Q_1, Q_2, Q_3 quartiles

$D_1, D_2, \ldots, D_9$ deciles

$P_1, P_2, \ldots, P_{99}$ percentiles

x value of a single score

f frequency with which a value occurs

Σ (capital sigma) summation

n number of scores in a sample

$n!$ factorial

N number of scores in a finite population; also used as the size of all samples combined

$\bar{x}$ mean of the scores in a sample

μ (mu) mean of all scores in a population

s standard deviation of a set of sample values

σ (lower case sigma) standard deviation of all values in a population

s^2 variance of a set of sample values

σ^2 variance of all values in a population

z standard score

$z(\alpha/2)$ critical value of z

t t distribution

$t(\alpha/2)$ critical value of t

df number of degrees of freedom

F F distribution

χ^2 chi-square distribution

χ_R^2 right-tailed critical value of chi-square

χ_L^2 left-tailed critical value of chi-square

p probability of an event or the population proportion

q probability or proportion equal to $1 - p$

$\hat{p}$ sample proportion

$\hat{q}$ sample proportion equal to $1 - \hat{p}$

$\bar{p}$ proportion obtained by pooling two samples

$\bar{q}$ proportion or probability equal to $1 - \bar{p}$

$P(A)$ probability of event A

$P(A|B)$ probability of event A assuming event B has occurred

$n^P r$ number of permutations of n items selected r at a time

$n^C r$ number of combinations of n items selected r at a time

THIRD EDITION

Elementary
Statistics

THIRD EDITION

Elementary Statistics

Mario F. Triola

Dutchess Community College, Poughkeepsie, New York

The Benjamin/Cummings Publishing Company

Menlo Park, California • Reading, Massachusetts • Don Mills, Ontario

Wokingham, U. K. • Amsterdam • Sydney • Singapore • Tokyo

Mexico City • Bogota • Santiago • San Juan

To Marc and Scott

Sponsoring Editor: Craig Bartholomew
Production Editor: Cici Oremland
Production Service: Stacey C. Sawyer, San Francisco
Book and Cover Design: Albert Burkhardt
Artists: Mary Burkhardt/Judith L. McCarty
Copyeditor: Loralee Windsor
Composition: Interactive Composition Corporation

Library of Congress Cataloging-in-Publication Data

Triola, Mario F.
  Elementary statistics

  Bibliography: p.
  Includes index.
  1. Statistics I. Title.
QA276.12.T76 1986   519.5   85-20012
ISBN 0-8053-9327-7

EFGHIJKLM-DO-8987

The Benjamin/Cummings Publishing Company, Inc.
2727 Sand Hill Road
Menlo Park, California 94025

Preface

Purpose of This Book

With the proliferation and widespread availability of calculators and computers, the importance and use of statistics are continuously increasing. Each of our lives is now affected in many ways by the analysis and management of data. The educated citizen and worker can no longer evade the challenges posed by our data-oriented society. This book is designed to develop the insight and skills needed to confront those challenges.

Audience

This book is an introduction to elementary statistics for nonmathematics students. A strong mathematics background is not required, but students should have completed a high school algebra course. Although this book does include underlying theory, it does not include the mathematical rigor more suitable for mathematics majors.

In writing this book, strong emphasis was placed on interesting, clear, and readable writing. Because the many examples and exercises in this book cover a wide variety of applications, it is appropriate for many disciplines. The previous editions have been used successfully by majors in psychology, sociology, business, data processing, computer science, engineering technology, biology, education, nursing, health, economics, ecology, agriculture, and many others.

Changes in the Third Edition

This third edition of *Elementary Statistics* includes all of the basic features of the previous editions. In response to extensive reader surveys, some sections have been added and almost every section has been modified to some extent. Also, several new features have been added:

Beginning of Chapter Features

- (New) List of **chapter sections** along with brief descriptions of their contents
- (New) Presentation of a **chapter problem**
- **Overview** of chapter including statement of objectives

End of Chapter Features

- **Computer project**
- **Review** of chapter
- (New) List of **important formulas**
- (New) **Vocabulary list** of important terms introduced in the chapter
- **Review Exercises**
- (New) **Case study activity**

Major Content Changes

- New section in Chapter 1: *The Nature of Data.* Includes discussion of nominal, ordinal, interval, and ratio levels of measurement.
- New optional section in Chapter 3: *Counting.* Includes permutations and combinations.
- New optional section in Chapter 6: *P-Values.*
- Inclusion of *stem-and-leaf plots* in Section 2-3.
- Inclusion of *box-and-whisker diagrams* in Section 4-6.
- Analysis of variance expanded to include treatment of samples with unequal sizes.

Exercises

The number of exercises has been substantially increased. There are now more than 1400 exercises, an increase of 40%. Also, a major effort has been made to group and arrange exercises so that there is usually an even-numbered exercise similar to each odd-numbered exercise.

As in the previous editions, the exercises are arranged in order of increasing difficulty. The exercises are divided into groups A and B, with the B types involving larger data sets or concepts requiring a stronger mathematical background. In a few rare cases, the B exercises introduce a new concept.

The exercise sets apply to a variety of real situations including this small sampling of topics:

X-ray radiation	Nuclear power
Energy consumption in homes	Consumer credit
Gasoline rationing	The reliability of computers

Blood cholesterol levels
Reliability of instrument
  readings
I.Q. tests
Car fatality rates for different
  age groups
Effectiveness of fire detecting
  devices
Surveys, polls, and the census

Chemical reactions
Income taxes
Fuel consumption rates
Product testing
Extrasensory perception
Biofeedback experiments
Campaign strategy
Gun control
Police response time

Other Features

- The **flowcharts** help clarify the more complicated procedures.

- Appendix B contains an expanded **glossary** of important terms.

- Appendix C contains an updated **bibliography.**

- Appendix D contains **answers** to almost all of the odd-numbered exercises.

- A **symbol table** is included on the front inside cover for quick and easy reference to key symbols.

- There are many more of the very popular **margin essays.** These short essays illustrate uses of statistics in very real and practical applications. The following is a sample of some of the topics covered.

Advertising: How commercials are regulated.

Biology: Were Mendel's experimental data manipulated?

Business: How airlines used sampling to save money by determining revenues from split-ticket sales.

Criminology: How valid are crime statistics?

Economics: How unemployment figures are obtained.

Education: Whether SAT scores can predict career success.

Energy: Quality control in a nuclear power plant.

Entertainment: How the Nielsen T.V. rating system works.

Medicine: How the Salk vaccine was tested.

Psychology: How to measure a seemingly qualitative characteristic, such as disobedience.

Sociology: Code of ethics for survey research.

Sports: Statistics and baseball strategy.

Supplements

STATDISK: A Computer Supplement

Elementary Statistics does not require the use of computers. For those who choose to supplement their course with computers, we have included

computer projects at the end of Chapters 2 through 11. Also, there are two new supplements available with this third edition.

- STATDISK is an easy-to-use statistical software package that does not require any previous computer experience. Developed as a supplement to this textbook, STATDISK is available for the IBM PC and the Apple IIe systems.

- STATDISK MANUAL/WORKBOOK includes instructions on the use of the STATDISK software package. It also includes experiments to be conducted by students.

The STATDISK software and Manual/Workbook have been designed so that instructors can assign computer experiments without using classroom time that is already quite limited. STATDISK includes a wide variety of programs that can be used throughout the course, and the experiments are designed to do more than number crunch or duplicate text exercises. They include concepts that can be discovered through computer use. Also, the text includes several sample displays that result from the use of STATDISK.

Other Supplements

The **Instructor's Resource Guide** includes

- answers to almost all of the text exercises
- test bank with sample tests for each chapter and comprehensive final examinations
- data sets that can be assigned to individual students
- transparency masters

The **Student Solutions Manual** provides detailed worked-out solutions to selected text exercises.

Acknowledgments

I wish to extend my sincere thanks for the suggestions made by the following reviewers of this third edition:

Dale Everson
University of Idaho

Josephine L. Gervase
Manchester Community
College

Frank Gunnip
Oakland Community College

William Koellner
Montclair State University

Larry Morgan
Montgomery Community
College

Charles Phillips
Sonoma State University

Robert Salhany
Rhode Island College

John Skillings
Miami University

David Stout
Pensacola Junior College

Loyd Wilcox
Golden West College

I also wish to thank the following individuals who completed a research questionnaire for the third edition:

B. Michael Adams	James Michali
Graydon Bell	Goro Nagase
Susan Brockman	Nicholas Norgaard
Frank Caldwell	Eric Nummela
Rich Campbell	Paul Olsen
Matthew Castellucci	Larry J. Pitzner
Roger Champagne	Dennis M. Ragan
Daniel Cherwien	Robert Reid
Daryl Close	William Rinaman
Frank D'Amico	Warren Ruud
Edward J. Danial	Robert Salhany
Margaret Davis	Charles Schwarz
Normand A. Dion	S. A. Sikorski
Don L. Evans	L. D. Stevens
Leora Foegen	Dave Stout
Annalisa L. France	Robert Sutliff
Josephine L. Gervase	Kenneth Tekel
Tony Giangrasso	Froylan Tiscareno
Joel Greenstein	S. B. Travis
Brian Hayes	M. B. Ulmer
Jonathan Henry	Lowell T. VanTassel
Susan L. Horn	P. VanVeldhutzen
Ronald W. Jorgensen	Sam Weaver
Anand S. Katiyar	Loyd V. Wilcox
R. Koo	Richard E. Wolfe
L. L. Krajewski	E. A. Yfantis
Theodore Lai	Frank Young
Rachel La Roe	Christina M. Yuengling
Phillip McGill	Anne Zeigler
Raymond H. Medley, Jr.	

I would like to extend special thanks to Larry Morgan of Montgomery Community College for his valuable assistance in checking solutions and providing very useful suggestions. Frank Gunnip has also been most helpful with his comments and his work on the test items for the book.

Finally, I wish to extend my thanks to Craig Bartholomew, Stacey Sawyer, Sally Elliott, and the entire Benjamin/Cummings staff. I also thank my wife Ginny and my children for their support, encouragement, and assistance.

To the Student

I recommend the use of a hand calculator. You should get one that is capable of addition, subtraction, multiplication, division, and computation of square roots, and it should use algebraic logic instead of chain logic. You can identify the type of logic by pressing these buttons:

$$2 + 3 \times 4 =$$

If the result is 14, the calculator uses algebraic logic. If the result is 20, the calculator uses chain logic, and you will be likely to make gross errors. Some relatively inexpensive calculators automatically compute the mean and the standard deviation and the slope and intercept values of a regression line. Although a calculator may not be required for the course, one will certainly make your life easier.

I also recommend that you read the overview carefully when you begin a chapter. Read the next section quickly to get a general idea of the material, and then return for a careful second reading. Try the exercises. If you encounter difficulty, go back and work some of the examples in the text and compare your solutions with the ones in the text.

When working on assignments, first attempt the earlier odd-numbered exercises. Check your answers with Appendix D and verify that you are correct before moving on to the other exercises. Keep in mind that neat and well-organized written assignments tend to be viewed more favorably. When you finish a chapter, check the review section to make sure that you didn't miss any major topics. Before taking tests, do the review exercises at the end of the chapters. In addition to helping you review, this will help you with your approach to an assortment of problems.

You might consider buying the *Student Solutions Manual* for this text. Written by Donald K. Mason of Elmhurst College, it gives detailed solutions to many of the text exercises.

M.F.T.
LaGrange, New York
January, 1986

Contents

CHAPTER **1**

Introduction to Statistics   2

Chapter Problem	3
1-1 Overview	4
1-2 Background	4
1-3 Uses and Abuses of Statistics	7
1-4 The Nature of Data	11
Vocabulary List	16
Exercises	16
Case Study Activity	19

CHAPTER **2**

Descriptive Statistics   20

Chapter Problem	21
2-1 Overview	22
2-2 Summarizing Data	23
Exercises A	30
Exercises B	34
2-3 Pictures of Data	35
Pie Charts	35

Histograms	37
Frequency Polygons	38
Ogives	38
Stem-and-Leaf Plots	40
Miscellaneous Graphics	43
Exercises A	44
Exercises B	49
2-4 Averages	50
Mean	51
Median	53
Mode	54
Midrange	55
Weighted Mean	55
Exercises A	59
Exercises B	62
2-5 Dispersion Statistics	64
Range	65
Variance	66
Standard Deviation	70
Exercises A	73
Exercises B	77
2-6 Measures of Position	79
Exercises A	84
Exercises B	86
Computer Project: Descriptive Statistics	86
Review	87

Important Formulas 88
Vocabulary List 88
Review Exercises 89
Case Study Activity 90

C H A P T E R **3**

Probability 92

Chapter Problem 93
3-1 Overview 94
3-2 Fundamentals 96
Rounding Off Probabilities 103
Exercises A 104
Exercises B 108
3-3 Addition rule 108
Exercises A 115
Exercises B 118
3-4 Multiplication Rule 119
Exercises A 126
Exercises B 129
3-5 Complementary Events 130
Exercises A 133
Exercises B 135
3-6 Counting 136
Exercises A 143
Exercises B 145
Computer Project: Probability 146
Review 147
Important Formulas 149
Vocabulary List 150
Review Exercises 150
Case Study Activity 153

C H A P T E R **4**

Probability Distributions 154

Chapter Problem 155
4-1 Overview 156
4-2 Random Variables 156
Discrete Random Variables 158
Exercises A 162
Exercises B 164
4-3 Mean, Variance, and Expectation 165
Expected Value 167
Exercises A 169
Exercises B 172
4-4 Binomial Experiments 173
Exercises A 184
Exercises B 187
4-5 Mean and Standard Deviation for the Binomial Distribution 188
Exercises A 193
Exercises B 194
4-6 Distribution Shapes 195
Exercises A 203
Exercises B 205
Computer Project: Probability Distributions 206
Review 207
Important Formulas 207
Vocabulary List 208
Review Exercises 208
Case Study Activity 212

C H A P T E R **5**

Normal Probability Distributions 214

Chapter Problem 215
5-1 Overview 216
5-2 The Standard Normal Distribution 217
Exercises A 223
Exercises B 224
5-3 Nonstandard Normal Distributions 225
Exercises A 229
Exercises B 232
5-4 Finding Scores When Given Probabilities 233
Exercises A 236
Exercises B 238

5-5 Normal as Approximation to
Binomial 239
Exercises A 247
Exercises B 249
5-6 The Central Limit Theorem 249
Exercises A 258
Exercises B 261
Computer Project: Normal
Probability Distributions 261
Review 262
Important Formulas 263
Vocabulary List 264
Review Exercises 264
Case Study Activity 266

C H A P T E R **6**

Testing Hypotheses 266

Chapter Problem 267
6-1 Overview 268
6-2 Testing a Claim About a Mean 269
Exercises A 284
Exercises B 287
6-3 P-Values 289
Exercises A 293
Exercises B 294
6-4 t Test 295
Exercises A 304
Exercises B 307
6-5 Tests of Proportions 307
Exercises A 313
Exercises B 315
6-6 Tests of Variances 315
Exercises A 321
Exercises B 324
Computer Project: Testing
Hypotheses 325
Review 325
Important Formulas 326
Vocabulary List 327
Review Exercises 327
Case Study Activity 329

C H A P T E R **7**

**Estimates and
Sample Sizes** 330

Chapter Problem 331
7-1 Overview 332
7-2 Estimates and Sample Sizes of
Means 332
Determining Sample Size 337
Small Sample Cases 339
Exercises A 340
Exercises B 343
7-3 Estimates and Sample Sizes of
Proportions 345
Sample Size 348
Exercises A 352
Exercises B 354
7-4 Estimates and Sample Sizes of
Variances 355
Exercises A 360
Exercises B 361
Computer Project: Estimates and
Sample Sizes 362
Review 362
Important Formulas 363
Vocabulary List 364
Review Exercises 364
Case Study Activity 366

C H A P T E R **8**

**Tests Comparing
Two Parameters** 368

Chapter Problem 369
8-1 Overview 370
8-2 Tests Comparing Two Variances 370
Exercises A 375
Exercises B 376
8-3 Tests Comparing Two Means 377

Exercises A 392
Exercises B 398
8-4 Tests Comparing Two Proportions 399
Exercises A 405
Exercises B 407
Computer Project: Tests
  Comparing Two Parameters 408
Review 409
Important Formulas 410
Review Exercises 411
Vocabulary List 415
Case Study Activity 415

C H A P T E R 9

Correlation and Regression 416

Chapter Problem 417
9-1 Overview 418
9-2 Correlation 419
Exercises A 428
Exercises B 432
9-3 Regression 433
Exercises A 444
Exercises B 447
9-4 Variation 448
Exercises A 454
Exercises B 455
Computer Project: Correlation
  and Regression 456
Review 456
Important Formulas 457
Review Exercises 457
Vocabulary List 459
Case Study Activity 459

C H A P T E R 10

Chi-Square and Analysis of Variance 460

Chapter Problem 461
10-1 Overview 462
10-2 Multinomial Experiments 462
Exercises A 468
Exercises B 471
10-3 Contingency Tables 472
Exercises A 478
Exercises B 482
10-4 Analysis of Variance 483
Exercises A 492
Exercises B 496
Computer Project: Chi-Square
  and Analysis of Variance 497
Review 497
Important Formulas 498
Review Exercises 499
Vocabulary List 501
Case Study Activity 501

C H A P T E R 11

Nonparametric Statistics 502

Chapter Problem 503
11-1 Overview 504
Advantages of Nonparametric Methods 504
Disadvantages of Nonparametric Methods 504
11-2 Sign Test 506
Exercises A 513
Exercises B 516
11-3 Wilcoxon Signed-Ranks Test for
  Two Dependent Samples 517
Exercises A 521
Exercises B 524
11-4 Wilcoxon Rank-Sum Test for
  Two Independent Samples 525
Exercises A 529
Exercises B 532
11-5 Kruskal-Wallis Test 533
Exercises A 537
Exercises B 540
11-6 Rank Correlation 540
Exercises A 549
Exercises B 552

11-7 Runs Test for Randomness 553
Exercises A 562
Exercises B 565
Computer Project: Nonparametric
    Statistics 565
Review 565
Important Formulas 566
Review Exercises 567
Vocabulary List 570
Case Study Activity 570

C H A P T E R **12**

Design, Sampling, and Report Writing

572

Identifying Objectives 573
12-1 Designing the Experiment 574
12-2 Sampling and Collecting Data 574

Random Sampling 574
Stratified Sampling 575
Systematic Sampling 575
Cluster Sampling 576
Importance of Sampling 576
12-3 Analyzing Data and Drawing
    Conclusions 577
12-4 Writing the Report 577
12-5 A Do-It-Yourself Project 578
Exercises A 579
Exercises B 580

Appendix A: Tables A1
Appendix B: Glossary A25
Appendix C: Bibliography A30
Appendix D: Answers to Selected
    Exercises A32
Index A80

Essays

C H A P T E R 1

Airline Companies Save by Sampling   5
Hertz and AAA Disagree on Car Costs   8
The Mt. St. Helens Myth   10
How Do You Measure Disobedience?   14

C H A P T E R 2

The Census   25
Authors of the *Federalist* Papers
    Identified   28
Histograms Can Be Revealing   37
Children See Much Television Violence   41
Over Half Our Presidents Are Above
    Average   52
The Class Size Paradox   56
United States Government Eliminates
    500,000 Farms   70
Index Numbers   80

C H A P T E R 3

Long-Range Weather Forecasting   95
Should You Guess on SAT's?   98
Mathematician Predicts Date of His
    Death   101
Just How Probable Is Probable?   102

You Bet?   109
Monkeys Are Not Likely to Type
    Hamlet   113
Probability and Prosecution   114
Nuclear Power Plant Has Unplanned
    "Event"   120
Convicted by Probability   122
Multiplication Rule Presents Difficulty
    to FAA   124
Redundancy   125
Not Too Likely   130
Odds Can Be Odd   132
Safety in Numbers   139
The Number Crunch   142

C H A P T E R 4

How Not to Pick Lottery Numbers   159
Prophets for Profits   168
Is Parachuting Safe?   174
Detecting Syphilis   183
Who Is Shakespeare?   190
Clusters of Disease   202

C H A P T E R 5

Bullhead City Gets Hotter   227

Are Geniuses Peculiar?   235
Reliability and Validity   252

CHAPTER 6

Why Professional Articles Are Rejected   270
Drug Approval Requires Strict
   Procedure   273
Product Testing Is Big Business   279
Beware of P-Value Misuse   291
Commercials, Commercials, Commercials   302
Misleading Statistics   308
Quality Control in a Nuclear Power
   Plant   310
The Power of Your Vote   319

CHAPTER 7

Technology Clouds Television Ratings   333
The Nielsen Television Rating System   335
Excerpts from a Department of
   Transportation Circular   336
A Professional Speaks About Sampling
   Error   339
Large Sample Size Is Not Good Enough   349
How One Telephone Survey Was
   Conducted   351
TV Watching Sets Records   358

CHAPTER 8

Cheating Success   371
Exit Polls on Their Way Out?   372
More Police, Fewer Crimes?   374
The Gender Gap in Wages   378
How Valid Are Crime Statistics?   386
The Golden Gate to Heaven   400
Polio Experiment   404

CHAPTER 9

Correlation Shows Trend   420
Student Ratings of Teachers   422

Have Atomic Tests Caused Cancer?   425
Unusual Economic Indicators   436
Rising to New Heights   442
What SAT Scores Measure   450

CHAPTER 10

Did Mendel Fudge His Data?   464
Of Course, Your Mileage May Vary   466
Survey Solicits Contributions   474
High Hopes   476
Statistics and Baseball Strategy   484
ZIP Codes Reveal Much   486
Pollster Lou Harris   491

CHAPTER 11

Air Is Healthier Than Tobacco   508
How Valid Are Unemployment Figures?   520
The Case of Coke versus Pepsi   528
Seat Belts Save Lives   536
Does Television Watching Cause
   Violence?   541
Study Criticized as Misleading   544
Dangerous to Your Health   548
Magazine Survey Results Reflect
   Leadership   554
Uncle Sam Wants You, if You're
   Randomly Selected   557
Consumer Price Index as a Measure
   of Inflation   561

CHAPTER 12

Survey Medium Can Affect Results   575
Phone Surveys   575
Invisible Ink Deceives Subscribers   576
Code of Ethics for Survey Research   578
Ethics in Experiments   579

THIRD EDITION

Elementary Statistics

1-1   **Overview**

1-2   **Background**
The **beginning** of statistics and its general **nature** are presented.

1-3   **Uses and Abuses of Statistics**
Examples of **beneficial uses** of statistics are presented, along with some of the common ways that statistics is used to **deceive.**

1-4   **The Nature of Data**
Different ways of **arranging data** are discussed. The four **levels of measurement** (nominal, ordinal, interval, ratio) are also defined.

Introduction
to Statistics

In the 1936 presidential election, an extensive telephone survey indicated that Alfred M. Landon would defeat Franklin D. Roosevelt. This poll was sponsored by *Literary Digest* magazine. More than two million people were surveyed, and it appeared that Landon would defeat Roosevelt by 57% to 43%. When the actual votes were counted, Roosevelt won by a margin of 62% to 38%. What went wrong? We will learn about the flaw in this 1936 survey, as well as several other ways in which statistics have been misused, either intentionally or unintentionally.

1-1 Overview

There are two basic meanings of the word *statistics*. In its simple meaning, *statistics* refers to data themselves, such as I.Q. scores, final exam averages, last year's value of exports, or the fatality rate for drivers who drink. A second meaning refers to statistics as a subject, which goes beyond the collection, tabulation, and summarizing of data. In this introductory statistics text, we will encounter ways of forming inferences that go beyond the initial data and enable us to form broader and more meaningful conclusions. In this chapter, we describe the nature of statistics and present a small sample of beneficial uses as well as some common abuses.

Throughout this book, we present some of the many ways the methods and theories of statistics have been used for the betterment of humanity, as well as some of the ways in which statistics have been abused by those who either didn't know any better or intentionally set out to deceive. The discussion of the ways in which statistics has been used for deceptive purposes does not come close to being a complete compendium of deceptive practices, but it should help you to become critical and analytical when you are presented with statistical claims. As you acquire more knowledge about standard acceptable techniques, you will be better prepared to challenge misleading statistical statements.

1-2 Background

Where did it all begin? In the seventeenth century, a successful store owner named John Graunt (1620–1674) had enough spare time to pursue outside interests. His curiosity led him to study and analyze a weekly church publication, called "Bills of Mortality," which listed births, christenings, and deaths and their causes. Based upon these studies, Graunt published his observations and conclusions in a work with the catchy title of "Natural and Political Observations Made upon the Bills of Mortality." This 1662 publication comprised the first real interpretation of social and biological phenomena based on a mass of raw data, and many people consider this to be the birth of statistics.

Graunt made observations about the differences between the birth and mortality rates of men and women. He noted a surprising consistency among events that seem to occur by chance. These and other early observations led to conclusions or interpretations that were invaluable in planning, evaluating, controlling, predicting, changing, or simply understanding some facet of the world in which we live.

We say that statistics can be used to predict, but it is very important to understand that we cannot predict with absolute certainty. In fact,

Airline Companies Save by Sampling

In the past, airline companies used an extensive and expensive accounting system to appropriate correctly the revenues from tickets that involved two or more companies. Now, instead of accounting for every ticket involving more than one company, they use a sampling method whereby a small percentage of these split tickets is randomly selected and used as a basis for appropriating all such revenues. The error created by this approach can cause some companies to receive slightly less than their fair share, but these losses are more than offset by the clerical savings accrued by circumventing the 100% accounting method. This concept saves companies millions of dollars each year.

statistical conclusions involve an element of uncertainty that can (and often does) lead to incorrect conclusions. It is possible to get ten consecutive heads when an ordinary coin is tossed ten times. Yet a statistical analysis of that experiment would lead to the incorrect conclusion that the coin is biased. That conclusion is not, however, certain. It is only a "likely" conclusion, which reflects the very low chance of getting ten heads in ten tosses.

This element of uncertainty sets statistics apart from other areas of applied mathematics, which require conclusions that meet the rigid criterion of certainty. For example, in plane geometry we begin with a list of basic rules or axioms and we *prove* without doubt that the sum of the angles of a triangle is 180°. Similarly, in arithmetic we prove that the product of any number and zero equals zero. In algebra, we prove that if $ac = bc$ while c is not zero, then $a = b$. But in statistics, our ultimate conclusions are not proved at all; they are only shown to be likely or unlikely.

In general, mathematics tends to be **deductive** in nature, meaning that acceptable conclusions are deduced with certainty from previous conclusions or assumptions. Statistics is often **inductive** in nature, because inferences are basically generalizations that may or may not correspond to reality.

In statistics, we commonly use the terms *population* and *sample*.

> **D E F I N I T I O N**
>
> A **population** is the complete and entire collection of elements (scores, people, measurements, and so on) to be studied.

> **D E F I N I T I O N**
>
> A **sample** is a subset of a population.

Closely related to the concepts of population and sample are the concepts of *statistic* and *parameter*.

> **D E F I N I T I O N**
>
> A **parameter** is a numerical measurement describing some characteristic of a population.

DEFINITION

A **statistic** is a numerical measurement describing some characteristic of a sample.

Suppose, for example, we collect sample data consisting of the annual incomes of a thousand randomly selected female lawyers. We could calculate the mean income for this sample group by finding the total of the one thousand incomes and then dividing that total by 1000. Since it is based on sample data, the resulting mean is considered to be a statistic. In contrast, the annual mean income of all female lawyers would be an example of a parameter.

Statisticians make inferences about an entire population based upon the observed data in a sample. Thus statisticians infer a general conclusion from known particular cases in the sample. In most branches of mathematics the procedure is reversed. That is, we first prove the generalized result and then apply it to the particular case. Geometers first prove that, for the population of all triangles, all possess the property that the sum of their respective angles is 180°. They then apply that established result to specific triangles, and they can be certain that the general property will always hold.

Statisticians, on the other hand, begin with a randomly selected sample of specific triangles and, after extracting the relevant numbers from those triangles, they conclude that it is likely that all triangles have a sum of 180°. This example is, in one sense, unfair, since it is possible to prove the deductive conclusion, but such proof is not always possible. Flashbulb manufacturers cannot test every one of their products in order to begin with a generalized property; psychologists and educators cannot administer I.Q. tests to all adults; pollsters cannot survey all voters. There are many situations in which sampling must be the first step in an inductive reasoning process.

As we proceed with our study of statistics, you will learn how to extract pertinent data from samples and how to infer conclusions based on the results of those samples. You will also learn how to assess the reliability of conclusions. Yet you should always realize that, while the tools of statistics can enable you to infer information about a population, you can never predict the behavior of any one individual.

A unique aspect of statistics is its obvious applicability to real and relevant situations. In many branches of mathematics we deal with abstractions that may initially appear to have little or no direct use in the real world, but the elementary concepts of statistics do have direct and practical applications.

1-3 Uses and Abuses of Statistics

Short essays that use real-world examples to illustrate the uses and abuses of statistics appear throughout the book. Among the uses are many applications prevalent today in the fields of business, economics, psychology, biology, computer science, military intelligence, English, physics, chemistry, medicine, sociology, political science, agriculture, and education. Statistical theory applied to these diverse fields often results in changes that benefit humanity. Social reforms are sometimes initiated as a result of statistical analyses of factors such as crime rates and poverty levels. Large-scale population planning can result from projections devised by statisticians. Manufacturers can provide better products at lower costs through the effective use of statistics in quality control. Epidemics and diseases can be controlled and anticipated through application of standard statistical techniques. Endangered species of fish and other wildlife can be protected through regulations and laws that are decided upon in part by statistical conclusions. Educators may discard innovative teaching techniques if statistical analyses show that traditional techniques are more effective. Through lower fatality rates, legislators can better justify laws such as those governing air pollution, auto inspections, and minimum drinking ages.

Students choose an elementary statistics course for a variety of reasons. Some students plan to major in psychology, economics, sociology, or other fields that require a statistics course. Others can find no alternative mathematics course that fulfills some minimum degree requirement. Still others have been known to elect a statistics course simply because they heard that it was interesting!

Apart from job-motivated or discipline-related reasons, the study of statistics can help you become more critical in your analyses of information so that you are less susceptible to misleading or deceptive claims. You use external data to make decisions, form conclusions, and build your own warehouse of knowledge. If you want to build a sound knowledge base, make intelligent decisions, and form worthwhile opinions, you must be careful to filter out the incoming information that is erroneous or deceptive. As educated and responsible members of society, you should sharpen your ability to recognize distorted statistical data; in addition, you should also learn to interpret undistorted data intelligently.

One hundred years ago Benjamin Disraeli said that "there are three kinds of lies: lies, damned lies, and statistics." It has also been said that "figures don't lie; liars figure." Some people have even been accused of using statistics like a drunk uses a lamppost: "more for support than for illumination." All of these statements refer to the abuses of statistics in which data are presented in a way that is misleading. The typical abuser has personal objectives and is willing to suppress unfavorable data while

Hertz and AAA Disagree on Car Costs

In one year, the Hertz Corporation estimated the cost of owning and operating a car to be 49¢ per mile, while the American Automobile Association estimate was 24.8¢ per mile. Hertz developed its own estimate but the AAA estimate came from Runzheimer and Co., a separate consulting firm. A spokesman for Runzheimer charged that Hertz rents and leases cars and "so they have a commercial ax to grind. The more the public thinks it costs to own and operate a car, the easier it is for them to rent or lease." A Hertz spokesman says "no one should believe any cents-per-mile figures. Many drivers prefer them for tax and employer reimbursement estimates, but dollars per year is your best cost gauge."

emphasizing supportive data. Here are a few examples of the many ways that data can be distorted.

The term *average* refers to several different statistical measures. (These will be discussed and defined in Chapter 2.) To most people, average means the sum of all values divided by the number of values, but this is only one type of average; it is called the arithmetic mean. There are also the median, the mode, the midrange, and others. (For example, given the 10 annual salaries of $15,000, $15,000, $15,000, $16,000, $18,000, $20,000, $22,000, $24,000, $25,000, and $45,000, we can correctly claim that the average is either $21,500 (arithmetic mean), $19,000 (median), $15,000 (mode), or $30,000 (midrange).) There are no objective criteria that can be used to determine the specific average that is most representative. Consequently, the user of statistics is free to select the average that best supports a favored position.

A union contract negotiator can choose the lowest average in an attempt to emphasize the need for a salary increase, while the management negotiator can choose the highest average to emphasize the well-being of the employees. In actual negotiations both sides are usually adept at exposing such ploys, but the typical citizen often accepts the validity of an average without really knowing which specific average is being presented and without knowing how different the picture would look if another average were used. The educated and thinking citizen is not so susceptible to potentially deceptive information; the educated and thinking citizen analyzes and criticizes statistical data so that meaningless or illusory contentions are not part of the base upon which his or her decisions are made and opinions formulated. If you read that the average annual salary of an American is $14,487, you should attempt to find out which of the averages that figure represents. If no additional information is available, you should realize that the given figure may be very misleading. Conversely, as you present statistics on data you have accumulated, you should attempt to provide descriptions identifying the true nature of the data.

Many visual devices—such as bar graphs and pie charts—can be used to exaggerate or deemphasize the true nature of data. (These are also discussed in Chapter 2.) In Figure 1-1 we show two bar graphs that depict the *same data,* but part (*b*) is designed to exaggerate the decline in values. While both bar graphs represent the same set of scores, they tend to produce very different subjective impressions. As an actual example, consider the graph shown in Figure 1-2, which appeared in *Time* as part of an advertising supplement sponsored by the member companies of the Chemical Manufacturer's Association. Can you see anything wrong with the graph? Too many of us look at a graph superficially and develop intuitive impressions based upon the pattern we see. Instead, we should scrutinize the graph and search for distortions of the type found in these illustrations. We should analyze the numerical information contained in a graph instead of being impressed by its general shape.

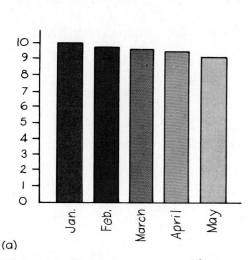

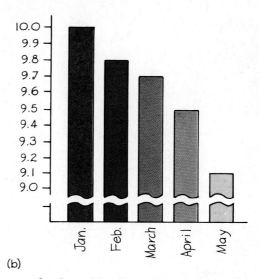

(a)

(b)

Figure 1-1 Both bar graphs depict the same data, but part (b) exaggerates what is in reality a modest decline.

Pictures of objects may also be misleading. If the dimensions of a two-dimensional figure (such as a square) are doubled, the area is increased by a multiple of four; if the dimensions of a solid figure are doubled, the volume is increased by a multiple of eight. Solids commonly used to depict data include moneybags, stacks of coins, army tanks, cows, and houses. Let's suppose that school taxes in a small community dou-

Figure 1-2

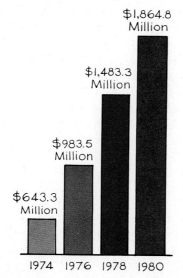

The Mt. St. Helens Myth

There was a popular belief that the state of Washington's birthrate jumped nine months after the eruption of Mt. St. Helens. However, a researcher for the state's Center for Health Statistics found no such increase. It was found that there were 189 births nine months after the eruption, compared to 201 births for the same period one year earlier. The birthrate did not jump; it showed a slight decline. Such rumors are often fostered by a common belief that the birthrate rises nine months after a major crisis.

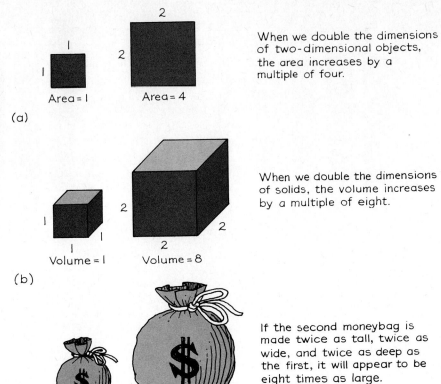

When we double the dimensions of two-dimensional objects, the area increases by a multiple of four.

(a)

When we double the dimensions of solids, the volume increases by a multiple of eight.

(b)

If the second moneybag is made twice as tall, twice as wide, and twice as deep as the first, it will appear to be eight times as large.

(c)

Figure 1-3

bled from one year to the next. By depicting the amounts of taxes as bags of money and by doubling the dimensions of the bag representing the second year, we can easily create the impression that taxes more than doubled (see Figure 1-3). Another variety of statistical "lying" is often inspired by small sample results. The toothpaste preferences of only ten dentists should not be used as a basis for a generalized claim such as "Caressed toothpaste is recommended by seven out of ten dentists." Even if the sample is large, it must be unbiased and representative of the population from which it comes. A popular illustration of a biased sample involves the 1936 presidential election in which an extensive telephone survey indicated that Landon would defeat Roosevelt. Obviously we haven't heard of President Landon, so we can conclude that the poll was misleading. Retrospective analysis reveals that a disproportionate number of affluent people had telephones in 1936, and they tended to favor Landon. Consequently the survey did not reach a representative sample of eligible voters. In this case, there was no intent to deceive. The faulty poll is easy to criticize in retrospect, but the basic flaw was not so obvious in 1936.

Sometimes the numbers themselves can be deceptive. A mean annual salary of $14,487.31 sounds precise and tends to instill a high degree of confidence in its accuracy. The figure of $14,500 doesn't convey the same sense of precision and accuracy.

Another source of statistical deception involves numbers that are ultimately guesses, such as the crowd count at a political rally, the dog population of New York City, and the amount of money bet illegally.

"Ninety percent of all our cars sold in this country in the last 10 years are still on the road." The typical consumer hears that commercial message and gets the impression that those cars must be well built in order to persist through those long years of driving. What the auto manufacturer failed to mention was that 90% of the cars they sold in this country were sold within the last 3 years!

The preceding examples comprise a small sampling of the ways in which statistics can be used deceptively. Entire books have been devoted to this subject, including Darrell Huff's *How to Lie with Statistics* and Robert Reichard's *The Figure Finaglers*. Understanding these practices will be extremely helpful in evaluating the statistical data found in everyday situations.

1-4 The Nature of Data

When thinking about collections of data, there is a common tendency to think of them as lists of numbers, such as the cholesterol levels of diseased rats (how disgusting) or the weights of freshly broiled Maine lobsters (that's better). Yet data may be nonnumerical, and even numerical data can belong to different categories with different characteristics. For example, a pollster may compile data consisting of nonnumeric data such as sex, race, and religion of voters in a sample. Numeric data, instead of being in an unordered list, might be data matched in pairs (discussed in Chapters 8, 9, 11) as in the two tables below.

Weight before diet (pounds)	110	115	139	164	176
Weight after diet (pounds)	106	115	132	155	163

Time required to sell car (minutes)	71	73	122	42	108
Cost of car (dollars)	11,418	11,872	12,364	10,050	12,500

Another very common arrangement for summarizing sample data is the contingency table (discussed in Section 10-3) such as the one given below.

	Identification of criminal	
	correct	wrong
Witness hypnotized	22	13
Witness not hypnotized	46	32

In this table the numbers are frequencies (counts) of sample results.

Clearly, the nature of the data can affect the nature of the central issue and the method used for anlaysis. With the paired diet data, a fundamental concern would be whether the diet is effective. Any analysis of these data should attempt to determine whether the "after" weights are significantly less than the corresponding "before" weights. With the paired sales data, the fundamental concern would be whether some relationship exists between sales time and car cost. This requires a different method of analysis. With the contingency table, the fundamental concern would be whether hypnotizing witnesses affects the quality of their ability to identify criminals. This requires yet another method of analysis. As we consider the topics of later chapters, we will see that the structure of the data does affect our choice of the method we will use. In addition to the structure of the data, our choice of analytical method is also affected by the nature of the data. It is common to classify data according to one of the following four **levels of measurement.**

Low
to
high

1. Nominal
2. Ordinal
3. Interval
4. Ratio

The **nominal level of measurement** is characterized by data that consist of names, labels, or categories only.

If we associate "nominal" with "name only," the meaning becomes easy to remember. An example of nominal data is the collection of sex classifications (male/female) for a sample of college graduates. Data at this nominal level of measurement cannot be arranged according to some ordering scheme. That is, there is no criterion by which values can be identified as greater than or less than other values.

EXAMPLE

The following are other examples of sample data at the nominal level of measurement.
(a) The sample consisting of 12 Democrats, 15 Republicans, and 9 Independents.
(b) The sample consisting of 14 students from New York, 17 from California, 8 from Connecticut, and 7 from Florida.
(c) The data summarized in the given contingency table.

	Identification of criminal	
	correct	wrong
Witness hypnotized	22	13
Witness not hypnotized	46	32

In the preceding example, it should be obvious that the data cannot be used for calculations since the categories lack any ordering or numerical significance. We cannot, for example, "average" 12 Democrats, 15 Republicans, and 9 Independents. Numbers are sometimes assigned to the different categories, especially when the data are processed by computer. We might find that Democrats are assigned 0, Republicans are assigned 1, and Independents are assigned 2. Even though we now have number labels, those numbers lack any real computational significance. The average of twelve 0s, fifteen 1s, and nine 2s might be 0.9, but that is a meaningless statistic. (0.9 does not represent a liberal Republican!)

The **ordinal level of measurement** involves data that may be arranged in some order, but differences between data values either cannot be determined or are meaningless.

EXAMPLE

The following are examples of data at the ordinal level of measurement.
(a) In a sample of 36 batteries, 12 were rated "good," 16 were rated "better," and 8 were rated "best."
(b) In a class of 19 students, 5 required remediation, 10 were average, and 4 were gifted.
(c) In a high school graduating class of 463 students, Sally ranked 12th, Allyn ranked 27th, and Mike ranked 28th.

How Do You Measure Disobedience?

The data you collect are at least as important as the statistical methodology you employ. However, it is often difficult to collect usable or relevant data. How do you collect data that relate to a characteristic that doesn't appear to be measurable, such as the level of disobedience in people? Stanley Milgram was a social psychologist who

devised a clever experiment which did just that. A researcher instructed a volunteer to operate a control board that gave increasingly painful "electric shocks" to a third person. Actually, no electric shocks were given and the third person was an actor who feigned increasing levels of pain and anguish. The volunteer began with 15 volts and was instructed to increase the shocks by increments of 15 volts up to a maximum of 450 volts. The disobedience level was the point at which the volunteer refused to follow the researcher's instructions to increase the voltage. Despite the actor's screams of pain and a feigned heart attack, two thirds of the volunteers continued to obey the researcher. Milgram was surprised by such a high level of obedience.

This ordinal level provides information about relative comparisons, but the degrees of differences are not available. We know that a low income worker earns less than a middle income worker, but we don't know how much less. Again, data at this level should not be used for calculations.

The **interval level of measurement** is like the ordinal level, with the additional property that we can determine meaningful amounts of differences between data. Data at this level may lack an inherent zero starting point.

Temperature readings of 25° F and 50° F are examples of data at this measurement level. Those values are ordered and we can determine their difference (often called the *distance* between the two values). However, there is no inherent zero point. The value of 0° F might seem like a starting point, but it is arbitrary and not inherent. The value of 0° F does not indicate "no heat" and it is incorrect to say that 50° F is twice as hot as 25° F.

> The **ratio level of measurement** is actually the interval level modified to include the inherent zero starting point.

E X A M P L E

Examples of data at this ratio level are as follows:
(a) Heights of pine trees around Lake Tahoe
(b) Volumes of helium in balloons
(c) Times (in minutes) of joggers in a marathon

In each of these data collections, values can be ordered, differences can be found, and there is an inherent zero starting point. This level is called the ratio level because the starting point makes ratios meaningful. For example, a tree 50 feet high is twice as tall as a tree 25 feet high, whereas 50° F is not twice as hot as 25° F.

The distinction between the interval and ratio levels of measurement is not always clear. Fortunately, the remainder of this text will generally treat the interval level and the ratio level in the same way. We

Levels of Measurement

Low

to

high

Level	Summary	Example
Nominal	Unordered categories.	Voter distribution: 45 Democrats, 80 Republicans, 90 Independents.
Ordinal	Categories are ordered but differences cannot be determined.	Voter distribution: 45 low income voters, 80 middle income voters, 90 upper income voters.
Interval	Differences between values can be found, but there may be no inherent starting point.	Temperatures of steel rods: 45° F, 80° F, 90° F
Ratio	Like interval, but with an inherent starting point.	Lengths of steel rods: 45 cm, 80 cm, 90 cm

should at least recognize the differences among these three categories: nominal, ordinal, and a combination of the interval and ratio levels of measurement.

A general and important guideline is that **the statistics based upon one level of measurement should not be used for a lower level (but can be used for a higher level).** We can, for example, calculate a mean (see Section 2-3) for data at the interval or ratio level, but not at the lower ordinal or nominal levels. An implication of this guideline is that data obtained from using a Likert scale, such as the one below, should not be used for calculations since these data are only at the ordinal level. This guideline is sometimes ignored and Likert scale results are often treated as interval or ratio level data, even though they are not. But serious errors may result from violating this guideline. If, for data-processing requirements, we assign the numbers of 0, 1, and 2 to Democrats, Republicans, and Independents (respectively) and proceed to calculate a mean, we are creating a meaningless statistic that can lead to incorrect conclusions.

Superior	Good	Average	Poor	Very Poor
1	2	3	4	5

In Chapter 2 we introduce basic important ways of dealing with data sets that are primarily at the interval or ratio levels of measurement.

Vocabulary List

Define and give an example of each term.

statistics (as a
   discipline)
population

sample
parameter
statistic
nominal

ordinal
interval
ratio

Exercises

1-1 John Graunt studied birth records during the seventeenth century and observed that "the number of male births exceeded the number of female births by about a thirteenth part." Recently, there were 1,791,000 male births and 1,703,000 female births in the United States. Restate Graunt's observation using the more recent data.

1-2 For 500 tosses of a coin, 240 heads turned up while 260 tails turned up.
(a) What percent of the total tosses were heads?
(b) We would expect that 50% of the results are heads. Does this mean that the coin is biased?

1-3 How does statistics differ from most other branches of applied mathematics?

1-4 What is the difference between a population and a sample?

1-5 How can the concepts of population and sample relate to deductive reasoning and inductive reasoning?

1-6 Give two reasons why a pollster cannot survey all voters.

1-7 A newspaper article reports that "this morning's demonstration was attended by 8755 students." Comment.

1-8 In 1936, why did voters with telephones tend to favor the Republican Landon for the presidency?

1-9 A district attorney claims that "organized crime paid $89,541 in bribes to local officials last year." Criticize that claim.

1-10 A survey was conducted in an attempt to estimate the number of men who watch a daytime soap opera on television. Why might the survey results indicate a much lower number than the correct one?

1-11 Census takers have found that in obtaining the ages of people, they would get more people of age 50 than of age 49 or of age 51. Can you explain how this might occur?

1-12 A pollster concluded from a survey that the average age of adult women in Cleveland was 31.2 years. Why might this figure be wrong?

1-13 A news report states that the police seized forged record albums that had a value of $1 million. How do you suppose the police computed the value of the forged albums, and in what other ways can that value be estimated? Why might the police be inclined to exaggerate the value of the albums?

1-14 A study of car accidents in New York City showed that more accidents occurred during the day than at night. Does this mean that it is safer to drive at night? How do you explain the greater number of accidents during the day?

1-15 A college conducts a survey of its alumni in an attempt to determine their median annual salary. Would alumni with very low salaries be likely to respond? How would this affect the result? Identify one other factor that might affect the result.

1-16 One study actually showed that smokers tend to get lower grades in college than nonsmokers. Does this mean that smoking causes lower grades? What other explanation is possible?

1-17 A report by the Nuclear Regulatory Commission noted that a particular nuclear reactor was being operated at "below-average standards." What is the approximate percentage of nuclear power plants that operate at below-average standards? Is a below-average standard necessarily equivalent to a dangerous or undesirable level?

1-18 An employee earning $400 per week was given a 20% cut in pay as part of her company's attempt to reduce labor costs. After a few weeks, this employee's dissatisfaction grew and her threat to resign caused her manager to offer her a 20% raise. The employee accepted this offer since she assumed that a 20% raise would make up for the 20% cut in pay.
  (a) What was the employee's weekly salary after she received the 20% cut in pay?
  (b) Use the salary figure from part (a) to find a 20% increase and determine the weekly salary after the raise.
  (c) Did the 20% cut followed by the 20% raise get the employee back to the original salary of $400 per week?

1-19 What differences are there between the following two statements, and which one do you believe is more accurate?
  (a) Drunken drivers cause more than half of all fatal car crashes.
  (b) Of all fatal motor vehicle crashes, more than 50% involve alcohol.

1-20 What is wrong with Figure 1-2?

1-21 The first edition of a textbook contains 1000 exercises. For the second edition, the author removed 100 of the original exercises and added 300 new exercises. Which of the following statements about the second edition are correct?
  (a) There are 1200 exercises.
  (b) There are 33% more exercises.
  (c) There are 20% more exercises.
  (d) 25% of the exercises are new.

1-22 In a typical year, about 56,000 deaths result from motor vehicle accidents.
  (a) How many deaths would result from motor vehicle accidents in a typical day?
  (b) How many deaths would result from motor vehicle accidents in a typical four-day period?
  (c) For the four days of the Memorial Day weekend (Friday through Monday), assume that driving increases by 25% and that there are 768 deaths resulting from motor vehicle accidents. Does it appear that driving is more dangerous over the Memorial Day weekend?

1-23 The Bureau of Census reports an average family size of 3.27 persons and an average household size of 2.73 persons. Explain the discrepancy.

1-24 A news article reports that women's salaries are typically 60% of the salaries of men in similar positions. One reason for this is that in recent years a larger proportion of women with no prior experience have entered the job market for the first time and their starting salaries are naturally lower. Apart from any discrimination based on sex, cite another reason that might help to explain women's lower salaries.

1-25   In each of the following, determine which of the four levels of measurement (nominal, ordinal, interval, ratio) is most appropriate.
  (a)   Upon completion of a course, an instructor is rated as being superior, above average, average, below average, or poor.
  (b)   The numbers 2, 3, 0, 4, 6 refer to the numbers of absences on the five work days in one week.
  (c)   Social security numbers.
  (d)   Zip codes.
  (e)   Annual incomes.
  (f)   Times in seconds required for a jogger to run around a track.
  (g)   Final course averages of A, B, C, D, F.
  (h)   Annual mean temperatures of the 50 state capitol cities.
  (i)   Years in which U.S. presidents were born.
  (j)   Temperatures at which various mixtures of water and alcohol will freeze.

1-26   In a final examination for a statistics course, one student received a grade of 50, and another student received a grade of 100.
  (a)   If we consider these numbers to represent only the points earned on the exam, then the score of 100 is twice that of 50. What is the corresponding level of measurement?
  (b)   If we consider these numbers to represent the amount of the subject learned in the course, it is wrong to conclude that the one student knows twice as much as the other. What is the level of measurement in this case?

1-27   Many people question what I.Q. scores actually measure. Assuming that I.Q. scores measure intelligence, is a person with an I.Q. score of 150 twice as intelligent as another person with an I.Q. score of 75?
  (a)   What does a yes answer imply about the level of measurement corresponding to I.Q. scores?
  (b)   What does a negative answer imply about the level of that data?

1-28   The years 1990, 1988, 1972, 1963, and 1984 form a collection of data at the interval level of measurement. Explain.

Chapter 1
Case Study Activity

Collect an example from a current newspaper or magazine in which data have been presented in a potentially deceptive manner. Identify the source from which the example was taken, explain briefly the way in which the data might be deceptive, and suggest how the data might be presented more fairly.

C H A P T E R 2

2-1 Overview

2-2 Summarizing Data
The construction of **frequency tables** is described.

2-3 Pictures of Data
Data is visualized through **pie charts, histograms, frequency polygons, ogives,** and **stem and leaf plots.**

2-4 Averages
These measures of central tendency are defined: **mean, median, mode, midrange,** and **weighted mean.**

2-5 Dispersion Statistics
These measures of dispersion are defined: **range, variance, standard deviation,** and **mean deviation.**

2-6 Measures of Position
For comparison purposes, the **standard score** (or **z score**) is defined. Also defined are **percentiles, quartiles,** and **deciles.**

Descriptive Statistics

Suppose that we have compiled the collection of data given in Table 2-1 on page 23 for the total number of hours of absence for 150 employees in one year. We might question whether these employees have an absence pattern that differs from that of similar firms. Perhaps a new sick leave policy has been implemented and we want to investigate its effects, or perhaps we would like to estimate next year's budget costs attributable to absences. These are only a few of the many possible reasons for wanting to know about this collection of data.

Surveying the list of 150 scores may reveal some characteristics of these data, but it is generally difficult to draw any meaningful conclusions from raw data of this type. An objective of this chapter is to develop a variety of methods that allow us to understand data sets such as the one in Table 2-1.

2-1 Overview

In analyzing a data set, we should first determine whether we know all values for a complete population, or whether we know only the values for some sample drawn from a larger population. That determination will affect both the methods we use and the conclusions we form.

We use methods of **descriptive statistics** to summarize or *describe* the important characteristics of a known set of data. If the 150 values given in Table 2-1 represent the absences for all employees in a firm, then we have known population data. We might then proceed to improve our understanding of this known population data by computing some average or by constructing a graph. We use descriptive statistics when we simply summarize known population data. In contrast, **inferential statistics** goes beyond mere description. We use inferential statistics when we use sample data to make *inferences* about a population.

Suppose we compute an average of the 150 scores (Table 2-1) and obtain a value of 87.9 hours. If the firm has 150 employees and our list is complete, that average of 87.9 hours is a descriptive statistic, which simply summarizes known data. Now suppose we treat those 150 scores as a sample drawn from a larger population of thousands of employees. If we conclude that the average annual absence time for an employee is 87.9 hours, we have made an *inference* that goes beyond the known data.

This chapter deals with the basic concepts of descriptive statistics, Chapter 3 includes an introduction to probability theory, and the subsequent chapters deal mostly with inferential statistics. Descriptive statistics and inferential statistics are the two basic divisions of the subject of statistics.

We use the tools of descriptive statistics in order to understand an otherwise unintelligible collection of data. The following three characteristics of data are extremely important, and they can give us much insight:

1. Representative score, such as an average.
2. Measure of scattering or variation.
3. Nature of the distribution, such as bell shaped.

We can learn something about the nature of the distribution by organizing the data and constructing graphs, as in Sections 2-2 and 2-3. In Section 2-4 we will learn how to obtain representative, or average, scores. We will measure the extent of scattering, or variation, among data as we use the tools found in Section 2-5. Finally, in Section 2-6 we will learn about measures of position so that we can better analyze or compare various scores. As we proceed through this chapter, we will refer to the 150 scores given in Table 2-1, and our insight into that data set will be increased as we reveal its characteristics.

Table 2-1					
Hours of Absence for 150 Employees					
85	192	13	27	165	87
129	4	141	4	105	102
119	145	189	12	44	157
96	8	142	41	79	21
47	51	38	111	88	67
122	92	87	46	221	111
89	66	57	47	136	90
43	14	39	163	149	150
81	140	78	99	134	116
21	159	105	118	79	138
73	4	103	98	134	36
4	32	116	78	0	73
71	80	76	45	85	15
154	74	77	96	6	66
15	55	108	106	91	4
87	9	151	118	57	120
131	0	88	171	49	92
80	58	65	143	55	64
84	70	71	73	8	80
130	167	92	115	167	15
94	221	114	35	80	182
75	57	67	147	0	22
164	154	104	110	102	234
76	116	100	127	128	0
127	36	98	133	72	122

2-2 Summarizing Data

Referring to the 150 absence times given in Table 2-1, we see that a survey of those values probably does not lead to any specific conclusions because the human mind usually cannot assimilate and organize that much data. In general, any large collection of raw data will remain unintelligible until it is organized and summarized. Because there are many reasons why an understanding of this data collection might be necessary, we will now proceed to consider ways of organizing the data.

A **frequency table** is an excellent device for making large collections of data much more intelligible. A frequency table is so named because it lists categories of scores along with their corresponding frequencies. The *frequency* for a category or class is the number of original scores that fall into that class.

This might seem complicated, but the construction of a frequency table is really a simple process. While construction may be time-consuming and monotonous, it is not very difficult. For extremely large

Table 2-2	
Score	Frequency
1–5	
6–10	
11–15	
16–20	
21–25	
26–30	

This frequency table has six classes.

The **lower class limits** are 1, 6, 11, 16, 21, 26.

The **upper class limits** are 5, 10, 15, 20, 25, 30.

The **class boundaries** are 0.5, 5.5, 10.5, 15.5, 20.5, 25.5, 30.5.

The **class marks** are 3, 8, 13, 18, 23, 28.

The **class width** is 5.

collections of scores, the data can be entered in a computer that is programmed to construct the appropriate frequency table automatically.

The following standard definitions formalize and identify some of the basic terminology associated with frequency tables. These definitions may seem difficult, so it will be helpful to examine Table 2-2, which illustrates them.

DEFINITION

Lower class limits are the smallest numbers that can actually belong to the different classes.

DEFINITION

Upper class limits are the largest numbers that can actually belong to the different classes.

DEFINITION

The **class boundaries** are obtained by increasing the upper class limits and decreasing the lower class limits by the same amount so that there are are no gaps between consecutive classes. The amount to be added or subtracted is one-half the difference between the upper limit of one class and the lower limit of the following class.

The Census

Every ten years, the United States Government undertakes a census intended to obtain information about each American. At stake are over $50 billion in government allocations, representation in Congress, redistricting of state and local governments, and business adjustments to changing populations. The Census Bureau once classified missed people and unanswered questions as unknown, but modern technology and sampling techniques now make possible the imputation of data, which means they make up their own answers. Millions of people are invented and given ages, religions, spouses, jobs, children, incomes, and so on.

There is much political pressure to correct for any undercount, since the error affects some regions and groups much more than others.

DEFINITION

The **class marks** are the midpoints of the classes.

DEFINITION

The **class width** is the difference between two consecutive lower class limits (or class boundaries).

The process of actually constructing a frequency table involves these key steps:

1. Decide on the number of classes your frequency table will contain. A class consists of one grouping of scores, such as 0–19. As a general guideline, the number of different classes in a frequency table should be between 5 and 20. The actual number of classes is arbitrary and may be affected by convenience or other subjective factors. We generally begin by examining the highest and lowest scores; they may suggest a convenient number of classes.

For example, if we have 500 I.Q. scores ranging from a low of 90 to a high of 119, there are only 30 possible scores; it is therefore natural to select either 10 classes of width 3 or 15 classes of width 2.

If we have 347 scores that represent the lives of transistors ranging from 217.2 hours to 638.7 hours, the range of 421.5 hours does not suggest a specific number of classes. There are no objective criteria for determining a number of distinct classes, so we must use our own subjective judgment. (Who says mathematics is not an art?)

2. Find the width of the classes by dividing the number of classes into the range (that is, the difference between the highest and lowest scores). For example, suppose we have I.Q. scores ranging from 90 to 119 and we want 15 class intervals. First we subtract 90 from 119 to get 29, the range. Next we divide 29 by 15 (since we want 15 class intervals) and round the answer of 1.933333 up to 2. This rounding *up* (not rounding *off*) is not only convenient—it also guarantees that all of the data will be included in the table.

DEFINITION

$$\text{class width} = \text{round } up \text{ of } \frac{\text{range}}{\text{number of classes}}$$

3. Select as a starting point either the lowest score or a value slightly less than the lowest score. This starting point is the lower class limit of the first class.

4. Add the class width to the starting point to get the second lower class limit. Add the class width to the second lower class limit to get the third, and so on.

5. List the lower class limits in a vertical column and enter the upper class limits, which can be easily identified at this stage.

EXAMPLE

We have a collection of scores representing the I.Q.'s of prison inmates. The highest and lowest scores are 148 and 70, respectively. Find the upper and lower limits of each class if we wish to have eight classes.

Solution

The difference between the highest and lowest scores is 78. Dividing 78 by 8 we get 9.75, which—for the sake of convenience—we round up to 10. Selecting 70 as the starting point is equivalent to selecting 70 as the lower limit of the first class. With a class width of 10, this means that the *next class* has 70 + 10, or 80, as its *lower limit*. The lower class limits are 70, 80, 90, . . . , 140. The first class must therefore be 70–79, while the second class becomes 80–89. The upper and lower limits of the 8 classes are as follows:

70–79
80–89
90–99
100–109
110–119
120–129
130–139
140–149

Once the class limits have been determined, it becomes relatively easy to identify the class boundaries and class marks. For the preceding list of class limits, the class boundaries are 69.5, 79.5, 89.5, . . . , 149.5. The class marks are 74.5, 84.5, 94.5, . . . , 144.5. In the next example we illustrate a case in which the use of whole numbers is not practical.

EXAMPLE

A large collection of college grade-point averages ranges from 0.00 to 3.96 and we want to develop a frequency table with ten classes. Find the upper and lower limits of each class.

Solution

The difference between the highest and lowest scores is 3.96. Dividing by the number of classes (ten), we get 0.396, which we round up to 0.40. (Rounding 0.396 up to 1 would cause all the data to fall in the first four classes with the last six classes all empty.) With a class width of 0.4, we get these lower class limits: 0.00, 0.40, 0.80, 1.20, 1.60, 2.00, 2.40, 2.80, 3.20, 3.60. From these lower class limits we can see that the upper class limits must be 0.39, 0.79, 1.19, . . . , 3.59, 3.99. The first class of 0.00–0.39 has a class mark of 0.195; the other class marks can be found by halving the sum of the upper and lower limits. The first class of 0.00–0.39 has an upper class boundary of 0.395 (the value midway between 0.39 and 0.40). Since the upper class boundary of 0.395 is 0.005 above the upper limit, it follows that the lower class boundary should be 0.005 below the lower class limit, so the value of −0.005 is determined.

Those lower class limits		suggest these upper class limits.

↘ ↙

0.00–0.39		−0.005–0.395
0.40–0.79		0.395–0.795
0.80–1.19	←Those class limits suggest these class boundaries.	0.795–1.195
1.20–1.59		1.195–1.595
1.60–1.99		1.595–1.995
2.00–2.39	→	1.995–2.395
2.40–2.79		2.395–2.795
2.80–3.19		2.795–3.195
3.20–3.59		3.195–3.595
3.60–3.99		3.595–3.995

It is usually desirable for all classes of a frequency table to have the same width, although it is sometimes impossible to avoid open-ended intervals such as "65 years or older." Also, the classes should be mutually exclusive; that is, each score will belong to exactly one class.

Common sense is often helpful in constructing frequency tables that are easier to use and understand. For the grade-point averages of the

Authors of the *Federalist* Papers Identified

In 1787–1788 Alexander Hamilton, John Jay, and James Madison anonymously published the famous *Federalist* papers in an attempt to convince New Yorkers that they should ratify the Constitution. The identity of most of the papers' authors became known, but the authorship of 12 of the papers was contested. Through statistical analysis of the frequencies of various words, we can now conclude that James Madison is the *likely* author of these 12 papers. For many of these disputed papers, the evidence in favor of Madison's authorship is overwhelming to the degree that we can be almost certain of being correct.

preceding example, we know that the highest and lowest possible values are 4.00 and 0.00, respectively. Also, we may know of special cutoff values like 2.00 (for probation) or 3.00 (for Dean's list). Based on this additional information we may be wise to adjust the frequency table so that it is easier to read and comprehend. We might, for example, use the following class limits:

    0.00–0.49
    0.50–0.99
    1.00–1.49
    1.50–1.99
    2.00–2.49
    2.50–2.99
    3.00–3.49
    3.50–3.99

Realizing that these adjusted class limits do not allow a perfect grade-point average of 4.00, we might also replace the last class of 3.50–3.99 with "3.50 and higher." We no longer have ten classes and the class widths may not be uniform throughout, but the resulting frequency table is much more readable and usable.

We will now consider the 150 employee absence times listed in Table 2-1 on page 23. The highest and lowest times are 234 hours and 0 hours, respectively. Assume that we want 12 classes in our frequency table; we divide 12 into 234 for a class width of 19.5, and, if we round that result up to 20, we get the following more convenient classes:

      0–19
     20–39
     40–59
       ⋮
    220–239

We can now complete the frequency table by recording a tally mark for each score in its proper class, and then we can find the total number of tally marks in each class (see Table 2-3). The first value of 85 would be recorded by a tally mark (|) in the 80–99 class. The second value in the first column (129) would be recorded with a tally mark in the 120–139 class, and so on. After four tally marks have been recorded in one class, the fifth mark would be drawn through the other four (|⧸||) for ease of counting by fives.

Table 2-3 provides a tremendous advantage by making intelligible the otherwise unintelligible list of 150 absence times. Yet this advantage is not gained without some cost. In constructing frequency tables, we may lose the accuracy of raw data. To see how this accuracy can be lost, consider the first class of 0–19. The table indicates that there are 19 values in that class, but there is no way to determine from the table

Table 2-3		
Absence time (hrs)	Tally marks	Frequency
0–19	⳾⳾ ⳾⳾ ⳾⳾ IIII	19
20–39	⳾⳾ ⳾⳾	10
40–59	⳾⳾ ⳾⳾ ⳾⳾	15
60–79	⳾⳾ ⳾⳾ ⳾⳾ ⳾⳾ II	22
80–99	⳾⳾ ⳾⳾ ⳾⳾ ⳾⳾ ⳾⳾	25
100–119	⳾⳾ ⳾⳾ ⳾⳾ ⳾⳾	20
120–139	⳾⳾ ⳾⳾ IIII	14
140–159	⳾⳾ ⳾⳾ III	13
160–179	⳾⳾ I	6
180–199	III	3
200–219		0
220–239	III	3

Table 2-4	
Absence time	Frequency
0–240	150

exactly what those original values are. We cannot reconstruct the original list of 150 absences from Table 2-3. The exact values have been compromised for the sake of comprehension. To take an extreme and somewhat absurd example, the 150 absence times could be put into a frequency table with one class, as in Table 2-4. Here the scores have been stripped of any semblance of precision. It is very easy to understand Table 2-4, but the data have lost almost all meaning.

Summarizing data generally involves a compromise between accuracy and simplicity. A frequency table with too few classes is simple but not accurate. A frequency table with too many classes is more accurate but not as easy to understand. The best arrangement is arrived at subjectively, usually in accordance with the common formats used in particular applications. Some of these difficulties will be overcome in the next section when stem-and-leaf plots are discussed.

A variation of the standard frequency table is used when cumulative totals are desired. The cumulative frequency for a class is the sum of the frequencies for that class and all previous classes. Table 2-5 is an example of a **cumulative frequency table,** and it corresponds to the same 150 absence times presented in Table 2-3. A comparison of the frequency column of Table 2-3 and the cumulative frequency column of Table 2-5 reveals that the latter values can be obtained from the former by starting at the top and adding on the successive values. For example, the cumulative frequency of 29 from Table 2-5 represents the sum of 19 and 10 from Table 2-3.

Table 2-5	
Absence time (hrs)	Cumulative frequency
Less than 20	19
Less than 40	29
Less than 60	44
Less than 80	66
Less than 100	91
Less than 120	111
Less than 140	125
Less than 160	138
Less than 180	144
Less than 200	147
Less than 220	147
Less than 240	150

In the next section, we will explore various graphic ways to depict data so that they are easy to understand. Frequency tables will be necessary for some of the graphs, and those graphs are often necessary for considering the way the scores are distributed. Frequency tables therefore become important prerequisites for later, more useful concepts.

2-2 Exercises A
Summarizing Data

2-1   What upper class limits correspond to lower class limits of 3, 6, 9, 12, 15?

2-2   What upper class limits correspond to lower class limits of 0, 10, 20, 30, 40, 50?

2-3   What upper class limits correspond to lower class limits of 0, 2.5, 5.0, 7.5, 10.0?

2-4   What upper class limits correspond to lower class limits of 50, 65, 80, 95, 110, 125?

2-5   For a certain frequency table, the lower limit of the first class is 10. The class width is 20 and there are five classes. Identify the other four lower class limits.

2-6   A frequency table has six classes and the class width is 6. If the lower limit of the first class is 8, identify the other five lower class limits.

2-7   In a frequency table with eight classes, the lower limit of the first class is 7.5, and the class width is 2.5. Identify the other lower class limits.

2-8   In a frequency table with six classes, the lower limit of the first class is 0, and the class width is 0.2. Find the other lower class limits.

In Exercises 2-9 through 2-12, find a suitable value for the class width given the information provided about the data set.

2-9   Minimum is 50, maximum is 99, 10 classes.

2-10   Minimum is 50, maximum is 91, 6 classes.

2-11   Minimum is 0.67, maximum is 3.24, 11 classes.

2-12   Minimum is 120, maximum is 239, 8 classes.

In Exercises 2-13 through 2-16, identify the class width for the frequency table identified.

2-13   Table 2-6   **2-14**   Table 2-7

2-15   Table 2-8   **2-16**   Table 2-9

Table 2-6	
IQ	Frequency
80–87	16
88–95	37
96–103	50
104–111	29
112–119	14

Table 2-7	
Time (hours)	Frequency
0.0–7.5	16
7.6–15.1	18
15.2–22.7	17
22.8–30.3	15
30.4–37.9	19

Table 2-8	
Weight (kg)	Frequency
16.2–21.1	16
21.2–26.1	15
26.2–31.1	12
31.2–36.1	8
36.2–41.1	3

Table 2-9	
Sales (dollars)	Frequency
0–21	2
22–43	5
44–65	8
66–87	12
88–109	14
110–131	20

In Exercises 2-17 through 2-20, identify the class marks for the frequency table identified.

2-17   Table 2-6   **2-18**   Table 2-7

2-19   Table 2-8   **2-20**   Table 2-9

In Exercises 2-21 through 2-24, identify the class boundaries for the frequency table identified.

2-21   Table 2-6   **2-22**   Table 2-7

2-23   Table 2-8   **2-24**   Table 2-9

In Exercises 2-25 through 2-28, construct the cumulative frequency table corresponding to the frequency table indicated.

2-25   Table 2-6   **2-26**   Table 2-7

2-27   Table 2-8   **2-28**   Table 2-9

In Exercises 2-29 through 2-32, modify the class limits of the frequency table identified so that the new limits make the table easier to read. Omit the frequencies since they cannot be determined. It may be necessary to change the number of classes.

2-29   Table 2-6   **2-30**   Table 2-7

2-31   Table 2-8   **2-32**   Table 2-9

In Exercises 2-33 through 2-36, use the given information to find the upper and lower limits of the first class.

2-33   Assume that you have a collection of scores representing the weights (rounded up to the nearest pound) of adult males. Also assume that the lowest and highest weights are 110 and 309, respectively. (You wish to construct a frequency table with 10 classes.)

2-34   You have been studying the jail sentences of those convicted of driving while intoxicated. You have a list of sentences ranging from 0 days in jail to 95 days in jail, and you intend to publish a frequency table with 16 classes.

2-35   A scientist is investigating the time required for a certain chemical reaction to occur. The experiment is repeated 200 times and the results vary from 17.3 to 42.7 seconds. (You wish to construct a frequency table with 12 classes.)

2-36   A track star has run the mile in 150 different events, and her times ranged from 238.7 seconds to 289.8 seconds. (You wish to construct a frequency table with 14 classes.)

2-37   The following scores represent the time (in hours) between failures of aircraft radios (Stewart model XK-84). Construct a frequency table with 10 classes.

108	168	152	74	85
136	52	34	62	137
150	175	121	136	126
120	143	174	164	48
42	127	148	81	137
243	74	122	86	128
139	123	86	85	125
103	197	132	49	72
110	123	111	137	77
129	115	215	154	130

2-38 A housing development is comprised of 50 homes that are identical in size. They have identical heating systems and lighting arrangements. In a study of energy consumption, the amounts of electricity (in kilowatt hours) used in the separate homes for a 1-month period are as follows. Construct a frequency table with 10 classes.

919	897	968	753	821	978	1021	837	852	691
967	829	1008	1040	924	902	884	836	788	968
950	821	951	815	806	923	926	778	825	851
948	739	909	823	903	1011	1000	894	777	994
945	896	903	932	976	841	806	741	820	1087

2-39 The Bureau of Weights and Measures is investigating consumer complaints that the Curtiss Sugar Company is cheating customers who buy sugar in 5-pound bags. The following scores represent the weights in pounds of 150 bags of sugar produced by the Curtiss Sugar Company. Construct a frequency table with 13 classes.

4.95	4.99	5.04	5.03	4.96	4.99
4.98	4.99	4.93	5.01	5.00	4.92
5.04	5.00	4.97	5.01	4.98	5.02
5.04	4.92	4.98	4.98	5.04	4.96
4.95	4.99	4.92	4.95	4.94	4.98
5.00	4.94	4.98	4.94	5.03	4.92
5.03	4.98	5.04	5.03	4.94	4.97
4.92	5.04	4.97	4.93	5.04	4.97
4.95	4.98	4.94	5.03	4.97	4.92
4.96	4.95	5.02	5.01	5.01	5.03
4.93	5.04	4.92	4.98	5.01	4.96
5.04	4.99	5.01	5.00	4.92	5.02
5.02	5.02	4.93	5.01	5.03	5.04
5.03	5.03	5.02	4.97	4.93	5.03
4.92	4.95	4.96	4.99	5.04	4.92
4.94	5.01	4.96	4.98	4.94	5.01
4.94	5.00	5.00	4.92	5.04	5.00
4.93	5.03	4.95	4.99	5.01	4.94
4.99	4.92	5.00	4.98	5.04	4.98
4.95	4.95	5.02	4.94	4.93	5.04
4.92	4.95	4.96	5.01	5.03	4.93
5.03	4.94	5.04	5.04	5.00	4.98
5.03	4.98	4.99	4.92	4.94	5.00
4.97	5.01	4.95	4.97	5.04	4.92
4.92	4.98	5.01	4.92	5.02	4.96

2-40 The Curtiss Sugar Company also distributes flour. The following scores represent the weights of 150 sacks of flour that bear the Curtiss label. Construct a frequency table with 12 classes.

1.09	0.92	1.00	1.12	0.89	0.98	1.09	1.14
0.75	1.03	1.09	0.95	0.92	1.06	1.20	0.95
1.12	0.95	1.00	1.06	0.95	0.98	0.89	0.98
0.92	1.06	0.84	0.89	0.92	0.78	0.95	1.00
0.92	0.98	0.92	0.84	1.17	1.03	1.00	0.98
0.89	0.75	1.09	1.03	0.98	0.89	0.95	0.92
0.98	1.20	1.00	1.03	0.95	0.95	0.95	1.03
1.14	1.00	0.95	0.95	1.06	1.03	0.95	0.95
1.00	0.92	0.87	0.89	1.00	0.87	0.98	0.92
1.00	1.03	0.98	0.98	1.00	1.00	0.87	0.87
0.89	0.98	1.00	0.87	0.98	0.95	1.00	1.00
0.95	1.00	1.17	0.98	0.87	1.00	1.09	0.92
0.89	0.89	1.00	1.09	1.06	0.87	0.89	0.98
0.92	1.00	0.98	0.89	1.09	0.87	0.89	1.09
1.06	0.92	1.06	1.00	1.03	1.00	1.06	1.06
0.95	0.89	0.92	0.95	0.75	1.09	1.09	0.87
1.03	0.89	1.03	0.87	0.84	0.78	1.09	0.84
0.98	0.95	0.98	0.95	0.98	1.03	1.03	1.00
0.98	1.00	1.00	1.09	1.00	1.03		

2-2 Exercises B
Summarizing Data

2-41 Compare frequency Tables 2-6, 2-7, 2-8, and 2-9. Are there any obvious differences in the *distribution* of the scores? Could such differences be readily observed by examining the raw scores?

2-42 Compare the frequency tables of Exercises 2-39 and 2-40. Is there any difference in the *distribution* of the scores? Can such a difference be readily observed by examining the raw scores? How can such a difference in distribution occur?

2-43 A frequency table has class marks of 7, 11, 15, 19, 23, 27. Find the class width and find class limits.

2-44 (a) Collect at least 100 scores from a source other than a textbook. Describe the source or situation used.
(b) Construct a frequency table representing the data from part (a).

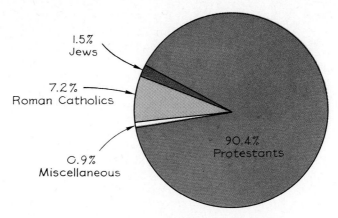

1.5%
Jews

7.2%
Roman Catholics

0.9%
Miscellaneous

90.4%
Protestants

Figure 2-1   Congregations in the United States

2-3   Pictures of Data

In the preceding section, we saw that frequency tables transform a disorganized collection of raw scores into an organized and understandable summary. In this section we consider ways of representing data in pictorial form. The obvious objective is to promote understanding of the data. We attempt to show that one graphic illustration can be a suitable replacement for hundreds of data.

Pie Charts

It is a simple and obvious characteristic of human physiology that people can comprehend one picture better than a more abstract collection of words or data. Consider this statement: Of the religious congregations in the United States, 1.5% are Jewish, 7.2% are Roman Catholic, 90.4% are Protestants, and 0.9% are of other denominations. What impressions did you develop? How much did you comprehend? Now examine the **pie chart** in Figure 2-1. While the pie chart depicts the same information as the quotation, it does a better job of dramatizing the data. The abstraction of numbers is overcome by the concrete reality of the "slices" of pie.

Pie charts are especially useful in cases involving data categorized according to attributes rather than numbers (data at the nominal level of measurement). The Jewish, Catholic, and Protestant religions are examples of attributes that do not correspond to numbers.

The construction of a pie chart from raw data is not too difficult. It simply involves the "slicing up" of the pie into the proper proportions. In Figure 2-2 we depict the pie chart representing a family budget. If the category of housing requires 24% of the total budget, then the housing slice should be 24% of the total pie.

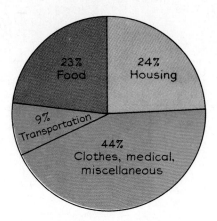

Figure 2-2 Family budget

Figure 2-3

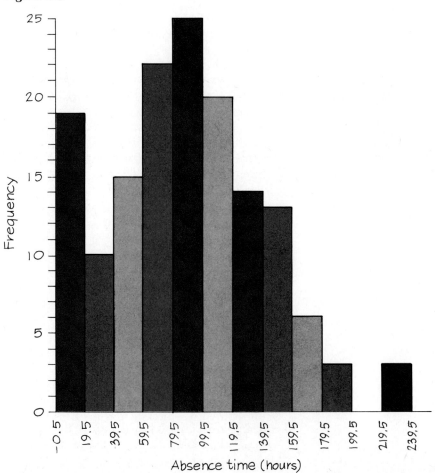

Histograms Can Be Revealing

In addition to revealing the nature of a distribution, histograms can sometimes reveal how the data were collected. One researcher was given a set of weights of adult American males. The data appeared to be reasonable, but a histogram of the last digit of each score showed that 0 and 5 occurred with much greater frequency than the other eight digits. From this, the analyst was able to conclude that the subjects were asked for their weights; they were not actually weighed. Many of the respondents apparently gave their weights in round figures, such as 170 and 185. After recognizing how the data were obtained, the analyst questioned their accuracy since there is a natural tendency for people to underestimate their true weight.

Histograms

A **histogram** is another common graphic way of presenting data. Histograms are especially important because they show the distribution of the data.

We generally construct a histogram to represent a set of scores after we have completed a frequency table. The standard format for a histogram usually involves a vertical scale that delineates frequencies and a horizontal scale that delineates values of the data being represented. We use shaded bars to represent the individual classes of the frequency table. Each bar extends from its lower class boundary to its upper class boundary so that we can mark the class boundaries on the horizontal scale. (However, improved readability is often achieved by using class limits or class marks instead of class boundaries.) As an example, Figure 2-3 is the histogram that corresponds directly to Table 2-3 in the previous section.

Before constructing a histogram from a completed frequency table, we must give some consideration to the scales used on the vertical and horizontal axes. We can begin by determining the maximum frequency, and the maximum frequency should suggest a suitable scale for the vertical axis. In Table 2-3 the maximum frequency of 25 will correspond to the tallest bar on the histogram, so 25 (or the next highest convenient number) should be located at the top of the vertical scale, with 0 at the bottom. In Figure 2-3, we designed the vertical scale to run from 0 to 25. The horizontal scale should be designed to accommodate all the classes of the frequency table. Consider the desired length of the horizontal axis along with the number of classes that must be incorporated. Both axes should be clearly labeled.

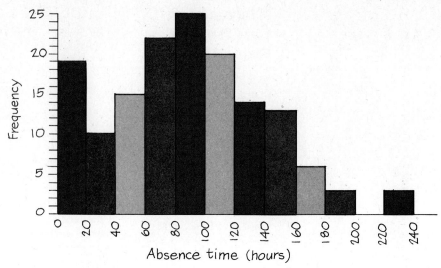

Figure 2-4   Histogram

The histogram of Figure 2-4 shows the same data as Table 2-3 and Figure 2-3, but with the class limits used on the horizontal scale for improved readability. The vertical axis is scaled down to create an impression of smaller differences. The flexibility in manipulating the scales for the vertical and horizontal axes can serve as a great device for deception. A clever person can exaggerate small differences or deemphasize large differences to suit his or her own purpose. However, conscientious and critical students of statistics become aware of such misleading presentations and carefully examine histograms by analyzing the objective numerical data, instead of simply glancing at the overall picture.

Frequency Polygons

A **frequency polygon** is a variation of the histogram in which the vertical bars are replaced by dots that are subsequently connected to form a line graph. When delineating the horizontal scale of a frequency polygon, we generally use class marks, so that the dots can be located directly above them. In Table 2-3 the class marks are 9.5, 29.5, 49.5, . . . , 229.5. In Figure 2-5 we identify those values along the horizontal axis, plot the points directly above them, connect those points with straight line segments, and then extend the resulting line graph at the beginning and end so that it starts and ends with a frequency of 0.

Ogives

Another common pictorial display is the **ogive** (pronounced "oh-jive"), or **cumulative frequency polygon.** This differs from an ordinary

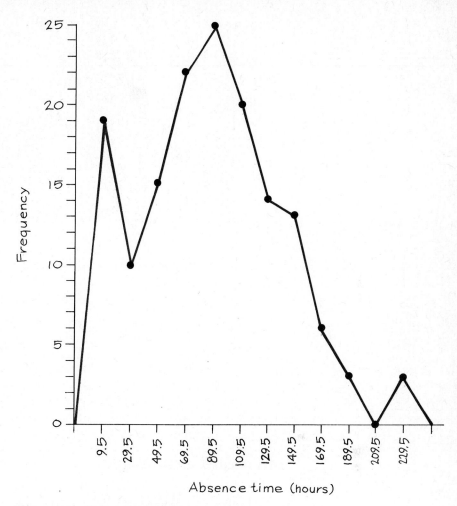

Figure 2-5 **Frequency polygon**

frequency polygon in that the frequencies are cumulative. That is, we add each class frequency to the total of all previous class frequencies, as in the cumulative frequency table described in Section 2-2. The vertical scale is again used to delineate frequencies, but we must now adjust the scale so that it accommodates the total of all individual frequencies. The horizontal scale should depict the class boundaries (see Figure 2-6). Ogives are useful when, instead of seeking the frequency for a given class interval, we want to know how many of our scores are above or below some level. For example, in doing an analysis of women's salaries, we might be more interested in the number of salaries below $20,000 than in the number of salaries between any two particular values.

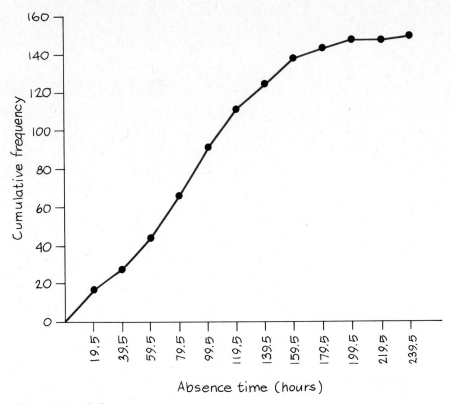

Figure 2-6 Ogive

Stem-and-Leaf Plots

Histograms, frequency polygons, and ogives are extremely useful for graphically displaying the distribution of data. By using those graphic devices, we are able to learn something about the data that is not apparent while the data remain in a list of values. This additional insight is clearly an advantage. However, in constructing histograms, frequency polygons, or ogives we also suffer the disadvantage of distorted data. When we transform raw data into a histogram, for example, we lose some information in the process. Generally we cannot reconstruct the original data set from the histogram, and this shows that some distortion has occurred. We will now introduce another device, which enables us to see the distribution of data without losing information in the process. The following concept is part of a relatively new branch of statistics often referred to as **exploratory data analysis** (EDA). See *Exploratory Data Analysis* by John W. Tukey for reference to other concepts in this field, which is gaining wider acceptance.

In a **stem-and-leaf plot** we sort data according to a pattern that reveals the underlying distribution. The pattern involves separating a number into two parts, usually the first digit and the other digits. The *stem* consists of the leftmost digits and *leaves* consist of the rightmost digits. The method is illustrated in the following example.

EXAMPLE

A librarian records the number of daily microfilm uses and compiles the sample data that follow. Construct a stem-and-leaf plot for this data.

| 10 | 11 | 15 | 23 | 27 | 28 | 38 | 38 | 39 | 39 |
| 40 | 41 | 44 | 45 | 46 | 46 | 52 | 57 | 58 | 65 |

Solution

We note that the numbers have first digits of 1, 2, 3, 4, 5, or 6 and we let those values become the stem. We then construct a vertical line and list the "leaves" as shown below. The first row represents the numbers 10, 11, 15. In the second row we have 23, 27, 28, and so on.

Stem	Leaves
1	015
2	378
3	8899
4	014566
5	278
6	5

By turning the page on its side we can see a distribution of these data, which, in this case, roughly approximates a bell shape. We have also retained all the information in the original list. We could reconstruct the original list of values from the stem-and-leaf plot. In the next example we illustrate the construction of a stem-and-leaf plot for data with three significant digits.

Children See Much Television Violence

The National Association for Better Broadcasting has estimated that between the ages of 5 and 15 a child will see about 13,000 violent deaths on television. This estimate was based on a one-week monitoring of Los Angeles television stations.

Results were projected according to the viewing habits of children in the 5 to 15 age bracket. This statistic has caused concern because many people believe that television violence has an adverse effect on child behavior.

E X A M P L E

An aeronautical research team is investigating the stall speed of an ultralight aircraft, and the following sample values are obtained (in knots). Construct a stem-and-leaf plot and the graph suggested by that plot.

21.7 24.0 22.4 22.4 24.3 22.3 22.6 25.2
24.1 21.8 23.2 23.9 23.5 23.2 23.9 23.8

Solution

Note that unlike the previous example, these data are not in order, so two sweeps will be necessary. For the first sweep, we record the data reading across one row at a time to get the following result.

Stem	Leaves
21.	78
22.	4436
23.	295298
24.	031
25.	2

For the second sweep we arrange each row of leaves in order from low to high as follows.

Stem	Leaves
21.	78
22.	3446
23.	225899
24.	013
25.	2

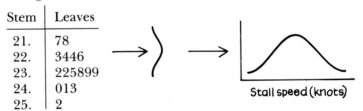

Stall speed (knots)

Here we let the stem consist of the two left digits while the leaf is the digit farthest to the right. Again, we can see the shape of the distribution by turning the page on its side and observing the columns of digits above the line.

For some data sets, simplified stem-and-leaf plots can be constructed by first rounding the values to two or three significant digits. If necessary, stem-and-leaf plots can be condensed by combining adjacent rows. They can also be stretched out by subdividing rows into those with the digits 0 through 4 and those with digits 5 through 9. Note that the following two stem-and-leaf plots represent the same set of data, but the plot on the right has been stretched out to include more rows. Also note that such changes may affect the apparent shape of the distribution.

$$
\begin{array}{c|c}
51 & 6899 \\
52 & 0347 \\
53 & 3788
\end{array}
\quad \xrightarrow{\text{expand}} \quad
\begin{array}{c|ll}
51 & & \text{(last digits of 0 through 4)} \\
51 & 6899 & \text{(last digits of 5 through 9)} \\
52 & 034 & \text{(last digits of 0 through 4)} \\
52 & 7 & \text{(last digits of 5 through 9)} \\
53 & 3 & \text{(last digits of 0 through 4)} \\
53 & 788 & \text{(last digits of 5 through 9)}
\end{array}
$$

In the preceding example we expanded the number of rows. We can also condense a stem-and-leaf plot by combining adjacent rows as shown below.

$$
\begin{array}{c|l}
50 & 01 \\
51 & 4 \\
52 & 56 \\
53 & 368 \\
54 & 2457 \\
55 & 3499 \\
56 & 0127 \\
57 & 358 \\
58 & 1269 \\
59 & 17
\end{array}
\quad \xrightarrow{\text{contract}} \quad
\begin{array}{c|l}
50\text{--}51 & 01*4 \\
52\text{--}53 & 56*368 \\
54\text{--}55 & 2457*3499 \\
56\text{--}57 & 0127*358 \\
58\text{--}59 & 1269*17
\end{array}
$$

$$ (581) \quad (591) $$

In the condensed plot, we separated digits in the "leaves" associated with the numbers in each stem by an asterisk. Every row in the condensed plot must include exactly one asterisk so that the shape of the plot is not distorted.

Another useful feature of stem-and-leaf plots is that their construction provides a fast and easy procedure for ranking data (putting data in order). Data must be ranked for a variety of statistical procedures, such as the Wilcoxon rank-sum test in Chapter 11 and the median in the next section.

Miscellaneous Graphics

Numerous pictorial displays other than the ones just described can be used to represent data dramatically and effectively. Some examples are soldiers, tanks, airplanes, stacks of coins, and moneybags. This list comprises a very meager sampling of the diverse graphic methods used to convey the nature of statistical data. In fact, there is almost no limit to the variety of different ways that data can be illustrated. However, pie charts, histograms, frequency polygons, and ogives are among the most common devices used (see Table 2-10 and Figures 2-7, 2-8, and 2-9). These graphic representations often convey the *distribution* of the data, and an understanding of the distribution is often critically important to the statistician. In the following section we consider ways of measuring other characteristics of data.

Table 2-10	
Score	Frequency
1–5	1
6–10	0
11–15	3
16–20	5
21–25	2

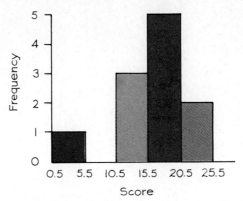

Figure 2-7 **Histogram** representing the data summarized in Table 2-10. The horizontal axis is delineated with **class boundaries.**

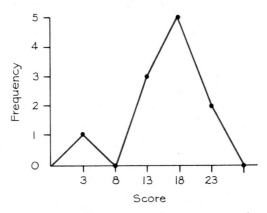

Figure 2-8 **Frequency polygon** representing the data summarized in Table 2-10. The horizontal axis is delineated with **class marks.**

2-3 Exercises A
Pictures of Data

2-45 Given the following frequency table, identify the values that should be entered along the horizontal scale. (See Figures 2-7, 2-8, and 2-9.) Assume that the frequency table will be used to construct
(a) a histogram
(b) a frequency polygon
(c) an ogive

Score	Frequency
50–59	2
60–69	0
70–79	6
80–89	3
90–99	1

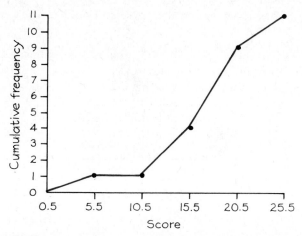

Figure 2-9 Ogive representing the data summarized in Table 2-10. The horizontal axis is delineated with **class boundaries.**

2-46 Construct a histogram that corresponds to the frequency table in Exercise 2-45.

2-47 Construct a frequency polygon that corresponds to the frequency table in Exercise 2-45.

2-48 Construct an ogive that corresponds to the frequency table in Exercise 2-45.

2-49 Given the following frequency table, identify the values that should be entered along the horizontal scale. (See Figures 2-7, 2-8, and 2-9.) Assume that the frequency table will be used to construct
(a) a histogram
(b) a frequency polygon
(c) an ogive

Score	Frequency
1–10	8
11–20	12
21–30	9
31–40	10
41–50	15
51–60	18

The scores represent the number of aircraft operations in one day for a particular airport.

2-50 Construct a histogram that corresponds to the frequency table in Exercise 2-49.

2-51 Construct a frequency polygon that corresponds to the frequency table in Exercise 2-49.

2-52   Construct an ogive that corresponds to the frequency table in Exercise 2-49.

2-53   Given the following frequency table, identify the values that should be entered along the horizontal scale. (See Figures 2-7, 2-8, and 2-9.) Assume that the frequency table will be used to construct
(a) a histogram
(b) a frequency polygon
(c) an ogive

Score	Frequency
1–30	15
31–60	22
61–90	30
91–120	35
121–150	30
151–180	35
181–210	20
211–240	10

The scores represent the time (in seconds) required for mice to work their way through a special type of maze.

2-54   Construct a histogram that corresponds to the frequency table in Exercise 2-53.

2-55   Construct a frequency polygon that corresponds to the frequency table in Exercise 2-53.

2-56   Construct an ogive that corresponds to the frequency table in Exercise 2-53.

2-57   Given the following frequency table, identify the values that should be entered along the horizontal scale. (See Figures 2-7, 2-8, and 2-9.) Assume that the frequency table will be used to construct
(a) a histogram
(b) a frequency polygon
(c) an ogive

Score	Frequency
0–99	55
100–199	45
200–299	60
300–399	70
400–499	75
500–599	60
600–699	80
700–799	50
800–899	75
900–999	40

The scores represent numbers randomly generated by a computer for use in a simulation.

2-58   Construct a histogram that corresponds to the frequency table in Exercise 2-57.

2-59   Construct a frequency polygon that corresponds to the frequency table in Exercise 2-57.

2-60   Construct an ogive that corresponds to the frequency table in Exercise 2-57.

In Exercises 2-61 through 2-64, list the original numbers in the data set represented by the given stem-and-leaf plots.

2-61

Stem	Leaves
2	00358

2-62

Stem	Leaves
1	001112278
2	3444569
3	013358

2-63

Stem	Leaves
40	6678
41	09999
42	13466
43	088

2-64

Stem	Leaves			
68	45	45	47	86
69	33	38	89	
70	52	59	93	
71	27			

In Exercises 2-65 through 2-68, construct the stem-and-leaf plots for the given data sets.

2-65   Hours of sleep for various subjects:

6.8   6.8   7.3   7.0   7.1   6.9   7.1   7.4   7.6   8.0
8.1   9.2   9.4   9.7   8.7

2-66   Freezing levels (in degrees Celsius below 0° C) of aircraft fire extinguishers:

37.6   41.1   41.3   42.7   38.0   38.0   38.9   40.0
40.6   40.7   40.9   39.8   39.2   39.0   39.4   39.0

2-67   Lap times of a race car (in seconds):

56.4   52.6   57.0   57.9   53.3   53.6   54.0   55.2   52.5   52.8
52.3   55.4   55.6   54.1   55.3   54.6   55.8   54.1

2-68   Takeoff distances (in feet) required for a light aircraft:

717   716   736   772   740   741   735   735   710   753   757   747
756   715   718   720   726   721   760   769   771   738   721

2-69   Construct the frequency table corresponding to the stem-and-leaf plot given on the next page. The data represents sales in an auto parts store. All values have been rounded to the nearest dollar.

Stem	Leaves
0	00011222233333344555677788999
1	0011122333334455555667
2	0000111133333346678
3	0011111223334445567899
4	001122233334555667789
5	01334678999
6	11355779
7	
8	0245
9	3
10	26

2-70 (a) Using the frequency table from Exercise 2-69, construct the corresponding histogram.

(b) Compare the histogram from part (a) to the shape of the stem-and-leaf plot given in Exercise 2-69.

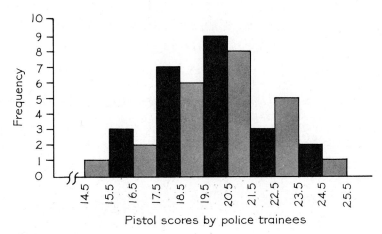

Figure 2-10 Histogram

2-71 Use the histogram in Figure 2-10.

(a) Construct the corresponding frequency table. Use the 11 classes 15, 16, 17, . . . , 25.

(b) Construct the corresponding ogive.

(c) Refer to the frequency table of part (a) to determine how many police trainees scored below 21.

(d) If 17 is a passing score, what percentage of trainees failed this test?

2-72 Use the frequency polygon in Figure 2-11.

(a) Construct the corresponding frequency table.

(b) Construct the corresponding ogive.

(c) How many subjects had a pulse rate between 75 and 84?

(d) What percentage of subjects had a pulse rate below 90?

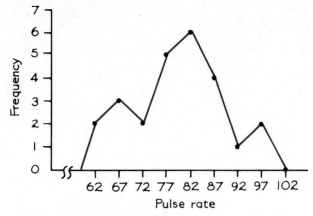

Figure 2-11 Frequency polygon

2-3 Exercises B
Pictures of Data

2-73 (a) Reconstruct the histogram of Exercise 2-50 so that the differences in frequencies are exaggerated.

(b) Reconstruct the histogram of Exercise 2-50 so that the differences in frequencies are deemphasized.

2-74 (a) Use a computer or telephone directory to obtain 100 random digits between 0 and 9, inclusive.

(b) Before doing any analysis, what do you predict for the shape of the histogram?

(c) Construct a frequency table with 10 classes.

(d) Construct a histogram with 10 classes.

2-75 Construct a stem-and-leaf plot of the employee absence times listed in Table 2-1.

2-76 Given the graphs of the following frequency polygons, construct the graphs of the corresponding ogives.

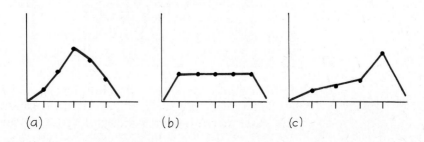

2-4 Averages

In this chapter we describe some of the fundamental methods and tools generally categorized as descriptive statistics. We began with frequency tables that are used to summarize raw data (Section 2-2) and then considered a variety of graphic devices that represent the data in pictorial form (Section 2-3). However, summary and graphic methods cannot be used easily if we are limited to verbal exchanges. In addition they are very difficult to use when we want to make statistical inferences.

Suppose, for example, that we wish to compare the reaction times of

Figure 2-12

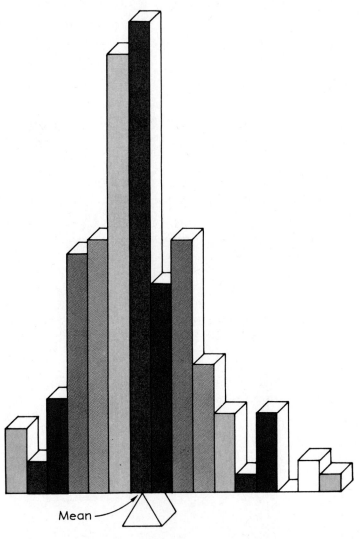

Mean

If a fulcrum is placed at the position of the mean, it will balance the histogram.

two different groups of people. We could collect the two sets of raw data, organize them into two frequency tables, and then construct the appropriate histograms. Next we could compare the relative shapes and positions of the two histograms, but we would be evaluating the strengths of their similarities subjectively. Our conclusions and inferences would be much more reliable if we could develop more objective procedures. Fortunately, we can.

The more objective procedures do require numerical measurements that describe certain characteristics of the data. In this and the following section we are concerned with some of these quantitative measurements. This section deals mainly with **averages,** which are also called **measures of central tendency** since they are measurements that are intended to capture the value at the center of the scores.

An important and basic point to remember is that there are different ways to compute an average. We have already stated that, given a list of scores and instructions to find the average, most people will obligingly proceed first to total the scores and then to divide that total by the number of scores. However, this particular computation is only one of several procedures that are classified under the umbrella description of an average. The most commonly used averages are mean, median, mode, and midrange.

Mean

The **arithmetic mean** is the most important of all numerical descriptive measurements, and it corresponds to what most people call an *average*. The arithmetic mean of a list of scores is obtained by adding the scores and dividing the total by the number of scores. This particular average will be employed frequently in the remainder of this text, and it will be referred to simply as the **mean** (see Figure 2-12).

E X A M P L E

Find the mean of the following quiz scores:

$$2, 3, 6, 7, 7, 8, 9, 9, 9, 10$$

Solution

First add the scores.

$$2 + 3 + 6 + 7 + 7 + 8 + 9 + 9 + 9 + 10 = 70$$

Then divide the total by the number of scores present.

$$\frac{70}{10} = 7$$

The mean score is therefore 7.

Over Half Our Presidents Are Above Average?

Historians were surveyed in an effort to rate the different presidents. The top four were Lincoln, Franklin D. Roosevelt, Washington, and Jefferson. Warren Harding was at the bottom of the list. One historian involved in the survey tabulations reported that "almost one out of every four (presidents) has been great or near great, and over half are above average." How can more than half our presidents be above average?

In many cases, we can provide a formula for computing some measurement. In statistics, these formulas often involve the Greek letter Σ (capital sigma) which is intended to denote the summation of a set of values. Σx means *sum the values that* x *can assume.* If we are working with a specific set of scores, the letter x is used as a variable that can assume the value of any one of those scores. Thus Σx means the sum of all of the scores. In the preceding example, the variable x assumes the values of 2, 3, 6, 7, 7, 8, 9, 9, 9, and 10, so Σx is

$$2 + 3 + 6 + 7 + 7 + 8 + 9 + 9 + 9 + 10, \text{ or } 70.$$

In future formulas, we may encounter expressions such as Σx^2. Standard usage of the symbol Σ requires that the values to be added must be in the form of the variable following Σ. Thus Σx^2 indicates that the individual scores must be squared *before* they are summed. Referring to the preceding scores, Σx^2 is

$$2^2 + 3^2 + 6^2 + 7^2 + 7^2 + 8^2 + 9^2 + 9^2 + 9^2 + 10^2$$
$$= 4 + 9 + 36 + 49 + 49 + 64 + 81 + 81 + 81 + 100$$
$$= 554$$

NOTATION

Σ	denotes **summation** of a set of values.
x	is the **variable** usually used to represent the individual raw scores.
n	represents the **number** of scores being considered.
$\bar{x}$	denotes the **mean** of a set of **sample** scores.
μ	denotes the **mean** of all scores in some **population.**

Since the mean is the sum of all scores (Σx) divided by the number of scores (n), we have the following formula.

Formula 2-1
$$\text{mean} = \frac{\Sigma x}{n}$$

The result can be denoted by $\bar{x}$ if the available scores are samples from a larger population; if all scores of the population are available, then we can denote the computed mean by μ (mu). Applying the formula to the preceding quiz scores, we get

$$\text{mean} = \frac{\Sigma x}{n} = \frac{70}{10} = 7$$

According to the definition of mean, 7 is the central value of the ten given quiz scores. Other definitions of averages involve different perceptions of how the center is determined. The median reflects another approach.

Median

The **median** of a set of scores is the middle value when the scores are arranged in order of increasing magnitude.

After first arranging the original scores in increasing (or decreasing) order, the median will be either of the following:

1. The number which is exactly in the middle of the list (if the number of scores is odd).
2. The mean of the two middle numbers (if the number of scores is even).

EXAMPLE

Find the median of the scores 7, 2, 3, 7, 6, 9, 10, 8, 9, 9, 10.

Solution

Begin by arranging the scores in increasing order.

$$2, 3, 6, 7, 7, 8, 9, 9, 9, 10, 10$$

With these eleven scores, the number 8 is located in the exact middle so that 8 is the median.

EXAMPLE

Find the median of the scores 7, 2, 3, 7, 6, 9, 10, 8, 9, 9.

Solution

Again, begin by arranging the scores in increasing order.

$$2, 3, 6, 7, 7, 8, 9, 9, 9, 10$$

With these ten scores, no single score is at the exact middle. Instead, the two scores of 7 and 8 share the middle. We therefore find the mean of those two scores.

$$\frac{7 + 8}{2} = \frac{15}{2} = 7.5$$

The median is 7.5.

EXAMPLE

Find the median of the 150 absence times listed in Table 2-1.

Solution

After arranging the scores in increasing order, we find that the 75th and 76th scores are both 87, so the median is 87. The scores can be arranged in increasing order by constructing a stem-and-leaf plot or by using a computer program such as STATDISK.

Mode

The **mode** is obtained from a collection of scores by selecting the score that occurs most frequently. In those cases where no score is repeated, we stipulate that there is no mode. In those cases where two scores both occur with the same greatest frequency, we say that each one is a mode and we refer to the data set as being **bimodal.** If more than two scores occur with the same greatest frequency, each is a mode and the data set is said to be **multimodal.**

EXAMPLE

The scores 1, 2, 2, 2, 3, 4, 5, 6, 7, 9 have a mode of 2.

EXAMPLE

The scores 2, 3, 6, 7, 8, 9, 10 have no mode since no score is repeated.

EXAMPLE

The scores 1, 2, 2, 2, 3, 4, 5, 6, 6, 6, 7, 9 have modes of 2 and 6, since 2 and 6 both occur with the same highest frequency.

EXAMPLE

A town meeting is attended by 40 Republicans, 25 Democrats, and 20 Independents. While we cannot numerically average these party affiliations, we can report that the mode is Republican, since that party had the highest frequency.

E X A M P L E

For the 150 absence times listed in Table 2-1, the value of 4 occurs five times and no other score occurs more than four times so that the mode is 4.

Midrange

The **midrange** is that average obtained by adding the highest score to the lowest score and then dividing the result by 2.

$$\text{midrange} = \frac{\text{highest score} + \text{lowest score}}{2}$$

E X A M P L E

Find the midrange of the scores 2, 3, 6, 7, 7, 8, 9, 9, 9, 10.

$$\text{midrange} = \frac{10 + 2}{2} = 6$$

R U L E

A simple rule for rounding answers is to carry one more decimal place than was present in the original data. We should round only the final answer and not intermediate values. For example, the mean of 2, 3, 5 is expressed as 3.3. Since the original data were whole numbers, we rounded the answer to the nearest tenth. The mean of 2.1, 3.4, 5.7 is rounded to 3.73.

You should know and understand the preceding four averages (mean, median, mode, and midrange). Other averages (geometric mean, harmonic mean, and quadratic mean) are not used as often, and they will be included only in Exercises B at the end of this section (see Exercises 2-106, 2-107, and 2-108).

Weighted Mean

The **weighted mean** is useful in many situations where the scores vary in their degree of importance. An obvious example occurs frequently in the determination of a final average for a course that includes four tests plus a final examination. If the respective grades are 70, 80, 75,

The Class Size Paradox

Students might prefer colleges with smaller classes. But there are at least two distinct ways to obtain mean class size, and the results can be dramatically different. In one college of the State University of New York, if we take the numbers of students in the 737 classes, we get a mean of 40 students. But if we were to compile a list of the class sizes for each student and use these numbers, we would get a mean class size of 147. This large discrepancy is due to the fact that there are many students in large classes, while there

are few students in small classes.

Suppose a college had only two classes, with 10 students in one and 90 students in the other. One obvious way to compute the mean class size is to compute $(10 + 90)/2 = 50$. But if we ask the 100 students for their class sizes we would get ten 10s and ninety 90s, and those 100 numbers would produce a mean of 82. The college would experience a mean class size of 50, but the students would experience a mean class size of 82.

Without changing the number of classes or faculty, the mean class size experienced by students can be reduced by making all classes close to the same size. This would also improve attendance, if we accept research evidence that attendance is better in smaller classes.

85, and 90, the mean of 80 does not reflect the greater importance placed on the final exam. Let's suppose that the instructor counts the respective tests as 15%, 15%, 15%, 15%, and 40%. The weighted mean then becomes

$$\frac{(70 \times 15) + (80 \times 15) + (75 \times 15) + (85 \times 15) + (90 \times 40)}{100}$$

$$= \frac{1050 + 1200 + 1125 + 1275 + 3600}{100}$$

$$= \frac{8250}{100} = 82.5$$

This computation suggests a general procedure for determining a weighted mean. Given a list of scores $x_1, x_2, x_3, \ldots, x_n$ and a corresponding list of weights $w_1, w_2, w_3, \ldots, w_n$, the weighted mean is obtained by computing

Table 2-11			
Absence time	Frequency f	Class mark x	$f \cdot x$
0–19	19	9.5	180.5
20–39	10	29.5	295.0
40–59	15	49.5	742.5
60–79	22	69.5	1529.0
80–99	25	89.5	2237.5
100–119	20	109.5	2190.0
120–139	14	129.5	1813.0
140–159	13	149.5	1943.5
160–179	6	169.5	1017.0
180–199	3	189.5	568.5
200–219	0	209.5	0
220–239	3	229.5	688.5
Total	150		13205.0

Formula 2-2 $$\text{weighted mean} = \frac{\Sigma(w \cdot x)}{\Sigma w}$$

That is, first multiply each score by its corresponding weight; then find the total of the resulting products thereby evaluating Σwx. Finally, add the values of the weights to find Σw and divide the latter value into the former.

Formula 2-2 can be modified so that we compute the mean from a frequency table. The absence data from Table 2-1 have been entered in Table 2-11 where we use the class marks as representative scores and the frequencies as weights. Then the formula for the weighted mean leads directly to Formula 2-3, which can be used to compute the mean of a set of scores in a frequency table.

Formula 2-3 $$\bar{x} = \frac{\Sigma(f \cdot x)}{\Sigma f}$$

where x = class mark and f = class frequency.

Recall that the class mark of a class interval is simply the midpoint value. For example, the class interval of 0–19 has 9.5 as its class mark. Thus Table 2-10 has class marks of 9.5, 29.5, 49.5, . . . , 229.5.

We can now compute the weighted mean.

$$\bar{x} = \frac{\Sigma(f \cdot x)}{\Sigma f} = \frac{13205.0}{150} = 88.0$$

If we had used the original collection of scores to calculate the mean directly, we would have obtained a mean of 87.9, so the value of the weighted mean obtained from the frequency table is quite accurate. The procedure we use is justified by the fact that a class like 0–19 can be represented by its class mark of 9.5, and the frequency number indicates that the representative score of 9.5 occurs 19 times. In essence, we are treating Table 2-11 as if it contained 19 values of 9.5 each, 10 values of 29.5 each, and so on. In examining Figure 2-3, the histogram of the data, the value of 88 for the mean does seem to indicate a central or balance point of the data, so the results of the calculations are at least reasonable.

We have stressed that the four basic measures of central tendency (mean, median, mode, midrange) can all be properly called averages and that the freedom to select a particular average can be a source of deception. Consider the extreme example of a class of students from which the following statistics have been computed.

mean I.Q. = 110.0
median I.Q. = 120.0
mode I.Q. = 95.0
midrange I.Q. = 100.0

Technically, the teacher of this class can use any of the preceding figures as the average I.Q. At the end of the academic year, the teacher could refer to the median value of 120.0 so that the teacher's own achievements and performance are enhanced. Even when we are not trying to enhance our achievements, the selection of the best average is not always easy. The different averages are characterized by different advantages and disadvantages. Consequently there are no objective criteria that determine the most representative average for a given set of data.

The mean has the distinct advantage of being the most familiar of the four basic averages. Also, for any finite set of scores, the mean always exists and is a single unique number that can be used in further statistical analyses. As another advantage, the mean takes every score into account. One serious disadvantage of the mean is that it is affected by exceptional extremes (high or low scores).

For any finite set of scores, the median always exists as a single unique number that is relatively easy to find. Unlike the mean, the median has the advantage of not being affected by some exceptional extreme values. However, a disadvantage is that the median is not sensitive to the value of every score.

When it exists, the mode is the easiest average to find, but there are many collections of data for which the mode is not unique or does not exist. A most serious disadvantage of the mode is its total insensitivity to all other scores. The mode is a useful measure in certain cases, but it may be a poor choice in many cases. A unique advantage of the mode is its applicability to data at the nominal level of measurement. For example, take a survey of 70 artists to determine their favorite colors. How can you average 18 reds, 23 blues, and 29 greens? The modal choice is green, but the other averages are useless in this example.

For any finite set of scores, the midrange always exists as a unique number that is easy to compute, but it is too strongly affected by extreme values and it does not take every score into account.

The preceding list of advantages and disadvantages emphasizes the lack of objectivity in the selection of a particular average. An understanding of this situation is a prerequisite for avoiding intended and accidental statistical deceptions that arise from the use of averages.

In the next section, we explore **measures of dispersion** that give information about variation within sets of data.

2-4   Exercises A
Averages

In Exercises 2-77 through 2-92, find the (a) mean, (b) median, (c) mode, (d) midrange for the given sample data.

2-77   In attempting to decipher transmitted data, the following frequencies were obtained for the letter *m*.

$$16, 7, 23, 5, 4, 23, 10$$

2-78   Eight leukemic mice were observed, and the numbers of days they survived without treatment were recorded as follows:

$$6, 7, 7, 7, 8, 8, 9, 9$$

2-79   The numbers of calories of "light" beer were recorded for 12-ounce samples of different brands, and the results are given below.

$$106, 99, 101, 103, 108, 107, 107, 107, 106$$

2-80   In a test of hearing, subjects estimate the loudness (in dBA) of sound, and their results are as follows.

$$68, 72, 70, 71, 68, 68, 75, 62, 80, 73, 68$$

2-81   Ten cars were tested for fuel consumption, and the following mpg ratings were obtained:

$$26, 23, 23, 23, 28, 24, 27, 25, 25, 27$$

2-82   A group of 12 subjects was tested on a device used in a biofeedback experiment, and the following readings (in microvolts) were obtained:

$$10.8, 6.7, 5.3, 7.1, 5.3, 7.9, 6.1, 4.9, 2.5, 5.3, 11.6, 6.8$$

2-83   Eight snow tires were tested, and their braking distances (in feet) on ice were

$$150, 138, 138, 165, 135, 148, 141, 125$$

2-84   A survey was made of 15 homemakers to determine the number of hours worked each week, and the following results were obtained:

$$52, 50, 46, 41, 45, 46, 47, 50, 42, 56, 61, 39, 57, 44, 45$$

2-85　In Section 1 of a statistics class, ten test scores were randomly selected, and the following results were obtained:

$$74, 73, 77, 77, 71, 68, 65, 77, 67, 66$$

2-86　In Section 2 of a statistics class, ten test scores were randomly selected, and the following results were obtained:

$$42, 100, 77, 54, 93, 85, 67, 77, 62, 58$$

2-87　The reaction times (in seconds) of a group of adult men were found to be

$$0.74, 0.71, 0.41, 0.82, 0.74, 0.85, 0.99, 0.71, 0.57, 0.85, 0.57, 0.55$$

2-88　The college entrance examination scores of 15 randomly selected high school seniors are

$$478, 433, 308, 478, 352, 433, 565, 240, 352, 455, 338, 400, 455, 350, 662$$

2-89　A farmer measured the heights (in centimeters) of 12 seedlings that had been allowed to grow for two years, and the following results were obtained:

$$64, 78, 91, 58, 81, 69, 103, 65, 51, 78, 76, 63$$

2-90　Job applicants are given an aptitude test, and their results are

$$186, 159, 173, 176, 216, 197, 206, 168, 204, 213, 240, 190, 180, 233, 202$$

2-91　A study was made of a regional center of the Internal Revenue Service to determine the number of income tax returns opened and sorted in 1 hour. The results for 11 employees were

$$248, 260, 259, 240, 248, 226, 250, 266, 255, 256, 234$$

2-92　A study was made to determine the number of defective items produced by 15 different employees in 1 day, and the results were

$$1, 1, 1, 2, 3, 7, 7, 8, 8, 9, 11, 12, 18, 19, 500$$

In Exercises 2-93 through 2-100, use the given frequency table. (a) Identify the class mark for each class interval. (b) Find the mean using the class marks and frequencies.

2-93

Index	Frequency
0.10–0.11	9
0.12–0.13	16
0.14–0.15	12
0.16–0.17	3
0.18–0.19	1

Frequency table of the blood alcohol concentrations of arrested drivers.

2-94

Number	Frequency
1–5	19
6–10	24
11–15	18
16–20	21
21–25	23
26–30	20
31–35	16
36–40	15

Frequency table of lottery numbers selected in a 26-week period.

2-95

Time	Frequency
0.0–4.9	12
5.0–9.9	3
10.0–14.9	8
15.0–19.9	15
20.0–24.9	10
25.0–29.9	5

Frequency table of computer connect times (in hours) for students in one semester.

2-96

Dosage	Frequency
10.0–10.4	5
10.5–10.9	10
11.0–11.4	20
11.5–11.9	15
12.0–12.4	8
12.5–12.9	2

Frequency table of radiation dosage (in milliroentgens) from an x-ray machine.

2-97

Time	Frequency
0–59	1
60–119	2
120–179	5
180–239	14
240–299	7
300–359	12
360–419	9

Frequency table of the time (in seconds) it takes the victim of a crime to call the police.

2-98

Number of trials	Frequency
1–5	2
6–10	1
11–15	12
16–20	10
21–25	5

Frequency table of the number of trials required by 30 different monkeys to learn a certain task.

2-99

Number	Frequency
0–9	5
10–19	19
20–29	32
30–39	14
40–49	1

Frequency table of the number of violent deaths seen on television by children in one week.

2-100

Index	Frequency
41–50	12
51–60	10
61–70	8
71–80	18
81–90	27
91–100	3

Frequency table of the pollution indices for a certain city.

2-4 Exercises B
Averages

2-101 (a) Find the mean, median, mode, and midrange for the following barometric pressure readings (in millibars):

$$1023.6, \ 1024.2, \ 1026.7, \ 1033.5, \ 1040.3$$

(b) Subtract 1000 from each score in part (a) and then find the mean, median, mode, and midrange.

(c) Is there any relationship between the answer set to part (a) and the answer set to part (b)? If so, what is it?

(d) In general, if a constant k is added to (or subtracted from) some set of scores, what happens to the mean, median, mode, and midrange?

(e) Do these results suggest a way of simplifying the computations for certain types of inconvenient numbers?

2-102 (a) Find the mean, median, mode, and midrange for the following wavelengths (in kilometers):

$$0.0236, \ 0.0242, \ 0.0267, \ 0.0335, \ 0.0403$$

(b) Multiply each score from part (a) by 10,000 and then find the mean, median, mode, and midrange.

(c) Is there any relationship between the answer set to part (a) and the answer set to part (b)? If so, what is it?

(d) In general, if every score in a data set is multiplied by some constant k, what happens to the mean, median, mode, and midrange?

(e) Do these results suggest a way of simplifying the computations for certain types of inconvenient numbers?

2-103 Compare the averages that comprise the solutions to Exercises 2-85 and 2-86. Do these averages discriminate or differentiate between the two lists of scores? Is there any apparent difference between the two data sets that is not reflected in the averages?

2-104 Using the data from Exercise 2-85, change the first score (74) to 900 and find the mean, median, mode, and midrange of this modified data set. How does the extreme value affect these averages?

2-105 Collapse the frequency table in Exercise 2-100 by using the three class intervals of 41–60, 61–80, and 81–100. Calculate the mean of the collapsed table and compare the result to that obtained in Exercise 2-100. Which result is likely to be better?

2-106 To obtain the **harmonic mean** of a set of scores, divide the number of scores n by the sum of the reciprocals of all scores.

$$\text{harmonic mean} = \frac{n}{\sum \frac{1}{x}}$$

For example, the harmonic mean of 2, 3, 6, 7, 7, 8 is

$$\frac{6}{\frac{1}{2} + \frac{1}{3} + \frac{1}{6} + \frac{1}{7} + \frac{1}{7} + \frac{1}{8}} = \frac{6}{1.4} = 4.3$$

(Note that 0 cannot be included in the scores.)
(a) Find the harmonic mean of 2, 3, 6, 7, 7, 8, 9, 9, 9, 10.
(b) A group of students drives from New York to Florida (1200 miles) at a speed of 40 miles per hour and returns at a speed of 60 miles per hour. What is their average speed for the round trip? (The harmonic mean is used in averaging speeds.)
(c) A dispatcher of charter buses calculates the average round trip speed (in miles per hour) for a certain route. The results obtained for 14 different runs are listed below. Based on these values, what is the average speed of a bus assigned to this route?

> 42.6 41.3 38.2 42.9 43.4 43.7 40.8 39.6
> 34.2 40.1 41.2 40.5 41.7 39.8

2-107 Given a collection of n scores (all of which are positive), the **geometric mean** is the nth root of their product. For example, to find the geometric mean of 2, 3, 6, 7, 7, 8, 9, 9, 9, 10, first multiply the scores.

$$2 \cdot 3 \cdot 6 \cdot 7 \cdot 7 \cdot 8 \cdot 9 \cdot 9 \cdot 9 \cdot 10 = 102,876,480$$

Then take the tenth root of the product since there are ten scores.

$$\sqrt[10]{102,876,480} = 6.3$$

(*cont. on p. 64*)

The geometric mean is often used in business and economics for finding average rates of change, average rates of growth, or average ratios. The *average growth factor* for money compounded at annual interest rates of 10%, 8%, 9%, 12%, and 7% can be found by computing the geometric mean of 1.10, 1.08, 1.09, 1.12, and 1.07. Find that average growth factor.

2-108   The **quadratic mean** (or root mean square, or R.M.S.) of a set of scores is obtained by squaring each score, adding the results, dividing by the number of scores n, and then taking the square root of that result.

$$\text{quadratic mean} = \sqrt{\frac{\Sigma x^2}{n}}$$

For example, the quadratic mean of 2, 3, 6, 7, 7, 8, 9, 9, 9, 10 is

$$\sqrt{\frac{4 + 9 + 36 + 49 + 49 + 64 + 81 + 81 + 81 + 100}{10}}$$

$$= \sqrt{\frac{554}{10}}$$

$$= \sqrt{55.4} = 7.4$$

The quadratic mean is usually used in physical applications. In power distribution systems, for example, voltages and currents are usually referred to in terms of their R.M.S values. Find the R.M.S. of the following power supplies (in volts): 151, 162, 0, 81, −68.

2-5 Dispersion Statistics

The preceding section was concerned with averages, or measures of central tendency. Those statistics were designed to reflect a certain characteristic of the data from which they came. That is, the averages are supposed to be *central* scores. However, other features of the data may not be reflected at all by the averages. Suppose, for example, that two different groups of ten students are given identical quizzes, with these results.

Group A	Group B
65	42
66	54
67	58
68	62
71	67
73	77
74	77
77	85
77	93
77	100

Computing the averages, we get

Group A		Group B	
mean	= 71.5	mean	= 71.5
median	= 72.0	median	= 72.0
mode	= 77	mode	= 77
midrange	= 71.0	midrange	= 71.0

From looking at these averages, we can see no difference between the two groups. Yet an intuitive perusal of both groups shows an obvious difference: The scores of Group B are much more widely scattered than those of Group A. This variability among data is one characteristic to which averages are not sensitive. Consequently, statisticians have tried to design statistics that measure this variability, or dispersion. The three basic measures of dispersion are range, variance, and standard deviation.

Range

The **range** is simply the difference between the highest value and the lowest value. For Group A, the range of 12 is the difference between 77 and 65. The range of Group B is 58 (100 − 42). This much larger range suggests greater dispersion. Be sure you avoid confusion between the midrange (an average) and the range (a measure of dispersion). The range is extremely easy to compute, but it's often inferior to other measures of dispersion. The rather extreme example of Groups C and D should illustrate this point.

Group C	Group D
1	2
20	3
20	4
20	5
20	6
20	14
20	15
20	16
20	17
20	18
Range = 19	Range = 16

The larger range for Group C suggests more dispersion than in Group D. But the scores of Group C are very close together while those of Group D are much more scattered. The range may be misleading in this case (and in many other circumstances) because it depends only on the maximum and minimum scores. Better measures of dispersion have been developed, but the improvement is not achieved without some loss. The loss occurs in the ease of computation. The variance and standard deviation measures generally have greater meaning, but they require computations that are much more difficult.

Variance

The **variance** is computed by applying the following formula to the given set of values.

Formula 2-4 $$\text{variance} = \frac{\Sigma(x - \bar{x})^2}{n - 1}$$

(Some authors define the variance with a denominator of n, but $n - 1$ provides a better estimate of the variance for a population when the formula is applied to a random sample from that population. Consequently, if the data are an entire population of scores, divide by n, but divide by $n - 1$ when dealing with sample data. For large values of n (such as those greater than 30) it really doesn't make too much difference which choice is made. For example, computations involving 100 scores will result in division by 100 or by 99, which gives values that are essentially the same.)

N O T A T I O N

σ^2 (where σ is the lowercase Greek sigma) denotes the variance of all scores in a *population*. (If all scores in a population are known, $\bar{x}$ in Formula 2-4 actually becomes μ.)

s^2 denotes the variance of a set of *sample* scores.

To use Formula 2-4 you should follow this procedure.

Procedure for Using Formula 2-4

   1. Find the mean of the scores ($\bar{x}$).
   2. Subtract the mean from each individual score ($x - \bar{x}$).
   3. Square each of the differences obtained from Step 2. That is, multiply each value by itself. (This produces numbers of the form $(x - \bar{x})^2$.)
   4. Add all of the squares obtained from Step 3 to get $\Sigma(x - \bar{x})^2$.
   5. Divide the preceding total by the number ($n - 1$); that is, one less than the total number of scores present. The result is the variance.

Upon students' initial exposure to the variance, there is sometimes a feeling of awe, mystery, or even an urge to acquire a withdrawal slip. These feelings can, of course, be overcome. Computation of the variance involves only the simple arithmetic operations of addition, subtraction, multiplication, and division. The mystery associated with this particular procedure must remain for some time, since a complete explanation of it is not possible at this stage.

We can point out that after the mean $\bar{x}$ has been obtained, any individual score x will differ or deviate from that mean by an amount of $(x - \bar{x})$. The sum of all such deviations, denoted by $\Sigma(x - \bar{x})$, might initially seem like a reasonable measure of dispersion or variance, but that sum will *always* equal 0. Scores greater than the mean will cause $(x - \bar{x})$ to be positive, while scores below the mean will cause $(x - \bar{x})$ to be negative values since the mean $\bar{x}$ is at the center of the scores.

How do we prevent this undesirable canceling out of positive and negative values so that the measure of dispersion is not always zero? One alternative is to use absolute values as in $\Sigma|x - \bar{x}|$. This leads to the **mean deviation,** which will be mentioned later, but this approach tends to be unsuitable for the important methods of statistical inference. Instead, we make all of the terms $(x - \bar{x})$ nonnegative by squaring them. The variance is therefore an average of the squared deviations from the mean.

E X A M P L E

Find the variance for these quiz scores: 2, 3, 5, 6, 9, 17.

Solution

Step 1: The mean is obtained by adding the scores (42) and then dividing by the number of scores present (6). The mean is 7.0.

Step 2: Subtracting the mean of 7.0 from each score, we get -5, -4, -2, -1, 2, and 10. (As a quick check, these numbers must always total 0.)

Step 3: Squaring each value obtained from Step 2, we get 25, 16, 4, 1, 4, and 100.

Step 4: The sum of all the preceding squares is 150.

Step 5: There are six scores, so we divide 150 by one less than 6. That is, $150 \div 5 = 30$. Using the rounding rule from the previous section, we report the variance as 30.0

Here is a helpful hint: Organize the computations for variance in the form of a table like Table 2-12 (see the following page).

Table 2-12

x	$(x - \bar{x})$	$(x - \bar{x})^2$
2	−5	25
3	−4	16
5	−2	4
6	−1	1
9	2	4
17	10	100
Totals: 42		150

$$\downarrow \qquad\qquad\qquad\qquad \downarrow$$

$$\bar{x} = \frac{42}{6} = 7.0 \qquad\qquad s^2 = \frac{150}{6-1} = \frac{150}{5} = 30.0$$

Formula 2-4 can be expressed in an equivalent form.

Formula 2-5 $$s^2 = \frac{n(\Sigma x^2) - (\Sigma x)^2}{n(n-1)}$$

Formulas 2-4 and 2-5 are equivalent in the sense that they will always produce the same results. This means that we can select the more convenient of the two formulas when we must compute a variance. Formula 2-5 will probably be more convenient if the mean hasn't yet been calculated or if it is known to be something other than a whole number. If the mean is a known whole number, Formula 2-4 will probably be more convenient.

E X A M P L E

Use Formula 2-5 to find the variance of these reaction times: 3.72, 4.63, 5.81, 4.73, 2.93, 5.13, 3.84.

Solution

Because there are seven scores, $n = 7$. The sum of the scores is 30.79, so $\Sigma x = 30.79$. Squaring each score and then adding the resulting squares gives $\Sigma x^2 = 141.0517$, so that

$$s^2 = \frac{n(\Sigma x^2) - (\Sigma x)^2}{n(n-1)} = \frac{7(141.0517) - (30.79)^2}{7(7-1)}$$

$$= \frac{39.3378}{42} = 0.937$$

We now have two formulas—2-4 and 2-5—for finding the variance and each has its own advantages. Formula 2-4 is as direct and "natural" as any formula for variance can be. But in the preceding example, we used Formula 2-5 since Formula 2-4 would have been too messy. (Try it and see for yourself!) Also, several brands of inexpensive pocket calculators are designed to compute variances automatically. A calculator that computes variances according to Formula 2-5 will require only three memory registers, in which n, Σx, and Σx^2 can be stored as the data are entered. (These are the required components of Formula 2-5.) If a calculator were to use Formula 2-4, it would need a separate memory register for each score (since all scores must be known before the mean can be obtained), and that is obviously impractical.

In the previous section we developed a procedure for computing the mean when the scores are incorporated in a frequency table. The fundamental importance of measures of dispersion encourages us to develop a procedure for finding the variance of scores summarized in a frequency table. In some cases, we may not have access to the original list of raw scores and the frequency table may have the only data available. In other cases, the number of raw scores may be large enough to render any individual treatment of the scores impractical. We again use the class marks and class frequencies and we obtain Formula 2-6, the variance computed from a frequency table.

Formula 2-6 $$s^2 = \frac{n[\Sigma(f \cdot x^2)] - [\Sigma(f \cdot x)]^2}{n(n-1)}$$

where x = class mark, f = class frequency, and n = sample size.

Table 2-13 summarizes the various steps in the use of Formula 2-6. The data in that table are the data consisting of the 150 absence times listed in Table 2-1.

Table 2-13

Absence time	Frequency f	Class mark x	$f \cdot x$	$f \cdot x^2$ or $f \cdot x \cdot x$
0–19	19	9.5	180.5	1714.75
20–39	10	29.5	295.0	8702.5
40–59	15	49.5	742.5	36753.75
60–79	22	69.5	1529.0	106265.5
80–99	25	89.5	2237.5	200256.25
100–119	20	109.5	2190.0	239805.0
120–139	14	129.5	1813.0	234783.5
140–159	13	149.5	1943.5	290553.25
160–179	6	169.5	1017.0	172381.5
180–199	3	189.5	568.5	107730.75
200–219	0	209.5	0	0
220–239	3	229.5	688.5	158010.75
Total	150		13205.0	1556957.5

United States Government Eliminates 500,000 Farms

When analyzing data from one year to the next, we must be wary of changing definitions that artificially alter the data. In the last century, the Department of Agriculture changed the official definition of a farm at least eight times. One change in the definition resulted in decreasing the number of farms by about 500,000. This change involved replacing annual sales of at least $250 (for less than 10 acres) or at least $50 (for more than 10 acres) to the new criterion of $1000 in annual sales. If we look for trends in the number of farms and are not aware of these changing definitions, we can easily form invalid conclusions.

$$s^2 = \frac{n[\Sigma(f \cdot x^2)] - [\Sigma(f \cdot x)]^2}{n(n-1)}$$

$$= \frac{150(1556957.5) - (13205.0)^2}{150(150-1)}$$

$$= \frac{59171600}{22350}$$

$$= 2647.5$$

According to the frequency table, the variance is 2647.5. If we compute the variance using the original list of scores, we get 2638.9; the variance obtained from the frequency table is quite accurate.

There are calculators and computer programs that will calculate the mean and variance from a frequency table. One calculator, for example, requires that class marks be entered along with the corresponding frequencies. The keys labeled $\Sigma+$ and Frq are used for the entries while keys labelled MEAN and VAR are used for the results.

Standard Deviation

DEFINITION

The **standard deviation** of a set of scores is the square root of the variance of those scores. That is, the standard deviation s is found by

$$s = \sqrt{\frac{\Sigma(x - \overline{x})^2}{n-1}} \quad \text{or} \quad \sqrt{\frac{n(\Sigma x^2) - (\Sigma x)^2}{n(n-1)}}$$

If the standard deviation is to be found for a set of scores summarized in a frequency table, we can simply compute the variance using Formula 2-6 and then take the square root of that result to obtain the corresponding standard deviation. For the job absence times of Table 2-13, the variance was found to be 2647.5. Consequently the standard deviation is $\sqrt{2647.5} = 51.5$. If we calculate the standard deviation from the original list of scores, we get 51.4. The result obtained from the frequency table is quite accurate.

Recall that the square root of a number is another number with the property that, when multiplied by itself, the result is the original number. Three is the square root of 9 because multiplying 3 by itself produces 9.

How do we interpret the values we compute for ranges, variances, and standard deviations? The following chart summarizes the measures of dispersion for the quiz scores of Group A and Group B on page 64. What do these numbers really mean?

	Group A	Group B
Range	12	58
Variance	22.7	331.8
Standard deviation	4.8	18.2

It is fairly easy to develop a strong intuitive sense for relating to the range. A range of 12 indicates data that span a much smaller gap than comparable data with a range of 58. Variances provide good comparisons, but their numerical values alone probably don't mean very much at this point. The Group A variance of 22.7 is less than the Group B variance of 331.8, so we conclude that Group A has less dispersion according to the criterion of variance. But if we try to see the link between the original Group A scores and the variance of 22.7, we will probably be unsuccessful at this stage. Subsequent discussions will shed more light on the interpretation of variances and standard deviations.

Perhaps some feeling for the concept of variance can be cultivated through a comparative study of the five examples of Figure 2-13. From that figure we should see that as the data spread farther apart, the corresponding values of the variance and standard deviation increase. This should help develop the intuitive sense that larger values of variance and standard deviation reflect greater amounts of scattering among data.

Another consideration that might be helpful in developing a sense for the standard deviation is a result from Chebyshev's theorem (see Exercise 2-147). One approximate result of that theorem can be stated as follows: *At least 89% of the values will fall within 3 standard deviations of the mean.* Suppose we know that a personality test produces scores with a mean of 300 and a standard deviation of 20. The values falling within 3 standard deviations of the mean are the values between 240 and 360. We therefore conclude that *at least* 89% of all results will lie between 240 and 360; values below 240 or above 360 are relatively rare. Now we have a better sense for how those scores are dispersed.

So far in this section we have discussed the three most common measures of dispersion, but there are others. The **mean deviation,** for example, is obtained by evaluating $\Sigma|x - \bar{x}| \div n$, where the absolute value signs (those vertical lines) require that each value of $(x - \bar{x})$ be recorded as the *positive* (or zero) difference between the score and the mean. The interquartile range, semi-interquartile range, and 10–90 percentile range are other measures of dispersion mentioned briefly in the following section.

Except for variance, the measures of dispersion are expressed in the same units as the original data. The standard deviation for a list of heights might be 3 inches. But the variance is in square units; for a set of heights, the variance might be 9 square inches. In comparison to the variance, a definite advantage of the standard deviation is that it is measured in the same units as the original data.

Increasing the
spread of the data
leads to larger values
of variance and
standard deviation.

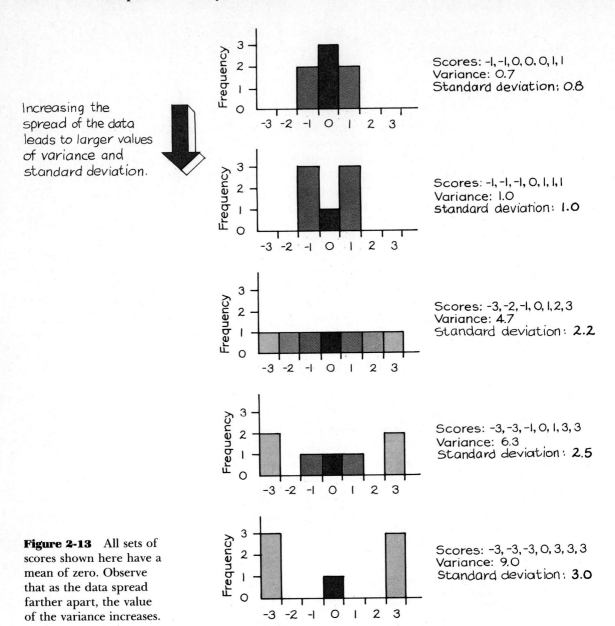

Scores: -1, -1, 0, 0, 0, 1, 1
Variance: 0.7
Standard deviation: 0.8

Scores: -1, -1, -1, 0, 1, 1, 1
Variance: 1.0
standard deviation: 1.0

Scores: -3, -2, -1, 0, 1, 2, 3
Variance: 4.7
Standard deviation: **2.2**

Scores: -3, -3, -1, 0, 1, 3, 3
Variance: 6.3
Standard deviation: **2.5**

Scores: -3, -3, -3, 0, 3, 3, 3
Variance: 9.0
Standard deviation: **3.0**

Figure 2-13 All sets of
scores shown here have a
mean of zero. Observe
that as the data spread
farther apart, the value
of the variance increases.

Measures of dispersion are extremely important in many practical
circumstances. Manufacturers interested in producing items of consistent quality are very concerned with statistics such as standard deviations.
A producer of car batteries might be pleased to learn that a product has
a mean life of 4 years, but that pleasure would become distress if the
standard deviation indicated a very large dispersion, which would corre-

spond to many battery failures long before the mean of 4 years. Quality control requires consistency, and consistency requires a relatively small standard deviation.

Some exercises of the previous section showed that if a constant is added to (or subtracted from) each value in a set of data, the measures of central tendency change by that same amount. In contrast, the measures of dispersion do not change when a constant is added to (or subtracted from) each score. To find the standard deviation for 8001, 8002, . . . , 8009, we could subtract 8000 from each score and work with more manageable numbers without changing the value of the standard deviation, which is a definite advantage. But now suppose a manufacturer learns that the average (mean) readings of a thousand pressure gauges are 10 pounds per square inch (psi) too low. A relatively simple remedy would be to rotate all of the dials to read 10 psi higher. That remedy would correct the mean, but the standard deviation would remain the same. If the gauges are inconsistent and have errors that vary considerably, then the standard deviation of the errors might be too large. If that is the case, each gauge must be individually calibrated and the entire manufacturing process might require an overhaul. Machines might require replacement if they do not meet the necessary tolerance levels. We can see from this example that measures of central tendency and dispersion have properties that lead to very practical and serious implications.

2-5 Exercises A
Dispersion Statistics

In Exercises 2-109 through 2-124, find the range, variance, and standard deviation for the given data.

2-109 In attempting to decipher transmitted data, the following frequencies were obtained for the letter *m*:

$$16, 7, 23, 5, 4, 23, 10$$

2-110 Eight leukemic mice were observed, and the numbers of days they survived without treatment were recorded as follows:

$$6, 7, 7, 7, 8, 8, 9, 9$$

2-111 The numbers of calories of "light" beer were recorded for 12 ounce samples of different brands, and the results are

$$106, 99, 101, 103, 108, 107, 107, 107, 106$$

2-112 In a test of hearing, subjects estimate the loudness (in dBA) of a sound, and their results are

$$68, 72, 70, 71, 68, 68, 75, 62, 80, 73, 68$$

2-113   Ten cars were tested for fuel consumption, and the following mpg ratings were obtained:

26, 23, 23, 23, 28, 24, 27, 25, 25, 27

2-114   For 12 subjects tested on a device used in a biofeedback experiment, the following readings (in microvolts) were obtained:

10.8, 6.7, 5.3, 7.1, 5.3, 7.9, 6.1, 4.9, 2.5, 5.3, 11.6, 6.8

2-115   Eight snow tires were tested, and their braking distances (in feet) on ice were

150, 138, 138, 165, 135, 148, 141, 125

2-116   A survey was made of 15 homemakers to determine the number of hours worked each week, and the following results were obtained:

52, 50, 46, 41, 45, 46, 47, 50, 42, 56, 61, 39, 57, 44, 45

2-117   In Section 1 of a statistics class, ten test scores were randomly selected, and the following results were obtained:

74, 73, 77, 77, 71, 68, 65, 77, 67, 66

2-118   In Section 2 of a statistics class, ten test scores were randomly selected, and the following results were obtained:

42, 100, 77, 54, 93, 85, 67, 77, 62, 58

2-119   The reaction times (in seconds) of a group of adult men were found to be

0.74, 0.71, 0.41, 0.82, 0.74, 0.85, 0.99, 0.71, 0.57, 0.85, 0.57, 0.55

2-120   The college entrance examination scores of 15 randomly selected high school seniors are

478, 433, 308, 478, 352, 433, 565, 240, 352, 455, 338, 400, 455, 350, 662

2-121   A farmer measured the heights (in centimeters) of 12 seedlings that had been allowed to grow for two years, and the following results were obtained:

64, 78, 91, 58, 81, 69, 103, 65, 51, 78, 76, 63

2-122   Job applicants are given an aptitude test, and their results are

186, 159, 173, 176, 216, 197, 206, 168, 204, 213, 240, 190, 180, 233, 202

2-123   A study was made of a regional center of the Internal Revenue Service to determine the number of income tax returns opened and sorted in 1 hour. The results for 11 employees were

248, 260, 259, 240, 248, 226, 250, 266, 255, 256, 234

2-124   A study was made to determine the number of defective items produced by 15 different employees in 1 day, and the results were

1, 1, 1, 2, 3, 7, 7, 8, 8, 9, 11, 12, 18, 19, 500

2-125   Add 15 to each score given in Exercise 2-124 and then find the range, variance, and standard deviation. Compare the results to those of Exercise 2-124.

2-126   Subtract 5 from each score given in Exercise 2-124 and then find the range, variance, and standard deviation. Compare the results to those of Exercise 2-124.

2-127   Multiply each score given in Exercise 2-124 by 10 and then find the range, variance, and standard deviation. Compare the results to those of Exercise 2-124.

2-128   Double each score in Exercise 2-124 and then find the range, variance, and standard deviation. Compare the results to those of Exercise 2-124.

In Exercises 2-129 through 2-136, find the variance and standard deviation for the given data.

2-129

Index	Frequency
0.10–0.11	9
0.12–0.13	16
0.14–0.15	12
0.16–0.17	3
0.18–0.19	1

Frequency table of the blood alcohol concentrations of arrested drivers.

2-130

Number	Frequency
1–5	19
6–10	24
11–15	18
16–20	21
21–25	23
26–30	20
31–35	16
36–40	15

Frequency table of lottery numbers selected in a 26-week period.

2-131

Time	Frequency
0.0–4.9	12
5.0–9.9	3
10.0–14.9	8
15.0–19.9	15
20.0–24.9	10
25.0–29.9	5

Frequency table of computer connect times (in hours) for students in one semester.

2-132

Dosage	Frequency
10.0–10.4	5
10.5–10.9	10
11.0–11.4	20
11.5–11.9	15
12.0–12.4	8
12.5–12.9	2

Frequency table of radiation dosage (in milliroentgens) from an x-ray machine.

2-133

Time	Frequency
0–59	1
60–119	2
120–179	5
180–239	14
240–299	7
300–359	12
360–419	9

Frequency table of the time (in seconds) it takes the victim of a crime to call the police.

2-134

Number of trials	Frequency
1–5	2
6–10	1
11–15	12
16–20	10
21–25	5

Frequency table of the number of trials required by 30 different monkeys to learn a certain task.

2-135

Number	Frequency
0–9	5
10–19	19
20–29	32
30–39	14
40–49	1

Frequency table of the number of violent deaths seen on television by children in one week.

2-136

Index	Frequency
41–50	12
51–60	10
61–70	8
71–80	18
81–90	27
91–100	3

Frequency table of pollution indices for a certain city.

2-137 Which would you expect to have a higher variance: the I.Q. scores of a class of 25 statistics students or the I.Q. scores of 25 randomly selected adults? Explain.

2-138 Which would you expect to have a higher standard deviation: the reaction times of 30 sober adults or the reaction times of 30 adults who had recently consumed three martinis each?

2-139 Is it possible for a set of scores to have a variance of zero? If so, how? Is it possible for a set of scores to have a negative variance?

2-140 Test the effect of an *outlier* (an extreme value) by changing the 500 in Exercise 2-124 to 25. Calculate the range, variance, and standard deviation for the modified set and compare the results to those originally obtained in Exercise 2-124.

2-5 Exercises B
Dispersion Statistics

2-141 Find the range, variance, and standard deviation for each of the following two groups. Which group has less dispersion according to the criterion of the range? Which group has less dispersion according to the criterion of the variance? Which measure of dispersion is "better" in this situation: the range or the variance?

Group C: 1, 20, 20, 20, 20, 20, 20, 20, 20, 20

Group D: 2, 3, 4, 5, 6, 14, 15, 16, 17, 18

2-142 The **coefficient of variation**, expressed in percent, is used to describe the standard deviation relative to the mean. It is calculated as

$$\frac{s}{\bar{x}} \cdot 100 \quad \text{or} \quad \frac{\sigma}{\mu} \cdot 100$$

(a) Find the coefficient of variation for I.Q. tests having a mean of 100 and a standard deviation of 15.

(b) Find the coefficient of variation for reading tests having a mean of 250 and a standard deviation of 50.

(c) Find the coefficient of variation for the following sample scores: 2, 2, 2, 3, 5, 8, 12, 19, 22, 30.

2-143 Given a collection of temperatures in degrees Fahrenheit, the following statistics are calculated:

$$\bar{x} = 40.2 \qquad s = 3.0 \qquad s^2 = 9.0$$

Find the values of $\bar{x}$, s, and s^2 for the same data set after each temperature has been converted to the Celsius scale.

$$\left[C = \frac{5}{9}(F - 32) \right]$$

2-144   If we consider the values 1, 2, 3, . . . , n to be a population, the variance can be calculated by the formula

$$\sigma^2 = \frac{n^2 - 1}{12}$$

This formula is equivalent to Formula 2-4 modified for division by n instead of $n - 1$, where the values are 1, 2, 3, . . . , n.
(a) Find the standard deviation of the population 1, 2, 3, . . . , 100.
(b) Find an expression for calculating the *sample* variance s^2 for the sample values 1, 2, 3, . . . , n.

2-145   Find the mean and standard deviation for the scores 18, 19, 20, . . . , 182. Treat those values as a population (see Exercise 2-144).

2-146   Find the variance and standard deviation for the set of sample data given below. The scores represent the numbers of letters mailed by a firm in different weeks.

6	33	46	50	56
104	121	140	158	171
186	190	192	206	221
229	236	245	254	259
262	290	340	346	368
371	376	377	408	421
430	442	496	501	535
540	554	566	600	614
622	623	627	636	638
682	684	695	711	733

2-147   Chebyshev's theorem states that the *proportion* of any set of data lying within K standard deviations of the mean is always *at least* $1 - 1/K^2$, where K is any positive number greater than 1.
(a) Given a mean I.Q. of 100 and a standard deviation of 15, what does Chebyshev's theorem say about the number of scores within two standard deviations of the mean (that is, 70–130)?
(b) Given a mean of 100 and a standard deviation of 15, what does Chebyshev's theorem say about the number of scores between 55 and 145?
(c) The mean score on the College Entrance Examination Board Scholastic Aptitude Test is 500 and the standard deviation is 100. What does Chebyshev's theorem say about the number of scores between 300 and 700?
(d) Using the data of part (c), what does Chebyshev's theorem say about the number of scores between 200 and 800?

2-148   Computers commonly use a random number generator, which produces values between 0 and 1. In the long run, all values occur with the same relative frequency. Find the mean and standard deviation for the numbers between 0.00000000 and 0.99999999. (Hint: See Exercises 2-127 and 2-144.)

2-6 Measures of Position

There is often a need to compare scores taken from two separate populations with different means and standard deviations. The standard score (or z score) can be used to help make such comparisons.

DEFINITION

The **standard score,** or **z score,** is the number of standard deviations that a given value is above or below the mean, and it is found by

$$z = \frac{x - \overline{x}}{s}$$

For example, which is better: a score of 65 on Test A or a score of 29 on Test B? The class statistics for the two tests are as follows.

Test A	Test B
$\overline{x} = 50$	$\overline{x} = 20$
$s = 10$	$s = 5$

For the score of 65 on Test A we get a z score of 1.5, since

$$z = \frac{x - \overline{x}}{s} = \frac{65 - 50}{10} = \frac{15}{10} = 1.5$$

For the score of 29 on Test B we get a z score of 1.8, since

$$z = \frac{x - \overline{x}}{s} = \frac{29 - 20}{5} = \frac{9}{5} = 1.8$$

That is, a score of 65 on Test A is 1.5 standard deviations above the mean, while a score of 29 on Test B is 1.8 standard deviations above the mean. This implies that the 29 on Test B is the better score. While 29 is below 65, it has a better *relative* position when considered in the context of the other test results. Later, we will make extensive use of these standard, or z, scores.

The z score provides a useful measurement for making comparisons between different sets of data. Quartiles, deciles, and percentiles are measures of position useful for comparing scores within one set of data or between different sets of data.

Just as the median divides the data into two equal parts, the three **quartiles,** denoted by Q_1, Q_2, and Q_3, divide the ranked scores into four equal parts. Roughly speaking, Q_1 separates the bottom 25% of the ranked scores from the top 75%, Q_2 is the median, and Q_3 separates the top 25% from the bottom 75%. To be more precise, at least 25% of the

1967 | 1984

$100 | $312.20

Index Numbers

The Consumer Price Index (CPI) is one of a family of yardsticks called *index numbers*. An index number allows us to compare the value of some variable relative to its value at some base period. The net effect is that we measure the change of the variable over some time span. We find the value of an index number by evaluating

$$\frac{\text{current value}}{\text{base value}} \times 100$$

The CPI is based on a weighted average of the costs of specific goods and services. Using 1967 as a base year with index 100, the CPI in 1984 was 312.2. This means that a combination of goods and services that cost $100 in 1967 would cost $312.20 in 1984. Many labor contracts have wage adjustments linked to changes in the CPI.

data will be less than or equal to Q_1 and at least 75% will be greater than or equal to Q_1. At least 75% of the data will be less than or equal to Q_3, while at least 25% will be equal to or greater than Q_3. Also, Q_2 is actually the median.

Similarly, there are nine **deciles,** denoted by $D_1, D_2, D_3, \ldots, D_9$, which partition the data into 10 groups with about 10% of the data in each group. There are also 99 **percentiles,** which partition the data into 100 groups with about 1% of the scores in each group. A student taking a competitive college entrance examination might learn that he or she scored in the 92nd percentile. This does not mean that the student received a grade of 92% on the test; it indicates roughly that whatever score he or she did achieve was higher than 92% of peers who took a similar test (and also lower than 8% of his or her colleagues). Percentiles are useful for converting meaningless raw scores into meaningful comparative scores. For this reason, percentiles are used extensively in educational testing. A raw score of 750 on a college entrance exam means nothing to most people, but the corresponding percentile of 93% provides useful comparative information.

The process of finding the percentile that corresponds to a particular score is fairly simple.

DEFINITION

$$\text{percentile of score } x = \frac{\text{number of scores less than } x}{\text{total number of scores}} \cdot 100$$

EXAMPLE

The following list contains 60 scores, which are already arranged in order from lowest to highest. Find the percentile corresponding to 12.0.

Waiting Time (Minutes)
of Customers Requesting Automobile Repairs

3.6	20.0	39.5	52.3	60.0	68.9
4.6	23.4	39.6	53.2	60.4	72.1
7.9	24.5	39.9	55.9	60.5	74.5
8.4	25.7	41.5	56.6	60.9	81.1
9.3	28.3	43.8	58.0	61.7	84.2
9.9	30.8	44.0	58.2	62.2	86.1
12.0	33.0	47.6	58.3	63.8	90.0
14.1	35.9	48.0	58.4	65.9	93.5
18.1	36.7	49.1	59.3	66.9	95.9
19.2	37.5	50.1	59.6	66.9	97.4

> **Solution**
>
> ---
>
> There are 6 scores less than 12.0 so we get
>
> $$\text{percentile of } 12.0 = \frac{6}{60} \cdot 100 = 10$$
>
> The score of 12.0 is the tenth percentile.

There are several different methods for finding the score corresponding to a particular percentile, but the process we will use is summarized in Figure 2-14 on p. 83.

> **E X A M P L E**
>
> Use the set of 60 scores from the last example to find the score corresponding to the 37th percentile.
>
> **Solution**
>
> ---
>
> We refer to Figure 2-14 and observe that the data are already ranked from lowest to highest. We now compute
>
> $$L = \left(\frac{k}{100}\right)n = \left(\frac{37}{100}\right)60 = 22.2$$
>
> We answer no when asked if $L = 22.2$ is a whole number, so we are directed to round L *up* (not off) to 23. The 37th percentile, denoted by P_{37}, is the 23rd score, starting with the lowest. Beginning with the lowest score of 3.6, we count down the list to find the 23rd score of 39.9, so that $P_{37} = 39.9$.

Note that there are 22 scores below 39.9 so that by the definition on page 80, the percentile of 39.9 is $(22/60) \cdot 100 = 36.7$, which is approximately 37. As the amount of data increases, these differences become smaller.

In the preceding example, we found the 37th percentile. Because of the sample size, the location L first became 22.2, which was rounded up to 23 because L was not originally a whole number. In the next example, we illustrate a case in which L does begin as a whole number. This condition will cause us to branch to the right in Figure 2-14.

EXAMPLE

Use the same set of 60 scores to find P_{25}, which denotes the 25th percentile.

Solution

Referring to Figure 2-14 and noting the data are already ranked from lowest to highest, we compute

$$L = \left(\frac{k}{100}\right) n = \left(\frac{25}{100}\right) 60 = 15$$

We now answer yes when asked if $L = 15$ is a whole number, and we then find that P_{25} is the value midway between the 15th and 16th scores. Since the 15th and 16th scores are 28.3 and 30.8, we get

$$P_{25} = \frac{28.3 + 30.8}{2} = 29.55$$

Once these calculations with percentiles are mastered, the analogous calculations for quartiles and deciles can be done with the same procedures by noting the following relationships.

Quartiles	Deciles
$Q_1 = P_{25}$	$D_1 = P_{10}$
$Q_2 = P_{50}$	$D_2 = P_{20}$
$Q_3 = P_{75}$	$D_3 = P_{30}$
	$\vdots$
	$D_9 = P_{90}$

Finding Q_3, for example, is equivalent to finding P_{75}.

Other statistics are sometimes defined using quartiles, deciles, or percentiles. For example, the **interquartile range** is a measure of dispersion obtained by evaluating $Q_3 - Q_1$. The **semi-interquartile range** is $(Q_3 - Q_1)/2$, the **midquartile** is $(Q_1 + Q_3)/2$, and the **10–90 percentile range** is defined to be $P_{90} - P_{10}$.

When dealing with large collections of data, more reliable results are obtained with greater ease when statistics software packages are used. See page 84 for the STATDISK computer display for the 150 job absence times introduced in Table 2-1. The user loads the software package and enters the 150 values. The STATDISK program provides all the results shown. This software package, like many others, is also capable of providing a histogram.

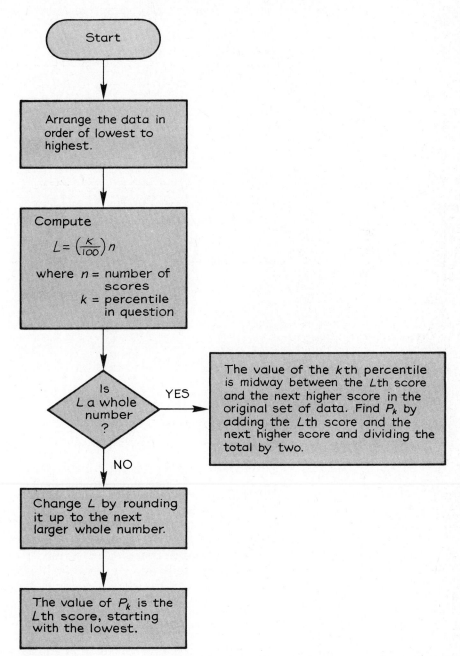

Figure 2-14 Procedure for finding the value of the kth percentile P_k

```
Sample Size = 150 Minimum = 0.0 Maximum = 234.0

          MEASURES OF CENTRAL TENDENCY

Mean = 87.9               Median = 87.0
Geometric Mean = 0.0    Harmonic Mean = [None]

              Midrange = 117.0
              Quadratic Mean = 101.7

          MEASURES OF DISPERSION

Sample Standard Deviation = 51.4
Population Standard Deviation = 51.2
    Range = 234.0

      Sample Variance = 2638.9
      Population Variance = 2621.3
        Standard Error = 4.2

      MEASURES OF POSITION - Quartiles

      Q1 = 51.0  Q2 = 87.0  Q3 = 122.0
```

2-6 Exercises A
Measures of Position

2-149   The times (in minutes) required for students to write a certain program are found to have a mean of 74 and a standard deviation of 19. For each of the following times, find the corresponding z score. (a) 93. (b) 110. (c) 55.

2-150   The sales receipts for auto parts are found to have a mean of $15.70 and a standard deviation of $17.30. Find the z score corresponding to each of the following sales totals. (a) $33.00. (b) $50.00. (c) $7.55.

2-151   A college entrance exam has a mean of 440 and a standard deviation of 100. For each of the given test scores, find the corresponding z score. (a) 540. (b) 340. (c) 640. (d) 290. (e) 740.

2-152   The scores on a test have a mean of 72 and a standard deviation of 8. For each of the given test scores, find the corresponding z score. (a) 80. (b) 84. (c) 60. (d) 52.

2-153 For a sample of record albums, the mean and standard deviation for the playing times of various songs are found to be 3.05 min and 0.877 min, respectively. Find the z score corresponding to each of the following playing times. (a) 4.00 min. (b) 2.50 min. (c) 5.00 min.

2-154 Assume that I.Q. scores have a mean of 100 and a standard deviation of 15. For each of the given I.Q. scores, find the corresponding z score. (a) 120. (b) 100. (c) 70. (d) 127. (e) 82.

2-155 Assume that an industrial test of memory yields results with a mean of 450 and a standard deviation of 40. For each of the given test scores, find the corresponding z score. (a) 530. (b) 500. (c) 400. (d) 300. (e) 450.

2-156 A screening test is used to measure the articulation of entering first-grade students. The results have a mean of 80 and a standard deviation of 5. For each of the given test scores, find the corresponding z score. (a) 83. (b) 75. (c) 72. (d) 80. (e) 100.

2-157 Which of the following two scores has the better relative position?
(a) A score of 53 on a test for which $\bar{x} = 50$ and $s = 10$.
(b) A score of 53 on a test for which $\bar{x} = 50$ and $s = 5$.

2-158 Two students took different language facility tests. Which of the following results indicates the higher level of language facility?
(a) A score of 60 on a test for which $\bar{x} = 70$ and $s = 10$.
(b) A score of 480 on a test for which $\bar{x} = 500$ and $s = 50$.

2-159 Three prospective employees take different tests of communicative ability. Which of the following scores corresponds to the highest level?
(a) A score of 60 on a test for which $\bar{x} = 50$ and $s = 5$.
(b) A score of 230 on a test for which $\bar{x} = 200$ and $s = 10$.
(c) A score of 540 on a test for which $\bar{x} = 500$ and $s = 15$.

2-160 Three students take different tests of neuroticism with the given results. Which is the highest relative score?
(a) A score of 3.6 on a test for which $\bar{x} = 4.2$ and $s = 1.2$.
(b) A score of 72 on a test for which $\bar{x} = 84$ and $s = 10$.
(c) A score of 255 on a test for which $\bar{x} = 300$ and $s = 30$.

2-161 Arrange the given data in order from lowest to highest. Then find the percentile corresponding to the score of 35.

35	40	41	42	65	55	88	97
16	18	21	93	81	71	91	86
47	46	85	77	22	25	79	83
69	70	75	51	73	59	56	60
62	63	67	53	28	29	38	39

2-162 Using the data in Exercise 2-161, find the percentile corresponding to the score of 41.

2-163   Using the data in Exercise 2-161, find the percentile corresponding to the score of 93.

2-164   Using the data in Exercise 2-161, find the percentile corresponding to the score of 29.

In Exercises 2-165 through 2-168, use the data in Exercise 2-161 to find the indicated percentile.

2-165 P_{50} **2-166** P_{37} **2-167** P_{67} **2-168** P_{84}

2-169   Use the data in Exercise 2-161 to find Q_1.

2-170   Use the data in Exercise 2-161 to find Q_3.

2-171   Use the data in Exercise 2-161 to find D_3.

2-172   Use the data in Exercise 2-161 to find D_7.

2-6   Exercises B
Measures of Position

2-173   (a)   Using the data in Exercise 2-161, find the interquartile range.
        (b)   Using the data in Exercise 2-161, find the midquartile.
        (c)   Using the data in Exercise 2-161, find the 10–90 percentile range.

2-174   (a)   For the data in Exercise 2-161, does $P_{50} = Q_2$? Does P_{50} always equal Q_2?
        (b)   For the data in Exercise 2-161, does $Q_2 = (Q_1 + Q_3)/2$?

2-175   Find Q_1, Q_2, and Q_3 for the data summarized in the given stem-and-leaf plot.

5	00011122
6	001233457899
7	4444455678899
8	012344567899
9	0001177

2-176

Construct a collection of data consisting of 50 scores for which $Q_1 = 20$, $Q_2 = 30$, $Q_3 = 70$.

Computer Project
Descriptive Statistics

(a)   Many computer systems are supported by software consisting of program packages for various uses. Statistical programs are common among such packages. For example, STATDISK is a statistical package designed as a supplement to this text. Use STATDISK or any other statistics software package to obtain descriptive statistics (mean, variance, standard deviation, and so on) for the 150 job absence times listed in Table 2-1.

(b) Develop a computer program that will take a set of data as input. Output should consist of the mean, variance, standard deviation, and the number of scores. Run the program using the set of 150 job absence times listed in Table 2-1.

Review

Chapter 2 deals mainly with the methods and techniques of descriptive statistics. The main objective of this chapter is to develop the ability to organize, summarize, and illustrate data and to extract from data some meaningful measurements. In Section 2-2 we considered the **frequency table** as an excellent device for summarizing data, while Section 2-3 dealt with graphic illustrations, including **pie charts, histograms, frequency polygons, ogives,** and **stem-and-leaf plots.** These visual illustrations help us to determine the position and distribution of a set of scores. In Section 2-4 we defined the common **averages,** or measures of central tendency. The **mean, median, mode,** and **midrange** represent different ways of characterizing the central value of a collection of data. The **weighted mean** is used to find the average of a set of scores that may vary in importance. In Section 2-5 we presented the usual **measures of dispersion,** including the **range, variance,** and **standard deviation;** these descriptive statistics are designed to measure the variability among a set of scores. The **standard score,** or **z score,** was introduced in Section 2-6 as a way of measuring the number of standard deviations by which a given score differs from the mean. That section also included the common measures of position: **quartiles, percentiles,** and **deciles.**

By now we should be able to organize, present, and describe collections of data composed of single scores. We should be able to compute the key descriptive statistics that will be used in later applications.

Using the concepts developed in this chapter, we now have a better understanding of the 150 absence times given in Table 2-1. The histogram allowed us to see the shape of the distribution of those values. We know that the mean is 87.9, the median is 87, the mode is 4, the standard deviation is 51.4, and the variance is 2638.9. The data are summarized in a frequency table and depicted in a histogram, frequency polygon, and ogive. By using these descriptive characteristics, a manager could form some meaningful and practical conclusions. Subsequent chapters will consider other ways of using sample statistics.

IMPORTANT FORMULAS

$$\overline{x} = \frac{\Sigma x}{n}$$

Mean

$$\overline{x} = \frac{\Sigma(f \cdot x)}{\Sigma f}$$

Computing the mean when the data is in a frequency table

$$s^2 = \frac{\Sigma(x - \overline{x})^2}{n - 1}$$

Variance

or

$$s^2 = \frac{n(\Sigma x^2) - (\Sigma x)^2}{n(n - 1)}$$

Short-cut formula for variance

$$s^2 = \frac{n[\Sigma(f \cdot x^2)] - [\Sigma(f \cdot x)]^2}{n(n - 1)}$$

Computing the variance when the data is in a frequency table

$$s = \sqrt{s^2}$$

The standard deviation is the square root of the variance.

$$z = \frac{x - \overline{x}}{s}$$

Standard, or z, score

Vocabulary List

Define and give an example of each term.

descriptive statistics
inferential statistics
frequency table
lower class limits
upper class limits
class boundaries
class marks
class width
cumulative
   frequency table
pie chart
histogram
frequency polygon
ogive

exploratory data
   analysis
stem-and-leaf plot
average
measure of central
   tendency
mean
median
mode
bimodal
multimodal
midrange
weighted mean
measure of
   dispersion

range
variance
standard deviation
mean deviation
standard score
z score
quartiles
deciles
percentiles
interquartile range
semi-interquartile
   range
midquartile
10–90 percentile
   range

Review Exercises

In Exercises 2-177 through 2-180, use the given sample data to find the: (a) mean, (b) median, (c) mode, (d) midrange, (e) range, (f) variance, (g) standard deviation.

2-177 Ages (in years) of a computer assembly team:

$$26, 26, 28, 32, 37, 40$$

2-178 Distances (in miles) traveled by commuting students:

$$8.4, 9.2, 28.6, 31.5, 10.1, 14.3, 8.4, 12.3$$

2-179 Diameters (in micrometers) of virus samples:

$$175, 183, 168, 191, 187, 183, 170, 174, 184, 181, 182$$

2-180 Voltage readings taken in an electrical circuit at different times:

$$7.1, 6.4, 6.7, 6.7, 6.7, 7.5, 7.7, 7.8, 7.9$$

2-181 A supplier constructs a frequency table for the number of car stereo units sold daily. Use that table to find the mean and standard deviation.

Number sold	Frequency
0–3	5
4–7	9
8–11	8
12–15	6
16–19	3

2-182 Using the frequency table given in Exercise 2-181, construct the corresponding histogram.

2-183 Using the frequency table given in Exercise 2-181, construct the corresponding frequency polygon.

2-184 Using the frequency table given in Exercise 2-181, construct the corresponding ogive.

2-185 Construct the stem-and-leaf plot for the data in Exercise 2-177.

2-186 Construct the stem-and-leaf plot for the data in Exercise 2-179.

2-187 For the data of Exercise 2-177, find the z score corresponding to 40.

2-188 For the data of Exercise 2-177, find the z score corresponding to 28.

2-189   The given scores represent the number of cars rejected in one day at an automobile assembly plant. The 50 scores correspond to 50 different randomly selected days. Construct a frequency table with 10 classes.

29	58	80	35	30	23	88	49	35	97
12	73	54	91	45	28	61	61	45	81
83	23	71	63	47	87	36	8	94	26
95	63	86	42	22	44	8	27	20	33
28	91	87	15	67	10	45	67	26	19

2-190   Construct a histogram that corresponds to the frequency table from Exercise 2-189.

2-191   For the data in Exercise 2-189, find (a) Q_1, (b) P_{45}, and (c) the percentile corresponding to the score of 30.

2-192   Use the frequency table from Exercise 2-189 to find the mean and standard deviation for the number of rejects.

2-193   (a) A set of data has a mean of 45.6. What is the mean if 5.0 is added to each score?
   (b) A set of data has a standard deviation of 3.0. What is the standard deviation if 5.0 is added to each score?
   (c) You just completed a calculation for the variance of a set of scores, and you got an answer of -21.3. What do you conclude?
   (d) True or false: If set A has a range of 50 while set B has a range of 100, then the standard deviation for data set A must be less than the standard deviation for data set B.
   (e) True or false: In proceeding from left to right, the graph of an ogive can never follow a downward path.

2-194   A psychologist gives a subject two different tests designed to measure spatial perception. The subject obtained scores of 66 on the first test and 223 on the other test. The first test is known to have a mean of 75 and a standard deviation of 15, while the second test has a mean of 250 and a standard deviation of 25. Which result is better? Explain.

Chapter 2
Case Study Activity

Through observation or experimentation, compile a list of sample data. Obtain at least 40 values, and try to select data from an interesting or meaningful population.
(a) Describe the nature of the data. That is, what do the values represent?
(b) What method was used to collect the values?

(c) What are some possible reasons why the data might not be representative of the population? That is, what are some possible sources of bias?

(d) Find the value of each of the following: sample size, minimum, maximum, mean, median, midrange, range, standard deviation, variance, and the quartiles Q_1 and Q_3.

(e) Construct a frequency table and histogram.

3-1 **Overview**

We identify chapter **objectives** and compare probability and statistics.

3-2 **Fundamentals**

We introduce the **basic definitions** of probability theory.

3-3 **Addition Rule**

We present the rule for finding the probability that one event *or* another event will occur.

3-4 **Multiplication Rule**

We present the rule for finding the probability that two events will *both* occur.

3-5 **Complementary Events**

We present the rule for finding the probability that an event does *not* occur.

3-6 **Counting (Optional)**

We present techniques for determining the total number of different ways that various events can occur.

Probability

Suppose that a political analyst plans to do a study of voting behavior in a county of 200,000 voters, of which 30% are Republican. He begins with a simple pilot study by paying a pollster $1200 for in-depth surveys of 12 randomly selected voters. But the pollster returns and reports that each of the 12 voters is a Republican. The political analyst protests and says that the sample is not random. The pollster replies that with such a small sample, it is easy to get 12 Republicans since there are 60,000 of them available. This argument might seem reasonable, but we will be able to analyze it more effectively by the techniques developed in this chapter. We will then learn how to determine whether or not the pollster is correct.

3-1   Overview

Suppose you discover a way to mark each molecule in an 8-ounce glass of water so that each one is recognizable. You then proceed to the nearest ocean beach and pour the water into the first wave. After waiting 20 years for the waters of the world to mix, you board a plane for Venice, Italy. Upon your arrival, you scoop a glass of water from the first canal you come to and examine the water in search of one of the molecules that you dumped in 20 years ago. Can you really expect to find one of the molecules in Venice? As absurd as this problem seems, the answer is actually yes. (If you find this incredible, you are not alone.) In fact, you can expect about 1000 of the original molecules to reappear in Venice.

In a class of 25 students, each is asked to identify the month and day of his or her birth. What are the chances that at least two students will share the same birthday? Again our intuition is misleading; it happens that at least two students will have the same birthday in more than half of all classes with 25 students.

The preceding conclusions are based upon simple principles of probability, which play a critical role in the theory of statistics. These introductory examples are not at all intuitively obvious, and they do require nontrivial calculations.

All of us now form simple probability conclusions in our daily lives. Sometimes these determinations are based on fact, while other probability determinations are subjective. In addition to being an important aspect of the study of statistics, probability theory is playing an increasingly important role in a society that must attempt to measure uncertainties.

In Chapter 1 we stated that inferential statistics involves the use of sample evidence in formulating inferences or conclusions about a population. These inferential decisions are based on probabilities or likelihoods of events. As an example, suppose that a statistician plans to test a coin for fairness. She tosses this coin 100 times, and the result is heads each time. Either the coin is fair and an *extremely* rare event has occurred, or the coin is not fair and it favors heads. What should we infer? The statistician, along with most reasonable people, would conclude that the coin is not fair. This decision is based on the very low probability of getting 100 consecutive heads with a normal coin.

In subsequent chapters we develop methods of statistical inference that rely on this type of thinking. It is therefore important to acquire a basic understanding of probability theory. We want to cultivate some intuitive feeling for what probabilities are, and we want to develop some very basic skills in calculating the probabilities of certain events (see Figure 3-1).

While probability theory is important for statistical inferences, we should note a fundamental difference between probability theory and

a. Duncan County voters   b. Jackson County voters

Figure 3-1   (a) Probability. With a *known population,* we can see that if one voter is randomly selected, there are 60,000 chances out of 200,000 of picking a Republican. This corresponds to a probability of 60,000/200,000 or 0.3. The population of all voters in the county is known, and we are concerned with the likelihood of obtaining a particular sample (a Republican). We are making a conclusion about a sample based on our knowledge of the population.
(b) Statistics. With an *unknown population,* we can obtain a fairly good idea of voter preferences by randomly selecting 500 voters and assuming that this sample is representative of the whole county population. After making the random selections, our sample is known and we can use it to make inferences about the population of all voters in the county. We are making a conclusion about the population based on our knowledge of the sample.

Long-Range Weather Forecasting

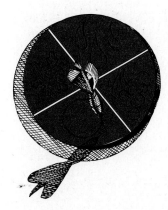

A high school teacher in Millbrook, New York, created a controversy when he conducted a classroom experiment in which long-range weather forecasts were made by throwing darts. The dart method resulted in 7 correct predictions among 20 winter storms, while a nearby private forecasting service correctly predicted only 6 of the storms. The teacher claimed that while short-range weather forecasts are usually quite accurate, long-range forecasts are not.

the theory of statistical inferences. **In a typical probability problem, the population is known and we want to determine the likelihood of observing a sample; in a typical statistical inference problem, a sample is given or obtained and we want to form a conclusion about the population.** In this sense, probability and statistics are opposites.

In this chapter we begin by introducing the fundamental concept of mathematical probability and we then proceed to investigate the basic rules of probability: the addition rule, the multiplication rule, and the rule of complements. We also consider techniques of counting the number of ways an event can occur. The primary objective of this chapter is to develop a sound understanding of probability values that will be used in subsequent chapters. A secondary objective is to develop the ability to solve simple probability problems, which are valuable in their own right as they are used to make decisions and better understand our world.

3-2   Fundamentals

In considering probability problems, we deal with experiments and events.

> **DEFINITION**
>
> An **experiment** is any process that allows us to obtain observations.

> **DEFINITION**
>
> An **event** is any collection of results or outcomes of an experiment.

For example, the random selection of a letter from a computer data bank is an experiment and the result of a vowel is an event. In this case, the event of a vowel will occur if the letter is a, e, i, o, or u. We might think of the event of getting a vowel as being a combination of the simpler outcomes corresponding to the individual vowel letters. It is often necessary to consider events that cannot be decomposed into simpler outcomes.

> **DEFINITION**
>
> A **simple event** is an outcome or an event that cannot be broken down any further.

When conducting the experiment of randomly selecting a letter from a data bank, the occurrence of an e is a simple event, but the occurrence of a vowel is not a simple event since it can be broken down into the five simple events of a, e, i, o, and u.

DEFINITION

The **sample space** for an experiment consists of all possible simple events. That is, the sample space consists of all outcomes that cannot be broken down any further.

There is no universal agreement for the actual definition of the probability of an event, but among the various theories and schools of thought, two basic approaches emerge most often. The approaches will be embodied in two rules for finding probabilities. The notation employed will relate P to probability, while capital letters such as A, B, and C will denote specific events. For example, A might represent the event of winning a million-dollar state lottery; $P(A)$ denotes the probability of event A occurring.

RULE 1

Empirical (relative frequency) approximation of probability. Conduct (or observe) an experiment a large number of times and count the number of times that event A actually occurs. Then $P(A)$ is estimated as follows.

$$P(A) = \frac{\text{number of times } A \text{ occurred}}{\text{number of times experiment was repeated}}$$

RULE 2

Classical approach to probability. Assume that a given experiment has n different simple events, each of which has an **equal chance** of occurring. If event A can occur in s of these n ways, then

$$P(A) = \frac{s}{n}$$

The first rule, often referred to as the **empirical,** or **relative frequency,** interpretation of probability, gives us only approximations or estimates of $P(A)$. The second rule, often referred to as the **classical**

Should You Guess on SAT's?

Students preparing for the SAT exams are often advised to avoid guessing, but is that good advice? Not necessarily! The correction for guessing is this: "For questions with five answer choices, one-fourth of a point is subtracted for each incorrect response; one-third of a point is subtracted for incorrect responses to questions with four answer choices." While that is a *correction* for guessing, it is not a *penalty*. Principles of probability can be used to show that in the long run, pure random guessing will neither raise nor lower the exam score. Also, students often can eliminate at least one choice and frequently have some intuitive sense for a correct answer. These factors suggest that the best strategy is to guess at every question for which the answer is not known. However, students should not waste too much time on such questions and, if possible, they should try to avoid tricky questions with attractive wrong answers.

approach, is useful in the analysis of simple experiments that involve equally likely outcomes, such as flipping coins, rolling dice, and so on (see Figure 3-2).

Complicated experiments, such as those involving events that are not equally likely, all require the relative frequency approximation. For example, to determine the probability of an 18-year-old male living to

Figure 3-2

When trying to determine $P(2)$ on a "shaved" die, we must repeat the experiment of rolling it many times and then form the ratio of the number of times 2 occurred to the number of rolls. That ratio is our estimate of $P(2)$.

(a) Empirical approach (Rule 1)

With a balanced and fair die, each of the six faces has an equal chance of occurring.

$$P(2) = \frac{\text{number of ways 2 can occur}}{\text{total number of simple events}} = \frac{1}{6}$$

(b) Classical approach (Rule 2)

age 65, we must study past records in order to arrive at a reasonable approximation. The events of living to age 65 and dying before age 65 are not equally likely, so we cannot use Rule 2.

Many experiments involving equally likely outcomes are so complicated that Rule 1 is used as a practical alternative to extremely complex computations. Consider, for example, the probability of winning at solitaire. In theory, Rule 2 applies, but in reality the better approach requires that we settle for an approximation obtained through Rule 1. That is, we should play solitaire many times and record the number of wins along with the number of attempts. The ratio of wins to trials is the approximation we seek.

The following examples are intended to illustrate the use of Rules 1 and 2. In some of these examples we use the term *random*. Throughout this chapter, a random selection means that each element that can be chosen has an equal chance of being chosen.

EXAMPLE

If one person is randomly selected from a group of 20 psychologists and 30 sociologists, what is the probability of getting a sociologist?

Solution

In selecting one person from this group, we note that there are 50 different possible people (simple events) available, of which 30 are sociologists. Assuming that each person has the same chance of being selected, Rule 2 applies and we get

$$P(\text{sociologist}) = \frac{30}{50} = 0.6$$

EXAMPLE

On a college entrance examination, each question has four possible answers. If an examinee makes a random guess on the last question, what is the probability that the response is incorrect?

Solution

There are four possible outcomes or answers, and there are three ways to answer incorrectly. Random guessing implies that the outcomes are equally likely, so we apply Rule 2 to get

$$P(\text{wrong answer}) = \frac{3}{4} \quad \text{or} \quad 0.75$$

EXAMPLE

Find the probability of a 20-year-old student living to be 21 years of age.

Solution

Here the two outcomes of living and dying are not equally likely, so the relative frequency approximation must be used. This requires that we observe a large number of 20-year-old students and then count those who live to be 21. Suppose that we survey 10,000 20-year-old students and find that 9961 lived to be 21 (these are realistic figures). Then the empirical approximation becomes

$$P(\text{20-year-old student living to 21}) = \frac{9{,}961}{10{,}000} \quad \text{or} \quad 0.996$$

This is the basic approach used by insurance companies in the development of mortality tables.

EXAMPLE

A magazine is developing a telephone solicitation department. What is the probability of getting an order on any given call?

Solution

Again the outcomes of getting or not getting an order are not equally likely, so the relative frequency approximation applies. In essence, past experience is projected, so we need to accumulate some experience before making any conclusions. Let's suppose that on the first 100 calls, there are 17 orders. We therefore estimate that

$$P(\text{order}) = \frac{17}{100} \quad \text{or} \quad 0.17$$

As the number of trials increases, the approximation becomes better.

EXAMPLE

Find the probability that a couple with three children will have exactly two boys. (Assume that boys and girls are equally likely and that the sex of any child is independent of any brothers or sisters.)

Mathematician Predicts Date of His Death

Jerome Cardan (1501–1576) was a doctor, mathematician, compulsive gambler, and author of *The Book on Games of Chance*, the first book on probability. His book, which included tips on how to cheat and how to catch others cheating, introduced some concepts of probability theory, such as the concept of expected value. Cardan correctly predicted the date on which he would die. However, he helped his prediction somewhat by applying a nonmathematical technique commonly known as suicide.

Solution

If the couple has exactly two boys in three births, there must be one girl. The possible outcomes are listed and each is assumed to be equally likely.

$$
\begin{aligned}
&\text{boy-boy-boy}\\
&\text{boy-boy-girl}\\
&\text{boy-girl-boy}\\
\text{two boys}\Big\langle\ &\text{boy-girl-girl}\\
&\text{girl-boy-boy}\\
&\text{girl-boy-girl}\\
&\text{girl-girl-boy}\\
&\text{girl-girl-girl}
\end{aligned}
$$

Of the eight different possible outcomes, three correspond to exactly two boys, so that

$$P(2 \text{ boys in 3 births}) = \frac{3}{8} = 0.375$$

EXAMPLE

If a year is selected at random, find the probability that February 30 is a Friday.

Solution

No year ever contains a February 30, so the given event cannot possibly occur. When an event is impossible, we say that its probability is zero.

The probability of any impossible event is 0.

EXAMPLE

If a year is selected at random, find the probability that Thanksgiving Day will be on a Thursday.

Solution

Thanksgiving Day is defined so that it will always be on a Thursday. The given event is therefore certain to occur. When an event is certain to occur, we say that its probability is 1.

The probability of any event that is certain to occur is 1.

Since any event imaginable is either impossible, certain, or somewhere in between, it is reasonable to conclude that the mathematical probability of any event is either 0, 1, or a number between 0 and 1 (see Figure 3-3). This property can be expressed as follows:

$$0 \leq P(A) \leq 1 \qquad \text{for any event } A$$

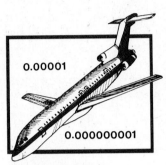

0.00001

0.000000001

Just How Probable Is Probable?

The mathematical definition requires that probability values must be between 0 and 1, with 1 corresponding to an event that is certain to occur and 0 corresponding to an impossible event. But what probability values correspond to terms such as *probable, improbable,* or *extremely probable?* In discussing systems design analysis of aircraft, the Federal Aviation Administration (FAA) has defined these terms. The following definitions are taken from an FAA publication.

1. Probable—Probable events may be expected to occur several times during the operational life of each airplane. A probability on the order of 0.00001 or greater (for each hour of flight).

2. Improbable—Improbable events are not expected to occur during the total operational life of a random single airplane of a particular type, but may occur during the total operational life of all airplanes of a particular type. A probability on the order of 0.00001 or less.

3. Extremely improbable—Extremely improbable events are so unlikely that they need not be considered to ever occur, unless engineering judgment would require their consideration. A probability on the order of 1×10^{-9} or less. (Note: 1×10^{-9} = 0.000000001.)

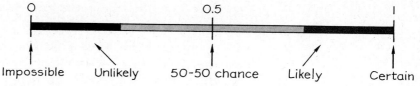

Figure 3-3 Possible values for probabilities

In Figure 3-3, the scale of 0 through 1 is shown on top, whereas the more familiar and common expressions of likelihood are shown on the bottom.

EXAMPLE

Your six-cylinder car is running roughly because of one faulty spark plug, and you have enough time to check and replace only one spark plug. What is the probability that you will select the defective one?

Solution

There are six possible outcomes corresponding to the six different spark plugs. Since one plug is defective, we conclude that

$$P(\text{defective plug}) = \frac{1}{6} = 0.167$$

Rounding Off Probabilities

Although it is difficult to develop a universal rule for rounding off probabilities, the following guide will apply to most problems in this text.

> Either give the exact fraction representing a probability, or round off final decimal results to three significant digits.

All the digits in a number are significant except for the zeros that are included for proper placement of the decimal point. The probability of 0.00128506 can be rounded to three significant digits as 0.00129. The probability of 1/3 can be left as a fraction or rounded in decimal form to 0.333, but not 0.3.

EXAMPLE

An experiment in ESP involves blindfolding the subject and allowing the subject to make one selection from a standard deck of cards. The subject must then identify the correct suit (that is, clubs, diamonds, hearts, or spades). What is the probability of making a correct random guess?

Solution

The standard deck contains 52 cards with 13 of each suit. There are therefore 52 possible outcomes, with 13 outcomes corresponding to whatever suit the subject selects. Therefore

$$P(\text{correct guess}) = \frac{13}{52} = \frac{1}{4} = 0.25$$

The $\frac{1}{4}$ probability of the previous example implies that, on the average, for each four guesses, there will be one correct selection. If our ESP subject does *significantly* better than one correct response in four, then we can conclude that the selections are not made solely on the basis of chance. But what is significant? Thirty correct guesses in 100 trials? Fifty correct guesses in 100 trials? Ninety-nine correct guesses in 100 trials? This decision relates to a key concept of statistical inference and will be explored later in the book.

An important concept of this section is the mathematical expression of probability as a number between 0 and 1. This type of expression is fundamental and common in statistical procedures, and we will use it throughout the remainder of this text. A typical computer output, for example, may involve a P-value expression such as "Significance less than 0.001." We will discuss the meaning of P-values later, but they are essentially probabilities of the type discussed in this section. We should recognize that a probability of 0.001 (equivalent to 1/1000) corresponds to an event that is very rare in the sense that it occurs an average of only once in a thousand trials.

3-2   Exercises A
Fundamentals

3-1   Which of the following could *not* be a probability?

$$\frac{3}{7}, \ 2, \ -\frac{1}{2}, \ \frac{3}{4}, \ \frac{99}{101}, \ 0, \ 1, \ 5, \ 1.11, \ 1.0001, \ 0.001, \ 0.9999$$

3-2 On a college entrance exam, each question has five possible answers. If an examinee makes a random guess on the first question, what is the probability that the response is correct?

3-3 In a simulation experiment, a computer program generates random whole numbers between 1 and 20 inclusive. What is the probability that the first generated number is 17?

3-4 An influenza vaccine is administered to 92 subjects and is found to be effective in 86 cases. What is the approximate probability that the vaccine will be effective?

3-5 A traffic survey indicates that of 3756 cars approaching a shopping plaza, 857 turned into the parking lot. Find the empirical probability of a car *not* entering the parking lot.

3-6 A farmer plants 900 seeds of which 210 are productive. What is the empirical probability of a similar seed being nonproductive?

3-7 In a recent national election, there were 12,300,000 citizens in the 18-to-20-year age bracket. Of these, 4,400,000 actually voted. Find the empirical probability that a person randomly selected from this group of 18-to-20-year-old citizens did vote in that national election.

3-8 Find the probability of selecting a date in May if one date is randomly drawn from the 366 possible dates in a leap year.

3-9 Among eight helicopters sent to rescue American hostages in Iran, three helicopters failed to operate properly. Given the same conditions, what is the empirical probability of failure for a helicopter?

3-10 A jury of four men and eight women has been selected for a sex-discrimination case. If the foreman of this jury is selected at random, find the probability that the foreman will be a woman.

3-11 A computer is used to generate random telephone numbers. Of the numbers generated and in service, 56 are unlisted, and 144 are listed in the telephone directory. If one of these telephone numbers is randomly selected, what is the probability that it is unlisted?

3-12 Among certain stocks selected in one day, 332 declined and 668 rose. If one of these stocks is selected at random, what is the probability it rose that day?

3-13 In an experiment involving smoke detectors, an alarm was triggered after the subject fell asleep. Of the 95 sleeping subjects, 40 were awakened by the alarm, and 55 did not awake. If one subject is randomly selected, find the probability that he or she is among those who were not awakened by the alarm.

3-14 In a survey of New York City subway cars, it was found that 123 cars had interior graffiti and 37 cars did not. If a car is randomly selected, what is the probability it will have interior graffiti?

3-15   In a survey of teenagers, it is found that 39 smoke and 261 do not. If a teenager is randomly selected, find the probability that he or she smokes.

3-16   In a survey of cars originally equipped with catalytic converters, it was found that 260 cars still had their converters, and 14 cars had them removed. What is the approximate probability of selecting a car with a removed catalytic converter if the selection is random and is limited to cars originally assembled with converters?

3-17   In a recent presidential election, 44 million voters were in favor of the Republican, and 36 million voters were in favor of the Democratic candidate. Find the probability that a voter selected at random cast a ballot in favor of the Republican.

3-18   A particular recessive genetic characteristic is present in 235 subjects studied and absent in 755. If one of these subjects is randomly selected, find the probability that the genetic characteristic will be present.

3-19   Among 1200 radio listeners surveyed, 155 would give no opinion but 163 indicated a preference for country music. If a listener is randomly selected and found to have an opinion, what is the approximate probability that they favor country music?

3-20   Among male dieters following the same routine, 16 lost weight, 4 gained weight, and 2 remained the same. If one of these dieters is selected at random, what is the probability that he lost weight?

3-21   Blood types are determined for a sample of people, and the results are given in the accompanying table. If one person from this sample group is randomly selected, find the probability that the person has type O blood.

Blood type	Frequency
O	18
A	9
B	8

3-22   A study was made of serious crimes in one year, and the totals for four states are given in the accompanying table. If one of these crimes is randomly selected, find the probability it was committed in Nevada.

State	Number of crimes
N.Y.	6900
Calif.	7800
Mich.	6700
Nevada	8600

3-23   A study was conducted to collect data on suicides in one year for one region and the results are given in the table on the top of page 107. If one of these suicide cases is randomly selected, find the probability of getting a male in the 25–34 age group.

	Male	Female
15–24 years of age	35	9
25–34 years of age	52	14

3-24 In a study of the effects of cigarette smoking on children, the data in the accompanying table were compiled. If one of these mothers is randomly selected, find the probability that she is in the category of those who smoke two or more packs a day and has a hyperactive child.

	Child is hyperactive	Child is not hyperactive
Mother smokes 2 or more packs per day	64	437
Mother smokes less than 2 packs per day	32	475

3-25 A couple plans to have two children.
(a) List the different possible outcomes according to the sex of each child. Assume that these outcomes are equally likely.
(b) Find the probability of getting two boys.
(c) Find the probability of getting exactly one child of each sex.

3-26 A couple plans to have four children.
(a) List the 16 different possible outcomes according to the sex of each child. Assume that these outcomes are equally likely.
(b) Find the probability of getting all girls.
(c) Find the probability of getting *at least* one child of each sex.
(d) Find the probability of getting *exactly* two children of each sex.

3-27 A quick quiz consists of three true-false questions and an unprepared student must guess at each one. The guesses will be random.
(a) List the different possible solutions.
(b) What is the probability of answering the three questions correctly?
(c) What is the probability of guessing incorrectly for all three questions?
(d) What is the probability of passing by guessing correctly for at least two questions?

3-28 Both parents have the brown-blue pair of eye-color genes, and each parent contributes one gene to a child. Assume that if the child has at least one brown gene, that color will dominate and the eyes will be brown. (Actually, the determination of eye color is somewhat more complex.)
(a) List the different possible outcomes. Assume that these outcomes are equally likely.
(b) What is the probability that a child of these parents will have the blue-blue pair of genes?
(c) What is the probability that a child will have brown eyes?

3-2   Exercises B
Fundamentals

3-29   On a distant planet, one parent from each of three different sexes is necessary for reproduction. Each parent contributes one gene from a triplet of genes that determine the color of a child's antenna. Three parents have the following genes:

> red-white-blue
>
> red-white-white
>
> red-red-white

(a) List the 27 different possible outcomes. Assume that these outcomes are equally likely.

(b) What is the probability that a child receives the blue-red-red combination of genes?

(c) The color of a child's antenna is determined by majority. For example, a child receiving the blue-white-white combination would have a white antenna. If one of each color is present, the antenna will be blue. What is the probability that a child has a white antenna?

(d) What is the probability that a child has a blue antenna? (See part (c).)

(e) What is the probability that a child has an antenna that is not white? (See part (c).)

3-30   Someone has reasoned that since we know nothing about the presence of life on Pluto, either life exists there or it does not, so that P(life on Pluto) = 0.5. Is this reasoning correct? Explain.

3-31   The odds against event A are in the same ratio as $P(A$ does not occur$)$: $P(A)$. If the odds against an event are 5 : 2, find the probability of the event occurring.

3-32   If the simple event A happens n different ways (all equally likely) and there are m different equally likely ways that event A does not occur, find an expression for $P(A)$.

3-33   If two flies land on an orange, find the probability that they lie on the same hemisphere.

3-34   Two points along a straight stick are randomly selected. The stick is then broken at those two points. Find the probability that the three pieces can be arranged to form a triangle.

3-3   Addition Rule

In the preceding section we introduced the basic concept of probability and we considered simple experiments and simple events. Many real circumstances involve compound events.

You Bet?

Only 40% of the money bet in the New York State lottery is returned in prizes, so the "house" has a 60% advantage. The house advantage at race-tracks is usually about 15%. In casinos, the house advantage is 5.26% for roulette, 5.9% for blackjack, 1.4% for craps, and 3% to 22% for slot machines. Some professional gamblers can systematically beat the house at blackjack if they use sophisticated card-counting techniques. The basic idea is to keep a record of cards that have been played so that the player will know when the deck has a disproportionately large number of high cards. When this occurs, the casino is more likely to lose and the player begins to place large bets. Many casinos eject card counters or shuffle the decks more frequently.

DEFINITION

Any event combining two or more simple events is called a **compound event.**

Computers are often used to randomly generate telephone numbers of people to be called for a survey or poll. That approach prevents the exclusion of subjects with unlisted numbers (often around 28% of the population). Suppose we consider only the last digit of the randomly generated telephone number. The event of getting an odd last digit is an example of a compound event since it combines the simple events of 1, 3, 5, 7, 9.

In this section we want to develop a rule for finding $P(A \text{ or } B)$, the probability that for a single outcome of an experiment, either event A or event B occurs or both events occur. (Throughout this text we use the inclusive *or*, which means either one, or the other, or both. We will *not* consider the exclusive *or*, which means either one or the other, but not both.) For example, suppose we stipulate that an experiment consists of using a computer to randomly generate the last digit of a telephone number, and we let events A and B be defined as follows:

A: the digit is odd

B: the digit is less than 5

The occurrence of A or B is the occurrence of an odd number or a number less than 5. Similarly, $P(A \text{ or } B)$ is the probability of getting an odd number or a number less than 5. We will now determine the probability of getting A or B through a direct and simple analysis. We know that there are ten possible outcomes (0, 1, 2, . . . , 9), so we can identify those that are either odd or less than 5.

0 is odd or less than 5 (it's less than 5)
1 is odd or less than 5 (it's both)
2 is odd or less than 5 (it's less than 5)
3 is odd or less than 5 (it's both)
4 is odd or less than 5 (it's less than 5)
5 is odd or less than 5 (it's odd)
6 is neither odd nor less than 5
7 is odd or less than 5 (it's odd)
8 is neither odd nor less than 5
9 is odd or less than 5 (it's odd)

Examining the preceding list we see that of the ten possible outcomes, exactly eight of them are odd or less than 5. This means that of the ten simple events that comprise the sample space, there are exactly eight ways that A or B can occur. Since the ten simple events are all equally likely, Rule 2 from Section 3-2 can be applied and we get $P(A \text{ or } B) = 8/10$.

If all problems were as simple and direct as this, we would have no real need for more abstract rules and formulas. But important problems of greater complexity do require formal generalized rules.

Let's use this last example as a basis for constructing one such rule. We know that $P(A) = P(\text{odd number}) = 5/10$ and $P(B) = P(\text{number less than 5}) = 5/10$ from Section 3-2. We also know that $P(A \text{ or } B) = 8/10$ from the above analysis. It would be nice if we could express $P(A \text{ or } B)$ in terms of $P(A)$ and $P(B)$, but how can we do that?

$P(A \text{ or } B) = 8/10$ does not seem to suggest any combination of $P(A) = 5/10$ and $P(B) = 5/10$. The problem really reduces to that of counting the number of ways A can occur (5), the number of ways B can occur (5), and getting a total of 8.

The obstacle here is the overlapping of events A and B. There are two ways that A and B will both occur simultaneously. If either 1 or 3 is the outcome, then we have an odd number that is also less than 5. **When combining the number of ways event A can occur with the number of ways B can occur, we must avoid double counting of those outcomes in which A and B happen simultaneously.** If we were to state that $P(A \text{ or } B) = P(A) + P(B)$ in this example, we would be wrong because $8/10 \neq 5/10 + 5/10$. The error arises from the two outcomes of 1 and 3, which were each counted twice. The last equation is corrected by a subtraction to compensate for that double counting.

$$\frac{8}{10} = \frac{5}{10} + \frac{5}{10} - \frac{2}{10}$$

The right-hand side of this last equation reflects the five ways of getting an odd number, the five ways of getting a number below 5, and the compensating subtraction of the two ways of getting both. This corrected equation can be generalized in the addition rule, which follows.

ADDITION RULE

$$P(A \text{ or } B) = P(A) + P(B) - P(A \text{ and } B)$$

where $P(A \text{ and } B)$ denotes the probability that A and B both occur at the same time as an outcome in the experiment.

See Figure 3-4.

EXAMPLE

A pollster surveys 100 subjects consisting of 40 Democrats (of which half are female) and 60 Republicans (of which half are female). Find the probability of randomly selecting one of these subjects and getting a Democrat or a female (see Figure 3-5).

Solution

By the addition rule,

P(Democrat or female)

$$= P(\text{Democrat}) + P(\text{female}) - P(\text{Democrat and female})$$

We now evaluate the terms on the right side of this equation. Of the 100 subjects there are 40 Democrats, so P(Democrat) = 40/100. The total number of females is 50, so P(female) = 50/100. Also there are 20 Democratic females (half of the 40) so that P(Democrat and female) = 20/100. Therefore,

P(Democrat or female)

$$= P(\text{Democrat}) + P(\text{female}) - P(\text{Democrat and female})$$

$$= \frac{40}{100} + \frac{50}{100} - \frac{20}{100}$$

$$= \frac{70}{100}$$

The addition rule is simplified whenever A and B cannot occur simultaneously, so that $P(A \text{ and } B)$ becomes zero.

Total area = 1

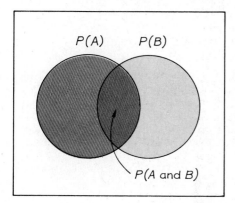

Figure 3-4 The probability of A or B equals the probability of A (left circle) plus the probability of B (right circle) minus the probability of A and B (football-shaped middle region). This figure should show that the addition of the areas of the two circles will cause a double counting of the football-shaped middle region. To compensate for that double counting, we subtract the area of that region. This is the basic concept that underlies the addition rule.

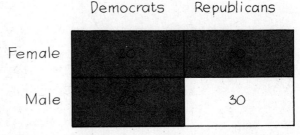

Figure 3-5 Shaded region represents "Democrats or females."

DEFINITION

Events *A* and *B* are **mutually exclusive** if they cannot occur simultaneously.

In the experiment of rolling one die, the event of getting a 2 and the event of getting a 5 are mutually exclusive events, since no outcome can be both a 2 and 5 simultaneously. The following pairs of events are other examples of mutually exclusive pairs in a single experiment. That is, within each of the following pairs of events, it is impossible for both events to occur at the same time.

$$\begin{cases} \text{Manufacturing a defective electronic component} \\ \text{Manufacturing a good electronic component} \end{cases}$$

$$\begin{cases} \text{Selecting a voter who is a registered Democrat} \\ \text{Selecting a voter who is a registered Republican} \end{cases}$$

$$\begin{cases} \text{Testing a subject with an I.Q. above 100} \\ \text{Testing a subject with an I.Q. below 95} \end{cases}$$

The following pairs of events are examples that are *not* mutually exclusive in a single trial. That is, within each of the following pairs of events, it is possible that both events occur at the same time.

$$\begin{cases} \text{Selecting a doctor who is a brain surgeon} \\ \text{Selecting a doctor who is a woman} \end{cases}$$

$$\begin{cases} \text{Selecting a voter who is a registered Democrat} \\ \text{Selecting a voter who is under 30 years of age} \end{cases}$$

$$\begin{cases} \text{Testing a subject with an I.Q. above 100} \\ \text{Testing a subject with an I.Q. above 110} \end{cases}$$

EXAMPLE

A computer randomly generates the last digit of a telephone number. Find the probability that the outcome is an 8 or 9.

Solution

The outcome of 8 and the outcome of 9 are mutually exclusive events. This means that it is impossible for both 8 and 9 to occur together when one digit is selected, so $P(8 \text{ and } 9) = 0$ and the addition rule is applied as follows:

$$P(8 \text{ or } 9) = P(8) + P(9) - P(8 \text{ and } 9)$$

$$= \frac{1}{10} + \frac{1}{10} - 0$$

$$= \frac{2}{10} \quad \text{or} \quad \frac{1}{5}$$

In Figure 3-6 we see that the addition rule can be simplified when the events in question are known to be mutually exclusive. The following examples involve application of the addition rule.

EXAMPLE

In a convention of 100 guests, there are 37 Democrats and 41 Republicans. Find the probability that a conventioneer selected at random is a Democrat or a Republican.

Solution

If D is the event of selecting a Democrat and R is the event of selecting a Republican, then D and R are mutually exclusive, so

$$P(D \text{ or } R) = P(D) + P(R) = \frac{37}{100} + \frac{41}{100} = \frac{78}{100}$$

We can ignore the subtraction of $P(D \text{ and } R)$. That probability is zero since D and R are mutually exclusive events.

Figure 3-6 If events A and B are known to be mutually exclusive, they are completely disjointed and involve no overlapping. When this occurs, the addition rule can be simplified to $P(A \text{ or } B) = P(A) + P(B)$.

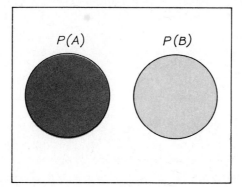

Monkeys Are Not Likely to Type *Hamlet*

A classical assertion holds that a monkey randomly hitting the keys on a typewriter will eventually produce the complete works of Shakespeare, if it continues to type year after year. Dr. William Bennet used the rules of probability to develop a computer simulation that addressed this problem, and he concluded that it would take a monkey about 1,000,000,000,000, 000,000,000,000,000,000, 000,000 years to reproduce Shakespeare's works.

Probability and Prosecution

Probability theory is used in forensic science as experts analyze criminal evidence. One particular area of rapid technological advancement is in the analysis of bloodstains. Several years ago, bloodstains could be categorized only according to the four basic blood types. Current capabilities allow scientists to identify at least a dozen different characteristics. It is estimated that the probability of two people having the same set of blood characteristics is about $\frac{1}{1000}$, and that evidence has already been used to help convict murderers.

EXAMPLE

In the same convention of 100 guests with 37 Democrats and 41 Republicans, there are 35 conventioneers under 30 years of age, including 17 of the Democrats and 13 of the Republicans. Find the probability that a conventioneer selected at random is a Democrat or is under 30 years of age.

Solution

Let Y denote the event of selecting someone under 30 years of age. We seek $P(D \text{ or } Y)$, but D and Y are not mutually exclusive because they do occur simultaneously whenever a Democrat under 30 is chosen. Consequently,

$$P(D \text{ or } Y) = P(D) + P(Y) - P(D \text{ and } Y)$$
$$= \frac{37}{100} + \frac{35}{100} - \frac{17}{100} = \frac{55}{100}$$

This example illustrates the method of compensating for the double counting that inevitably results when overlapping events are involved.

EXAMPLE

Men were once drafted into the U.S. Army according to the random selection of birthdays. If the 366 different possible birthdays are written on separate slips of paper and mixed in a bowl, find the probability of making one selection and getting a birthday in May or November.

Solution

Let M denote the event of drawing a May date, while N denotes the event of drawing a November date. Clearly, M and N are mutually exclusive because no date is in both May and November. Thus

$$P(M \text{ or } N) = P(M) + P(N) = \frac{31}{366} + \frac{30}{366} = \frac{61}{366}$$

Again, the subtraction of $P(M \text{ and } N)$ can be ignored since it is zero.

EXAMPLE

Using the same population of 366 different birthdays, find the probability of making one selection that is the first day of a month or a November date.

Solution

Let F denote the event of selecting a date that is the first of the month. Here F and N are not mutually exclusive because they can occur simultaneously (as on November 1). Thus

$$P(F \text{ or } N) = P(F) + P(N) - P(F \text{ and } N)$$

$$= \frac{12}{366} + \frac{30}{366} - \frac{1}{366}$$

$$= \frac{41}{366}$$

If there is any doubt about whether the subtraction of $P(A \text{ and } B)$ should be included, then include it. If it turns out that A and B are mutually exclusive, then $P(A \text{ and } B) = 0$, so the formula automatically reduces to $P(A \text{ or } B) = P(A) + P(B)$. Mistakes are made when the subtraction of $P(A \text{ and } B)$ is ignored in situations that require it—that is, when events that are not mutually exclusive are treated as being mutually exclusive! One positive indication of such an error is a total probability that exceeds 1, but errors involving the addition rule do not necessarily cause the probability to exceed 1.

3-3 Exercises A
Addition Rule

3-35 For each pair of events given, determine whether the two events are mutually exclusive for a single experiment.
(a) Selecting a dominant personality type.
    Selecting a submissive personality type.
(b) Selecting a voter who favors gun control.
    Selecting a Conservative.
(c) Selecting a blonde (natural or otherwise).
    Selecting a person with brown eyes.
(d) Selecting someone born in the United States.
    Selecting a citizen.
(e) Selecting an honest politician.
    Selecting an incumbent politician.
(f) Selecting an unmarried person.
    Selecting a person with an employed spouse.
(g) Selecting a registered voter.
    Selecting someone over 65 years of age.
(h) Selecting a required course.
    Selecting an elective course.

(i) Selecting a novel.
Selecting a biography.

(j) Selecting a summer day.
Selecting a legal holiday.

3-36   If a birth date is randomly selected from the 366 different possibilities (all equally likely), find the probability that it is a date in November or December.

3-37   If a computer randomly generates the last digit of a telephone number, find the probability that it is odd or greater than 2.

3-38   A sample group consists of 15 patients with type O blood, 8 patients with type A blood, and 7 patients with type B blood. If one of these patients is randomly selected, find the probability that their blood type is A or B.

3-39   A labor study involves a sample of 12 mining companies, 18 construction companies, 10 manufacturing companies, and 3 wholesale companies. If a company is randomly selected from this sample group, find the probability of getting a mining or construction company.

3-40   A convention of 100 guests consists of 37 Democrats, 41 Republicans, and 22 Independents. Find the probability of randomly selecting one guest and getting a Democrat or an Independent.

3-41   An empty pond is filled with water and stocked with 500 trout, 400 bass, and 300 sunfish. Find the probability of randomly catching the first fish and getting a bass or a sunfish.

3-42   A neighborhood is comprised of 82 Italians, 75 Irish, 40 Germans, and 2 Austrians. Find the probability of randomly selecting one resident and getting an Italian or an Austrian.

3-43   A statistics class is attended by 12 psychology majors, 8 business majors, 4 biology majors, and 6 education majors. Find the probability of randomly selecting one student and getting a biology or business major.

3-44   Using the statistics class of Exercise 3-43, find the probability of randomly selecting one student and getting a psychology or biology or education major.

3-45   An insurance firm serves 5700 clients, of which 4100 are males. There are 2600 clients under 30 years of age, and 1900 of these are males. Find the probability of making one random selection and getting a female or a client under 30 years of age.

3-46   Using the clients from Exercise 3-45, find the probability of making one random selection and getting a male or a client under 30 years of age.

3-47   A pollster surveys 100 subjects consisting of 40 Democrats (of which half are female) and 60 Republicans (of which half are female). Find the probability of randomly selecting one of these subjects and getting a Republican or a male.

3-48 Using the subjects from Exercise 3-47, find the probability of randomly selecting one subject and getting a Democrat or a male.

3-49 In a convention of 100 guests with 37 Democrats and 41 Republicans, there are 35 conventioneers under 30 years of age, including 17 of the Democrats and 13 of the Republicans. Find the probability that a conventioneer selected at random is a Republican or is under 30 years of age.

3-50 In one local survey, 100 subjects indicated their opinions on a zoning ordinance. Of the 62 favorable responses, there were 40 males. Of the 38 unfavorable responses, there were 15 males. Find the probability of randomly selecting one of these subjects and getting a male or a favorable response.

3-51 A psychology study related to personality characteristics involves 1000 subjects, of which 200 are neurotic, 800 are submissive, and 155 are both neurotic and submissive. What is the probability that a person selected at random from this group is either neurotic or submissive?

3-52 The 12 residents of one house include exactly 5 registered voters of which 4 are males; there are 3 other males in the house. If a resident is randomly selected, find the probability of getting a female who is a registered voter.

3-53 A manufacturer assembles cars and trucks at three different locations (A, B, C) according to the accompanying table, which represents daily output. If one vehicle is randomly selected, find the probability that it is either a car or was assembled at site B.

	A	B	C
Car	20	40	10
Truck	5	5	30

3-54 For the data of Exercise 3-53, find the probability that for a randomly selected vehicle, the assembly took place at site B.

3-55 The accompanying table summarizes the age distribution of the 7000 students attending a particular college. If one of these students is randomly selected, find the probability of getting a student less than 35 years of age.

Age	Number
15–22	3700
23–29	1500
30–34	600
35–44	800
over 44	400

3-56 Using the data from Exercise 3-55, find the probability that when one student is randomly selected, the student's age is either below 23 or over 44.

3-57   A survey included information about the sex of the respondent and the education of the father. The results are summarized in the accompanying table. If one of the respondents is randomly selected from this group, find the probability that either a male is selected or the respondent's father completed college.

	Below high school	High school	College
Male	115	140	35
Female	60	70	15

3-58   Using the data from Exercise 3-57, find the probability that if one of the respondents from the given group is randomly selected, his or her father completed either high school or college.

3-59   A statistics class contains 15 females and 12 males. Five of the females are business majors, while the remaining females are psychology majors. All of the males are psychology majors. Find the probability of randomly selecting a student who is a psychology major or a female.

3-60   Using the statistics class of Exercise 3-59, find the probability of randomly selecting a student who is a business major or a male.

3-61   Using the statistics class of Exercise 3-59, find the probability of randomly selecting a student who is a psychology major or a male.

3-62   Using the statistics class of Exercise 3-59, find the probability of randomly selecting a student who is a business major or a female.

3-3 Exercises B
Addition Rule

3-63   If $P(A \text{ or } B) = 1/3$, $P(B) = 1/4$, $P(A \text{ and } B) = 1/5$, find $P(A)$.

3-64   (a) If $P(A) = 0.4$ and $P(B) = 0.5$, what is known about $P(A \text{ or } B)$ if A and B are mutually exclusive events?
     (b) If $P(A) = 0.4$ and $P(B) = 0.5$, what is known about $P(A \text{ or } B)$ if A and B are not mutually exclusive events?

3-65 $P(A \text{ or } B) = 0.8$ while $P(A) = 0.4$. What is known about events A and B?

3-66   Find $P(B)$ if $P(A \text{ or } B) = 0.6$, $P(A) = 0.6$, and A and B are mutually exclusive events.

3-67   How is the addition rule changed if the exclusive *or* is used instead of the inclusive *or*? Recall that the exclusive *or* means either one or the other, but not both.

3-68   If $P(A \text{ or } B) = P(A) + P(B) - P(A \text{ and } B)$, develop a rule for $P(A \text{ or } B \text{ or } C)$.

3-4 **Multiplication Rule**

In Section 3-3 we developed a rule for finding the probability that A or B will occur in a given experiment. In the addition rule, we used the term $P(A$ and $B)$, which denoted the probability that A and B both occurred. Determination of a value for $P(A$ and $B)$ is often easy when we are dealing with a single simple experiment, but now we want to develop a more general rule for finding $P(A$ and $B)$, where A occurs in one experiment, while B occurs in another. This is an important distinction: the $P(A$ and $B)$ used in the addition rule is different from the $P(A$ and $B)$ used in this section. In the addition rule, $P(A$ and $B)$ means that A and B both occur simultaneously in a single event, but the $P(A$ and $B)$ we will now consider involves two separate events with the outcome of A followed by the outcome of B.

We begin with a simple example, which will suggest a preliminary multiplication rule. Then we use another example to develop a variation and ultimately obtain a generalized multiplication rule.

Probability theory is used extensively in the analysis and design of tests. Practical considerations often require that standardized tests allow only those answers that can be corrected easily, such as true–false or multiple choice. Let's assume that the first question on a test is a true–false type while the second question is multiple choice with five possible answers (a, b, c, d, e). We will use the following two questions. Try them!

1. "Jimmy Carter was the first U.S. president who was born in a hospital."   (True, False)
2. What foreign city receives the largest number of American tourists each year?

> (a) Montreal
> (b) London
> (c) Paris
> (d) Rome
> (e) Tijuana

We want to determine the probability of getting both answers correct by making random guesses. We begin by listing the different possible answers.

T,a	T,b	T,c	T,d	T,e
F,a	F,b	F,c	F,d	F,e

If the answers are random guesses, then the ten possible outcomes are equally likely. The correct answers are "true" and "e" so that

$$P(\text{both correct}) = P(\text{true and e}) = \frac{1}{10}$$

Considering the component answers of "True" and "e", respectively, we see that with random guesses we have $P(\text{True}) = 1/2$ while $P(e) = 1/5$.

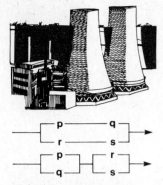

Nuclear Power Plant Has an Unplanned "Event"

In 1974 a team of scientists, led by an M.I.T. professor of nuclear engi-neering, estimated the likelihood of a serious nuclear accident to be about one in a million. At 4:00 A.M. on Wednesday, April 4, 1979, a siren signaled the beginning of an "event" at a nuclear power plant at Three Mile Island in Pennsylvania. Residents of the area prepared for a full-scale evacuation because of the danger of a major catastrophe, which fortunately never materialized. This accident proved that the backup and fail-safe systems were not as infallible as nuclear energy proponents had claimed. There were at least five separate equipment failures. The ensuing reassessment of the likelihood of a nuclear catastrophe is certain to have a strong impact on future energy policies. Scientists will undoubtedly conduct a careful review of the design of nuclear power plants, including the use of redundant, or backup, components. Compare the two systems at left and determine which is likely to be more reliable. Assume that p, q, r, and s are valves and that each system supplies water necessary for cooling purposes.

Recognizing that $1/10$ is the product of $1/2$ and $1/5$, we observe that $P(\text{True and e}) = P(\text{T}) \cdot P(\text{e})$ and we use this observation as a basis for formulating the following rule.

RULE

Multiplication rule (preliminary)

$$P(A \text{ and } B) = P(A) \cdot P(B)$$

EXAMPLE

Three floppy disks are produced and one of them is defective. Two disks are randomly selected for testing, but the first is replaced before the second selection is made. Find the probability that both disks are good.

Solution

Letting G represent the event of selecting a good disk, we want $P(G \text{ and } G)$. With $P(G) = 2/3$ we apply the multiplication rule to get

$$P(G \text{ and } G) = P(G) \cdot P(G) = \frac{2}{3} \cdot \frac{2}{3} = \frac{4}{9}$$

Although the preceding solution is mathematically correct, common sense suggests that product testing should be conducted without replacement of the items already tested. There is always the chance that you could test the same item twice. Also, we cannot be sure of the reliability of any generalization based on a specific case, such as our preliminary multiplication rule. We would be wise to test that rule in a variety of cases to see if any errors arise. We will see that the preliminary rule is sometimes inadequate, as in the improved testing procedure in the next example.

EXAMPLE

Let's again assume that we have three floppy disks of which one is defective. We will again randomly select two disks, but we will *not* replace the first selection. We will find the probability of getting two good disks with this improved testing procedure.

Solution

To understand the sample space better, we will represent the defective disk by D and the two good disks by G_1 and G_2. Assuming that the first selection is not replaced, we now list the different possible outcomes.

$G_1 \; G_2$
$G_1 \; D$
$G_2 \; G_1$ — Both disks are good.
$G_2 \; D$
$D \; \; G_1$
$D \; \; G_2$

By examining this list of six equally likely outcomes, we see that only two cases correspond to two good disks so that $P(G \text{ and } G) = 2/6$ or $1/3$. In contrast, our preliminary multiplication rule was used in the preceding example to produce a result of $4/9$. Here, the correct result is $1/3$, so the preliminary multiplication rule does not fit this case.

In this example, the preliminary multiplication rule does not take into account the fact that the first selection is not replaced. $P(G)$ again begins with a value of $2/3$, but after getting a good disk on the first selection, there would be one good disk and one defective disk remaining so that $P(G)$ becomes $1/2$ on the second selection.

Here is the key concept of this last example: **Without replacement of the first selection, the second probability is affected by the first**

Convicted by Probability

In 1964 Mrs. Juanita Brooks was robbed in Los Angeles. According to witnesses, the robber was a Caucasian woman with blond hair in a ponytail, who escaped in a yellow car driven by a black male with a mustache and beard. Janet and Malcolm Collins were arrested and convicted after a college mathematics instructor testified that there is only about one chance in 12 million that any couple would have the characteristics described by the witnesses. The following estimated probabilities were presented in court.

P(yellow car) $= \frac{1}{10}$
P(man with mustache) $= \frac{1}{4}$
P(girl with hair in a ponytail) $= \frac{1}{10}$
P(girl with blond hair) $= \frac{1}{3}$
P(black man with beard) $= \frac{1}{10}$
P(interracial couple in car) $= \frac{1}{1000}$

The convictions were reversed by the California Supreme Court, which noted that no evidence was presented to support the stated probabilities and the independence of the characteristics was not established.

result. Since this dependence of the second event on the first result is so important, we formulate a special definition.

DEFINITION

Two events A and B are **independent** if the occurrence of one does not affect the probability of the occurrence of the other. (Several events are similarly independent if the occurrence of any does not affect the probabilities of the occurrence of the others.) If A and B are not independent, they are said to be **dependent**.

The multiplication rule holds if the events A and B are independent; in that case $P(B)$ is not affected by the occurrence of A. But in our last example, where the first disk was not replaced before the second selection, $P(G)$ was affected by the result of the first selection. The correct result of $1/3$ could be obtained by examining the listed sample space or by calculating $\frac{2}{3} \cdot \frac{1}{2} = \frac{1}{3}$, where the probability of $1/2$ was affected by the first selection.

Apart from the concept of independence, this last example suggests an improved multiplication rule.

RULE

Let $P(B|A)$ represent the probability of B occurring after assuming that A has already occurred.

Multiplication rule for dependent events

$$P(A \text{ and } B) = P(A) \cdot P(B|A)$$

This improved multiplication rule seems to accommodate dependent events, but how do we treat independent events? If A and B are independent, $P(B|A)$ must equal $P(B)$, since the definition of independence states that the occurrence of A should not affect $P(B)$. That is, if A and B are independent, we can use the first multiplication rule. The following rule combines the previous results and summarizes the key concept of this section.

MULTIPLICATION RULE

$P(A \text{ and } B) = P(A) \cdot P(B)$ if A and B are **independent**.

$P(A \text{ and } B) = P(A) \cdot P(B|A)$ if A and B are **dependent**.

This multiplication rule is easily extended to several events. In general, **the probability of any sequence of independent events is simply the product of their corresponding probabilities.** The next two examples illustrate this extension of the multiplication rule.

E X A M P L E

A couple plans to have four children. Find the probability that all four children are girls. Assume that boys and girls are equally likely and that the sex of each child is independent of the sex of any other children.

Solution

The probability of getting a girl is $\frac{1}{2}$ and the four births are independent of each other, so

$$P(\text{four girls}) = \frac{1}{2} \cdot \frac{1}{2} \cdot \frac{1}{2} \cdot \frac{1}{2} = \frac{1}{16}$$

E X A M P L E

A pill designed to prevent certain physiological reactions is advertized as 95% effective. Find the probability that the pill will work in all of 15 separate and independent applications.

Solution

Since the 15 applications are independent, we get

$$P(15 \text{ successes}) = 0.95 \cdot 0.95 \cdot 0.95 \cdots 0.95 \ (15 \text{ times})$$
$$= 0.95^{15}$$
$$= 0.463$$

Whereas the preceding two examples illustrate an extension of the multiplication rule for independent events, the next example illustrates a similar extension of the multiplication rule for dependent events.

In this section, we have seen that the independence of events must be considered in the determination of probabilities like $P(A \text{ and } B)$, since the component probabilities may be affected. Replacement of randomly selected items usually involves independent events, while failure to replace selected items usually causes the events to be dependent.

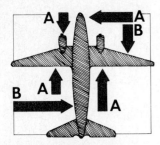

Multiplication Rule Presents Difficulty to FAA

In a publication discussing the analysis of component failures in aircraft, the Federal Aviation Administration (FAA) refers specifically to the multiplication rule as it relates to "system independence and redundancy." The FAA states that "the most often encountered difficulty with quantitative analyses presented to the FAA has been the improper treatment of events which are not mutually independent. The probability of occurrence of two events which are mutually independent may be multiplied to obtain the probability that both events occur using the formula: $P(A \text{ and } B) = P(A) \cdot P(B)$. This multiplication will produce an incorrect solution if A and B are not mutually independent.

EXAMPLE

A Congressional committee of four men and six women is appointed. If a reporter randomly selects three different committee members for interviews, find the probability that they are all women.

Solution

Letting A, B, and C represent the respective events of randomly selecting a woman on the first, second, and third attempts, we see that those events are dependent since successive probabilities are affected by previous results. The probability of getting two different women on the first two selections can be found through direct application of the multiplication rule for dependent events.

$$P(A \text{ and } B) = P(A) \cdot P(B|A)$$
$$= \frac{6}{10} \cdot \frac{5}{9} = \frac{30}{90}$$

If the probability of the first woman is $\frac{6}{10}$ and the probability of the second woman is $\frac{5}{9}$, then the probability of the third woman must be $\frac{4}{8}$. This seems reasonable since, after the first two women are selected, there would be eight remaining members, of which four are women. Therefore

$$P(A \text{ and } B \text{ and } C) = P(A) \cdot P(B|A) \cdot P(C|A \text{ and } B)$$
$$= \frac{6}{10} \cdot \frac{5}{9} \cdot \frac{4}{8}$$
$$= \frac{120}{720} = \frac{1}{6}$$

EXAMPLE

In Section 3-1 we described a pollster who supposedly surveyed 12 randomly selected voters and got 12 Republicans. The population consisted of 200,000 voters, of which 30% (or 60,000) are Republicans. The pollster argued that this could easily happen, but can it?

Solution

We seek the probability of getting 12 Republicans in 12 selections. The events of getting the 12 Republicans are dependent since the sampling was done without replacement. We get

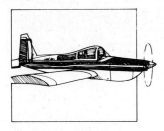

Redundancy

Reliability is often designed into systems through redundancy of critical components. For example, the typical single-engine light aircraft has only one engine, but each cylinder can be fired by either of two separate and independent spark plugs. There are two independent electrical systems so that a failure of one will not lead to engine failure. Aircraft used for instrument flight typically have two separate radios. Such redundancy is an application of the multiplication rule in probability theory. If the probability of failure of one component, such as a spark plug, is 0.001, the probability of failure of dual components, such as both plugs firing the same cylinder, is $0.001 \times 0.001 = 0.000001$. In this case, there is clearly safety in numbers.

$$P(12 \text{ Republicans}) = \frac{60,000}{200,000} \times \frac{59,999}{199,999} \times \cdots \times \frac{59,989}{199,989}$$
$$= 0.000000531$$

When small samples are drawn from large populations, the common practice is to treat the events as being independent so that the calculations are simplified. Taking advantage of the fact that each component fraction in the above expression is approximately 0.30, we get

$$P(12 \text{ Republicans}) = (0.30) \times (0.30) \times \cdots \times (0.30) \qquad (12 \text{ times})$$
$$= 0.30^{12} = 0.000000531$$

In any event, we see that the probability of getting 12 Republicans with random selections is incredibly small. The selection process used by the pollster can now be criticized with sound justification.

When using the multiplication rule for finding probabilities in compound events, a **tree diagram** is sometimes helpful in determining the number of possible outcomes. In a tree diagram, we depict schematically the possible outcomes of an experiment as line segments emanating from one starting point. In Figure 3-7 we show the tree diagram that summarizes the possibilities for parents who plan to have three children. Note that the tree diagram in Figure 3-7 presents the eight different possible outcomes as the eight different possible paths begin at the left and end at the right. Assuming that boys and girls are equally likely, we see that the eight paths are equally likely, so each of the eight possible outcomes has a probability of $\frac{1}{8}$. Such diagrams are helpful in counting

Figure 3-7

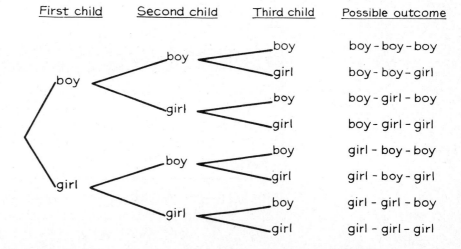

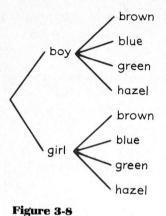

Figure 3-8

the number of possible outcomes if the number of possibilities is not too large. In cases involving large numbers of choices, the use of tree diagrams is impractical. However, they are useful as visual aids to provide insight into the multiplication rule. Suppose, for example, that we want to find the probability of a child being a girl with blue eyes. We will assume that boys and girls are equally likely. We also assume, for the sake of simplicity, that the possible eye colors of brown, blue, green, and hazel are all equally likely. (In reality, this will not usually be true.) In Figure 3-8, the possible outcomes are summarized and we can see that the outcome of a girl with blue eyes is only one of the eight branches. Since we are assuming that the branches are equally likely, we get P(girl with blue eyes) $= \frac{1}{8}$. Examining Figure 3-8 we see that for each sex there are four eye colors. The total number of outcomes is therefore 4 taken 2 times, or 8. The tree diagram illustrates the reason for multiplication.

3-4 Exercises A
Multiplication Rule

3-69 For each following pair of events, classify the two events as independent or dependent.

(a) Making a correct guess on the first question of a multiple choice quiz.
Making a correct guess on the second question of the same multiple choice quiz.

(b) Staying out of jail.
Being a criminal.

(c) Getting an unfavorable reaction when a certain drug is administered to a mouse.
Getting an unfavorable reaction when the same drug is administered to another mouse.

(d) Finding your television inoperable.
Finding your car inoperable.

(e) Finding your television inoperable.
Finding your kitchen light inoperable.

3-70 If 23.7% of all physicians are surgeons, find the probability that two different physicians selected at random will both be surgeons.

3-71 In a study of families with divorced parents, it has been found that $\frac{1}{3}$ of the children suffer depression. If four children with divorced parents are randomly selected from those in the study, find the probability that they all suffer from depression.

3-72 The first two answers to a true-false quiz are both true. If random guesses are made for those two questions, find the probability that both answers are correct.

3-73 There is a 0.17 probability that a telephone solicitation for a magazine will produce an order. Find the probability that two orders are obtained in two separate calls.

3-74 Special thermometers are used in a scientific experiment and there is a 50% chance that any such thermometer is in error by more than 2°. If ten thermometers are used, find the probability that they are all in error by more than 2°.

3-75 Four firms using the same auditor independently and randomly select a month in which to conduct their annual audits. What is the probability that all four months are different?

3-76 If a death is selected at random, assume that there is 0.057 probability that it was caused by an accident. Find the probability that three randomly selected deaths were all accidental.

3-77 A circuit requiring a 500 ohm resistance is designed with five 100 ohm resistors arranged in series. The proper resistance is achieved if all five resistors function correctly. There is a 0.992 probability that any individual resistor will not fail. What is the probability that all five of the resistors will work correctly to provide the 500 ohm resistance?

3-78 One couple attracted media attention when their three children, who were born in different years, were all born on July 4. Find the probability that three randomly selected people were all born on July 4.

3-79 If the probability of a certain pill being effective is 0.95, find the probability that it will be effective in each of 65 independent applications.

3-80 An experiment begins with six identical mice in the same cage. Each day one mouse is randomly selected, injected with a chemical, and then put back into the same cage. Find the probability of selecting the same mouse on each of the first three days.

3-81 An experiment begins with four female mice and two male mice in the same cage. One mouse is randomly selected each day and put in a separate cage. Find the probability that the first three removals are all females.

3-82 A car has eight spark plugs, of which two are defective. Find the probability of locating both defective spark plugs in only two random selections. Assume that the first selection is not replaced.

3-83 An insurance firm serves 5700 clients, of which 4100 are male. There are 2600 clients under 30 years of age, and 2000 of these are male. Find the probability of randomly selecting a female client under 30 years of age.

3-84 If two people are randomly selected, find the probability that the second person has the same birthday as the first.

3-85 A computer simulation involves the random generation of digits from 1 through 5. If two digits are generated, find the probability that they are both odd in each case.
(a) The first selected digit cannot be selected again.
(b) The first selected digit is available for the second selection.

3-86 A container holds 12 eggs, of which 5 are fertile.
(a) Find the probability of randomly selecting 3 eggs that are all fertile if each egg is replaced before the next selection is made.
(b) Find the probability of randomly selecting 3 eggs that are all fertile if the eggs selected are not replaced.

3-87 A governor's conference is attended by 30 Republican governors and 20 Democratic governors. Two prizes of state aid are to be raffled off by the President and each governor has one ticket. Find the probability that both winners are Republicans in each case.
(a) The first winning ticket is replaced.
(b) The first winning ticket is not replaced.

3-88 For the raffle of Exercise 3-87, find the probability that both winners belong to the same party in each case.
(a) The first winning ticket is replaced.
(b) The first winning ticket is not replaced.

3-89 Eight defective batteries are present in a bin of 100 batteries. The entire bin is approved for shipment if no defects show up when three randomly selected batteries are tested.
(a) Find the probability of approval if the selected batteries are replaced.
(b) Find the probability of approval if the selected batteries are not replaced.
(c) Comparing the results to parts (a) and (b), which procedure is more likely to reveal a defective battery? Which procedure do you think is better?

3-90 Of 2 million components produced by a manufacturer in one year, 5000 are defective. If two of these components are randomly selected and tested, find the probability that they are both good in each of the following cases.
(a) The first selected component is replaced.
(b) The first selected component is not replaced.

3-91 An approved jury list contains 20 men and 20 women. Find the probability of randomly selecting 12 of these people and getting an all-male jury.

3-92 An insurance investigator finds that for a year in Chicago, there were 3,900,000 motor vehicles registered and 70,000 of them were reported stolen. One family had five motor vehicles and all were reported stolen in separate incidents in one year. What is the probability that five randomly selected motor vehicles would all be stolen?

3-4 Exercises B
Multiplication Rule

3-93 Use a calculator to compute the probability that of 25 people, no two share the same birthday.

3-94 Use a calculator to compute the probability that of 50 people, no two share the same birthday.

3-95 A gasoline rationing scheme used in 1973 and again in 1979 mandated that cars with even-numbered license plates may get gas only on even-numbered days of the month. Similarly, cars with odd-numbered plates could refuel only on odd-numbered days of the month. In addition, all gas stations had to close every Sunday. Assume that this plan is now in effect and that you have an even-numbered license plate. Find the probability that you can get gas on a day selected at random from the current year.

3-96 A poll was taken on the campus of a large university in order to determine student attitudes about a variety of issues. Fifty students (40 males and 10 females) were polled on their involvement in campus activities, and the results showed 20 responses of yes and 30 responses of no. See the chart.

	Yes	No	
Male			40
Female			10
	20	30	

(a) If one of the 50 students is randomly selected, find the probability of getting a male.

(b) If one of the 50 students is randomly selected, find the probability of getting a female.

(c) If one of the 50 students is randomly selected, find the probability of getting a student who answered yes.

(d) If one of the 50 students is randomly selected, find the probability of getting a student who answered no.

(e) Assuming that the sex of the respondent has no effect on the response, find the probability of randomly selecting one of the 50 polled students and getting a male who answered yes.

(f) Using the probability from part (e) and the fact that there are 50 respondents, what would you expect to be the number of males who answered yes?

(g) If the number from part (f) is entered in the appropriate box in the chart, are the numbers in the other boxes then determined? If so, find them.

3-5   Complementary Events

In this section we begin by defining complementary events and present one last rule of probabilities that relates to these events.

DEFINITION

The **complement** of event A, denoted by $\overline{A}$, consists of all outcomes not in A.

The complement of event A is the event A does *not* occur. If A represents an outcome of answering a test question correctly, then the complementary event $\overline{A}$ represents a wrong answer.

The definition of complementary events implies that they must be mutually exclusive, since it is impossible for an event and its opposite to occur at the same time. Also, we can be absolutely certain that either A does or does not occur. That is, either A or $\overline{A}$ must occur. These observations enable us to apply the addition rule for mutually exclusive events as follows:

$$P(A \text{ or } \overline{A}) = P(A) + P(\overline{A}) = 1$$

We justify $P(A \text{ or } \overline{A}) = P(A) + P(\overline{A})$ by noting that A and $\overline{A}$ are mutually exclusive, and we justify the total of 1 by our absolute certainty that A either does or does not occur. This result of the addition rule leads to the following three *equivalent* forms.

Rule of Complementary Events

$$P(A) + P(\overline{A}) = 1$$
$$P(\overline{A}) = 1 - P(A)$$
$$P(A) = 1 - P(\overline{A})$$

The first form comes directly from our original result, while the second (see Figure 3-9) and third variations involve very simple equation manipulations. A major advantage of the rule of complementary events is that it can sometimes be used to significantly reduce the workload required to solve certain problems. As an example, let's consider a very ambitious couple planning to have seven children. We want to determine the probability of getting at least one boy among those seven children. The direct solution to this problem is messy, but a simple indirect approach is made possible by our rule of complementary events. Let's denote the event of getting at least one boy in seven by B. We will begin

Not Too Likely

N. C. Wickramashinghe of University College in Cardiff, Wales, stated that the chance of getting life from a random shuffling of amino acids is about 1 in $10^{40,000}$. He said that this is equivalent to a tornado blowing through a junkyard and creating a jumbo jet in the process.

Total area = 1

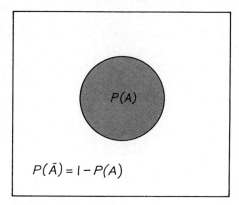

$P(\bar{A}) = 1 - P(A)$

Figure 3-9

by finding $P(\bar{B})$, the probability of *not* getting at least one boy (which is equivalent to getting seven girls). Now $P(\bar{B})$ is relatively easy to compute if we make two reasonable assumptions:

1. $P(\text{girl}) = \frac{1}{2}$.

2. The sexes of successive babies are independent of those of any younger children.

(Neither of these two assumptions is exactly correct, but they can be used with extremely good results.)

Applying the multiplication rule for independent events, we get

$$P(\bar{B}) = P(\text{7 consecutive girls}) = \frac{1}{2} \cdot \frac{1}{2} \cdot \frac{1}{2} \cdot \frac{1}{2} \cdot \frac{1}{2} \cdot \frac{1}{2} \cdot \frac{1}{2} = \frac{1}{128}$$

However, we are seeking $P(B)$ so that the rule of complementary events can be applied.

$$P(B) = 1 - P(\bar{B})$$
$$= 1 - \frac{1}{128}$$
$$= \frac{127}{128}$$

As complex as this solution may appear, it is trivial in comparison to the alternate solutions that involve a direct approach.

E X A M P L E

A conference is attended by 40 doctors and 10 psychologists. If three different participants are randomly selected for a panel discussion, find the probability that at least one is a psychologist.

Solution

Let A represent the event of selecting at least one psychologist when three participants are randomly selected. Then $\overline{A}$ becomes the event of not getting at least one psychologist. That is, $\overline{A}$ signifies that no psychologists are selected, so all three participants are doctors. We use the multiplication rule to get

$$P(\overline{A}) = P(\text{all 3 are doctors})$$
$$= \frac{40}{50} \cdot \frac{39}{49} \cdot \frac{38}{48}$$
$$= 0.504$$

We now use the rule of complementary events to find $P(A)$.

$$P(A) = 1 - P(\overline{A})$$
$$= 1 - 0.504$$
$$= 0.496$$

Odds Can Be Odd

Odds are used extensively by patrons of casinos and race tracks, but mathematicians and scientists tend to use mathematical probabilities (numbers between 0 and 1) instead. Odds facilitate trans- actions involving bets and payoffs. Someone betting on a horse with odds of 20 to 1 will make a profit of $20 for each $1 bet, assuming that the horse wins. However, odds are extremely awkward in mathematical and scientific calculations. If a device has a 0.2 proba- bility of being defective, the probability of two such items being de- fective is $0.2 \times 0.2 = 0.04$. The same calcu- lation using odds would begin with odds of 4 to 1. We can easily multiply 0.2 by 0.2 to get 0.04, but how do we combine 4 to 1 with 4 to 1 to get the correct result of 24 to 1? For calculations in mathe- matics and science, probabilities are more useful than odds. Odds are indeed odd.

The key concept employed in the two previous examples is equating "at least 1" with the *opposite* of "none." In an informal notation, this means that P(at least 1) $= 1 - P$(none).

The rule of complementary events enables us to make many other important conclusions. If, for example, we have a 0.05 probability of an error in a test, we can conclude that the probability of no error is $1 - 0.05$, or 0.95. If we know that the true probability of a baby being a boy is 0.512, we can conclude that P(girl) $= 1 - 0.512 = 0.488$. There are many cases when the probability of an event is known and we need the probability of the complementary event.

The concepts and rules of probability theory presented in this chapter consist of elementary and fundamental principles. A more complete study of probability is not necessary at this time since our main objective is to study the elements of statistics, and we have already covered the probability theory that we will need. We hope that this chapter generates some interest in probability for its own sake. The importance of probability is continuing to grow as it is used by more and more scientists, economists, politicians, biologists, insurance specialists, executives, and other professionals.

3-5 Exercises A
Complementary Events

In Exercises 3-97 through 3-104, determine the probability of the given event and the probability of the complementary event.

3-97   A defective capacitor is randomly selected from a box containing 20 capacitors of which 6 are defective. (Only one selection is made.)

3-98   A letter is randomly selected from the alphabet and the result is a vowel (a, e, i, o, u).

3-99   A number from 1 through 10 is randomly selected and the result is even.

3-100   A baby is born and that baby is a boy.

3-101   A day of the week is randomly selected and it is a Friday or a Saturday.

3-102   A tax return is randomly selected from a group of 20, and the return selected shows the largest income in that group.

3-103   In a class of 15 girls and 10 boys, one student is randomly called and that student is a girl.

3-104   Three television stations draw lots to determine which station will cover the local football game, and the station with the lowest budget wins.

In the remaining exercises, find the probability described.

3-105   If a husband and wife plan to have three children, find the probability that they have at least one boy. Assume that boys and girls are equally likely and that the births are independent.

3-106   A typing pool is made up of five men and five women. If three different typists are randomly selected from this pool, find the probability that at least one of the three is a man.

3-107   When correct procedures are followed, there is a 0.95 probability that a certain test will lead to the correct conclusion. If the correct procedures are used and this test is applied in five independent cases, what is the probability that the conclusion will be incorrect at least once?

3-108   A true-false test of four questions is given and an unprepared student must make random guesses in answering each question. What is the probability of at least one correct response?

3-109   In a neighborhood of 500 residents there are four illegal aliens. A random check is made of 20 of these residents. Find the probability that at least one illegal alien will be found among the 20 different residents.

3-110   In a certain state, 4% of all cars on the road are in obvious violation of some law. For a random inspection of 15 different cars, what is the probability that at least one car will be found in violation of a law?

3-111   A certain method of contraception is found to be 95% effective. What is the probability of at least one pregnancy among ten different couples using this method of contraception?

3-112   Three programmers are given a problem, which they work on independently. The probabilities of completing a working program by the end of the day are 1/5, 2/5, and 1/4, respectively. Find the probability that a working program will be available by the end of the day.

3-113   An employee needs to call any one of five colleagues. Assume that the five colleagues are random selections from a population in which 28% have unlisted numbers. Find the probability that at least one of the five fellow workers will have a listed number.

3-114   A circuit is designed so that a critical function is properly performed if at least one of four identical and independent components does not fail. The probability of failure for any one of these four components is 0.081. Find the probability that the critical function will be properly performed.

3-115   An investor estimates that for each of five oil stocks, there is a probability of 0.002 that there will be a substantial decline in value. If shares from all five companies are purchased, what is the probability that a substantial loss will be suffered from at least one of the stocks?

3-116 It is found that for medical students given selected case studies, there is a 0.850 probability of a correct diagnosis. For three students each given one case, what is the probability of at least one incorrect diagnosis?

3-117 Under certain conditions, there is a 0.015 probability that a randomly selected driver will fail a breathalyzer test for sobriety. If the police test 100 randomly selected drivers, what is the probability of finding at least one driver who fails the sobriety test?

3-118 A nuclear weapon will not be misfired if at least one of five separate and independent fail-safe mechanisms functions properly. The estimated likelihoods of failure for these fail-safe devices are 0.105, 0.200, 0.001, 0.115, and 0.340, respectively. What is the probability that a misfire will not occur?

3-119 A critical component in a circuit will work properly only if three other components all work properly. The probabilities of a failure for the three other components are 0.010, 0.005, and 0.012. Find the probability that at least one of these three components will fail.

3-120 A pollster must interview residents from a neighborhood of 80 adults. If there are 20 professionals, 50 skilled workers, and 10 unemployed residents, find the probability of randomly selecting four different adult residents and getting at least one professional.

3-5 Exercises B
Complementary Events

3-121 (a) If $P(A) = 1$ and A and B are complementary events, what is known about event B?

(b) If $P(A) = 0$ and A and B are complementary events, what is known about event B?

(c) If $P(A)$ is at least 0.7 and A and B are complementary events, what is known about event B?

3-122 All pairs of complementary events must be mutually exclusive pairs. Must all mutually exclusive pairs of events also be complementary? Support your response with specific examples.

3-123 (a) Develop a formula for the probability of not getting either A or B on a single trial. That is, find an expression for $P(\overline{A \text{ or } B})$.

(b) Develop a formula for the probability of not getting A or not getting B on a single trial. That is, find an expression for $P(\overline{A} \text{ or } \overline{B})$.

(c) Compare the results from parts (a) and (b). Are they the same or are they different?

3-124 Find the probability that, of 25 randomly selected people, at least 2 share the same birthday.

3-125   Find the probability that, of 50 randomly selected people, at least 2 share the same birthday.

3-126   A company has been manufacturing resistors at a cost of 50¢ each, and there is a probability of 0.900 that any such resistor will work. An employee suggests that another process will generate resistors in groups of four at a cost of 10¢ for each resistor. There is a 50% failure rate for the resistors produced by this second method. Is it better to produce four resistors at 10¢ each, or the one resistor at a cost of 50¢? Explain.

3-6   Counting (Optional)

In calculating probabilities, we often obtain the total number of different possible outcomes by counting them. However, it sometimes happens that the number of different possible outcomes is extremely large so that the usual counting procedure becomes impractical. In this section we introduce techniques that can be used to replace the standard counting process with fast and efficient procedures. In addition to their use in probability problems, these counting techniques have also grown in importance because of their applicability to problems relating to computers and programming.

Suppose that we use a computer to randomly select one of the two Rh factors (positive, negative) and one of the three blood types A, O, or B. Let's assume that we want the probability of getting a positive Rh factor and type A blood. We can represent this probability as P(positive and A) and apply the multiplication rule to get

$$P(\text{positive and } A) = \frac{1}{2} \cdot \frac{1}{3} = \frac{1}{6}$$

We could also arrive at the same probability by examining the tree diagram in Figure 3-10.

In Figure 3-10 we see that there are six branches that, by the random computer selection method, are all equally likely. Since only one branch

Figure 3-10

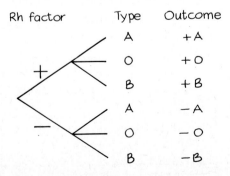

Rh factor	Type	Outcome
	A	+ A
+	O	+ O
	B	+ B
	A	− A
−	O	− O
	B	− B

corresponds to P(positive and A) we get a probability of 1/6. Apart from the calculation of the probability, this solution does reveal another principle, which is a generalization of the following specific observation: with *two* Rh factors and *three* blood types, there are *six* different possibilities for the compound event of selecting a factor and type. We now state this generalized principle.

FUNDAMENTAL COUNTING RULE

For a sequence of two events in which the first event can occur *m* ways and the second event can occur *n* ways, the events together can occur a total of $m \cdot n$ ways.

EXAMPLE

The first question on a standard test is true–false, while the second question is multiple choice with possible answers of a, b, c, d, e. How many different possible answer sequences are there for these two questions?

Solution

The first question can be answered two ways and the second question can be answered five ways so the total number of different possible answer sequences is given by $2 \cdot 5 = 10$.

The fundamental counting principle given above easily extends to situations involving more than two events, as illustrated in the following example.

EXAMPLE

In computer design, if a byte is defined to be a sequence of eight bits, and each bit must be a 0 or 1, how many different bytes are possible?

Solution

Since each bit can occur in two ways (0 or 1) and we have a sequence of eight bits, the total number of different possibilities is given by $2 \cdot 2 \cdot 2 \cdot 2 \cdot 2 \cdot 2 \cdot 2 \cdot 2 = 256$.

The next rule uses the factorial symbol ! which denotes the product of decreasing whole numbers as illustrated below. Many calculators have a factorial key, and Table A-1 in Appendix A lists factorials of values from 0 through 20.

$$5! = 5 \cdot 4 \cdot 3 \cdot 2 \cdot 1 = 120$$
$$4! = 4 \cdot 3 \cdot 2 \cdot 1 = 24$$
$$3! = 3 \cdot 2 \cdot 1 = 6$$
$$2! = 2 \cdot 1 = 2$$
$$1! = 1$$
$$0! = 1 \text{ (by definition)}$$
$$n! = n(n - 1)(n - 2) \cdots 1$$

Using the factorial symbol, we now present the factorial rule.

FACTORIAL RULE

n different items can be arranged in order $n!$ different ways.

This factorial rule reflects the fact that the first item may be selected n different ways, the second item may be selected $n - 1$ ways, and so on.

EXAMPLE

How many possible arrangements are there of the letters A, B, and C? Do any of these arrangements form a word?

Solution

From the diagram below we see that there are six arrangements, and one of the arrangements does form a word.

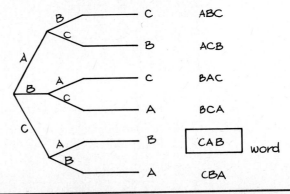

Safety in Numbers

Some hotels have abandoned the traditional room key in favor of an electronic key made of paper and aluminum foil. A central computer changes the access code to a room as soon as a guest checks out. A typical electronic key has 32 different positions that are either punched or left untouched. This configuration allows for 2^{32} or 4,294,967,296 different possible codes, so it is impractical to develop a complete set of keys or try to make an illegal entry by trial and error.

EXAMPLE

A presidential candidate plans to visit the capitol of each of the 50 states. How many different routes are possible?

Solution

The 50 state capitols can be arranged 50! ways, so that the number of different routes is 50! or 30,414 *followed by 60 zeros;* that is an incredibly large number! Now we can see why the symbol ! is used for factorials!

The preceding example is a variation of a classical problem called the traveling salesman problem. It is especially interesting because the large number of possibilities precludes a direct computation for each route, even if computers are used. The time for the fastest computer to directly calculate the shortest possible route is about

1,000,000,000,000,000,000,000,000,000,000,000,000,000 *centuries*!

Clearly, those who use computers to solve problems must be able to recognize when the number of possibilities is so large.

In the factorial counting rule, we determine the number of different possible ways we can arrange a number of items in some type of ordered sequence. Sometimes we don't want to include all of the items available. When we refer to arrangements, we imply that *order* is taken into account. Arrangements are commonly called permutations, which explains the use of the letter P in the following rule.

PERMUTATIONS RULE

The number of **permutations** (or arrangements) of r items selected from n available items is

$$_nP_r = \frac{n!}{(n - r)!}$$

It must be emphasized that in applying the preceding permutations rule, we must have a total of n items available, we must select r of the n items, and we must consider rearrangements of the same items to be different. In the following example we are asked to find the total number of different arrangements that are possible. That clearly suggests use of the permutations rule.

EXAMPLE

If an editor must arrange five articles in a magazine and there are eight articles available, how many different arrangements are possible?

Solution

Here we want the number of arrangements of $r = 5$ items selected from $n = 8$ available articles so that the number of different possible arrangements is

$$_8P_5 = \frac{8!}{(8 - 5)!} = \frac{8!}{3!} = 6720$$

This permutation rule can be thought of as an extension of the fundamental counting rule. We can solve the preceding problem by using the fundamental counting rule in the following way. With 8 articles available and with space for only 5 articles, we know that there are 8 choices for the first article, 7 choices for the second article, 6 choices for the third article, 5 choices for the fourth article, and 4 choices for the fifth article. The number of different possible arrangements is therefore $8 \cdot 7 \cdot 6 \cdot 5 \cdot 4 = 6720$, but $8 \cdot 7 \cdot 6 \cdot 5 \cdot 4$ is actually $8! \div 3!$. In general, whenever we select r items from n available items, the number of different possible arrangements is $n! \div (n - r)!$, and this is expressed in the permutations rule.

When we intend to select r items from n available items but do not take order into account, we are really concerned with possible combinations rather than permutations. That is, when different orderings of the same items are to be counted separately, we have a permutation problem, but when different orderings are *not* to be counted separately, we have a combination problem and may apply the following rule.

COMBINATIONS RULE

The number of **combinations** of r items selected from n available items is

$$_nC_r = \frac{n!}{(n - r)! \ r!}$$

The combinations rule makes sense when we think of it as a modification of the permutations rule. Note that

$$_nC_r = \frac{n!}{(n - r)! \ r!} = \frac{n!}{(n - r)!} \cdot \frac{1}{r!} = {}_nP_r \cdot \frac{1}{r!} = \frac{{}_nP_r}{r!}$$

Realizing that the combinations rule disregards the order of the r selected items while the permutations rule counts all of the different orderings separately, we see that $_nC_r$ will be a fraction of $_nP_r$. For any particular selection of r items, the factorial rule shows that there are $r!$ different arrangements. We know that any particular selection of r items would be counted as only one combination, but it would yield $r!$ different arrangements. As a result, the total number of combinations will be $1/r!$ of the total number of permutations, and that result is expressed in the combinations rule.

E X A M P L E

If 12 jurors are to be selected and there are 15 candidates available, how many different combinations of jurors are possible?

Solution

We want to select $r = 12$ jurors from $n = 15$ available candidates. We are not concerned with the order in which the jurors are selected; we are concerned only with the different possible combinations. Applying the last rule we get

$$_{15}C_{12} = \frac{15!}{(15 - 12)!\ 12!} = \frac{15!}{3!\ 12!}$$

$$= \frac{15 \cdot 14 \cdot 13 \cdot \cancel{12} \cdot \cancel{11} \cdot \cancel{10} \cdot \cancel{9} \cdot \cancel{8} \cdot \cancel{7} \cdot \cancel{6} \cdot \cancel{5} \cdot \cancel{4} \cdot \cancel{3} \cdot \cancel{2} \cdot \cancel{1}}{(3 \cdot 2 \cdot 1)\cancel{12} \cdot \cancel{11} \cdot \cancel{10} \cdot \cancel{9} \cdot \cancel{8} \cdot \cancel{7} \cdot \cancel{6} \cdot \cancel{5} \cdot \cancel{4} \cdot \cancel{3} \cdot \cancel{2} \cdot \cancel{1})}$$

$$= \frac{15 \cdot 14 \cdot 13}{3 \cdot 2 \cdot 1} = 455$$

We stated that the counting techniques of this section are sometimes used in probability problems. The following examples illustrate such applications.

E X A M P L E

In the New York State lottery, first prize is won if a player selects the correct six-number combination when six different numbers from 1 through 44 are drawn. If a player selects one particular six-number combination, find the probability of winning.

Solution

Since 6 different numbers are selected from 44 different possibilities, the total number of combinations is

$$_{44}C_6 = \frac{44!}{(44-6)!\,6!} = \frac{44!}{38!\,6!} = 7,059,052$$

With only one combination selected, the player's probability of winning is only 1/7,059,052.

The Number Crunch

Every so often telephone companies split regions with one area code into regions with two or more area codes because the increased number of telephones in the area have nearly exhausted the possible numbers that can be listed under a single code. A seven-digit telephone number cannot begin with a 0 or 1, but if we allow all other possibilities, we get $8 \times 10 \times 10 \times 10 \times 10 \times 10 \times 10 = 8,000,000$ different possible numbers! Even so, after surviving for 80 years with the single area code of 212, New York City was recently partitioned into the two area codes of 212 and 718. Los Angeles, Houston, and San Diego have also endured split area codes.

EXAMPLE

A home security device with ten buttons is disarmed when three different buttons are pushed in the proper sequence. (No button can be pushed twice.) If the correct code is forgotten, what is the probability of disarming this device by randomly pushing three of the buttons?

Solution

The number of different possible three-button sequences is

$$_{10}P_3 = \frac{10!}{(10-3)!} = 720$$

The probability of randomly selecting the correct three-button sequence is therefore 1/720.

EXAMPLE

A dispatcher sends a delivery truck to eight different locations. If the order in which the deliveries are made is randomly determined, find the probability that the resulting route is the shortest possible route.

Solution

With eight locations there are 8! or 40,320 different possible routes. Among those 40,320 different possibilities, only two routes will be shortest (actually the same route in two different directions). Therefore, there is a probability of only 2/40,320 or 1/20,160 that the selected route will be the shortest possible route.

In this last example, application of the appropriate counting technique made the solution easily obtainable. If we had to determine the number of routes directly by listing them, we would labor for over 11 hours while working at the rapid rate of one route per second! Clearly, these counting techniques are extremely valuable.

Since there is often confusion when choosing between the permutations rule and the combinations rule, we provide the following example, which is intended to emphasize the difference between them.

EXAMPLE

Five students (Al, Bob, Carol, Donna, and Ed) have volunteered for service to the student government.
(a) If three of the students are to be selected for a special *committee*, how many different committees are possible?
(b) If three of the students are to be nominated for the offices of president, vice president, and secretary, how many different *slates* are possible?

Solution

(a) When forming the committee, order of selection is irrelevant. The committee of Al, Bob, and Ed is the same as that of Bob, Al, and Ed. Therefore, we want the number of combinations of 5 students when 3 are selected. We get

$$_5C_3 = \frac{5!}{(5-3)!\,3!} = 10$$

There are 10 different possible committees.
(b) When forming slates of candidates, the order is relevant. The slate of Al for president, Bob for vice president, and Ed for secretary is different from the slate of Bob, Al, and Ed for president, vice president, and secretary, respectively. Here we want the number of permutations of 5 students when 3 are selected. We get

$$_5P_3 = \frac{5!}{(5-3)!} = 60$$

There are 60 different possible slates.

3-6 Exercises A
Counting

In Exercises 3-127 through 3-138, evaluate the given expression.

3-127	$6!$	**3-128**	$10!$	**3-129**	$\dfrac{40!}{39!}$
3-130	$(20-15)!$	**3-131**	$_5P_2$	**3-132**	$_5C_2$
3-133	$_{10}C_7$	**3-134**	$_{10}P_7$	**3-135**	$_nC_n$

3-136 $_nP_n$ **3-137** $_nP_0$ **3-138** $_nC_0$

3-139   Data are grouped according to sex (male, female) and according to income level (low, middle, high). How many different possible categories are there?

3-140   Text is coded according to a scheme in which the letters are rearranged. How many different ways can text with eight different letters be arranged?

3-141   A computer operator must select four jobs from among ten available jobs waiting to be completed. How many different arrangements are possible?

3-142   A computer operator must select four jobs from ten available jobs waiting to be completed. How many different combinations are possible?

3-143   An IRS agent must audit 12 returns from a collection of 22 flagged returns. How many different combinations are possible?

3-144   A health inspector has time to visit seven of the 20 restaurants on a list. How many different routes are possible?

3-145   How many ways can a station manager arrange nine commercials in a show?

3-146   How many different seven-digit telephone numbers are possible if the first digit cannot be 0 or 1?

3-147   An airline mail route must include stops at seven cities.
(a) How many different routes are possible?
(b) If the route is randomly selected, what is the probability that the cities will be arranged in alphabetical order?

3-148   A six-member FBI investigative team is to be formed from a list of 30 agents.
(a) How many different possible combinations can be formed?
(b) If the selections are random, what is the probability of getting the six agents with the most experience?

3-149   How many different social security numbers are possible? Each social security number is a sequence of nine digits.

3-150   A computer simulation involves the random selection of numbers. If the whole numbers from 1 to 25 are available and 12 of them are selected without repetition, how many different arrangements are possible?

3-151   A union must elect four officers from 16 available candidates. How many different slates are possible if one candidate is nominated for each office?

3-152   (a) How many different ZIP codes are possible if each code is a sequence of five digits?
(b) If a computer randomly generates five digits, what is the probability it will produce your ZIP code?

3-153 A television program director has 14 shows available for Monday night and five shows must be chosen.
(a) How many different possible combinations are there?
(b) If 650 different combinations are judged to be incompatible, find the probability of randomly selecting five shows that are compatible.

3-154 A telephone company employee must collect the coins at 40 different locations.
(a) How many different routes are possible?
(b) If two of the routes are the shortest, find the probability of randomly selecting a route and getting one of the two shortest routes.

3-155 A representative of an environmental protection agency plans to sample water at ten different ponds randomly selected from 25 available ponds.
(a) How many different combinations are possible?
(b) What is the probability of randomly selecting the ten ponds with the lowest pollution levels?

3-156 A multiple choice test consists of ten questions with choices a, b, c, d, e.
(a) How many different answer keys are possible?
(b) If all ten answers are random guesses, what is the probability of getting a perfect score?

3-157 The Bureau of Fisheries once asked Bell Laboratories for help in finding the shortest route for getting samples from locations in the Gulf of Mexico. How many different routes are possible if samples must be taken from 24 locations?

3-158 A manager must choose five secretaries from among 12 applicants and assign them to different stations.
(a) How many different arrangements are possible?
(b) If the selections are random, what is the probability of getting the five youngest secretaries selected in order of age.

3-6 Exercises B
Counting

3-159 The number of permutations of n items when x of them are identical to each other and the remaining $n - x$ are identical to each other is given by

$$\frac{n!}{(n - x)! \, x!}$$

If a sequence of ten trials results in only successes and failures, how many ways can three successes and seven failures be arranged?

3-160 If a couple with ten grandchildren is randomly selected, what is the probability that the ten grandchildren consist of two boys and eight girls? (See Exercise 3-159.)

3-161   Most calculators or computers cannot directly calculate 70! or higher. When n is large, $n!$ can be *approximated* by

$$n! = 10^K \quad \text{where } K = (n + 0.5)\log n + 0.39908993 - 0.43429448n$$

Evaluate 50! using the factorial key on a calculator and also by using the approximation given here.

3-162   The Bureau of Fisheries once asked Bell Laboratories for help in finding the shortest route for getting samples from 300 locations in the Gulf of Mexico. There are 300! different possible routes. If 300! is evaluated, how many digits are used in the result? (See Exercise 3-161.)

Computer Project
Probability

In many situations, probability problems can be solved by writing a computer program that simulates the relevant circumstances. In determining the probability of winning the game of solitaire, for example, it is easier to program a computer to play solitaire than it is to develop calculations using the rules of probability. The computer can then play solitaire 1000 times and the probability of winning is estimated to be the number of wins divided by 1000. Such simulations often involve the computer's ability to generate random numbers. The subroutine RND is available in many BASIC implementations. One common use causes an entry of RND(X) to return a value between 0 and 1, such as 0.23560387. That result can be manipulated to produce desired values.

```
10 RANDOMIZE
20 LET R = INT (100 * RND(X))
30 IF R > 37 THEN 20
40 PRINT R
```

In the short program above, we take a value like 0.23560387, multiply by 100 to get 23.560387, and then take only the integer part to get 23. If the result is greater than 37, we go back and try again. Those three lines produce randomly selected values from the list 0, 1, 2, . . . , 37. This short program simulates the spinning of a roulette wheel if we stipulate that 37 corresponds to 00.

(a)   Enter those lines and run the program 20 times. If you were betting on the number 7, how many times did you win?

(b)   In roulette, if you bet $1 on "odd," you win $1 if the result is an odd number. (Remember that in our program, 37 should not win since it represents 00.) Modify the program so that you begin with $10 and you bet $1 on each spin until you either reach $20 or go broke. The final output should indicate whether you reached $20 or went broke first.

(c) A gambler is in Las Vegas with only $10 but must have $20 for the return trip home, or else he must walk. Use computer simulations to determine which of the following two strategies is more likely to get him the $20 he needs for the return trip.

   i. Bet the entire $10 on "odd" for one spin of the roulette wheel.

   ii. Bet on odd, $1 at a time, until reaching $20 or going broke.

Review

This chapter introduced the basic concept of **probability.** We began with two rules for finding probabilities. Rule 1 represents the *empirical* approach, whereby the probability of an event is approximated by actually conducting or observing the experiment in question.

RULE 1

$$P(A) = \frac{\text{number of times } A \text{ occurred}}{\text{number of times experiment was repeated}}$$

Rule 2 is called the **classical** approach, and it applies only if all of the outcomes are equally likely.

RULE 2

$$P(A) = \frac{s}{n} = \frac{\text{number of ways } A \text{ can occur}}{\text{total number of different outcomes}}$$

We noted that the probability of any impossible event is 0, while the probability of any certain event is 1. Also, for any event A,

$$0 \le P(A) \le 1$$

In Section 3-3, we considered the **addition rule** for finding the probability that A **or** B will occur. In evaluating $P(A \text{ or } B)$, it is important to consider whether the events are *mutually exclusive*—that is, whether they can both occur at the same time (see Figure 3-11).

In Section 3-4 we considered the **multiplication rule** for finding the probability that A **and** B will occur. In evaluating $P(A \text{ and } B)$, it is important to consider whether the events are *independent*—that is whether

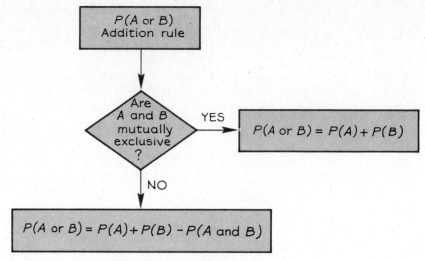

Figure 3-11   Finding the probability that for a single trial, either event A or B (or both) will occur

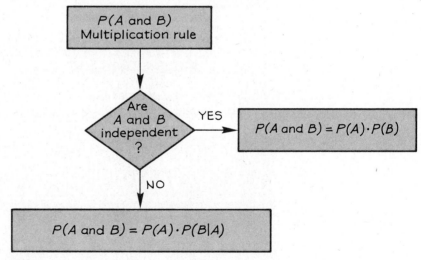

Figure 3-12   Finding the probability that event A will occur in one trial and event B will occur in another trial

the occurrence of one event affects the probability of the other event (see Figure 3-12).

In Section 3-5 we considered **complementary** events (opposites). Using the addition rule, we were able to develop the **rule of complementary events:** $P(A) + P(\overline{A}) = 1$. We saw that this rule can sometimes be used to simplify probability problems.

In Section 3-6 we considered techniques for determining the total number of different possibilities for various events. We presented the fundamental counting rule, the factorial rule, the permutations formula, and the combinations formula, all summarized below with the other important formulas from this chapter.

Most of the material that follows this chapter deals with statistical inferences based upon probabilities. As an example of the basic approach used, consider a test of someone's claim that a quarter is fair. If we flip the quarter 10 times and get 10 consecutive heads, we can make one of two inferences from these sample results:

1. The coin is actually fair and the string of 10 consecutive heads is a fluke.
2. The coin is not fair.

The statistician's decision is based upon the **probability** of getting 10 consecutive heads which, in this case, is so small (1/1024) that the inference of unfairness is the better choice. The purpose of this example is to emphasize the important role played by probability in the standard methods of statistical inference.

IMPORTANT FORMULAS

$0 \leq P(A) \leq 1$ for any event A

$P(A \text{ or } B) = P(A) + P(B)$ if A, B are mutually exclusive

$P(A \text{ or } B) = P(A) + P(B) - P(A \text{ and } B)$ if A, B are not mutually exclusive

$P(A \text{ and } B) = P(A) \cdot P(B)$ if A, B are independent

$P(A \text{ and } B) = P(A) \cdot P(B|A)$ if A, B are dependent

$P(\overline{A}) = 1 - P(A)$

$P(A) = 1 - P(\overline{A})$

$P(A) + P(\overline{A}) = 1$

$m \cdot n$ = the total number of ways two events can occur, if the first can occur m ways while the second can occur n ways

$n!$ = the number of ways n different items can be arranged

$$_nP_r = \frac{n!}{(n-r)!}$$ the number of *permutations* (arrangements) when r items are selected from n available items

$$_nC_r = \frac{n!}{(n-r)! \, r!}$$ the number of *combinations* when r items are selected from n available items

Vocabulary List

Define and give an example of each term.

experiment

event

simple event

sample space

empirical
approximation of
probability

classical approach to
probability

compound event

addition rule

mutually exclusive
events

multiplication rule

independent events

dependent events

tree diagram

complement of an
event

fundamental
counting rule

factorial rule

permutations rule

combinations rule

Review Exercises

3-163   A computer simulation requires the random generation of whole numbers from 1 through 15. If one such number is randomly selected, what is the probability that it is odd?

3-164   A research project involves one patient with type A blood, one with type B blood, one with type O blood, and one with type AB blood. If one of these patients is randomly selected, what is the probability of getting the patient with type O blood?

3-165   A hearing is attended by twelve people in favor of a construction moratorium and eight people opposed. If one person is randomly selected to be the first speaker, what is the probability that he or she is opposed?

3-166   A course is taken by seven engineering majors, five math majors, three computer science majors, and four physics majors. If one of these students is randomly selected as an assistant, what is the probability of getting a math major?

3-167   A medical study involves the subjects summarized in the accompanying table. If one subject is randomly selected, find the probability of getting someone with blood type A or B.

Blood type

	A	O	B	AB
Healthy	7	20	3	9
Ill	3	10	2	6

3-168   Using the medical data from Exercise 3-167, if one subject is randomly selected, find the probability of getting someone who is healthy or has type O blood.

3-169   Using the medical data from Exercise 3-167, if one subject is randomly selected, find the probability of getting someone who is ill or has type A blood.

3-170   Using the medical data from Exercise 3-167, if one subject is randomly selected, find the probability of getting someone who is healthy or has blood type A or O.

3-171   An experiment consists of randomly choosing a day of the week and a month of the year. Find the probability of getting Saturday-September.

3-172   If a couple plans to have five children, find the probability that they are all boys. Assume that boys and girls have the same chance of being born and that the sex of any child is not affected by the sex of any other children.

3-173   Of 120 auto ignition circuits, there are 18 defects. If two circuits are randomly selected, find the probability that they are both defective in each case.
(a) The first selection is replaced before the second selection is made.
(b) The first selection is not replaced.

3-174   In 60 patients given a vaccine, 35 unfavorable reactions occurred. If the records of two different patients are randomly selected, find the probability that they both had unfavorable reactions.

3-175   There is a probability of 0.030 that a power cell will fail. Find the probability that among five such cells there will be at least one cell that does *not* fail.

3-176   In a certain state, 30% of the voters are Republican. If four voters are randomly selected for an exit poll, what is the probability of getting at least one Republican?

3-177   One method has a 60% rate of success in helping people to discontinue cigarette smoking. If eight randomly selected smokers use this method, what is the probability that at least one of them will give up smoking?

3-178   It has been found that a particular psychological disorder is improved through group therapy with a success rate of 40%. If ten subjects with this disorder are randomly selected and treated with group therapy, what is the probability that at least one of them will improve?

3-179   If a couple plans to have four children, find the probability that they are all of the same sex. Assume that boys and girls are equally likely and that the sexes are independent.

3-180   Eight men and seven women have applied for a temporary job. If three different applicants are randomly selected from this group, find the probability of each event.
(a) All three are women.
(b) There is at least one woman.

3-181   (a) What is the probability of an event that is known to be impossible?
(b) What is the probability of an event that will definitely occur?
(c) What is $P(A)$ if $P(\overline{A}) = 0.22$?

(d) If events A and B are mutually exclusive, must they be complementary?

(e) If events A and B are complementary, must they be mutually exclusive?

3-182   A survey is made in a neighborhood of 65 Democrats and 15 Republicans. Of the Democrats 30 are women, while 10 of the Republicans are women. If one subject from this group is randomly selected, find the probability of getting each outcome.
(a) A male or a Democrat.
(b) A male Democrat.
(c) A Democrat or a Republican.

3-183   Evaluate the following.
(a) 8!   (b) $_8P_6$ (c) $_{10}C_8$ (d) $_{80}C_{78}$

3-184   In assigning flight numbers, a dispatcher can choose any one of six cities for departures and any one of four other cities for arrivals. How many different routes are possible?

3-185   If seven different customers arrive for service in random order, what is the probability that they will be in alphabetical order?

3-186   A question on a history test requires that five events be arranged in the proper chronological order. If a random arrangement is selected, what is the probability that it will be correct?

3-187   In one segment of a radio show, a disk jockey must select a program consisting of four songs arranged in some order. If twelve songs are available from the rotation, how many different programs are possible?

3-188   In attempting to gain access to a computer data bank, a computer is programmed to automatically dial every phone number with the prefix 478 followed by four digits. How many such telephone numbers are possible?

3-189   When buying a home computer system, a programmer can choose any one of six central processing units, any one of four disk drives, any one of eight printers, and any one of four modems. How many different configurations are possible?

3-190   A city council decides to appoint six members to a zoning committee. If the appointments are to be made from a list of 14 available candidates, how many different ways can the zoning committee be formed?

3-191   A psychologist conducts experiments in which a monkey must select the correct interconnecting parts. When five parts are selected from twelve available parts, how many different combinations are there?

3-192   In a test of sensory perception, a subject must select the three brightest colors and arrange them in the order of decreasing brightness. If eight colors are available, how many different arrangements are possible when three colors are selected?

Chapter 3
Case Study Activity

In Section 3-1 we stated that at least two students will have the same birthday in more than half of the classes with 25 students. Exercise 3-124 requires the theoretical solution to that problem, but many such problems can be solved by developing a **simulation.** Let the days of the year be represented by the numbers from 001 through 365. Now refer to a page randomly selected from a telephone book and record the last three digits of 25 telephone numbers, but ignore 000 and any cases above 365. Check to determine whether or not at least two of the dates (represented by the three digit numbers) are the same. It should take about 30 minutes to repeat this experiment ten times. Find the estimated probability, which is the number of times matched dates occurred, divided by the number of times the experiment is repeated.

CHAPTER

4

4-1 Overview
We identify chapter **objectives.**

4-2 Random Variables
We describe **discrete** and **continuous random variables** and **probability distributions.**

4-3 Mean, Variance, and Expectation
Given a probability distribution, we describe methods of determining the **mean, standard deviation,** and **variance,** and we define the **expected value** of a probability distribution.

4-4 Binomial Experiments
We learn how to calculate probabilities in **binomial experiments.**

4-5 Mean and Standard Deviation for the Binomial Distribution
We learn how to calculate the **mean, standard deviation,** and **variance** for a binomial distribution.

4-6 Distribution Shapes
We investigate the shape or nature of various probability distributions and establish a correspondence between probability and area.

Probability Distributions

Recently, the Federal Aviation Administration studied the possibility of allowing twin-engine commercial jets to make transatlantic flights previously open only to jets with at least three engines. The lowered requirement is of great interest to manufacturers of twin-engine jets (such as the Boeing 767). Also, the two-engine jets use about half the fuel of jets with three or four engines. Obviously, the key issue in approving the lowered requirement is the probability of a twin-engine jet making a safe transatlantic crossing. This probability should be compared to that of three- and four-engine jets. Clearly, such a study involves a thorough understanding of the related probabilities, and the contents of this chapter will enable us to develop such an understanding. We will return to this case study as we develop the relevant principles.

4-1 Overview

In Chapter 2 we discussed the histogram as a device for showing the frequency distribution of a set of data. In Chapter 3 we discussed the basic principles of probability theory. In this chapter we combine those concepts to develop probability distributions that are basically theoretical models of the frequency distributions we produce when we collect sample data. We construct frequency tables and histograms using *observed* real scores, but we construct probability distributions by presenting possible outcomes along with their *probable* frequencies.

Suppose a casino manager suspects cheating at a dice table. The manager can compare the frequency distribution of the actual sample outcomes to a theoretical model that describes the frequency distribution likely to occur with fair dice. In this case the probability distribution serves as a model of a theoretically perfect population frequency distribution. In essence, we can determine what the frequency table and histogram would be like for a pair of fair dice rolled an infinite number of times. With this perception of the population of outcomes, we can then determine the values of important parameters such as the mean, variance, and standard deviation.

The concept of a probability distribution is not limited to casino management. In fact, the remainder of this book and the very core of inferential statistics depend on some knowledge of probability distributions. To analyze the effectiveness of a new drug, for example, we must know something about the probability distribution of the symptoms the drug is intended to correct.

This chapter deals mostly with discrete cases, while subsequent chapters involve continuous cases. We begin by distinguishing between discrete and continuous random variables.

4-2 Random Variables

We have already stated that an experiment is a process that allows us to obtain observations. Let's assume that we are dealing with experiments of the type that will have some number associated with each outcome. Typically, that number will vary from trial to trial. For example, consider the experiment of launching two rockets. If we count the number of successful launches in those two attempts, we are associating a number with each outcome, and that number will vary from trial to trial. Since such outcomes vary and are determined by chance, they are represented by **random variables.** The values of a random variable are the numbers we associate with the different simple events that comprise the sample space.

EXAMPLE

A jet has two independently operating engines. Let the random variable represent the number of engines that fail during a flight. Since there can be 0 failure, 1 failure, or 2 failures, the values of the random variable are 0, 1, and 2.

EXAMPLE

A quiz consists of ten multiple-choice questions. Let the random variable represent the number of correct answers. This random variable can take on the values of 0, 1, 2, 3, 4, 5, 6, 7, 8, 9, and 10.

Random variables may be discrete or continuous. We know what a finite number of values is (1, or 2, or 3, and so on), but our definition of a discrete random variable also involves the concept of a **countable** number of values. As an example, suppose that a random variable represents the number of times a die must be rolled before a 6 turns up. This random variable can asume any one of the values 1, 2, 3, ... We now have an infinite number of possibilities, but they correspond to the counting numbers. Consequently this type of infinity is called countable. In contrast, the number of points on a continuous scale is not countable and represents a higher degree of infinity. There is no way to count the points on a continuous scale, but we can count the number of times a die is rolled, even if the rolling seems to continue forever.

DEFINITION

A **discrete random variable** has either a finite number of values or a countable number of values.

For example, suppose that a random variable can assume the value that represents the number of United States Senators present for a roll call. That random variable can assume only one of 101 different values (0, 1, 2, . . . , 99, 100) and, since 101 is a finite number, the random variable is discrete.

DEFINITION

A **continuous random variable** has infinitely many values, and those values can be associated with points on a continuous scale in such a way that there are no gaps or interruptions.

Just as count data are usually associated with discrete random variables, measurement data are usually associated with continuous random variables.

As an example, we stipulate that a random variable can assume the value representing the exact speed of a car at a particular instant. The car might be traveling 42.135724 . . . kilometers per hour. This random variable can assume any value that corresponds to a point on the continuous interval shown in Figure 4-1. (We assume that the car never exceeds 80 kilometers per hour.) We now have an infinite number of possible values, which are not countable since there is a correspondence with the continuous scale of Figure 4-1. This random variable is continuous, not discrete.

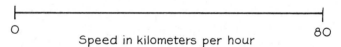

Speed in kilometers per hour

Figure 4-1

Random variables that represent heights, weights, times, and temperatures are usually continuous, as are those that represent speeds.

In reality, it is usually impossible to deal with exact values of a continuous random variable. Instead, we usually convert continuous values to discrete values by rounding off to a limited number of decimal places. If each speed between 0 kilometers per hour and 80 kilometers per hour is rounded off to the nearest integer value, we reduce the total number of possibilities to the finite number of 81. In this way, continuous random variables are made discrete. This chapter is involved almost exclusively with discrete random variables.

Discrete Random Variables

Inferential statistics is often used to make decisions in a wide variety of different fields. We begin with sample data and attempt to make inferences about the population from which the sample was drawn. If the sample is very large we may be able to develop a good estimate of the population frequency distribution, but samples are often too small for that purpose. The practical approach is to use information about the sample along with general knowledge about population distributions.

How Not to Pick Lottery Numbers

In the New York State lottery, a player makes an entry by selecting six different numbers between 1 and 44. After six numbers are randomly se-

lected, any entries with the correct six numbers share in the top prize, which has run as high as $22 million (won by a single person). Since the winning numbers are randomly selected, any combination of six numbers will have the same chance as any other combination. However, some numbers are better bets than others! The combination of 1, 2, 3, 4, 5, 6 is a poor choice because many people select that particular combination; if those numbers were selected, the top prize would be very small since it would be split among so many people. In a typ-

ical week, the first prize total was $2,775,282; if the winning numbers had been 1, 2, 3, 4, 5, 6, there would have been 5,930 winners who would have received only $468 each. Since the chances of winning with any six-number combination are the same, it makes sense to pick a combination not selected by many other players. Avoid patterns like 2, 4, 6, 8, 10, 12 or combinations that form a pattern on the entry card. Players would be wise to pick their numbers randomly, not according to some pattern that other players might also follow.

Much of this general information is included in this and the following chapters. Without a knowledge of probability distributions, users of statistics would be severely limited in the inferences they could make. We intend to develop the ability to work with discrete and continuous probability distributions, and we begin with the discrete case because it is simpler.

DEFINITION

A **probability distribution** is the collection of all values that a random variable can assume, along with the probabilities that correspond to these values.

Table 4-1

x	$P(x)$
0	$\frac{1}{4}$
1	$\frac{1}{2}$
2	$\frac{1}{4}$

For example, suppose a drug is administered to two patients where the random variable is the number of cures (0, 1, or 2), and assume that the probability of a cure is $\frac{1}{2}$. The probability of zero cures is $\frac{1}{4}$, the probability of one cure is $\frac{1}{2}$, while the probability of two cures is $\frac{1}{4}$. Table 4-1 summarizes the probability distribution for this situation. (The given probabilities can be easily verified by listing the four cases in the sample space, beginning with "cure-cure.")

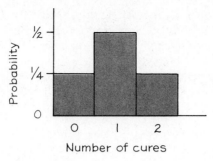

Figure 4-2

There are various ways of graphing these probability distributions, but we illustrate only the histogram, which was introduced in Chapter 2.

The horizontal axis delineates the values of the random variable, while the vertical scale represents probabilities. In Figure 4-2, we show the histogram representing the probability distribution of the last example. Note that along the horizontal axis, the values of 0, 1, and 2 are located at the centers of the rectangles. This implies that the rectangles are each 1 unit wide, so the areas of the three rectangles are $1 \cdot \frac{1}{4}, 1 \cdot \frac{1}{2}, 1 \cdot \frac{1}{4}$ (or $\frac{1}{4}, \frac{1}{2}$, and $\frac{1}{4}$).

In general, if we stipulate that each value of the random variable is assigned a width of 1 on the histogram, then the areas of the rectangles will total 1. **We can therefore associate the probability of each numerical outcome with the area of the corresponding rectangle.** This correspondence between probability and area is an important concept that will be used many times in later chapters.

Suppose we have an experiment with an identified discrete random variable. What do we know about any two events that lead to different values of the random variable? They must be mutually exclusive since their simultaneous occurrence would necessarily lead to the same value of the random variable. In the experiment of giving the drug to two patients, for example, the three values of the random variable (0, 1, 2) are mutually exclusive outcomes. (Among two patients, we cannot get exactly 1 cure and 2 cures at the same time.) Knowing that all the values of the random variable will cover all events of the entire sample space, and knowing that events that lead to different values of the random variable are mutually exclusive, we can conclude that the sum of $P(x)$ for all values of x must be 1. Also, $P(x)$ must be between 0 and 1 for any value of x.

Requirements for $P(x)$ to be a probability distribution:

1. $\Sigma P(x) = 1$ where x assumes all possible values
2. $0 \le P(x) \le 1$ for every value of x

These two requirements for probability distributions are actually direct descendents of the corresponding rules of probabilities (discussed in Chapter 3).

E X A M P L E

Does $P(x) = x/5$ (where x can take on the values of 0, 1, 2, 3) determine a probability distribution?

Solution

If a probability distribution is determined, it must conform to the preceding two requirements. But

$$\Sigma P(x) = P(0) + P(1) + P(2) + P(3)$$

$$= \frac{0}{5} + \frac{1}{5} + \frac{2}{5} + \frac{3}{5}$$

$$= \frac{6}{5}$$

so that the first requirement is not satisfied and a probability distribution is not determined.

E X A M P L E

Does $P(x) = x/10$ (where x can be 0, 1, 2, 3, or 4) determine a probability distribution?

Solution

For the given function we conclude that

$$P(0) = \frac{0}{10} = 0 \qquad P(3) = \frac{3}{10}$$

$$P(1) = \frac{1}{10} \qquad\qquad P(4) = \frac{4}{10}$$

$$P(2) = \frac{2}{10}$$

The sum of these probabilities is 1, and each $P(x)$ is between 0 and 1, so both requirements are satisfied. Consequently, a probability distribution is determined. The graph of this probability distribution is shown in Figure 4-3. Note that the sum of the areas of the rectangles is 1, and each rectangle has an area between 0 and 1.

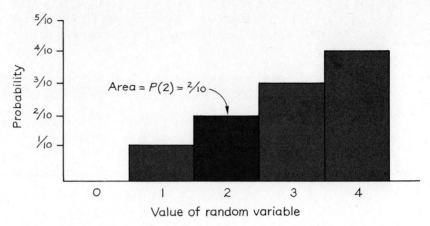

Figure 4-3　Value of random variable

We saw in the overview that probability distributions are extremely important in the study of statistics. We have just considered probability distributions of discrete random variables, and later chapters will consider fundamental probability distributions of continuous random variables.

4-2　Exercises A
Random Variables

In Exercises 4-1 through 4-12, determine whether a probability distribution is given. In those cases where P(x) does not determine a probability distribution, identify the requirement that is not satisfied.

4-1

x	$P(x)$
0	$\frac{1}{4}$
1	$\frac{1}{4}$
2	$\frac{1}{4}$
3	$\frac{1}{4}$
4	$\frac{1}{4}$

4-2

x	$P(x)$
5	0.1
10	0.2
15	0.3

4-3

x	$P(x)$
1	$\frac{1}{10}$
3	$\frac{2}{10}$
5	$\frac{3}{10}$
7	$\frac{2}{10}$
9	$\frac{1}{10}$

4-4

x	$P(x)$
1	$\frac{1}{12}$
2	$\frac{2}{12}$
3	$\frac{3}{12}$
4	$\frac{3}{12}$
5	$\frac{2}{12}$
6	$\frac{1}{12}$

4-5 $P(x) = x$ for x = 0.1, 0.3, 0.6

4-6 $P(x) = x$ for $x = 0, \frac{1}{2}, \frac{1}{4}$

4-7 $P(x) = x/10$ for x = 0, 1, 2, 3, 4, 5

4-8 $P(x) = x/20$ for x = 2, 3, 4, 5, 6

4-9 $P(x) = x - 0.5$ for x = 0.5, 0.6, 0.7, 0.8, 0.9

4-10 $P(x) = x - 2.5$ for x = 2, 3

4-11 $P(x) = \dfrac{1}{2(2 - x)!x!}$ for x = 0, 1, 2

4-12 $P(x) = \dfrac{3}{4(3 - x)!x!}$ for x = 0, 1, 2, 3

In Exercises 4-13 through 4-20, do each of the following.

(a) *List the values that the random variable* x *can assume.*

(b) *Determine the probability* P(x) *for each value of* x.

(c) *Summarize the probability distribution as a table that follows the format of Table 4-1.*

(d) *Construct the histogram that represents the probability distribution for the random variable* x. *(See Figure 4-3.)*

(e) *Indicate the area of each rectangle in the histogram of part (d) so that there is a correspondence between area and probability.*

4-13   A drug is administered to three patients and the random variable x represents the number of cures that occur. The probabilities corresponding to 0 cures, 1 cure, 2 cures, and 3 cures are found to be 0.125, 0.375, 0.375, and 0.125, respectively.

4-14   A manufacturer produces gauges in batches of three and the random variable x represents the number of defects in a batch. The probabilities corresponding to 0 defects, 1 defect, 2 defects, and 3 defects are found to be 0.70, 0.20, 0.09, and 0.01, respectively.

4-15   A computer simulation involves the repeated generation of a random number. The only possibilities are 0, 1, 2, and they are equally likely.

4-16   A computer is used to select the last digit of telephone numbers to be dialed for a poll. The possible values are 0, 1, 2, . . . , 9, and they are equally likely.

4-17   The random variable x represents the number of boys in a family of three children. (*Hint:* Assuming that boys and girls are equally likely, we get $P(2) = 3/8$ by examining the sample space of bbb, bbg, bgb, bgg, gbb, gbg, ggb, ggg.)

4-18   The random variable x represents the number of boys in a family of four children. (See Exercise 4-17.)

4-19   Two dice are rolled. The random variable x represents the total number of dots that turn up. (*Hint:* the sample space consists of 36 cases and $P(2) = 1/36$.)

4-20   The construction of a data security code involves the random generation of two digits where each digit can be 0, 1, 2, 3, or 4. Both digits can be the same, and all selections are equally likely. The random variable x represents the sum of the two digits.

4-2   Exercises B

4-21   Let $P(x) = 1/(2x)$ where $x = 1, 2, 3, \ldots$. Is $P(x)$ a probability distribution?

4-22   Let $P(x) = 1/2^x$ where $x = 1, 2, 3, \ldots$. Is $P(x)$ a probability distribution?

4-23   It has been determined that the proportion of record albums that are counterfeits is 1/5. Random samples of such albums are collected in groups of 4, and x represents the number of counterfeits in a group. Use the multiplication rule and rule of complements (Chapter 3) to complete the table so that the probability distribution is determined.

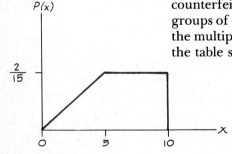

Figure 4-4

x	$P(x)$
0	
1	0.410
2	0.154
3	
4	

4-24   Given that x is a continuous random variable with the distribution shown in Figure 4-4, find each of the following.
(a) The total area of the enclosed region.
(b) The probability that x is less than 5.
(c) The probability that x is less than 7.
(d) The probability that x is between 6 and 7.
(e) The probability that x is between 3 and 7.

4-3 Mean, Variance, and Expectation

We know how to find the mean and variance for a given set of scores (see Chapter 2), but suppose we have scores that are "conceptualized" in the sense that we know the probability distribution instead of a sample of observations or measurements. For example, suppose that we want to determine the mean and variance for the numbers of boys that will occur in pairs of independent births. (Assume that boys and girls are equally likely.) We have no specific results to work with, but we do know that the outcomes of 0, 1, and 2 boys have probabilities of 1/4, 1/2, 1/4, respectively. One way to find the mean and variance for the numbers of boys is to pretend that theoretically ideal results actually occurred. We list the different possible outcomes.

$$
\begin{array}{ll}
\text{boy-boy} & (2) \\
\text{boy-girl} & (1) \\
\text{girl-boy} & (1) \\
\text{girl-girl} & (0)
\end{array}
$$

Since they are equally likely, we can pretend that the four results actually did occur. The mean of 2, 1, 1, and 0 is 1, while the variance is 0.5. (The variance is found by applying Formula 2-4. In this use of Formula 2-4 we divide by n instead of $n - 1$ because we assume that we have all scores of the population.) Instead of considering four trials that yield theoretically ideal results, we could pretend that a large number of trials (say 4000) yielded these theoretically ideal results

		Number of boys	Frequency
boy-boy	(2 boys 1000 times)		
boy-girl	(1 boy 1000 times)	2	1000
girl-boy	(1 boy 1000 times)	1	2000
girl-girl	(0 boys 1000 times)	0	1000

Our random variable represents the number of boys in two births, so the preceding results suggest a list of 4000 scores that consist of 1000 twos, a total of 2000 ones, and 1000 zeros. The mean of those 4000 scores is again 1, while the variance is again 0.5. In this example, it really makes no difference whether we presume 4 theoretically ideal trials or 4000.

Instead of pretending that theoretically ideal results have occurred, we can find the mean of a discrete random variable by Formula 4-1.

Formula 4-1 $$ \mu = \Sigma x \cdot P(x) $$

This formula is justified by relating $P(x)$ to its role in describing the relative frequency with which x occurs. Recall that the mean of any list of scores is the sum of those scores divided by the total number of scores n. We usually compute the mean in that order; that is, first sum the scores and then divide the total by n. However, we can obtain the same result by dividing each individual score by n and then summing the quotients. For example, to find the mean of 2, 1, 1, 0 we usually write

$$\frac{2 + 1 + 1 + 0}{4} = 1$$

but we can also compute the mean as

$$\frac{2}{4} + \frac{1}{4} + \frac{1}{4} + \frac{0}{4} = 1$$

Similarly, we can find the mean of the preceding 4000 scores by writing

$$\frac{(2 \cdot 1000) + (1 \cdot 2000) + (0 \cdot 1000)}{4000} = 1$$

but we can also compute that mean as

$$\mu = \left(2 \cdot \frac{1000}{4000}\right) + \left(1 \cdot \frac{2000}{4000}\right) + \left(0 \cdot \frac{1000}{4000}\right) = 1$$

The right side of this last equation corresponds to $\Sigma x \cdot P(x)$ in Formula 4-1.

For each specific value of x, $P(x)$ can be considered the relative frequency with which x occurs. $P(x)$ effectively accommodates the repetition of specific x scores when we compute the mean. It also incorporates division by n directly into the summation process.

Similar reasoning enables us to take the variance formula from Chapter 2 ($\sigma^2 = \Sigma(x - \mu)^2/N$) and apply it to a random variable of a probability distribution to get $\sigma^2 = \Sigma(x - \mu)^2 \cdot P(x)$. Again the use of $P(x)$ accommodates the repetition of specific x scores and simultaneously accomplishes the division by N. This latter formula for variance is usually manipulated into an equivalent form to facilitate computations, as shown in Formula 4-2.

Formula 4-2 $$\sigma^2 = [\Sigma x^2 \cdot P(x)] - \mu^2$$

To apply Formula 4-2 to a specific case, we square each value of x and multiply that square by the corresponding probability and then add all of those products. We then subtract the square of the mean. The standard deviation σ can be easily obtained by simply taking the square root of the variance. If the variance is found to be 0.5, the standard deviation is $\sqrt{0.5}$, or about 0.7.

EXAMPLE

Use Formulas 4-1 and 4-2 to find the mean, variance, and standard deviation of the random variable that represents the number of boys in two independent births. (Assume that a girl or a boy is equally likely to occur.)

Solution

In the following table, the two leftmost columns summarize the probability distribution. The three rightmost columns are created for the purposes of the computations that are required.

x	$P(x)$	$x \cdot P(x)$	x^2	$x^2 \cdot P(x)$
0	$\frac{1}{4}$	0	0	0
1	$\frac{1}{2}$	$\frac{1}{2}$	1	$\frac{1}{2}$
2	$\frac{1}{4}$	$\frac{2}{4}$	4	1
Total		1		1.5

From the table we see that $\mu = \Sigma x \cdot P(x) = 1$ and

$$\sigma^2 = [\Sigma x^2 \cdot P(x)] - \mu^2 = (1.5) - 1^2 = 0.5$$

The standard deviation is the square root of the variance, so that

$$\sigma = \sqrt{0.5} = 0.7$$

An important advantage of these techniques is that a probability distribution is actually a model of a theoretically perfect population frequency distribution. Since the probability distribution allows us to perceive the population, we are able to determine the values of important parameters such as the mean, variance, and standard deviation. This in turn allows us to make the inferences that are necessary for decision-making in a multitude of different professions.

Expected Value

We will now consider some applications of the expected value (or mean) of a discrete random variable. The uses of expected value (also called mathematical expectation) are extensive and varied, and they play a very important role in an area of application called *decision theory*. Gamblers use knowledge of expected values in their playing strategies.

Prophets for Profits

For the unreasonable price of $40 per year, subscribers can receive a weekly publication "designed to be of help in picking a winning set of Lotto or N.Y. State Lottery numbers." Apparently blind to the fact that the lottery numbers are randomly selected each week, this booklet recommends numbers to bet. Some numbers are recommended as "hot" because they have been coming up often, others are recommended as "due" because they haven't been coming up often. Other choices involve horoscopes, dreams, and numbers that have "appeared or talked to" a seeress!

Insurance firms compute expected values so that they can set their rates at profitable, but competitive, levels. Let's consider a very simple insurance problem involving a man who, on his 64th birthday, obtains a $1000, one-year life insurance policy at a cost of $50. Based on past mortality experience, the insurance company estimates that there is a 0.963 probability that this man will live for at least 1 year. It follows that there is a 0.037 probability that he will not survive for that year. How much can the insurance company expect to earn on this policy?

	x	$P(x)$
Man lives	$50	0.963
Man dies	−$950	0.037

The accompanying table summarizes the situation. If the man lives, the insurance company gains the premium cost of $50, and there is a 0.963 probability of this happening. If he dies, the company has already collected $50 but it must pay out $1000 for a net loss of $950, and there is a probability of 0.037 that this will occur. We represent the loss with a negative sign. We now define the expected value of a discrete random variable, since we can use it in this situation. Note that this definition refers to the same expression found in Formula 4-1 for the mean.

DEFINITION

The **expected value** of a discrete random variable is $E = \Sigma x \cdot P(x)$

For the example under consideration, we get

$$E = (\$50 \times 0.963) + (-\$950 \times 0.037) = \$13$$

The value of E is the average return the company can expect in the long run of many such trials. With $E = \$13$ we see that the company can expect to earn $13 for each such policy. Another way to view this result is to note that for each 1000 qualified policy holders, premiums paid to the company are $1000 \times \$50 = \$50,000$. If the expected 37 men die during the year, the company must pay benefits of $37 \times \$1000 = \$37,000$. The difference of $13,000 averages out to a net profit of $13 per policy. In reality, the true earnings must be adjusted for overhead costs that must be covered. But this example does illustrate the importance of expected value for the entire insurance industry.

EXAMPLE

A promoter for an outdoor rock concert will lose $4000 in expenses if the event is canceled because of rain or nonappearance of the performers. If the event is held, he will make a profit of $3000. The probability of a cancellation is estimated to be 0.180. What is the promoter's expected value?

Solution

In this example, x assumes the values of $-\$4000$ and $\$3000$, with the corresponding probabilities of 0.180 and 0.820. In Table 4-2 we multiply each value of x by the corresponding probability and then add the resulting products. Note that this is the same process used for finding the mean of a discrete random variable.

Table 4-2		
x	$P(x)$	$x \cdot P(x)$
-4000	0.180	-720
3000	0.820	2460
Total		1740

The result indicates that the promoter has an expected value of $1740. This value represents the promoter's average return if he were to run many such concerts under the same conditions.

In Chapter 10 we will use the concept of expected value to compare actual survey results to expected results; the degree of similarity or disparity will enable us to form some meaningful conclusions.

4-3 Exercises A
Mean, Variance, and Expectation

In Exercises 4-25 through 4-32, find the mean, variance, and standard deviation of the random variable x.

4-25

x	$P(x)$
0	0.2
1	0.5
2	0.3

4-26

x	$P(x)$
0	0.2
1	0.7
2	0.1

4-27

x	$P(x)$
1	0.15
2	0.45
3	0.35
4	0.05

4-28

x	$P(x)$
5	0.20
6	0.10
7	0.45
8	0.25

4-29

x	$P(x)$
5	$\frac{1}{4}$
10	$\frac{1}{4}$
20	$\frac{1}{4}$
50	$\frac{1}{4}$

4-30

x	$P(x)$
5	$\frac{1}{20}$
10	$\frac{3}{20}$
20	$\frac{7}{20}$
50	$\frac{9}{20}$

4-31

x	$P(x)$
2	$\frac{4}{30}$
4	$\frac{6}{30}$
6	$\frac{10}{30}$
8	$\frac{6}{30}$
10	$\frac{4}{30}$

4-32

x	$P(x)$
2	$\frac{1}{5}$
4	$\frac{1}{5}$
6	$\frac{1}{5}$
8	$\frac{1}{5}$
10	$\frac{1}{5}$

4-33　A cab driver determines the probability distribution (below) for the number of passengers per trip. Find the mean, variance, and standard deviation for the number of passengers per trip.

4-34　A pilot determines the probability distribution (below) for the number of radio frequency changes that she must make in a 4-hour flight. Find the mean, variance, and standard deviation for the number of frequency changes required.

4-35　A telephone company survey studies the number of times a telephone will ring before it is answered. Assume that the probability distribution is as given below. Find the mean, variance, and standard deviation for the number of rings.

4-36　An automatic device is used to set newsprint. The probability distribution for the number of errors per column is given below. Find the mean, variance, and standard deviation for the number of errors per column.

4-33

x	$P(x)$
1	0.6
2	0.2
3	0.1
4	0.1

4-34

x	$P(x)$
6	0.05
7	0.10
8	0.05
9	0.05
10	0.10
11	0.20
12	0.20
13	0.15
14	0.05
15	0.05

4-35

x	$P(x)$
1	0.10
2	0.05
3	0.27
4	0.34
5	0.12
6	0.08
7	0.03
8	0.01

4-36

x	$P(x)$
0	0.03
1	0.04
2	0.09
3	0.08
4	0.15
5	0.18
6	0.12
7	0.10
8	0.13
9	0.06
10	0.02

4-37 On any given weekday, the probabilities of 0, 1, or 2 accidents on a certain highway are 0.650, 0.300, and 0.050 respectively. Find the mean, variance, and standard deviation for the number of accidents in a day.

4-38 A commuter airline company finds that for a certain flight, the probabilities of 0, 1, 2, or 3 vacant seats are 0.705, 0.115, 0.090, and 0.090 respectively. Find the mean, variance, and standard deviation for the number of vacant seats.

4-39 There is a probability of 0.18 that a World Series will last four games, a 0.18 probability that it will last five games, a 0.20 probability that it will last six games, and a 0.44 probability that it will last seven games. Find the mean, variance, and standard deviation for the number of games that World Series' contests last.

4-40 A computer store finds that the probabilities of selling 0, 1, 2, 3, or 4 microcomputers in one day are 0.245, 0.370, 0.210, 0.095, and 0.080 respectively. Find the mean, variance, and standard deviation for the number of microcomputer sales in one day.

4-41 If you have a $\frac{1}{4}$ probability of gaining $500 and a $\frac{3}{4}$ probability of gaining $200, what is your expected value?

4-42 If you have a $\frac{1}{10}$ probability of gaining $1000 and a $\frac{9}{10}$ probability of losing $300, what is your expected value?

4-43 If you have a $\frac{1}{10}$ probability of gaining $200, a $\frac{3}{10}$ probability of losing $300, and a $\frac{6}{10}$ probability of breaking even, what is your expected value?

4-44 A car wash loses $30 on rainy days and gains $120 on days when it does not rain. If the probability of rain is 0.15, what is the expected value?

4-45 A contractor bids on a job to construct a building. There is a 0.7 probability of making a $175,000 profit, and there is a probability of 0.3 that the contractor will break even. What is the expected value?

4-46 A contractor bids on a job to construct a building. There is a 0.7 probability of making a $175,000 profit and a 0.3 probability of losing $250,000. What is the expected value?

4-47 A flight-training school makes a profit of $200 on fair days but loses $150 for each day of bad weather. If the probability of bad weather is 0.4, what is the expected value?

4-48 A popular magazine runs a sweepstakes as an advertising method. The various prizes are listed, along with their approximate numerical odds of winning.

First prize:	$25,000	(one chance in 15,660,641)
Second prize:	$24,000	(one chance in 5,455,950)
Third prize:	$5,000	(one chance in 3,132,128)

Fourth prize: $1,000   (one chance in 313,213)

Fifth prize: $25   (one chance in 7830)

(a)   Compute the expected value.

(b)   If the only cost of entering this contest is a stamp, how much does a person win or lose in the long run?

4-3   Exercises B
Mean, Variance, and Expectation

4-49   A couple plans to have five children. Let the random variable be the number of boys that will occur. Assume that a boy or a girl is equally likely to occur and that the sex of any successive child is unaffected by previous brothers or sisters.

x	$P(x)$
0	
1	
2	
3	
4	
5	

(a)   List the 32 different possible simple events.

(b)   Enter the probabilities in the table at left, where x represents the number of boys in the five births.

(c)   Graph the histogram for the probability distribution from (b).

(d)   Find the mean number of boys that will occur among the five births.

(e)   Find the variance for the number of boys that will occur.

(f)   Find the standard deviation for the number of boys that will occur.

(g)   On the histogram in part (c), identify the location of the mean from part (d). Also identify the location of the value that is exactly one standard deviation above the mean. Then identify the location of the value that is exactly one standard deviation below the mean.

4-50   The variance for the discrete random variable x is 1.25.

(a)   Find the variance of the random variable $5x$.

(b)   Find the variance of the random variable $x/5$.

(c)   Find the variance of the random variable $x + 5$.

(d)   Find the variance of the random variable $x - 5$.

4-51   A discrete random variable can assume the values $1, 2, \ldots, n$ and those values are equally likely.

(a)   Show that $\mu = (n + 1)/2$.

(b)   Show that $\sigma^2 = (n^2 - 1)/12$.

(*Hint:* $1 + 2 + 3 + \cdots + n = n(n + 1)/2$.

$1^2 + 2^2 + 3^2 + \cdots + n^2 = n(n + 1)(2n + 1)/6$.)

4-52   Verify that $\sigma^2 = [\Sigma x^2 \cdot P(x)] - \mu^2$ is equivalent to $\sigma^2 = \Sigma(x - \mu)^2 \cdot P(x)$.

(*Hint:*   For constant c, $\Sigma cx = c\Sigma x$. Also, $\mu = \Sigma x \cdot P(x)$.)

4-4 Binomial Experiments

Many real and important applications require that we find the probability that some event will occur x times out of n different trials. In any one trial, either the event occurs or it does not, so we are dealing with an element of "twoness" present in situations such as these:

- Elections: voters choosing between two candidates.
- Manufacturing: machines producing acceptable or defective items.
- Education: subjects passing or failing a test.
- Medicine: new drugs being effective or ineffective.
- Psychology: mental health treatment being effective or ineffective.
- Agriculture: crops being profitable or nonprofitable.
- Advertising: television viewers recalling or not recalling a sponsor's name.

All these situations exhibit the element of twoness that results in a special type of discrete probability distribution called the *binomial distribution*. A binomial distribution is a list of outcomes and probabilities for a binomial experiment.

> ### DEFINITION
>
> A **binomial experiment** is one that meets all the following requirements:
>
> **1.** The experiment must have a fixed number of trials.
> **2.** The trials must be independent.
> **3.** Each trial must have all outcomes classified into two categories.
> **4.** The probabilities must remain constant for each trial.

This definition indicates that in a binomial experiment, we have a fixed number of independent trials where each outcome has only two classifications. The term independent simply means that the outcome of one trial will not affect the probabilities of outcomes on subsequent trials. The birth of a girl does not affect the sex of the next baby that is born, so these events are independent.

If a real experimental trial has many outcomes, each such outcome must be classified into one of two categories to use the binomial distribution. In a multiple-choice question, there may be five possible selections, but the two categories of "right" and "wrong" are suitable for a binomial model. In randomly selecting a number from 1 to 100, we have 100 possible choices, but we can devise the two categories of "8" and "not 8" for binomial purposes. In short, the property of twoness applies to possible categories and not to specific individual outcomes or simple events.

Is Parachuting Safe?

About 35 people die each year from injuries sustained while parachuting. In comparison, a typical year includes about 200 scuba diving fatalities, 7000 drownings, 900 bicycle deaths, 800 lightning deaths, and 1150 deaths from bee stings.

Of course, this does not necessarily mean that it's safer to parachute than to ride a bicycle or swim. A fair comparison should include fatality *rates*, not just the number of fatalities. It has been estimated that in parachuting, 1 fatality will occur in about 25,000 jumps.

The author, with much trepidation, made two parachute jumps but gave up the sport after missing the spacious drop zone both times.

NOTATION

S and F (success and failure) denote the two possible categories of all outcomes. p and q will denote the probabilities of S and F, respectively, so that

$$P(S) = p$$
$$P(F) = 1 - p = q$$

n will denote the fixed number of trials.

x will denote a specific number of successes in n trials so that x can be any whole number between 0 and n, inclusive.

p denotes the probability of success in *one* of the n trials.

q denotes the probability of failure in *one* of the n trials.

$P(x)$ denotes the probability of getting exactly x successes among the n trials.

The word *success* as used here does not necessarily correspond to a good event. Selecting a defective parachute may be classified a success, even though the results of such a selection may be less than pleasant. Either of the two possible categories may be called the success S as long as the corresponding probability is identified as p. The value of q can always be found by subtracting p from 1. If $p = 0.4$, then $q = 1 - 0.4$, or 0.6.

We will soon introduce a formula for computing probabilities in a binomial experiment, but we begin with an example intended to illustrate the reasoning underlying that formula. Try to follow this example carefully so that the ensuing formula will not seem to be a mysterious revelation.

Assume that a drug is given to five patients. From past experience, the drug has been found to be effective in 10% of the cases for which it is used as a treatment. If we want to find the probability that the drug is successful in all five cases, we can use the multiplication rule for independent events (see Section 3-4) to compute

$P(5 \text{ successes})$

$$= P(\text{success}) \cdot P(\text{success}) \cdot P(\text{success}) \cdot P(\text{success}) \cdot P(\text{success})$$
$$= 0.1 \times 0.1 \times 0.1 \times 0.1 \times 0.1 = 0.00001$$

The probability of 0.1 remained constant for each of the five trials because this example involves five separate and independent patients.

Let's again consider the same drug with the same 10% success rate and the same five patients, but this time we want the probability that the drug is successful in exactly two of the five patients. That is, we want the probability of two successes and three failures. Since $P(\text{success}) = 0.1$, it follows that $P(\text{failure}) = 0.9$; we may be very tempted to solve this problem by computing

$$0.1 \times 0.1 \times 0.9 \times 0.9 \times 0.9 = 0.00729$$

This solution is very common and natural, but it is incomplete because it contains an implicit assumption that the first two patients were successfully treated by the drug, while the last three patients were not. But we are seeking the probability of exactly two successes with no stipulation that the failures must be endured by the last three patients. If we represent success by S and failure by F, the product of 0.00729 corresponds to the compound event S-S-F-F-F. However, there are other ways or orders of listing two successes and three failures. There are actually ten possible arrangements, and each one has a probability of 0.00729. We list them at left.

S-S-F-F-F
F-F-F-S-S
F-F-S-F-S
F-S-F-S-F
S-F-F-F-S
F-F-S-S-F
F-S-S-F-F
S-F-F-S-F
S-F-S-F-F
F-S-F-F-S

Since there are ten ways of getting exactly two cures among the five patients and each different way has a probability of 0.00729, there is a total probability of $10 \cdot 0.00729$, or 0.0729, that two of the five subjects will be cured by the drug. We can generalize this specific result by stating that with n independent trials in which $P(S) = p$ and $P(F) = q$, the multiplication rule indicates that the probability of the first x cases being successes while the remaining cases are failures is

$$\underbrace{p \cdot p \cdots p}_{x \text{ times}} \cdot \underbrace{q \cdot q \cdots q}_{(n-x) \text{ times}}$$

or

$$p^x \cdot q^{n-x}$$

But $P(x)$ denotes the probability of x successes among n trials in *any* order, so that $p^x \cdot q^{n-x}$ must be multiplied by the number of ways the x successes and $n - x$ failures can be arranged. The number of ways in which it is possible to arrange x successes and $n - x$ failures is shown in Formula 4-3.

Formula 4-3
$$\frac{n!}{(n-x)!\, x!}$$

The factorial symbol, introduced in Section 3-6, denotes the product of decreasing factors. Many calculators have a factorial key, and the values of factorials for 0 through 20 can be found in Table A-1 of Appendix A. Some sample factorials are listed here.

$$3! = 3 \cdot 2 \cdot 1 = 6$$
$$2! = 2 \cdot 1 = 2$$
$$1! = 1$$
$$0! = 1 \text{ (by definition)}$$
$$n! = n \cdot (n-1)(n-2) \ldots 1$$

The expression given as Formula 4-3 does correspond to $_nC_r$, also introduced in Section 3-6. (Coverage of Section 3-6 is not necessary for this chapter.) We will not derive Formula 4-3, but its role should become clear in the example that follows. Formula 4-3 can be derived by using the binomial theorem for the expansion of $(a + b)^n$.

We can now combine this counting device (Formula 4-3) with the direct application of the multiplication rule for independent events to get a general formula for computing what we will call binomial probabilities.

Formula 4-4 **Binomial probability formula**

$$P(x) = \frac{n!}{(n - x)! \, x!} \cdot p^x \cdot q^{n-x}$$

Let's use this formula to repeat the solution to our original problem by finding the probability of exactly two cures among five randomly selected patients where $P(\text{cure}) = P(S) = p = 0.1$. In this problem, recall that $n = 5$, $p = 0.1$, $q = 0.9$, and $x = 2$ so that

$$P(2) = \frac{5!}{(5 - 2)! \, 2!} \cdot 0.1^2 \cdot 0.9^{5-2}$$

$$= \frac{120}{6 \cdot 2} \cdot 0.1^2 \cdot 0.9^3$$

$$= 10 \cdot 0.01 \cdot 0.729$$

$$= 0.0729$$

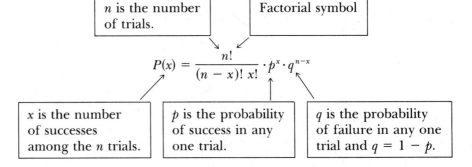

This specific example can help us see the rationale underlying Formula 4-4. The component $n!/(n - x)! \, x!$ counts the number of ways of getting x successes in n trials. In this example we have ten ways of getting two successes in five trials. The product $p^x \cdot q^{n-x}$ gives us the probability of getting exactly x successes among n trials in one particular order. The product of those factors is the total probability representing all ways of getting exactly x successes in n trials. The following examples further illustrate the use of the binomial probability formula.

EXAMPLE

A genetic trait of one family manifests itself in 25% of the offspring. If eight offspring are randomly selected, find the probability that the trait is manifested in exactly three cases.

Solution

The experiment is binomial because of the following:

1. We have a fixed number of trials (8).
2. The trials are independent, since the offspring don't affect each other.
3. There are two categories, since each offspring either manifests the genetic trait or does not.
4. The probability of 0.25 remains constant for each trial.

The immediate objective is to identify the values of n, p, q, and x correctly so that we can apply the binomial probability formula. We have

$$n = 8 \qquad \text{(number of trials)}$$
$$p = 0.25 \qquad \text{(probability of success)}$$
$$q = 0.75 \qquad \text{(probability of failure)}$$
$$x = 3 \qquad \text{(desired number of successes)}$$

We should check for consistency by verifying that what we call a success, as counted by x, is the same success with probability p. That is, we must be sure that x and p refer to the same concept of success. Using the values for n, p, q, and x in the binomial probability formula, we get

$$P(3) = \frac{8!}{(8-3)!\ 3!} \cdot 0.25^3 \cdot 0.75^{8-3}$$

$$= \frac{40{,}320}{(120)(6)} \cdot 0.015625 \cdot 0.237305$$

$$= 0.208$$

There is a probability of 0.208 that the genetic trait will be manifested in exactly three of the eight offspring.

EXAMPLE

A fire science specialist has determined that there is a 0.310 probability of an alarm being false. Find the probability of exactly one false alarm among the next four alarms reported.

Solution

With $n = 4$, $P(S) = p = 0.310$, and $P(F) = q = 1 - 0.310 = 0.690$, we want $P(1)$. Applying the binomial probability formula directly we get

$$P(1) = \frac{4!}{(4 - 1)! \; 1!} \cdot 0.310^1 \cdot 0.690^3$$

$$= \frac{24}{6 \cdot 1} \cdot 0.310^1 \cdot 0.690^3$$

$$= 4 \cdot 0.310 \cdot 0.3285$$

$$= 0.407$$

Table 4-3

x	$P(x)$
0	0.227
1	0.407
2	0.275
3	0.082
4	0.009

In this example, we can also use the binomial probability formula to find $P(0)$, $P(2)$, $P(3)$, and $P(4)$ so that the complete probability distribution for the number of false alarms will be known. These results are shown in Table 4-3, where x denotes the number of false alarms among the next four alarms.

We can depict a table such as 4-3 in the form of a histogram, as in Figure 4-5. The shape of a probability distribution is often a critically important feature. (We consider that characteristic in more detail in Section 4-6.)

In many cases, the requirement of independent trials is not strictly satisfied, but—for all practical purposes—independence can be assumed. For example, when sampling ten voters from a population of 50,000, sampling would normally be done without replacement, so that the ten trials lack independence. But the number sampled is so small

Figure 4-5

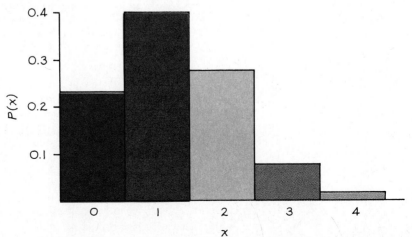

relative to the population size that we can assume independence, with negligible error. This idea is used extensively by professional statisticians in industry and research.

EXAMPLE

In a certain county, 30% of the voters are Republicans. If ten voters are randomly selected, find the probability that six of them will be Republicans.

Solution

Assuming that the total number of voters in the county is large, the binomial probability model can be used since the necessary conditions are essentially met. From the given information we see that we must find $P(6)$ while $n = 10$, $x = 6$, $P(S) = p = 0.3$, and $P(F) = q = 1 - 0.3 = 0.7$. Through direct application of the binomial probability formula we get

$$P(6) = \frac{10!}{(10 - 6)! \, 6!} \cdot 0.3^6 \cdot 0.7^{10-6}$$
$$= 210 \cdot 0.000729 \cdot 0.2401$$
$$= 0.0368$$

An alternative to computing with the binomial probability formula involves the use of the table of binomial probabilities (Table A-2 in Appendix A). To use this table, first locate the relevant value of n in the leftmost column and then locate the corresponding value of x that is desired. At this stage, one row of numbers should be isolated. Now align that row with the proper probability of p by using the column across the top. The isolated number represents the desired probability (missing its decimal point at the beginning). A very small probability such as 0.000000345 is indicated by 0+. For example, the table indicates that for $n = 10$ trials, the probability of $x = 2$ successes when $P(S) = p = 0.05$ is 0.075. (A more precise value is 0.0746348, but the table values are approximate.)

EXAMPLE

Use either the binomial probability formula or the binomial probability table to solve the following problem: A manufacturer has produced 100,000 radios of which 5% are defective. In a random sample of six radios, find the probability of getting exactly two defective radios.

Solution

The large population size and the small number of selections will, in effect, cause the trials to be essentially independent. With $n = 6$, $P(S) = p = 0.05$ and $x = 2$, we compute

$$P(2) = \frac{6!}{(6 - 2)! \, 2!} \cdot 0.05^2 \cdot 0.95^4$$

$$= \frac{720}{24 \cdot 2} \cdot 0.0025 \cdot 0.8145$$

$$= 0.0305$$

The probability of randomly selecting six radios, of which two are defective, is therefore 0.031. (Using the table we get 0.031.)

EXAMPLE

Find the probability that of five babies, there are exactly four girls. (Assume that a boy or a girl is equally likely to occur.)

Solution

This is a binomial experiment with $n = 5$, $P(S) = p = 0.5$, and $x = 4$. Applying the binomial probability formula we get

$$P(4) = \frac{5!}{(5 - 4)! \, 4!} \cdot 0.5^4 \cdot 0.5^{5-4}$$

$$= \frac{120}{1 \cdot 24} \cdot 0.0625 \cdot 0.5$$

$$= 0.15625$$

Thus the probability of getting four girls in five births is 0.156. (Using the table we also get 0.156.)

Although the values in Table A-2 are approximate and only selected values of p are included, the use of such a table often provides quick and easy results. The next example illustrates situations in which the use of Table A-2 can replace several computations with the binomial probability formula.

EXAMPLE

The operation of a component in a computer system is so critical that five of these components are built in for backup protection. The latest quality control tests indicate that this type of component has a 30% failure rate. (For optimists, that's a 70% success rate.) Assume that this is a binomial experiment and use Table A-2 to find the probabilities of the following events.

(a) Among the five components, at least two continue to work.
(b) Among the five components, fewer than four continue to work.
(c) Among the five components, at least one continues to work.

Solution

For all three parts of this problem, we stipulate that a success corresponds to a component that continues to work. With five components and a 30% failure rate, we have $n = 5$, $p = 0.7$, and $q = 0.3$.

(a) At least two of the five components continue to work, so that the number of successes is two or more. Thus $x = 2$ or 3 or 4 or 5. We use Table A-2 with $n = 5$, $p = 0.7$, and $x = 2, 3, 4, 5$ to get the probabilities of 0.132, 0.309, 0.360, 0.168. We total these probabilities, since the addition rule for mutually exclusive events applies. That is,

$$P(2 \text{ or } 3 \text{ or } 4 \text{ or } 5) = P(2) + P(3) + P(4) + P(5)$$

The probability of at least two working components is therefore 0.969.

(b) If fewer than four components continue to work, we have $x = 0$ or 1 or 2 or 3. With $n = 5$ and $p = 0.7$, we get the corresponding probabilities of 0.002, 0.028, 0.132, 0.309, which give a total of 0.471.

(c) If at least one component continues to work, we have $x = 1$ or 2 or 3 or 4 or 5. With $n = 5$ and $p = 7$, we get the corresponding probabilities of 0.028, 0.132, 0.309, 0.360, 0.168, which give a total of 0.997. (With some cleverness we might note that "at least one working component" is equivalent to *not* having $x = 0$, so that the probability we seek can be found by excluding only the value 0.002 corresponding to $x = 0$; that approach suggests an answer of 0.998, which differs from 0.997 only because of rounding errors.)

To appreciate the ease of using Table A-2, solve the preceding problem by using the binomial probability formula!

Many computer statistics packages include an option for generating binomial probabilities. The following sample output from STATDISK illustrates one such program. The user simply enters n and p and the entire probability distribution table is displayed. The column labeled "Cum. Prob." represents cumulative probabilities by adding up the values of $P(x)$ as it goes down the column.

```
This program uses the BINOMIAL PROBABILITY FORMULA
to find probabilities and cumulative probabilities.

You must enter:

n = the number of trials or sample size (Max. 125)
p = the probability of success on a single trial

                    n = 5

                    p = 0.30000

              Mean = 1.50000
         St. Dev. = 1.02470
         Variance = 1.05000


     X          P(X)          Cum. Prob.

     0        0.16807          0.16807
     1        0.36015          0.52822
     2        0.30870          0.83692
     3        0.13230          0.96922
     4        0.02835          0.99757
     5        0.00243          1.00000
```

In the beginning of this chapter, we briefly discussed the Federal Aviation's consideration of allowing twin engine jets to make transatlantic flights. (Existing regulations required at least three engines.) A realistic estimate for the probability of an engine failing on a transatlantic flight is 1/14,100. Using that probability and the binomial probability formula, we can develop the probability distributions as summarized in Tables 4-4 and 4-5, where x represents the number of engines that fail on a transatlantic flight.

Detecting Syphilis

In the 1940s, the United States Army included the Kahn-Wasserman test in its physical examination of inductees. This test, designed to detect the presence of syphilis, required blood samples that were given chemical analyses. Since the testing procedure was time-consuming and expensive, a more efficient one was sought. One researcher noted that if samples from a large number of blood specimens were mixed in pairs and then tested, the total number of chemical analyses would be greatly reduced. Syphilitic inductees could be identified by retesting the few blood samples that were included in the pairs indicating the presence of syphilis. But if the total number of analyses was reduced by pairing blood specimens, why not put them in groups of three or four or more? Probability theory was used to find the most efficient group size, and a general theory was developed for detecting the defective members of any population. Bell Laboratories, for example, used this approach to identify defective condensers by putting groups of them in a vacuum and testing for leakage.

Table 4-4	Three Engines
x	$P(x)$
0	0.9997872491
1	0.0002127358
2	0.0000000151
3	0.0000000000

Table 4-5	Two Engines
x	$P(x)$
0	0.9998581611
1	0.0001418339
2	0.0000000050

Using those tables and assuming that a flight will be completed if at least one engine works, we get

P(safe flight with twin-engine jet) $= P$(0 or 1 engine failures)

$$= 0.9998581611 + 0.0001418339 = 0.9999999950$$

P(safe flight with three-engine jet) $= $ (0, 1, or 2 engine failures)

$$= 0.9997872491 + 0.0002127358 + 0.0000000151$$

$$= 1.0000000000 \text{ (rounded to 10 decimal places)}$$

The result of 1.0000000000 is actually a rounded off form of the more precise result of 0.9999999999996433. Comparing the two resulting probabilities, we see that the three-engine jet would be safer, as expected, but the difference doesn't seem to be too significant. All of those leading nines in the results suggest that both configurations are quite safe. We could analyze the significance of those final probabilities further, but this illustration does show a very useful application of the binomial probability formula.

To keep this section in perspective, remember that the binomial probability formula is but one of many probability formulas that can be used for different situations. It is, however, among the most important and most useful of all discrete probability distributions. In practical cases it is often used in problems such as quality control, voter analysis, medical research, military intelligence, and advertising.

4-4   Exercises A
Binomial Experiments

4-53   Which of the following are binomial experiments or can be treated as binomial experiments with negligible error?
(a) Testing a sample of five condensors (with replacement) from a population of 20 condensors, of which 40% are defective.
(b) Polling 500 voters on the presidential election from a population of 200,000 voters if 35% are Republicans and 65% are Democrats.
(c) Testing a sample of eight drug dosages from a population of 5000, of which 2% are contaminated.
(d) Polling 1000 voters in the presidential election from a population of 8 million voters, of which 40% are Democrats, 35% are Republicans, and 25% are Independents.
(e) Firing 20 missiles at a target with a hit rate of 90%.
(f) Tossing an unbiased coin 500 times.
(g) Tossing a biased coin 500 times.
(h) Surveying 500 consumers to find the brands of toothpaste they prefer.
(i) Surveying 500 consumers to determine whether their preferred brand of toothpaste is Brand X.
(j) Administering a driving test to 50 license applicants with a passing rate of 72%.

4-54   In a binomial experiment, a trial is repeated n times. Find the probability of x successes if $P(S) = p$. (Use the given values of n, x, and p and the table of binomial probabilities (Table A-2).)
(a) $n = 10, x = 3, p = 0.5$ (b) $n = 10, x = 3, p = 0.4$
(c) $n = 7, x = 0, p = 0.01$ (d) $n = 7, x = 0, p = 0.99$
(e) $n = 7, x = 7, p = 0.01$ (f) $n = 15, x = 5, p = 0.7$
(g) $n = 12, x = 11, p = 0.6$ (h) $n = 9, x = 6, p = 0.1$
(i) $n = 8, x = 5, p = 0.95$ (j) $n = 14, x = 14, p = 0.9$

4-55   For each of the following, evaluate $\dfrac{n!}{(n-x)!\,x!}$

to find the number of ways you can arrange x successes and $n-x$ failures.
(a) $n = 5, x = 3$ (b) $n = 5, x = 0$
(c) $n = 8, x = 7$ (d) $n = 8, x = 3$
(e) $n = 8, x = 8$

4-56   For each of the following, evaluate p^x, q^{n-x}, and then $p^x \cdot q^{n-x}$.
(a) $p = 0.8, q = 0.2, n = 4, x = 1$
(b) $p = 0.3, q = 0.7, n = 5, x = 0$
(c) $p = 0.4, n = 5, x = 2$
(d) $p = 0.6, n = 5, x = 3$
(e) $p = 0.2, n = 5, x = 5$

In Exercises 4-57 through 4-80, identify the values of n, x, p, *and* q, *and find the value requested.*

4-57   Find the probability of getting exactly four boys in ten births. (Assume that male and female births are equally likely.)

4-58   Find the probability of getting exactly six girls in seven births. (Assume that male and female births are equally likely.)

4-59   One treatment has a 60% success rate in helping people stop smoking. Find the probability of successfully treating three smokers in a group of eight smokers.

4-60   One prison experiences a 70% return rate for released prisoners. Find the probability of getting 12 returning prisoners from the next group of 15 that are released.

4-61   Of the subjects taking a certain civil service examination, 30% fail. Find the probability that of seven subjects, exactly four pass the test.

4-62   The probability of a computer component being defective is 0.01. Find the probability of getting exactly 2 defective components in a sample of 12.

4-63   In a certain college, 60% of the entering freshmen graduate. Find the probability that of ten randomly selected entering students, exactly seven will graduate.

4-64   A company produces batteries in batches of 500. In each batch, a sample of five batteries is tested and, if more than one defect is found, the entire batch is tested. Assume a defect rate of 1%.
(a)   Find the probability of getting no defects in the sample of five.
(b)   Find the probability of getting exactly one defect in the sample of five.
(c)   Use the results of parts (a) and (b) to find the probability of getting more than one defect in the sample of five batteries.

4-65   Police find that one patrol unit gets a 2% arrest rate when it sets up a checkpoint for drunk drivers. Find the probability that of 20 drivers checked, there will be exactly one arrest.

4-66   A survey has shown that of the people who have a personal computer, 18% include word processing among its uses. Find the probability that of 15 randomly selected people with personal computers, exactly 3 use it for word processing.

4-67   A study has revealed that 29% of all programmers are women. Find the probability that among ten randomly selected programmers, exactly five are women.

4-68   In an experiment on genetics, it has been found that a certain type of plant has a 25% chance of yielding a white flower and a 75% chance of yielding a red flower. Find the probability that among 16 such plants, exactly 12 plants will yield a red flower.

4-69   A multiple-choice test has 30 questions, and each one has five possible answers, of which one is correct. If all answers are guesses, find the probability of getting exactly four correct answers.

4-70   A poll shows that 16.9% of a student population indicate that their likely career will be computer related. Find the probability that among eight randomly selected students, exactly three give that indication.

4-71   Of all adult Americans, 35% exercise almost every day. Find the probability that among 14 randomly selected Americans, there are 7 who exercise almost every day.

4-72   A survey shows that for a certain brand of designer sunglasses, 25% are counterfeits. Find the probability that in a sample of ten such sunglasses, there will be four counterfeits.

4-73   In Table A-2, the probability corresponding to $n = 3$, $x = 2$, and $p = 0.01$ is shown as 0+. Find the exact probability represented by this 0+.

4-74   In Table A-2, the probability corresponding to $n = 5$, $x = 4$, and $p = 0.10$ is shown as 0+. Find the exact probability represented by this 0+.

4-75   A binomial experiment consists of three trials with a 1/3 probability of success in each trial. Construct the probability distribution table for this experiment. (Use the same format as Table 4-3.)

4-76   A binomial experiment consists of four trials with a 1/4 probability of success in each trial. Construct the probability distribution table for this experiment. (Use the same format as Table 4-3.)

4-77   The rhythm method of birth control is rated as 60% effective. Find the probability of at least one conception in ten separate trials.

4-78   It is found that for one section of Interstate 80, 92% of all vehicles are traveling at speeds greater than 55 mph. Find the probability that among five randomly selected vehicles, at least four are going faster than 55 mph.

4-79   A large housing complex contains thousands of apartments and the average vacancy rate is 6%. Find the probability that of 20 apartments in various units, there are fewer than 3 vacancies.

4-80   If 11% of all American families have incomes below the poverty level, find the probability that among ten randomly selected families, fewer than three will be below the poverty level.

4-4   Exercises B
Binomial Experiments

4-81   Suppose that an experiment meets all conditions to be binomial except that the number of trials is not fixed. Then the **geometric distribution,** which gives us the probability of getting the first success on the xth trial, is described by $P(x) = p(1 - p)^{x-1}$, where p is the probability of success on any one trial. Assume that the probability of a defective computer component is 0.2. Find the probability that the first defect is the seventh component tested.

4-82   The **Poisson distribution** is used as a mathematical model describing the probability distribution of the arrivals of entities requiring service (such as cars arriving at a gas station, planes arriving at an airport, or moviegoers arriving at a theater). The Poisson distribution is defined by the equation

$$P(x) = \frac{\mu^x \cdot e^{-\mu}}{x!}$$

where x represents the number of arrivals during a given time interval (such as 1 hour), μ is the mean number of arrivals during that same time interval, and e is a constant approximately equal to 2.718. Assume that $\mu = 15$ cars per hour for a gas station and find the probability of each event.

(a) No arrivals in an hour.
(b) One arrival in an hour.
(c) Ten arrivals in an hour.
(d) Fifteen arrivals in an hour.
(e) Twenty arrivals in an hour.
(f) Thirty arrivals in an hour.

4-83   If we sample from a small finite population without replacement, the binomial distribution should not be used because the events are not independent. If sampling is done without replacement and the outcomes belong to one of two types, we can use the **hypergeometric distribution.** If a population has A objects of one type, while the remaining B objects are of the other type, and if n samples are drawn without replacement, then the probability of getting x objects of Type A and $n - x$ objects of type B is

$$P(x) = \frac{A!}{(A - x)! \, x!} \cdot \frac{B!}{(B - n + x)! \, (n - x)!} \div \frac{(A + B)!}{(A + B - n)! \, n!}$$

Five people are randomly selected (without replacement) from a population of seven men and three women. Find the probability of getting four men and one woman.

4-84   The binomial distribution applies only to cases involving two types of outcomes, whereas the **multinomial distribution** involves more than two categories. Suppose we have three types of mutually exclusive outcomes denoted by A, B, and C. Let $P(A) = p_1$, $P(B) = p_2$, and $P(C) = p_3$. In n independent trials, the probability of x_1 outcomes of type A, x_2 outcomes of type B, and x_3 outcomes of type C is given by

$$\frac{n!}{(x_1!)(x_2!)(x_3!)} \cdot p_1^{x_1} \cdot p_2^{x_2} \cdot p_3^{x_3}$$

(a) Extend the result to cover six types of outcomes.

(b) A genetics experiment involves six mutually exclusive genotypes identified as A, B, C, D, E, and F, and they are all equally likely. If 20 offspring are tested, find the probability of getting exactly five A's, four B's, three C's, two D's, three E's, and three F's.

4-5   Mean and Standard Deviation for the Binomial Distribution

The binomial distribution is a probability distribution, so the mean, variance, and standard deviation for the appropriate random variable can be found from the formulas presented in Section 4-3.

$$\mu = \Sigma x \cdot P(x)$$
$$\sigma^2 = [\Sigma x^2 \cdot P(x)] - \mu^2$$

However, these formulas, which apply to all probability distributions, can be significantly simplified for the special case of binomial distributions. Given the binomial probability formula and the above general formulas for μ and σ^2, we can pursue a series of somewhat complicated algebraic manipulations that ultimately lead to the following desired result.

For a **binomial experiment,**

Formula 4-5 $\mu = n \cdot p$

Formula 4-6 $\sigma^2 = n \cdot p \cdot q$

Formula 4-7 $\sigma = \sqrt{n \cdot p \cdot q}$

The formula for the mean does make sense intuitively. If we were to analyze 100 births, we would expect to get about 50 girls, and np in this experiment becomes $100 \cdot \frac{1}{2}$, or 50.

The variance is not so easily justified, and we prefer to omit the complicated algebraic manipulations that lead to the second formula. Instead, we will show that these simplified formulas do lead to the same results as the more generalized formulas from Section 4-3.

EXAMPLE

A dealer considers a car to be successful if it does not require repairs covered by the warranty. If the probability of a car being successful is 0.8, find the mean, variance, and standard deviation for the number of successful cars among the three cars sold in one day.

Solution

In this binomial experiment we are given $n = 3$ and $P(S) = p = 0.8$. It follows that $P(F) = q = 0.2$. We now use Formulas 4-5, 4-6, and 4-7.

$$\mu = n \cdot p$$
$$= 3 \cdot 0.8 = 2.4$$
$$\sigma^2 = n \cdot p \cdot q$$
$$= 3 \cdot 0.8 \cdot 0.2 = 0.48$$
$$\sigma = \sqrt{n \cdot p \cdot q}$$
$$= \sqrt{0.48} = 0.69$$

We have completed the solution since the values of μ, σ^2, and σ have been determined. However, we want to show that these same values will result from the use of the more general formulas established in Section 4-3. We begin by computing the mean using $\mu = \Sigma x \cdot P(x)$. The possible values of x are 0, 1, 2 and 3, but we also need the values of $P(0)$, $P(1)$, $P(2)$, and $P(3)$. We get these values from Table A-2 or the binomial probability formula and enter them in the table. (We let $n = 3$, $p = 0.8$, and $q = 0.2$.)

x	$P(x)$	$x \cdot P(x)$	x^2	$x^2 \cdot P(x)$
0	0.008	0	0	0
1	0.096	0.096	1	0.096
2	0.384	0.768	4	1.536
3	0.512	1.536	9	4.608
Total		2.4		6.24

Who Is Shakespeare?

A poll of 1553 randomly selected adult Americans revealed that:

- 89% could identify Shakespeare.

- 58% could identify Napoleon.

- 47% could identify Freud.

- 92% could identify Columbus.

- 71% knew what happened in 1776.

Ronald Berman, a past director of the National Endowment for the Humanities, said "I don't worry about them (those who don't know what happened in 1776). I worry about the people who take polls." He goes on to say that the poll should ask substantive questions, such as: What is the difference between democracy and totalitarianism?

We now use results from the table to apply the general formulas from Section 4-3, as follows.

$$\mu = \Sigma x \cdot P(x) = 2.4$$
$$\sigma^2 = [\Sigma x^2 \cdot P(x)] - \mu^2$$
$$= 6.24 - 2.4^2$$
$$= 0.48$$
$$\sigma = \sqrt{0.48} = 0.69$$

The two different sets of computations produced the same results, but it should be noted that the first set of computations were much simpler than the second set. That difference becomes more dramatic as the number of trials, n, becomes larger.

The preceding computations should lead to two conclusions. First, the use of $\mu = n \cdot p$ and $\sigma^2 = n \cdot p \cdot q$ in binomial experiments will produce the same results as the more general formulas of $\mu = \Sigma x \cdot P(x)$ and $\sigma^2 = [\Sigma x^2 \cdot P(x)] - \mu^2$. Second, the latter formulas tend to be much more complicated, so if we know an experiment is binomial, we should use the simplified formulas. They will conserve time and effort and, at the same time, reduce the chance for arithmetic errors.

EXAMPLE

On a four-engine jet, the probability of any engine failing on a transatlantic flight is 0.0000709. Find the mean number of engine failures per flight for such aircraft. Also find the standard deviation. (Assume that the engines operate independently.)

Solution

With $n = 4$ and P(engine failure) $= p = 0.0000709$ we get

$$\mu = n \cdot p$$
$$= 4 \cdot 0.0000709 = 0.000284$$
$$\sigma = \sqrt{n \cdot p \cdot q}$$
$$= \sqrt{4 \cdot 0.0000709 \cdot 0.9999291}$$
$$= \sqrt{0.0002836} = 0.0168$$

E X A M P L E

A quality control manager knows that her company manufactures integrated circuit chips in daily batches of 2100 with a 5% rate of defects. Find the mean and standard deviation for the number of defects per batch.

Solution

With $n = 2100$, $P(\text{defect}) = p = 0.05$, we compute

$$\mu = n \cdot p$$
$$= 2100 \cdot 0.05 = 105$$
$$\sigma = \sqrt{n \cdot p \cdot q \cdot}$$
$$= \sqrt{2100 \cdot 0.05 \cdot 0.95}$$
$$= \sqrt{99.75} = 9.99$$

Formulas 4-5, 4-6, and 4-7 enable us to calculate the mean, variance, and standard deviation for the number of successes in a binomial experiment. We will see later that in many situations, it is highly unlikely that a score will differ from the mean by more than three standard deviations. In our last example of integrated circuit chips, it is very unlikely that the number of defects will differ from the mean of 105 by more than 30 (or about three standard deviations). Therefore, the number of defects is usually between 75 and 135. If the quality control manager gets more than 135 defects in a batch, she should suspect that something has changed and production may be going out of control. This illustrates the importance of Formulas 4-5, 4-6, and 4-7.

When n is large and p is close to 0.5, the binomial distribution tends to resemble the smooth curve that approximates the histograms in Figures 4-6 and 4-7. Note that the data tend to form a bell-shaped curve. In Chapter 5, we will use this property for solving certain applied problems.

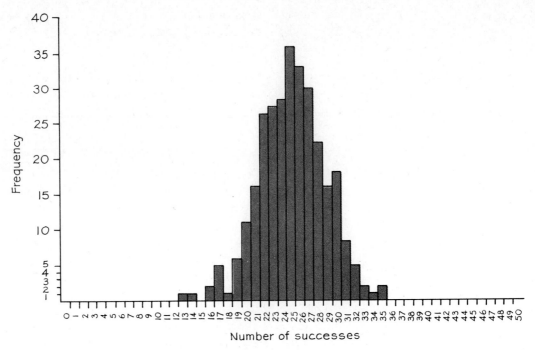

Figure 4-6   Summary of 300 real experiments. Each experiment has 50 trials ($n = 50$) with $p = 0.5$.

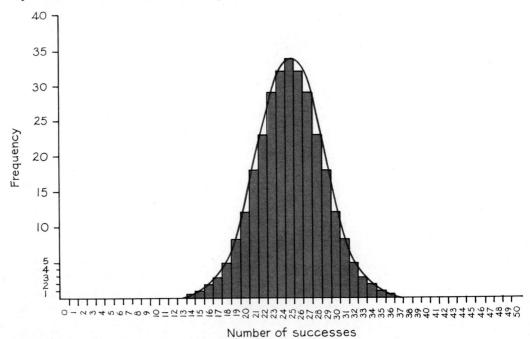

Figure 4-7   Summary of 300 ideal theoretical experiments. Each experiment has 50 trials ($n = 50$) with $p = 0.5$.

4-5 Exercises A
Mean and Standard Deviation for the Binomial Distribution

In Exercises 4-85 through 4-96, find the mean μ, variance σ^2, and standard deviation σ for the given values of n *and* p. *Assume that the binomial conditions are satisfied in each case.*

4-85 $n = 16, p = 0.5$ **4-86** $n = 36, p = 0.5$

4-87 $n = 10, p = 0.3$ **4-88** $n = 15, p = 0.2$

4-89 $n = 20, p = 0.9$ **4-90** $n = 11, p = 0.1$

4-91 $n = 104, p = 0.6$ **4-92** $n = 25, p = 0.25$

4-93 $n = 18, p = \frac{1}{3}$ **4-94** $n = 500, p = 0.85$

4-95 $n = 300, p = \frac{3}{4}$ **4-96** $n = 600, p = \frac{3}{8}$

4-97 Find the mean, variance, and standard deviation for the number of boys in families with five children. (Assume that a boy or a girl is equally likely to occur and that the sex of any child is independent of any brothers or sisters.)

4-98 A certain computer component is manufactured in lots of 64. If 1% of the components are defective, find the mean and standard deviation for the number of defects in each lot.

4-99 Of the subjects taking a certain civil service examination, 30% fail. If the test is given to groups of 20, find the mean and standard deviation for the number of failures in each group.

4-100 A certain type of seed has a 75% germination rate. If the seeds are planted in rows of 20, find the mean and standard deviation for the number of seeds that germinate in each row.

4-101 An aircraft rated for flight in inclement weather is normally equipped with two independent radios that can be used for communications. If radios of this type have a 1.5% failure rate per flight, find the mean and standard deviation for the number of failures on a flight of an aircraft equipped with two radios.

4-102 A certain pollution-measuring instrument gives accurate readings 95% of the days that it is used. Find the mean and standard deviation for the number of days in one week for which the readings are not accurate.

4-103 Of all individual tax returns, 37% include errors made by the taxpayer. If IRS examiners are assigned randomly selected returns in batches of 12, find the mean and standard deviation for the number of erroneous returns per batch.

4-104   A pathologist knows that 14.9% of all deaths are attributable to a myocardial infarction. Find the mean and standard deviation for the number of such deaths that will occur in typical communities of 5000 people.

4-105   In standard English text, the letter *e* occurs with a relative frequency of 0.130. Find the mean and standard deviation for the number of times *e* will be found on standard pages of 2600 letters.

4-106   Women comprise 43.0% of the civilian labor force of workers aged 16 years and over. Find the mean and standard deviation for the number of women in groups of 250 randomly selected members of the civilian labor force.

4-107   In planning for no smoking areas on an aircraft, it is estimated that 37% of all adults smoke. Find the mean and standard deviation for the number of smokers in groups of 240 adults.

4-108   A survey has shown that 43% of all unregistered voters prefer the Democratic party. A follow-up study is to be conducted with 1200 randomly selected unregistered voters. Find the mean and standard deviation for the numbers of Democrats in groups of 1200.

4-109   Among all military personnel, 13.8% are officers. A study of military personnel involves random selections in groups of 50. Find the mean and standard deviation for the number of officers per group.

4-110   Among Americans aged 20 years or over, 12.5% sleep at least nine hours each night. A study of dreams requires 36 volunteers who are at least 20 years old. Find the mean and standard deviation for the number of such people in groups of 36 who sleep at least nine hours each night.

4-111   A psychologist studying suicides found that 63.8% of the men and 38.0% of the women used firearms. A sample of 200 cases is randomly selected from the population of male suicides. For such groups, find the mean and standard deviation for the number of cases in which firearms are used.

4-112   Of cars originally built with catalytic converters, 4.4% have had them removed. One point of a certain highway is passed by 12,600 cars per hour. Assume that all of those cars were originally equipped with catalytic converters. What is the mean and standard deviation for the hourly numbers of cars with their catalytic converters removed?

4-5 Exercises B
Mean and Standard Deviation for the Binomial Distribution

4-113   The most profitable production arrangement for a certain transistor results in a 40% yield, meaning that 40% of the transistors produced are acceptable. The transistors are produced in groups of ten and Table 4-6

Table 4-6

x	Frequency
0	1
1	17
2	62
3	112
4	142
5	95
6	45
7	22
8	3
9	1
10	0

describes the yield for 500 groups. Compute the theoretical mean and standard deviation for the number of acceptable transistors in a group of ten. Compute the mean $\bar{x}$ and standard deviation s based on the sample summarized in the table. Compare the results. (x represents the number of acceptable transistors in a group of ten.)

In Exercises 4-114 through 4-116, consider as unusual anything that differs from the mean by more than twice the standard deviation. That is, unusual values are either less than $\mu - 2\sigma$ or greater than $\mu + 2\sigma$.

4-114   Is it unusual to get 450 girls and 550 boys in 1000 independent births?

4-115   Is it unusual to find 5 defective transistors in a sample of 20 if the defective rate is 20%?

4-116   A company manufactures an appliance, gives a warranty, and 95% of its appliances do not require repair before the warranty expires. Is it unusual for a buyer of ten such appliances to require warranty repairs on two of the items?

4-6   Distribution Shapes

We have already stated that the *distribution* of data is an extremely important characteristic and may strongly affect the methods we use or the conclusions we draw. As we consider distributions of data in this section, we have two main objectives:

1. We will identify some of the more common distributions and show how such identification often becomes critical.

2. We will introduce a method of working with uniform distributions that are continuous instead of discrete. This method will show how to obtain probabilities by finding areas, and it will be a good preparation for a very important concept to be introduced in Chapter 5.

We noted in the first chapter that one of the most fascinating aspects associated with a study of statistics is the surprising regularity and predictability of events that seem to happen by chance. In this section we consider some common collections of data, along with their corresponding histograms. Examination of a histogram representing data is extremely helpful in characterizing the way that the data are distributed. An understanding of the distribution may in turn lead to helpful and valuable inferences. Let's begin with a simple example involving the analysis of the last question on an I.Q. test. Assume that this is a multiple-choice question with five possible answers and that the responses of 100 subjects are summarized in Table 4-7. The histogram that corresponds to Table 4-7 is Figure 4-8. You do not have to be an expert statistician to recognize that Figure 4-8 depicts a distribution that is essentially even, or uniform.

Table 4-7	
Response	Frequency
a	20
b	22
c	19
d	18
e	21

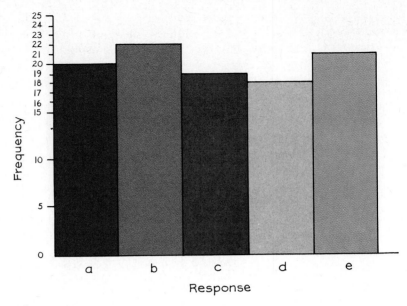

Figure 4-8

If we determine that the correct response is *b*, we might expect a higher frequency for that correct response. The nature of the distribution should tell us something about the validity or usefulness of that last test question. Because the responses are uniformly distributed, there is good reason to believe that the subjects are guessing. Perhaps there is insufficient time to consider all questions carefully and the last few responses are last-minute guesses, or perhaps this last question is so incredibly difficult that everybody makes random guesses. In any event, the distribution of responses suggests that the test can be improved by changing the last question.

Let's consider a second example involving a manufacturer of compact cars who is concerned with the comfort of very tall drivers. The manufacturer seeks information about the distribution of heights and compiles the sample data for car buyers summarized in Table 4-8.

Examination of the sampling distribution leads the manufacturer to conclude that relatively few buyers (about 3%) are more than 76 inches tall (see Figure 4-9). Since the accommodation of these towering torsos would demand expensive design changes, the car manufacturer elects to sacrifice that tall share of the market that equals 3%. Had the distribution been uniform, as in Figure 4-8, then the manufacturer would have proceeded with the necessary design changes, thereby avoiding the sacrifice of about 20% of the potential market.

The key observation at this stage is the essential difference between the distributions of Figures 4-8 and 4-9. Both distributions are very real and arise naturally in various circumstances, yet their inherent differences lead to different inferences. Figure 4-10 presents some of the

Table 4-8	
Height (inches)	Frequency
61–64	4
65–68	27
69–72	46
73–76	20
77–80	3

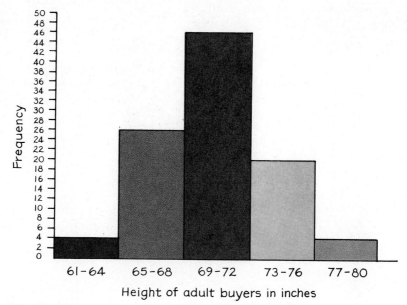

Figure 4-9

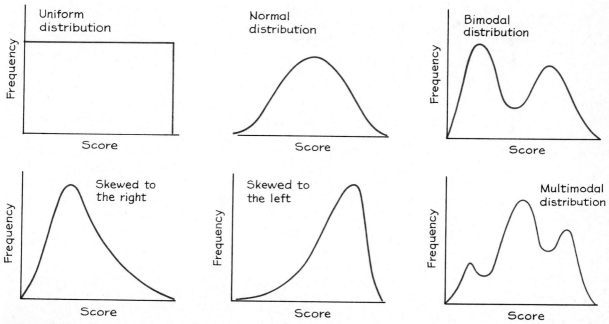

Figure 4-10

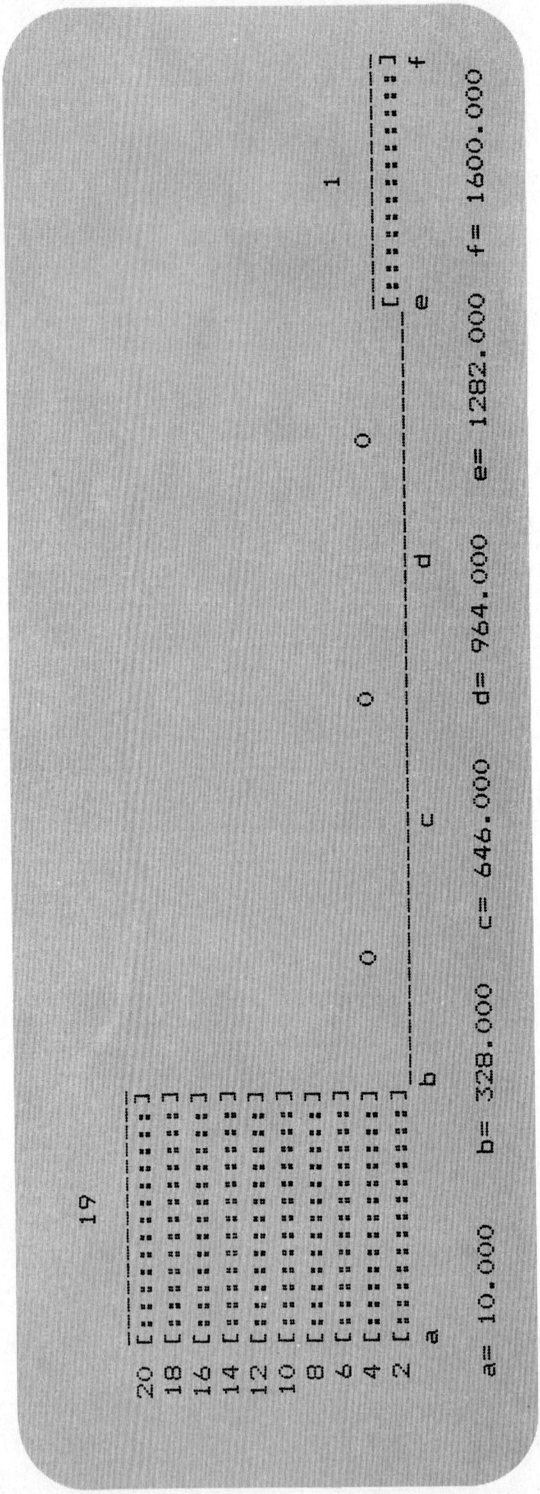

198

Figure 4-11(a)

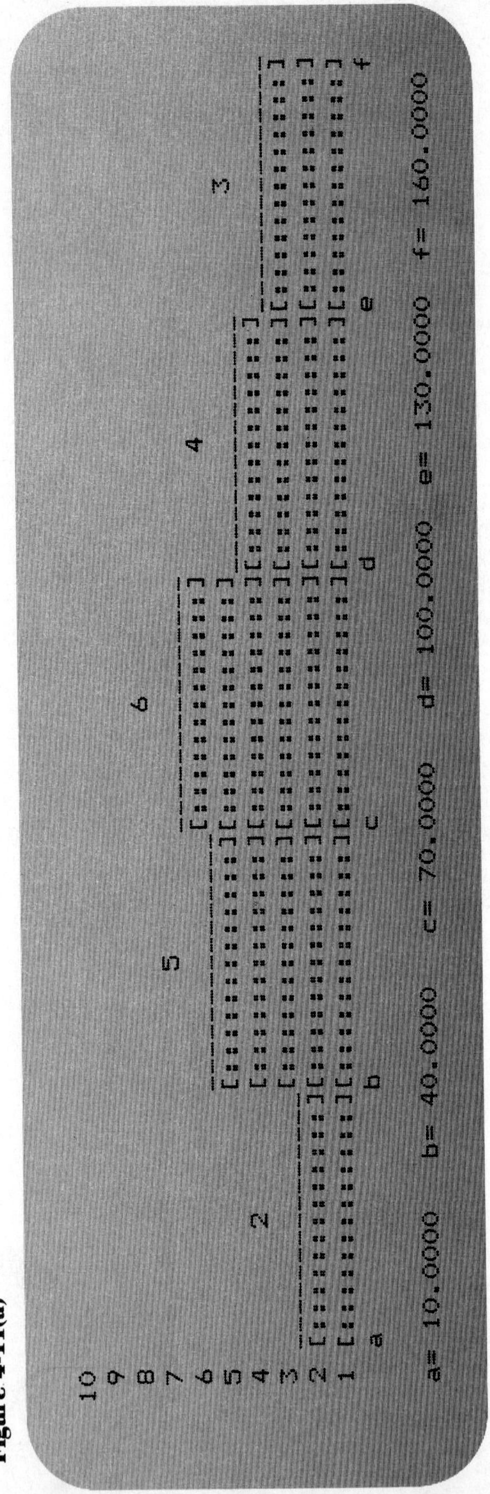

(b)

common shapes of histograms that arise in real data problems. The uneven bars of histograms have been smoothed to form continuous curves as was done in Figure 4-7.

We must be careful when using histograms to determine the nature of the distribution. In Figure 4-11(a), we show a computer printout of a histogram in which the triglyceride levels of 20 subjects were entered. The last score was incorrectly entered with an extra zero so that 1600 was used instead of 160. In Figure 4-11(b) we show the computer printout for the corrected data set. From Figure 4-11(b) we conclude that the data are normally distributed, but that was not at all apparent from Figure 4-11(a), even though only one score was incorrect. In this case, the outlier caused a severe distortion of the histogram. In other cases, outliers may be correct values of data but may continue to disguise the true nature of the distribution through histograms such as the one shown in Figure 4-11(a).

The example of Figure 4-11(a) suggests that we should avoid making important conclusions about the distribution of data without exploring and verifying the values themselves. In addition to the histogram, we might also construct a **box-and-whisker diagram** (sometimes called a **boxplot**), which reveals more information about how the data are spread out. In order to construct such a diagram, we must first determine the minimum, maximum, and median values. We must also obtain the values of the two **hinges.** After arranging the data in increasing order, we find the left hinge by finding the median of the bottom half of the data. The right hinge is the median of the top half of the data. This definition of hinges may vary with textbooks. The 20 scores depicted in Figure 4-11(a) are arranged in increasing order and listed below.

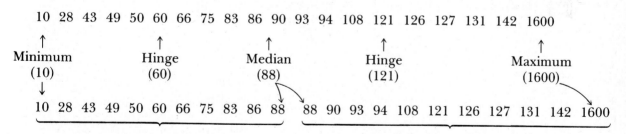

The left hinge is found by determining the median of this bottom half of the data.

The right hinge is found by determining the median of this top half of the data.

Note that the hinges are different from the quartiles. For this data set, $Q_1 = 55$ and $Q_3 = 123.5$ while the hinges are 60 and 121. To construct a box-and-whisker diagram, we now begin with a horizontal scale as in Figure 4-12(a). We "box" in the hinges as shown and we use "whiskers" to connect the minimum score to a hinge and the maximum score to a

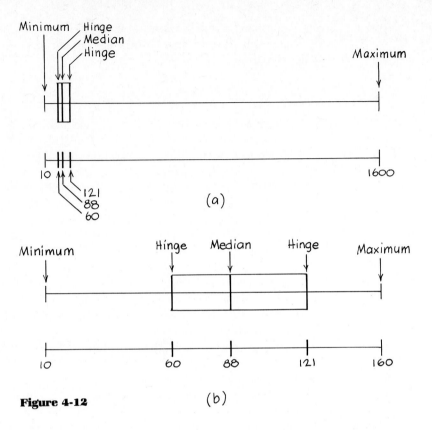

Figure 4-12

hinge. Figure 4-12(b) shows the box-and-whisker diagram for the data set after the incorrect score of 1600 has been corrected to 160.

Note that by the procedures used here, approximately one-fourth of the values should fall between the low score and the left hinge, the two middle quarters are in the boxes, and approximately one-fourth of the values are between the right hinge and the maximum. The diagram therefore shows how the data are spread out. The uneven spread shown in Figure 4-12(a) is in strong contrast to the even spread shown in Figure 4-12(b). Suspecting that triglyceride levels are normally distributed, we would expect to see a box-and-whisker diagram like the one in Figure 4-12(b), whereas the diagram in Figure 4-12(a) would raise suspicion and instigate further investigation.

Whether we use histograms or box-and-whisker plots, an understanding of the data requires some knowledge of the distribution. Armed with that knowledge and a knowledge of the key parameters (like the mean and standard deviation), we can often formulate useful inferences. The following simple example illustrates an important concept, which will subsequently be applied to more useful, more realistic, and more difficult situations.

Let's assume that the Curtiss Sugar Company packs sugar in bags

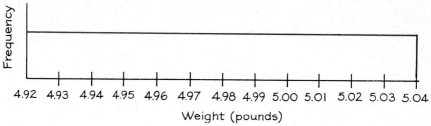

Figure 4-13

labeled 5 pounds. However, the packaging machine isn't perfect and the actual weights are *uniformly* distributed with a mean of 4.98 pounds and a range of 0.12 pound. Using only the mean, range, and type of distribution, we are able to construct the frequency polygon shown in Figure 4-13. The mean will be at the center, while the maximum is above the mean by an amount equal to one-half of the range. The minimum is below the mean by an amount equal to one-half the range. (If we had been given the values of the minimum and maximum in a uniform distribution, the mean would be one-half their sum.)

E X A M P L E

Given the uniform distribution of Figure 4-13, find the probability that a randomly selected bag of sugar will weigh less than 4.98 pounds.

Solution

Since all bags weigh between 4.92 and 5.04 pounds, we can state that for the weight X of one randomly selected bag, $P(4.92 \leq X \leq 5.04) = 1$. That is, we assign 1 to the area under the line in Figure 4-13 so that we can establish a natural and usable correspondence between area and probability. Since exactly one-half the area is associated with the stipulated weight of less than 4.98 pounds, we get a probability of $\frac{1}{2}$ as a solution.

E X A M P L E

Given the same distribution as in Figure 4-13, find the probability of randomly selecting a weight between 4.93 and 5.03 pounds.

Solution

The area associated with the interval between 4.93 and 5.03 is $\frac{10}{12}$ of the total area, so the probability we seek is $\frac{10}{12}$, or $\frac{5}{6}$.

Clusters of Disease

Periodically, much media attention is given to a cluster of cases of a disease in a given community. Typical of this was a New Jersey community's cluster of 13 leukemia cases in a 5-year period. The normal leukemia rate for a comparable community would be only 1 case in 10 years.

Research studies of such clusters can lead to valuable conclusions. For example, such a study led to the discovery that asbestos fibers can be carcinogenic. That con-clusion was reached through analysis of a cluster of cancer cases near African asbestos mines.

When we analyze diseases and deaths in an attempt to identify clusters, however, we must be careful to avoid the error of positioning a cluster boundary in a deceiving way. We can sometimes create an artificial cluster that doesn't really exist by locating the outer boundary so that it just barely includes cases of disease or death. That would be somewhat like gerrymandering, even though it might be unintentional.

EXAMPLE

If weights less than 4.96 pounds are unacceptable, find the probability that a randomly selected bag of sugar is acceptable.

Solution

Again using Figure 4-13, we see that $\frac{4}{12}$, or $\frac{1}{3}$, of the total area corresponds to unacceptable weights (less than 4.96), so the probability of selecting an acceptable weight must be $\frac{8}{12}$, or $\frac{2}{3}$.

This concept of assigning 1 to the area under a curve works well for uniform distributions, since the associated rectangles are easily partitioned into smaller units. Normal distributions similar to the one depicted in Figure 4-14 occur often in reality, but the area computations are much more difficult because of the curve involved. (Methods for normal distribution computations are pursued in Chapter 5.)

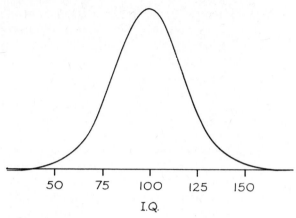

I.Q.

Figure 4-14

4-6   Exercises A
Distribution Shapes

4-117   A traffic engineer analyzes a model in which vehicle speeds are uniformly distributed between a low of 30 mph and a high of 40 mph. Find the probability that the speed of a randomly selected vehicle is
(a)   Less than 35 mph.
(b)   Greater than 32 mph.
(c)   Between 33 mph and 39 mph.

4-118   Technicians are trained with a computer simulation that randomly generates temperatures for one sector of a chemical reactor. Those temperatures are uniformly distributed between 400°C and 500°C. Find the probability that a randomly selected temperature is
(a)   Greater than 480°C.
(b)   Between 410°C and 490°C.
(c)   Between 450°C and 600°C.

4-119   A computer simulation involves the random generation of whole numbers between 5001 and 10,000, inclusive. If the distribution of these numbers is uniform, find the probability that a randomly selected number is:
(a)   Less than 3000.
(b)   Greater than 8000.
(c)   Less than 7501.
(d)   Between 6000 and 7508 (inclusive).
(e)   Greater than 8000 but less than 9000.

4-120   A set of sample data represents salaries of sales personnel employed by a vacuum cleaner manufacturer. Assuming that those salaries are uniformly distributed with a mean of $13,000 and a range of $5000, find each of the following.
(a) The maximum salary.
(b) The minimum salary.
(c) The probability of randomly selecting a salary between $9500 and $14,000.

4-121   The heights of 3-year-old pine trees in New York State follow a normal bell-shaped distribution, but an experiment begins with the formation of a uniformly distributed population having a mean of 25 inches and a range of 3 inches. For the experimental group, find each of the following.
(a) The height of the tallest tree.
(b) The height of the shortest tree.
(c) The probability of randomly selecting a tree taller than 24.5 inches.

4-122   Cholesterol levels are artificially introduced into blood samples so that the resulting mixtures are uniformly distributed with a mean of 295 and a range of 110. Find each of the following.
(a) The highest cholesterol level.
(b) The lowest cholesterol level.
(c) The probability of randomly selecting a mixed sample and getting a cholesterol level between 253 and 387, inclusive.

4-123   A machine is designed to pour 50 cubic centimeters of a drug into a bottle. Because of a flaw, the machine pours amounts that are uniformly distributed with a mean of 51.5 cubic centimeters and a range of 1.2 cubic centimeters.
(a) What is the maximum volume of the drug poured by the machine?
(b) What is the minimum volume of the drug poured by the machine?
(c) If a poured sample is randomly selected, find the probability that it is between 51.0 cubic centimeters and 52.0 cubic centimeters.

4-124   Pollution levels are uniformly distributed over a city. If the mean pollution index is 53.0 with a range of 7.5, find each of the following.
(a) The maximum pollution level.
(b) The minimum pollution level.
(c) The probability of randomly selecting a place where the pollution index is between 46.0 and 50.0.

In Exercises 4-125 through 4-132, use the given data to construct box-and-whisker diagrams. Identify the values of the minimum, maximum, median, and hinges.

4-125   Ages of selected full-time undergraduate students (in years):

17.2   17.9   18.6   18.8   19.3   19.3   20.0   20.1   23.4   26.3

4-126   Monthly rental costs of apartments in one region (in dollars):

420   420   420   430   450   450   465   470   490   600

4-127   Blood alcohol contents of drivers given breathalyzer tests:

0.02   0.04   0.08   0.08   0.09   0.10   0.10   0.12   0.13   0.19

4-128   Time intervals (in months) between adjacent births for selected families:

12.9   13.4   18.3   24.7   31.2   31.3   32.0   32.1   33.4   33.8   34.1
36.2   41.7   41.9   52.5

4-129   Blood pressure levels (in mm of mercury) for patients who have taken 25 mg of the drug captopril.

198   180   142   157   181   183   162   130   170   164
170   173   173   175   195   190   193   157   159   138

4-130   Number of words typed in a five-minute civil service test taken by twenty-five different applicants.

174   181   219   213   213   207   106   111   143   160   166   350   183
198   193   190   190   185   220   221   229   257   243   281   308

4-131   Time (in hours of operation) between failures for prototypes of computer printers:

34   22   4   9   27   36   12   40   29   32
35   25   7   9   26   36   45   43   41   2
31   31   30   14   15   18   10   27   38   21

4-132   Time (in seconds) required for long distance telephone calls:

45   12   21   180   27   33   35   38   41   42
43   65   63   60   60   152   126   121   49   50
56   59   59   59   68   73   85   98   107

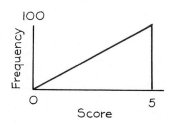

4-6 Exercises B
Distribution Shapes

4-133   Given the triangular distribution (see figure above left), find the probability of randomly selecting a score that is:
(a) Less than 3.
(b) Between 2 and 4.
(*Hint:* The area of any triangle $= \frac{1}{2} \times$ base $\times$ height.)

4-134   Given the accompanying distribution, find the probability of randomly selecting a score that is:
(a) Less than 12.
(b) Between 5 and 15.
(c) Between 5 and 19.

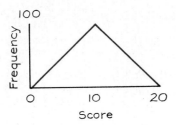

4-135   For the data set found in Exercise 4-132, find Q_1 and Q_3 and compare those quartile values to the values of the hinges.

4-136   Make sketches of histograms that correspond to the given box-and-whisker diagrams.

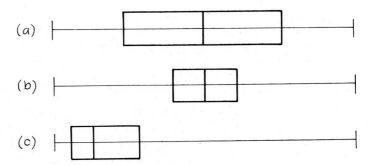

Computer Project
Probability Distributions

(a) Develop your own program to compute probabilities using the binomial probability formula. The program should take the values of n, x, and p as input, and output should be the corresponding probability of x successes in n trials of a binomial experiment. Run the program to generate all of the probabilities represented by 0+ in the section of Table A-2 for which $n = 10$.
(b) Use an existing software package that is capable of producing binomial probabilities. With $p = 0.5$, find the largest value of n that will not lead to an error. Also, find the probabilities represented by 0+ in the section of Table A-2 for which $n = 10$.

Review

The central concern of this chapter was the concept of a probability distribution. Here we dealt with **discrete** probability distributions, while successive chapters deal with continuous probability distributions.

In an experiment yielding numerical results, the **random variable** can take on those different numerical values. A **probability distribution** consists of all values of a random variable, along with their corresponding probabilities. By constructing a histogram of a probability distribution, we can see a useful correspondence between those probabilities and the areas of the rectangles in the histogram.

Of the infinite number of different probability distributions, special attention is given to the important and useful **binomial probability distribution,** which is characterized by these properties:

1. There is a fixed number of trials (denoted by n).
2. The trials must be independent.
3. Each trial must have outcomes that can be classified in *two* categories.
4. The probabilities involved must remain constant for each trial.

We saw that probabilities for the binomial distribution can be computed by using Table A-2 or by using the binomial probability formula, where n is the number of trials, x is the number of successes, p is the probability of a success, and q is the probability of a failure.

For the special case of the binomial probability distribution, the mean, variance, and standard deviations of the random variable can be easily computed by using the formulas given in the summary below.

We concluded the chapter with a section emphasizing the importance of determining the shape of a distribution and the usefulness of making a correspondence between probability and area in a histogram. We also introduced box-and-whisker diagrams as a way of obtaining information about the distribution and spread of data sets.

IMPORTANT FORMULAS

Requirements for a discrete probability distribution:

1. $\Sigma P(x) = 1$ for all possible values of x.
2. $0 \leq P(x) \leq 1$ for any particular value of x.

For *any* discrete probability distribution:

$$\text{Mean} \quad \mu = \Sigma x \cdot P(x)$$
$$\text{Variance} \quad \sigma^2 = [\Sigma x^2 \cdot P(x)] - \mu^2$$

Expected value of discrete random variable:

$$E = \Sigma x \cdot P(x)$$

(continued)

Binomial probability formula:

$$P(x) = \frac{n!}{(n-x)!\,x!} \cdot p^x \cdot q^{n-x}$$

For *binomial* probability distributions:

Mean $\mu = n \cdot p$

Variance $\sigma^2 = n \cdot p \cdot q$

Standard deviation $\sigma = \sqrt{n \cdot p \cdot q}$

Vocabulary List

Define and give an example of each term.

random variable

discrete random
  variable

continuous random
  variable

probability
  distribution

expected value

binomial experiment

binomial probability
  formula

box-and-whisker
  diagram

hinge

uniform distribution

Review Exercises

4-137   If $P(x)$ is described by the accompanying table, does $P(x)$ form a probability distribution? Explain.

x	$P(x)$
0	0.4
1	0.4
2	0.4

4-138   If $P(x)$ is described by the accompanying table, does $P(x)$ form a probability distribution? Explain.

x	$P(x)$
-1	0.35
0	0.15
1	0.40
2	0.10

4-139   Does $P(x) = (x + 1)/5$ (for $x = -1, 0, 1, 2$) determine a probability distribution? Explain.

4-140 Does $P(x) = x/12$ (for $x = 1, 2, 3, 4$) determine a probability distribution? Explain.

4-141 A probability distribution $P(x)$ is described by the accompanying table.

x	$P(x)$
5	0.6
10	
20	0.1

(a) Fill in the missing probability.
(b) Find the mean of the random variable x.
(c) Find the variance of the random variable x.

4-142 A probability distribution $P(x)$ is described by the accompanying table.
(a) Complete the table.
(b) Find the mean of the random variable x.
(c) Find the standard deviation of the random variable x.

x	$P(x)$
0	0.20
1	0.70
5	

4-143 An experiment in parapsychology involves the selection of one of the numbers 1, 2, 3, 4, and 5, and they are all equally likely.
(a) Summarize the corresponding probability distribution.
(b) Find the mean of the random variable x.
(c) Find the variance of the random variable x.

4-144 It has been found that the probabilities of 0, 1, 2, 3, and 4 wrong answers on a quiz are 0.20, 0.35, 0.30, 0.10, and 0.05, respectively.
(a) Summarize the corresponding probability distribution.
(b) Find the mean of the random variable x.
(c) Find the standard deviation of the random variable x.

4-145 Among the voters in a certain region, 40% are Democrats, and five voters are randomly selected.
(a) Find the probability that all five are Democrats.
(b) Find the probability that exactly three of the five are Democrats.
(c) Find the mean number of Democrats that would be selected in such groups of five.
(d) Find the variance of the number of Democrats that would be selected in such groups of five.
(e) Find the standard deviation for the number of Democrats that would be selected in such groups of five.

4-146 In a quality control study, it was found that 5% of all auto frames manufactured by one company had at least one defective weld.
(a) Find the probability of getting three frames with defective welds when 20 frames are randomly selected.
(b) If frames are examined in groups of 20, find the mean number of defective frames in each group.
(c) Find the standard deviation for the number of defective frames in groups of 20.

4-147 A recessive genetic trait is known to occur in 50% of all offspring born within a certain population. A study involves the random selection of offspring in groups of 16.
(a) Find the probability of selecting a group and getting exactly eight offspring with the recessive trait.
(b) Find the mean number of offspring having the recessive trait in such groups of 16.
(c) Find the variance for the number of offspring having the recessive trait in such groups of 16.

4-148 A study of college enrollments shows that 40% of all full-time undergraduates are under 20 years of age.
(a) If ten full-time undergraduates are randomly selected, find the probability that at least half of them are under 20.
(b) Find the mean number of full-time undergraduates that would be found in groups of ten.
(c) Find the standard deviation for the number of full-time undergraduates that would be found in groups of ten.

4-149 The annual divorce rate is described as follows: "There are 5.3 divorced people per 1000 population."
(a) Find the probability that among ten randomly selected people, exactly one person is divorced.
(b) Find the standard deviation for the numbers of divorced people in groups of 1000.

4-150 In a study of sleep patterns of adults, it has been found that 23% of the sleep time is spent in the REM (rapid eye movement) stage.
(a) If a sleeping adult is observed at five randomly selected times, find the probability that exactly one of the five observations will be made during a REM stage.
(b) Find the standard deviation for the number of REM stages observed in groups of five observations.

4-151 A solid-state device requires 15 electronic components, two of which cost less than 50¢ each. One component is randomly selected from each of eight such devices.

(a) Find the probability of getting at least one of the components costing less than 50¢.

(b) Find the mean number of components costing under 50¢ in such groups of eight.

(c) Find the standard deviation for the numbers of components costing under 50¢ in such groups of eight.

4-152 In a study of middle-aged adults (40 to 65 years), it was found that 7.8% suffer from hypertension. A follow-up study begins with the random selection of 20 middle-aged adults.

(a) Find the probability that exactly one-fourth of the selected subjects suffer from hypertension.

(b) Find the mean number of hypertension cases found in such groups of 20.

(c) Find the standard deviation for the numbers of hypertension cases found in groups of 20.

4-153 Temperatures (in degrees Fahrenheit) are uniformly distributed with a minimum of 40° F and a maximum of 80° F. If a temperature reading is randomly selected, find the probability that it is between 50° F and 75° F.

4-154 Refer to the same temperature data described in Exercise 4-153. If one temperature reading is randomly selected, find the probability that it will be between 45° F and 90° F.

4-155 A voltage regulator is designed to produce uniformly distributed voltage levels with a minimum of 7.00 volts and a maximum of 9.00 volts. If a reading is taken, find the probability that the voltage is below 8.55 volts.

4-156 Grade stakes used in surveying have lengths that are uniformly distributed between 85 cm and 95 cm. If one such stake is randomly selected, find the probability that its length differs from the mean by more than 2 cm.

4-157 Construct the box-and-whisker diagram for the following weights of subjects about to begin a diet program. Label the hinges, the median, and the minimum and maximum.

| 155 | 153 | 171 | 142 | 142 | 165 | 167 | 163 |

4-158 Construct the box-and-whisker diagram for the following times (in months) served by convicted felons.

| 26 | 48 | 20 | 36 | 32 | 32 | 33 | 30 | 29 | 24 |

4-159 Construct the box-and-whisker diagram for the following daily customer counts achieved by one store:

| 42 | 45 | 48 | 59 | 67 | 72 | 74 | 75 | 75 | 77 |
| 78 | 78 | 78 | 81 | 82 | 82 | 88 | 94 | 97 | 99 |

4-160   Construct the box-and-whisker diagram for the following scores achieved by subjects on a personality test.

95	95	91	51	51	103	120	112	74	68	60	59
58	53	83	105	106	106	108	54	57	56	56	96

Chapter 4
Case Study Activity

For families of four children, assume that girls and boys are equally likely. Find the probabilities of 0, 1, 2, 3, and 4 girls, respectively, and summarize the results in the form of a probability distribution table. Calculate the mean and standard deviation for the numbers of girls in families with four children. Now proceed to collect data from 100 simulated families as follows. Refer to a telephone book and record the last four digits of 100 telephone numbers. Letting the digits 0, 1, 2, 3, 4 represent girls and letting 5, 6, 7, 8, 9 represent boys, find the number of girls in each of the 100 "families." Summarize the results in a frequency table and find the mean and standard deviation. Compare the results from the observed data to the theoretical results.

5-1 Overview
We identify chapter **objectives** and describe the **normal distribution.**

5-2 The Standard Normal Distribution
We define the **standard normal distribution** and describe methods for determining probabilities by using that distribution.

5-3 Nonstandard Normal Distributions
We use the z score (or standard score) to work with normal distributions in which the mean is not zero or the standard deviation is not one.

5-4 Finding Scores When Given Probabilities
With normal distributions, we determine the values of scores that correspond to various given probabilities.

5-5 Normal as Approximation to Binomial
We can sometimes use the **normal** distribution to determine probabilities in a **binomial** experiment.

5-6 The Central Limit Theorem
The sampling distribution of sample means tends to be a normal distribution with mean μ and standard deviation $\sigma/\sqrt{n}$.

Normal Probability Distributions

A manufacturer produces shear pins, which are used to protect heavy construction machinery by snapping when the load reaches 3000 pounds. When the pin snaps, the blade disengages so that the gears and engine are not damaged. Using the principles covered in Chapters 2 and 4, a quality control analyst has determined that the actual breaking points of the shear pins are normally distributed with a mean of 2860 pounds and a standard deviation of 52 pounds. A parts buyer complains that many of the pins are defective because they break only when the load is well above 3000 pounds. In this chapter we will learn how to find the actual percentage of such defective pins so that our quality control analyst can respond to the buyer. This will be one of many applications of the principles covered in this chapter.

5-1   Overview

Chapter 4 explained the difference between discrete random variables and continuous random variables and dealt with discrete probability distributions. This and subsequent chapters consider continuous probability distributions. Specifically, this chapter involves the normal distribution, which is the most important of all continuous probability distributions because it has such a wide range of applications to real circumstances.

The normal distribution was originally developed as a result of studies dealing with errors that occur in various experiments. We now recognize that many real and natural occurrences, as well as many physical measurements, have frequency distributions that are approximately normal. Blood cholesterol levels, heights of adult women, weights of 5-year-old boys, diameters of New York McIntosh apples, and the lengths of newborn sharks are all examples of collections of values whose frequency polygons will closely resemble the normal probability distribution (see Figure 5-1).

Scores on standardized tests, such as I.Q. tests or college entrance examinations, also tend to be normally distributed. We will see in this chapter that sample means tend to be normally distributed, even if the underlying distribution is not. Also, the normal distribution is often used as a good approximation to other types of distributions—both discrete and continuous.

The smooth bell-shaped curve shown in Figure 5-1 can be algebraically described by the equation

$$y = \frac{e^{-(x-\mu)^2/2\sigma^2}}{\sigma\sqrt{2\pi}} \tag{5-1}$$

where μ represents the mean score of the entire population, σ is the standard deviation of the population, π is approximately 3.142, and e is approximately 2.718. Equation 5-1 relates the horizontal scale of x values to the vertical scale of y values. We will not use this equation in actual computations since we would need some knowledge of calculus to do so.

In reality, collections of scores may have frequency polygons that approximate the normal distribution illustrated in Figure 5-1, but we cannot expect to find perfect normal distributions that conform to the precise relationship of equation 5-1. As a realistic example, consider the 150 I.Q. scores summarized in Table 5-1 and illustrated in Figure 5-2. The histogram of Figure 5-2 closely resembles the smooth bell-shaped curve of Figure 5-1, but it does contain some imperfections. In a theoretically ideal normal distribution, the tails extend infinitely far in both directions as they get closer to the horizontal axis. But this property is usually inconsistent with the limitations of reality. For example, I.Q. scores cannot be less than zero, so no histogram of I.Q. scores can extend infinitely far to the left. Similarly, objects cannot have negative weights

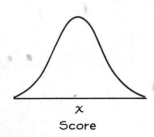

x
Score

Figure 5-1
The normal distribution

Table 5-1	
Frequency Table	
I.Q.	Frequency
55.0–64.9	1
65.0–74.9	5
75.0–84.9	15
85.0–94.9	31
95.0–104.9	39
105.0–114.9	36
115.0–124.9	15
125.0–134.9	4
135.0–144.9	3
145.0–154.9	1

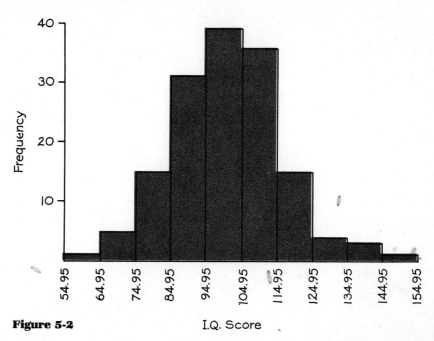

Figure 5-2   I.Q. Score

nor can they have negative distances. Limitations such as these require that frequency distributions of real data can only approximate a normal distribution.

This chapter presents the standard methods used to work with normally distributed scores, and it includes applications. In addition to the importance of the normal distribution itself, the methods are important in establishing basic patterns and concepts that will apply to other continuous probability distributions.

5-2   The Standard Normal Distribution

There are actually many different normal probability distributions, each dependent on only two parameters: the population mean μ and the population standard deviation σ. Figure 5-3 shows two different normal distributions of I.Q. scores where the difference is due to a change in the standard deviation. A change in the value of the population mean μ would cause the curve to be moved to the right or left. Among the infinite possibilities, one particular normal distribution is of special interest.

DEFINITION

The **standard normal distribution** is a normal probability distribution that has a mean of 0 and a standard deviation of 1.

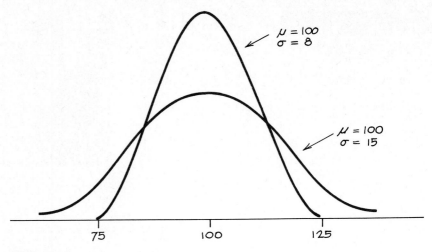

Figure 5-3

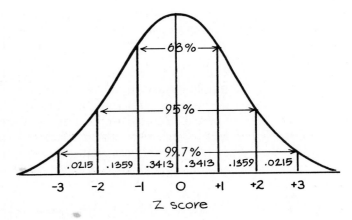

Figure 5-4   Standard normal distribution

The choices of $\mu = 0$ and $\sigma = 1$ enable us to greatly simplify the general algebraic equation (5-1) of normal distributions. Working with this simplified form, mathematicians are able to perform various analyses and computations. Figure 5-4 shows a graph of the standard normal distribution with some of the computed results. For example, the area under the curve and bounded by scores of 0 and 1 is 0.3413. The area under the curve and bounded by scores of -1 and -2 is 0.1359. The sum of the six known areas in Figure 5-4 is 0.9974, but if we include the small areas in the two tails we get a total of 1. (In any probability distribution, the total area under the curve is 1.) Nearly all (99.7%) of the values lie within three standard deviations from the mean.

Figure 5-4 illustrates only six probability values, but a more complete

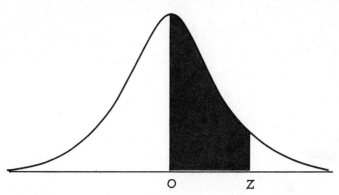

Figure 5-5   The standard normal distribution. The area of the shaded region bounded by the mean of zero and the positive number z can be found in Table A-3.

table has been compiled to provide more precise data. Table A-3 (in Appendix A) gives the probability corresponding to the area under the curve bounded on the left by a vertical line above the mean of zero and bounded on the right by a vertical line above any specific positive score denoted by z (see Figure 5-5). Note that when you use Table A-3, the hundredths part of the z score is found across the top row. To find the probability associated with a score between 0 and 1.23, for example, begin with the z score of 1.23 by locating 1.2 in the left column. Then find the value in the adjoining row of probabilities that is directly below 0.03. There is a probability of 0.3907 of randomly selecting a score between 0 and 1.23. It is essential to remember that this table is designed only for the standard normal distribution, which has a mean of 0 and a standard deviation of 1. Nonstandard cases will be considered in the next section.

Since normal distributions originally resulted from studies of experimental errors, the following examples that deal with errors in measurements should be helpful.

EXAMPLE

A manufacturer of scientific instruments produces thermometers that are supposed to give readings of 0°C at the freezing point of water. Tests on a large sample of these instruments reveal that some readings are too low (denoted by negative numbers) and that some readings are too high (denoted by positive numbers). Assume that the mean reading is 0° while the standard deviation of the readings is 1.00°C. Also assume that the frequency distribution of errors closely resembles the normal distribution. If one thermometer is randomly selected, find the probability that, at the freezing level of water, the reading is between 0° and +1.58°.

Solution

We are dealing with a standard normal distribution and we are looking for the area of the shaded region in Figure 5-5 with z = 1.58. We find from Table A-3 that the shaded area is 0.4429. The probability of randomly selecting a thermometer with an error between 0° and +1.58° is therefore 0.4429.

The solutions to this and the following examples are contingent on the values listed in Table A-3. But these values did not appear spontaneously. They were arrived at through calculations that relate directly to equation 5-1. Table A-3 therefore serves as a convenient means of circumventing difficult computations involving that equation.

EXAMPLE

With the thermometers from the preceding example, find the probability of randomly selecting one thermometer that reads (at the freezing point of water) between 0° and −2.43°.

Solution

We are looking for the region shaded in Figure 5-6(a), but Table A-3 is designed to apply only to regions to the right of the mean (zero) as in Figure 5-5. However, by observing that the normal probability distribution possesses symmetry about zero, we see that the shaded regions in parts (a) and (b) of Figure 5-6 have the same area. Referring to Table A-3 we can easily determine that the shaded area of Figure 5-6(b) is 0.4925, so the shaded area of Figure 5-6(a) must also be 0.4925. That is, the probability of randomly selecting a thermometer with an error between 0° and −2.43° is 0.4925.

Figure 5-6

(a)

(b)

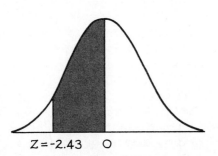

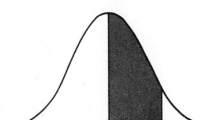

Z = −2.43   O

O   Z = 2.43

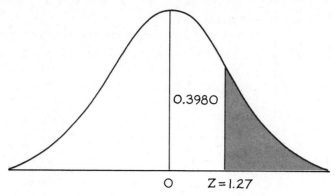

Figure 5-7

We incorporate an obvious but useful observation in the following example, but first go back for a minute to Figure 5-5. A vertical line directly above the mean of zero divides the area under the curve into two equal parts, each containing an area of 0.5. Since we are dealing with a probability distribution, the total area under the curve must be 1.

EXAMPLE

With these same thermometers, we again make a random selection. Find the probability that the chosen thermometer reads (at the freezing point of water) greater than +1.27°.

Solution

We are again dealing with normally distributed values having a mean of 0° and a standard deviation of 1°. The probability of selecting a thermometer which reads above +1.27° corresponds to the shaded area of Figure 5-7. Table A-3 cannot be used to find that area directly, but we can use the table to find the adjacent area of 0.3980. We can now reason that since the total area to the right of zero is 0.5, the shaded area is 0.5 − 0.3980, or 0.1020. Thus there is a 0.1020 probability of randomly selecting one of the thermometers with a reading greater than +1.27°.

We are able to determine the area of the shaded region in Figure 5-7 by an *indirect* application of Table A-3. The following example illustrates yet another indirect use.

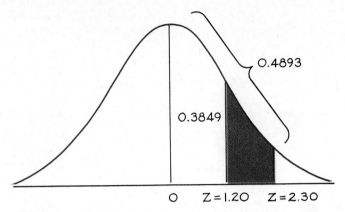

Figure 5-8

EXAMPLE

Back to the same thermometers: Assume that one thermometer is randomly selected and find the probability that it reads (at the freezing point of water) between 1.20° and 2.30°.

Solution

The probability of selecting a thermometer that reads between +1.20° and +2.30° corresponds to the shaded area of Figure 5-8. However, Table A-3 is designed to provide only for regions bounded on the left by the vertical line above zero. We can use the table to find the areas of 0.3849 and 0.4893 as shown in this figure. If we denote the area of the shaded region by A, we can see from the figure that

$$0.3849 + A = 0.4893$$

so that

$$A = 0.4893 - 0.3849$$
$$= 0.1044$$

The probability we seek is therefore 0.1044.

The previous examples were contrived so that the mean of 0 and the standard deviation of 1 coincided exactly with the values of the standard normal distribution described in Table A-3. In reality, it would be unusual to find such a nice relationship, since typical normal distributions involve means different from 0 and standard deviations different from 1.

These nonstandard normal distributions introduce another problem. What table of probabilities can be used since Table A-3 is designed around a mean of 0 and a standard deviation of 1? For example, I.Q. scores are normally distributed with a mean of 100 and a standard deviation of 15. Scores in this range are far beyond the scope of Table A-3. Section 5-3 examines these nonstandard normal distributions and the methods used in dealing with them.

5-2 Exercises A
The Standard Normal Distribution

In Exercises 5-1 through 5-36, assume that the readings on the thermometers are normally distributed with a mean of 0° and a standard deviation of 1.00°. A thermometer is randomly selected and tested. Find the probability that the reading in degrees is:

5-1 Between 0 and 0.50.   **5-2** Between 0 and 2.00.

5-3 Between 0 and 1.40.   **5-4** Between 0 and 1.64.

5-5 Between −1.00 and 0.   **5-6** Between −1.82 and 0.

5-7 Between 0 and −2.04.   **5-8** Between 0 and −1.72.

5-9 Greater than 1.00.   **5-10** Greater than 0.25.

5-11 Greater than 2.19.   **5-12** Greater than 1.08.

5-13 Less than −1.00.   **5-14** Less than −0.30.

5-15 Less than −0.66.   **5-16** Less than −1.37.

5-17 Greater than −1.00.   **5-18** Greater than −0.55.

5-19 Less than 1.17.   **5-20** Less than 0.85.

5-21 Between −1.00 and 2.00.   **5-22** Between −0.25 and 0.75.

5-23 Between −2.00 and 1.50.   **5-24** Between −1.29 and 1.37.

5-25 Between 1.00 and 2.00.   **5-26** Between 0.50 and 1.65.

5-27 Between 1.20 and 1.33.   **5-28** Between 1.05 and 2.05.

5-29 Between −0.60 and −0.20.   **5-30** Between −2.00 and −1.50.

5-31 Between −0.25 and −1.35.   **5-32** Between −1.07 and −2.11.

5-33 Greater than 0.   **5-34** Less than 0.

5-35 Either less than −0.50 or greater than 1.50.

5-36 Either less than −1.96 or greater than 1.96.

5-2 Exercises B
The Standard Normal Distribution

5-37   With $\mu = 0$ and $\sigma = 1$ in a normally distributed population, find the percentage of values between $\mu - 3\sigma$ and $\mu + 3\sigma$.

5-38   Assume that we have the same normally distributed thermometer readings with a mean of $0°$ and a standard deviation of $1.00°$. If 5% of the thermometers are rejected because they read too high and another 5% are rejected because they read too low, what is the maximum error that will not lead to rejection?

5-39   Assume that we have the same normally distributed thermometer readings with a mean of $0°$ and a standard deviation of $1.00°$. If a buyer establishes a maximum acceptable error and rejects 2% of these thermometers, what are the highest and lowest acceptable readings?

5-40   (a) In a normally distributed collection of scores with mean 0 and standard deviation 1, find the area under the curve bounded by the lines above $z = 1.40$ and $z = 1.80$.

    (b) Suppose we let $z = 1.80 - 1.40 = 0.40$. We then find the corresponding area from Table A-3. Does this result agree with the answer to part (a)? Why or why not? Explain.

5-41   In equation 5-1, if we let $\mu = 0$ and $\sigma = 1$ we get

$$y = \frac{e^{-x^2/2}}{\sqrt{2\pi}}$$

which can be approximated by

$$y = \frac{2.7^{-x^2/2}}{2.5}$$

Graph the last equation after finding the y coordinates that correspond to the following x coordinates: -4, -3, -2, -1, 0, 1, 2, 3, and 4. (A calculator capable of dealing with exponents will be helpful.) Attempt to determine the approximate area bounded by the curve, the x-axis, the vertical line passing through 0 on the x-axis, and the vertical line passing through 1 on the x-axis. Compare this result to Table A-3.

5-42   Suppose we begin with a population of normally distributed values with a mean of 0 and a standard deviation of 1.00, and we add 5 to every score. If a value is randomly selected from this modified population, find the probability that it is between 4.00 and 7.00.

5-43   Suppose we begin with a population of normally distributed values with a mean of 0 and a standard deviation of 1.00, and we add the positive number k to every score. If a value is randomly selected from this population, find the probability that it is greater than $k + 1$.

5-44 A population has a mean of 0 and a standard deviation of 1.00 but is uniformly distributed instead of being normally distributed. Such a distribution has minimum and maximum values of $-\sqrt{3}$ and $\sqrt{3}$, respectively. Find the probability of randomly selecting a value between 0 and 1.00 with this uniform distribution, and compare it to the area between 0 and 1.00 for a normal distribution with a mean of 0 and a standard deviation of 1.00.

5-3 Nonstandard Normal Distributions

In Section 5-2 we considered only the standard normal distribution, but in this section we will extend the same basic concepts to include nonstandard normal distributions. This inclusion will greatly expand the variety of practical applications we can make since, in reality, most normally distributed populations will have either a nonzero mean and/or a standard deviation different from 1.

We continue to use Table A-3, but we require a way of standardizing these nonstandard cases. This is done by letting

$$z = \frac{x - \mu}{\sigma} \tag{5-2}$$

so that z is the number of standard deviations that a particular score x is away from the mean. We call z the z *score* or *standard score* and it is used in Table A-3 (see Figure 5-9).

Suppose, for example, that we are considering a normally distributed collection of I.Q. scores known to have a mean of 100 and a standard deviation of 15. If we seek the probability of randomly selecting one I.Q. score that is between 100 and 130, we are concerned with the area shown in Figure 5-10. The difference between 130 and the mean of 100 is 30 I.Q. points, or exactly two standard deviations. The shaded area in Figure 5-10 will therefore correspond to the shaded area of Figure 5-5 where $z = 2$. We get $z = 2$ either by reasoning that 130 is two standard deviations above the mean of 100 or by computing

$$z = \frac{x - \mu}{\sigma} = \frac{130 - 100}{15} = \frac{30}{15} = 2$$

Figure 5-9

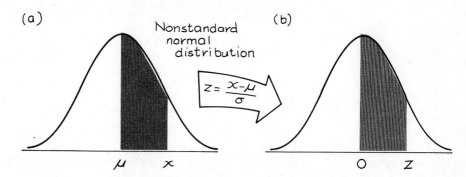

(a)

Nonstandard normal distribution

$z = \frac{x - \mu}{\sigma}$

(b)

μ x

O z

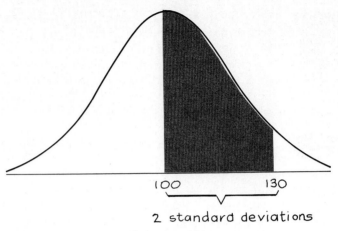

Figure 5-10   I.Q. score

With $z = 2$, Table A-3 indicates that the shaded region we seek has an area of 0.4772 so that the probability of randomly selecting an I.Q. score between 100 and 130 is 0.4772. Thus Table A-3 can be indirectly applied to any normal probability distribution if we use equation 5-2 as the algebraic way of recognizing that the z score is actually the number of standard deviations that x is away from the mean. The following examples illustrate that observation. Before considering the next example we should note that I.Q. and many other types of test scores are discrete whole numbers, while the normal distribution is continuous. We ignore that conflict in this and the following section since the results are minimally affected, but we will introduce a "continuity correction factor" in Section 5-5 when it really becomes necessary. However, we might note for the present that finding the probability of getting an I.Q. score of *exactly* 130 would require that the discrete value of 130 be represented by the continuous interval from 129.5 to 130.5. From that brief example we can see how the continuity correction factor can be used to deal with discrete data.

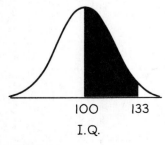

Figure 5-11

EXAMPLE

If I.Q. scores are normally distributed with a mean of 100 and a standard deviation of 15, find the probability of randomly selecting a subject with an I.Q. between 100 and 133.

Solution

Referring to Figure 5-11, we seek the probability associated with the shaded region. In order to use Table A-3, the nonstandard data must be standardized by applying equation 5-2.

$$z = \frac{x - \mu}{\sigma} = \frac{133 - 100}{15} = \frac{33}{15} = 2.20$$

The score of 133 therefore differs from the mean of 100 by 2.20 standard deviations. Corresponding to a z score of 2.20, Table A-3 indicates a probability of 0.4861. There is therefore a probability of 0.4861 of randomly selecting a subject having an I.Q. between 100 and 133.

Note that when Table A-3 is used in conjunction with equation 5-2, the nonstandard population mean corresponds to the standard mean of zero. As a result, probabilities extracted directly from Table A-3 must represent regions whose left boundary is the line denoting the mean.

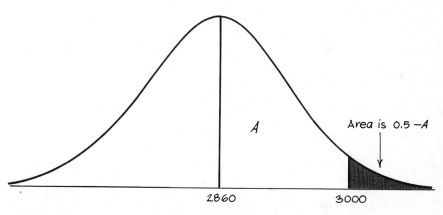

Figure 5-12   Breaking point (pounds)

Bullhead City Gets Hotter

A fireman in Bullhead City, Arizona, was given the responsibility of reporting daily weather statistics to the National Weather Service. One day, a Weather Service representative demanded that the thermometer be moved from the firehouse lawn to a more natural setting. The fireman selected a dry and dusty area 100 yards away from the cooler lawn. This move resulted in readings about 5 degrees higher, which often made Bullhead City the hottest place in the United States. As Bullhead City gained a newly found prominence in many television weather reports, some residents denounced the notoriety as a handicap to business, while other residents felt that business was helped. Under more standardized conditions, measuring instruments—such as thermometers and scales—tend to produce errors that are normally distributed.

EXAMPLE

In the beginning of this chapter, we described the production of shear pins with normally distributed breaking points. The mean is 2860 pounds, the standard deviation is 52 pounds, and a pin is considered defective if its breaking point is above 3000 pounds. A buyer complains that many of the pins are defective. Find the percentage of defective pins.

Solution

In Figure 5-12, the shaded region corresponds to the area representing shear pins with a breaking point of 3000 pounds or more. We cannot find the area of that shaded region directly, but we can use equation 5-2 to find the adjacent area immediately to the left of the region we seek.

$$z = \frac{x - \mu}{\sigma} = \frac{3000 - 2860}{52}$$

$$= \frac{140}{52} = 2.69$$

We now use Table A-3 to get an area of 0.4964 for the region bounded by 2860 and 3000. Since the total area to the right of 2860 is 0.5, the shaded region must be $0.5 - 0.4964$ or 0.0036. The proportion of defective shear pins is 0.0036, which is equivalent to 0.36%. Since only 0.36% of the pins are defective, the buyer is wrong in claiming that many of the pins are defective.

E X A M P L E

A college entrance examination is designed so that the scores are normally distributed with a mean of 500 and a standard deviation of 100. Find the percentage of subjects with scores between 550 and 675.

Solution

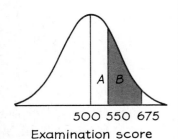

A   B

500 550 675

Examination score

Figure 5-13

Figure 5-13 depicts the relevant region (B), but we cannot obtain the desired probability directly since the left boundary is not the mean. Instead we use equation 5-2 to find the total probability for regions A and B combined and then proceed to subtract the probability of region A. To get the probability for regions A and B combined, we let $\mu = 500$, $\sigma = 100$, and $x = 675$, which gives

$$z = \frac{x - \mu}{\sigma} = \frac{675 - 500}{100} = \frac{175}{100} = 1.75$$

From Table A-3 with $z = 1.75$ we get a probability of 0.4599. To get the probability representing region A, we let $\mu = 500$, $\sigma = 100$, and $x = 550$, so

$$z = \frac{x - \mu}{\sigma} = \frac{550 - 500}{100} = \frac{50}{100} = 0.50$$

From Table A-3 with $z = 0.50$ we get a probability of 0.1915.

$$P(\text{regions } A \text{ and } B \text{ combined}) - P(\text{region } A)$$
$$= 0.4599 - 0.1915$$
$$= 0.2684$$

Since there is a probability of 0.2684 that a randomly selected subject will score between 550 and 675, we conclude that 26.84% of the subjects will score in that range.

In this section we extended the concept of Section 5-2 to include more realistic nonstandard normal probability distributions. We noted that the formula $z = (x - \mu)/\sigma$ algebraically represents the number of standard deviations that a particular score x is away from the mean. However, all the examples we have considered so far are of the same general type: a probability (or percentage) determined by using the normal distribution (described in Table A-3) when given the values of the mean, standard deviation, and relevant score(s). In some practical cases, the probability (or percentage) is known and we must determine the relevant score(s). Problems of this type are considered in the following section.

5-3 Exercises A
Nonstandard Normal Distributions

5-45 In a study of hypertense patients, it was found that systolic blood pressure readings (in mm of mercury) are normally distributed with a mean of 174.0 and a standard deviation of 21.1. If a hypertense patient is randomly selected, find the probability that their reading is between 174.0 and 185.0.

5-46 In studying ocean conditions, the Bureau of Fisheries found that for one location, the August water temperatures (in degrees Fahrenheit) were normally distributed with a mean of 83.6 and a standard deviation of 2.4. For a randomly selected time in August, find the probability that the water temperature is between 83.6 and 87.0.

5-47 A librarian records the numbers of books checked out for various days and finds that they are normally distributed with a mean of 230.0 and a standard deviation of 65.5. For a randomly selected day, find the probability that the number of books checked out is between 200.0 and 230.0.

5-48 A company manager does a study of the lengths of time clients are kept on hold when calling for information. Those times are found to be normally distributed with a mean of 72.0 seconds and a standard deviation of 15.0 seconds. Find the probability that a call will be kept on hold between 60.0 seconds and 72.0 seconds.

5-49 In a certain country, heights of adult males are normally distributed with a mean of 70.2 inches and a standard deviation of 4.2 inches. If cars are designed to accommodate a maximum height of 6 feet 4 inches, what percentage of adult males will find these cars unsuitable?

5-50 Scores on a college entrance examination are normally distributed with a mean of 500 and a standard deviation of 100. One college gives priority acceptance to subjects scoring above 650. What percentage of subjects are eligible for priority acceptance?

5-51   In tests conducted on jet pilots, it has been found that blackout thresholds are normally distributed with a mean of 4.7 G and a standard deviation of 0.8 G. Find the probability of randomly selecting a jet pilot with a blackout threshold that is less than 3.5 G.

5-52   A manufacturer of bulbs for movie projectors advertises a life of 50 hours. A study of these bulbs indicates that their lives are normally distributed with a mean of 61 hours and a standard deviation of 6.3 hours. What is the percentage of bulbs that fail to last as long as the manufacturer claims?

5-53   A standard I.Q. test produces normally distributed results with a mean of 100 and a standard deviation of 15. If an average I.Q. is defined to be any I.Q. between 90 and 109, find the probability of randomly selecting an I.Q. that is average.

5-54   Assume that weights of newborn children are normally distributed with a mean of 116 ounces and a standard deviation of 12 ounces. Find the probability of randomly selecting a newborn child whose weight is between:
(a)   108 ounces and 124 ounces.
(b)   110 ounces and 122 ounces.

5-55   Certain trees grow in such a way that their maximum heights are normally distributed with a mean of 12.0 meters and a standard deviation of 2.0 meters. Find the percentage of trees having maximum heights:
(a)   Between 10.0 meters and 15.0 meters.
(b)   Above 13.0 meters.

5-56   A machine fills sugar boxes in such a way that the weights of the contents are normally distributed with a mean of 2260 grams and a standard deviation of 20 grams. If a box is randomly selected, find the probability that the weight of the contents is between:
(a)   2250 grams and 2275 grams.
(b)   2225 grams and 2300 grams.

5-57   A test for reaction times produces normally distributed results with a mean of 0.78 second and a standard deviation of 0.06 second. Find the percentage of subjects with reaction times between 0.80 second and 0.85 second.

5-58   A scientist repeats an experiment and records the times required for a certain chemical reaction. The times are normally distributed with a mean of 44.35 seconds and a standard deviation of 0.30 second. Find the percentage of times between 44.00 seconds and 44.30 seconds.

5-59   In one college, it has been found that first semester students have grade-point averages that are normally distributed with a mean of 2.46 and a standard deviation of 0.65. What percentage of students will have grade-point averages between 1.00 and 2.00?

5-60 A sociologist must rely on mailed questionnaires for required information. He sends questionnaires in batches of 1000 and finds that the mean number of responses per batch is 168 while the standard deviation is 26.0. Also, the numbers of responses per batch are normally distributed. For a typical batch, find the probability that the number of responses is between 175 and 200.

5-61 A traffic study conducted at one point on an interstate highway shows that vehicle speeds (in mph) are normally distributed with a mean of 61.3 and a standard deviation of 3.3. If a vehicle is randomly checked, what is the probability that its speed is between 55.0 and 60.0?

5-62 A consumer product testing team analyzes the energy consumed by color television sets and finds that the consumption levels (in kwh) are normally distributed with a mean of 320 and a standard deviation of 7.5. For a randomly selected set, find the probability that the level is between 325 and 335.

5-63 A particular x-ray machine gives radiation dosages (in milliroentgens) that are normally distributed with a mean of 4.13 and a standard deviation of 1.27. For a randomly selected x-ray procedure, find the probability that the radiation dosage is between 5.00 and 6.00.

5-64 The times required for nurses to learn a particular task are found to be normally distributed with a mean of 18.4 minutes and a standard deviation of 2.7 minutes. What percentage of nurses will take 15 minutes to 18 minutes to learn this task?

5-65 A manufacturer contracts to supply ball bearings with diameters between 24.60 millimeters and 25.40 millimeters. Product analysis indicates that the ball bearings manufactured have diameters that are normally distributed with a mean of 25.10 millimeters and a standard deviation of 0.20 millimeter. What percentage of ball bearings fail to satisfy the contract specifications?

5-66 Solve Exercise 5-65 after changing the standard deviation to 0.15. The standard deviation can be decreased by improving the precision of the machinery through parts replacements, lubrication, and other maintenance methods.

5-67 In a study of the coliform contamination in streams, an environmentalist finds that one region has normally distributed coliform levels (number of cells per 100 ml) with a mean of 122 and a standard deviation of 14. For a randomly selected sample, find the probability that the coliform level differs from the mean by more than 25.

5-68 A sociologist studies the rate of drug abuse among youths (12 to 17 years of age) and obtains a number of "blocks" with 100 randomly selected youths in each block. The mean number of hallucinogen users per block is 4.36 and the standard deviation is 0.84. Find the percentage of blocks in which the number of hallucinogen users is below 4 or above 5.

5-3 Exercises B
Nonstandard Normal Distributions

5-69   Assume that the following scores are representative of a normally distributed population.
(a) Find the mean $\bar{x}$ of this sample.
(b) Find the standard deviation s of this sample.
(c) Find the percentage of these sample scores that are between 51 and 54, inclusive.
(d) Find the percentage of *population* scores between 51 and 54. Use the sample values of $\bar{x}$ and s as estimates of μ and σ.

50	51	52	51	47	57
51	50	50	51	50	47
51	48	48	50	49	52
49	50	53	47	53	52
51	48	51	54	51	50

5-70   The accompanying frequency table summarizes the age distribution for a number of students randomly selected from the population of students at a large university. Although the data do not appear to be normally distributed, assume that they are and find $\bar{x}$ and s. Then use those statistics as estimates of μ and σ in order to find the proportion of students over 21 years of age. How does the result compare to the sample statistics summarized in the table?

Age	f
15–19	220
20–22	173
23–24	55
25–29	98
30–34	62
35–44	85
45–59	36
60–65	4

5-71   A population has a mean of 100 and a standard deviation of 15 but is uniformly distributed. Such a distribution has a minimum of 74 and a maximum of 126. Find the probability of randomly selecting a value between 80 and 110 with this uniform distribution, and compare it to the area between 80 and 110 for a normal distribution with the same mean of 100 and standard deviation of 15.

5-72   Suppose we begin with a population of normally distributed values with mean μ and standard deviation σ, and we proceed to add k to every score. If a value is randomly selected from this modified population, find the probability that it is between $\mu + k$ and $\mu + k + \sigma/2$.

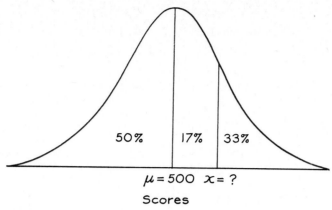

$\mu = 500 \quad x = ?$

Scores

Figure 5-14

5-4 Finding Scores When Given Probabilities

All the examples and problems in Sections 5-2 and 5-3 involved the determination of a probability or percentage based on the normal distribution data of Table A-3, a given mean, standard deviation, and relevant score(s). In this section we consider the same types of circumstances, but we will alter the known data so the computational procedure will change. These techniques will closely parallel some of the important procedures that will be introduced later in the book.

Let's begin with a practical problem. We have an intelligence test that produces normally distributed scores with a mean of 500 and a standard deviation of 100. Assuming that we want to identify the top 33% so that we can develop a specialized experimental learning program, what specific score serves as a cutoff that separates the top 33% from the lower 67%? Figure 5-14 depicts the relevant normal distribution.

We can find the z score that corresponds to the x value we seek after first noting that the region containing 17% corresponds to a probability of 0.1700. Looking at Table A-3 we can see that a probability of 0.1700 corresponds to a z score of 0.44. This means that the desired value of x is 0.44 standard deviation away from the mean. Since the standard deviation is given as 100, we can conclude that 0.44 standard deviation is equivalent to 44. The score x is above the mean and 44 units away. Since the mean is given as 500, the score x must be 500 + 44, or 544. That is, the test score of 544 separates the top 33% from the lower 67%. Thus, 544 is the 67th percentile, or P_{67}.

We could have achieved the same results by noting that

$$z = \frac{x - \mu}{\sigma} \text{ becomes } 0.44 = \frac{x - 500}{100}$$

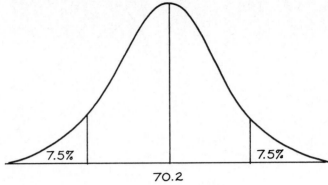

Figure 5-15   Height (inches)

when we substitute the given values for the mean μ, the standard deviation σ, and the z score corresponding to a probability of 0.1700. We can solve this last equation for x by multiplying both sides by 100 and then adding 500 to both sides. We get $x = 544$.

EXAMPLE

A clothing manufacturer finds it unprofitable to make clothes for very tall or very short adult males. The executives decide to discontinue production of goods for the tallest 7.5% and the shortest 7.5% of the adult male population. Find the minimum and maximum heights they will continue to serve. Assume that heights of adult males are normally distributed with a mean of 70.2 inches and a standard deviation of 4.2 inches (see Figure 5-15).

Solution

We note that the two outer regions total 15%, so the two equal inner regions must comprise the remaining 85%. This implies that each of the two inner regions must represent 42.5%, which is equivalent to a probability of 0.425. We can use Table A-3 to find the z score that yields a probability of 0.425. With $z = 1.44$, the corresponding probability of 0.4251 is close enough so we conclude that the upper and lower cutoff scores will correspond to $z = 1.44$ and $z = -1.44$. With $\mu = 70.2$ and $\sigma = 4.2$, we can use both values of z in $z = (x - \mu)/\sigma$ to get

$$1.44 = \frac{x - 70.2}{4.2} \quad \text{and} \quad -1.44 = \frac{x - 70.2}{4.2}$$

Solving both of these equations results in the values of 76.248 and 64.152. That is, the company will make clothing only for adult males between 64.152 inches and 76.248 inches in height.

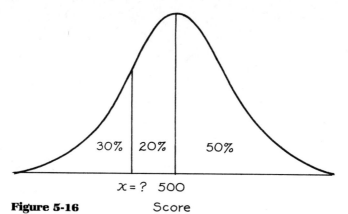

30% 20% 50%

$x = ?$ 500

Figure 5-16   Score

Are Geniuses Peculiar?

A 35-year study of children with extremely high I.Q.'s showed, through statistical analysis, that a disproportionately large number of these gifted children came from families of high social class, while relatively few came from lower social and economic classes. Also, the gifted children in this sample achieved greater than average success in later years. This study further showed that the appearances and personalities of geniuses did not tend to be peculiar. In fact, these people exhibited a lower incidence of maladjustment and unusual appearance than those having average I.Q. scores.

In the preceding example we can see that the thoughtless use of equation 5-2 will produce the maximum height of 76.248 inches, but determination of the minimum height requires an adjustment that comes only with an understanding of the whole situation. The moral is clear: Don't plug in numbers blindly. Instead, develop an understanding of the underlying meaning. Always draw a graph of the normal distribution with the relevant labels and apply common sense to guarantee that the results are reasonable. The graph and the common-sense check help reduce the incidence of errors.

The following example shows how blind application of equation 5-2 leads to serious errors in the case of a negative z score.

EXAMPLE

As one of its admissions criteria, a college requires an entrance examination score that is among the top 70% of all scores. Assuming a normal distribution with $\mu = 500$ and $\sigma = 100$, find the minimum acceptable score.

Solution

From Figure 5-16 we see that the area bounded by the unknown relevant score x and the mean comprises 20% of the total area. If that 20% region were to the right of the mean, we could use Table A-3 directly with no difficulty. Let's proceed by pretending that this is the case. Twenty percent, or 0.20, is approximated by a z score of 0.52 (see Table A-3). We can now reason as follows. The value of x differs from the mean of 500 by 0.52 standard deviation. With $\sigma = 100$, 0.52 standard deviation becomes $0.52 \cdot 100 = 52$, so x must be 52 *below* 500. That is, x must be 448. We can see from Figure 5-16 that x is less than 500, so the difference of 52 must be subtracted from 500 instead of added to 500. If this solution were attempted through a superficial application of equation 5-2, we would get the erroneous answer of 552. But 552 is unreasonable because the given data require a result less than 500. This problem could have been quickly solved by simply letting $z = -0.52$ as follows:

$$z = \frac{x - \mu}{\sigma} \text{ becomes } -0.52 = \frac{x - 500}{100}$$

which implies that $x = 448$. However, it would be easy to make the mistake of forgetting the negative sign in -0.52. This illustrates the importance of drawing a picture when dealing with the normal distribution. In solving problems of this type, we should *always* draw a figure—such as Figure 5-16—so that we can see exactly what is happening.

5-4 Exercises A
Finding Scores When
Given Probabilities

In Exercises 5-73 through 5-84, assume that the readings on a scale are normally distributed with a mean of 0 meters and a standard deviation of 1 meter.

5-73 Ninety-five percent of the errors are below what value?

5-74 Ninety-nine percent of the errors are below what value?

5-75 Ninety-five percent of the errors are above what value?

5-76 Ninety-nine percent of the errors are above what value?

5-77 If the top 5% and the bottom 5% of all errors are unacceptable, find the minimum and maximum acceptable errors.

5-78 If the top 0.5% and the bottom 0.5% of all errors are unacceptable, find the minimum and the maximum acceptable errors.

5-79 If the top 10% and the bottom 5% of all errors are unacceptable, find the minimum and maximum acceptable errors.

5-80 If the top 15% and the bottom 20% of all errors are unacceptable, find the minimum and maximum acceptable errors.

5-81 Find the value that separates the top 40% of all errors from the bottom 60%.

5-82 Find the value that separates the top 82% of all errors from the bottom 18%.

5-83 Find the value of the third quartile (Q_3), which separates the top 25% of all errors from the bottom 75%.

5-84   Find the value of P_{18} (18th percentile), which separates the bottom 18% of all errors from the top 82%.

5-85   A population consists of patients classified as moderately hypertense, and their systolic blood pressure readings (in mm of mercury) are normally distributed with a mean of 174.0 and a standard deviation of 21.2. Find the 95th percentile for these blood pressure levels. That is, find the blood pressure level separating the top 5% from the lower 95%.

5-86   In studying ocean conditions, the Bureau of Fisheries found that for one location, the August water temperatures (in degrees Fahrenheit) were normally distributed with a mean of 83.6 and a standard deviation of 2.4. Find the temperature that is exceeded 10% of the time.

5-87   A librarian records the number of books checked out for various days and finds that they normally are distributed with a mean of 230.0 and a standard deviation of 65.5. Find Q_3, the third quartile. That is, find the number that separates the upper 25% from the lower 75%.

5-88   A company manager finds that the lengths of times clients are kept on telephone hold are normally distributed with a mean of 72.0 seconds and a standard deviation of 15.0 seconds. Find D_7, the seventh decile. That is, find the time that separates the top 30% from the lower 70%.

5-89   If heights of adult males are normally distributed with a mean of 70.2 inches and a standard deviation of 4.2 inches, and 95% of all adult males satisfy a minimum height requirement for police officers, what is that minimum height requirement?

5-90   Scores on a college entrance examination are normally distributed with a mean of 500 and a standard deviation of 100. If 70% of all examinees pass, find the passing grade.

5-91   In tests conducted on jet pilots, it has been found that their blackout thresholds are normally distributed with a mean of 4.7 G and a standard deviation of 0.8 G. If the Air Force were to establish the rule that a pilot's blackout threshold must be in the top 67%, what would be the lowest acceptable level?

5-92   A manufacturer of bulbs for movie projectors finds that the lives of the bulbs are normally distributed with a mean of 61.0 hours and a standard deviation of 6.3 hours. The manufacturer will guarantee the bulbs so that only 3% will be replaced because of failure before the guaranteed number of hours. For how many hours should the bulbs be guaranteed?

5-93   A standard I.Q. test produces normally distributed results with a mean of 100 and a standard deviation of 15. A class of high school science students is grouped homogeneously by excluding students with I.Q. scores in either the top 5% or the bottom 5%. Find the lowest and highest possible I.Q. scores of students remaining in the class.

5-94 A machine fills sugar boxes in such a way that the weights of the contents are normally distributed with a mean of 2260 grams and a standard deviation of 20 grams. Another machine checks the weights and rejects packages in the top 1% or bottom 1%. Find the minimum and maximum acceptable weights.

5-95 Reaction times in a certain test are normally distributed with a mean of 0.78 second and a standard deviation of 0.06 second. Find the value of the 10–90 percentile range. That is, find the difference between the 10th percentile and the 90th percentile.

5-96 In one college, grade-point averages of first semester students are normally distributed with a mean of 2.46 and a standard deviation of 0.65. Find the interquartile range. That is, find the difference between the first quartile Q_1 and the third quartile Q_3.

5-97 A manufacturer has contracted to supply ball bearings. Product analysis reveals that the diameters are normally distributed with a mean of 25.1 millimeters and a standard deviation of 0.2 millimeter. The largest 7% of the diameters and the smallest 13% of the diameters are unacceptable. Find the limits for the diameters of the acceptable ball bearings.

5-98 A manufacturer of color television sets tests competing brands and finds that the amounts of energy they require are normally distributed with a mean of 320 kwh and a standard deviation of 7.5 kwh. If the lowest 30% and the highest 20% are not included in a second round of tests, what are the limits for the energy amounts of the remaining sets?

5-99 A particular x-ray machine gives radiation dosages (in milliroentgens) that are normally distributed with a mean of 4.13 and a standard deviation of 1.27. A dosimeter is set so that it displays yellow for radiation levels that are not in the top 10% or bottom 30%. Find the lowest and highest "yellow" radiation levels.

5-100 In a study of coliform contamination in streams, an environmentalist finds that one region has normally distributed coliform levels (number of cells per 100 ml) with a mean of 122 and a standard deviation of 14. The contamination levels are classified into three equal groups of low, medium, and high. Find the minimum and maximum levels for the "medium" category.

5-4 Exercises B
Finding Scores When Given Probabilities

5-101 A city sponsored cross-country race has 4830 applicants, but only 200 are allowed to run in the final race. A qualifying run was held two weeks before the final, and the times are normally distributed with a mean of 36.2 minutes and a standard deviation of 3.8 minutes. If the 200 fastest times qualify, what is the cutoff time?

5-102   A teacher gives a test and gets normally distributed results with a mean of 50 and a standard deviation of 10. Grades are to be assigned according to the following scheme. Find the numerical limits for each letter grade.

A: Top 10%.
B: Scores above the bottom 70% and below the top 10%.
C: Scores above the bottom 30% and below the top 30%.
D: Scores above the bottom 10% and below the top 70%.
F: Bottom 10%.

5-103   Assume that the following scores are representative of a normally distributed list of test scores.
(a)   Find the mean of $\bar{x}$ of this sample.
(b)   Find the standard deviation s of this sample.
(c)   If a grade of A is given to the top 5%, find the minimum numerical score that corresponds to A in this sample.
(d)   Find the theoretical score that separates the top 5% by using the sample mean and standard deviation as estimates for the population mean and standard deviation.

59	69	64	60	77	72	74	64
76	61	77	47	69	82	76	69
60	72	66	92	80	74	78	54
82	59	64	77	69	66	56	53
72	79	58	59	59	50	72	76

5-104   The accompanying stem-and-leaf plot represents randomly selected test scores of applicants for a civil service job. Use the sample data to find $\bar{x}$ and s, then use those values as estimates of μ and σ in order to determine the value of the first quartile Q_1. Compare that result to the value of Q_1 found by directly using the sample values.

4	0445
5	355
6	446899
7	00356689
8	0244578
9	1335

5-5   Normal as Approximation to Binomial

The title of this section should really be "The Normal Distribution Used as an Approximation to the Binomial Distribution," but we wanted something a little snappier. This longer version does better express the true intent of this section: We want to solve certain binomial probability problems by approximating the binomial probability distribution by the normal probability distribution. Consider the following situation.

In a study of retention among employees, one firm found that 40% of its workers stayed with the company for at least five years. Assuming that each new employee has a 40% chance of staying for at least five years, find the probability that, among the next 100 new employees, at least 45 will last five years or more.

The preceding situation can be properly considered as a binomial experiment with $n = 100$, $p = 0.40$, $q = 0.60$, and x assuming the values of $45, 46, 47, \ldots, 99, 100$. Table A-2 stops at $n = 25$, so it cannot be used here because we have $n = 100$. The binomial probability formula

$$P(x) = \frac{n!}{(n - x)! \, x!} \cdot p^x \cdot q^{n-x}$$

can be applied 56 times beginning with

$$P(45) = \frac{100!}{(100 - 45)! \, 45!} \cdot 0.40^{45} \cdot 0.60^{100-45}$$

The resulting 56 probabilities can be totaled to produce the correct result. However, these computations would be *extremely* time-consuming, tedious, and generally hazardous to mental health. Fortunately, we will use a much simpler method.

In some cases, which will be described soon, the normal distribution serves as a good approximation to the binomial distribution. Reexamine the graph of Figure 4-7, where we illustrate the binomial distribution for $n = 50$ and $p = 0.5$. The graph of that figure strongly resembles a normal distribution. Similar observations have led mathematicians to recognize that the normal distribution may be used as an approximation to the binomial distribution. Although results will be approximate, they are usually good and we can circumvent the lengthy computation of the binomial probability function and use normal distribution computations instead. This explains why we bother with an approximation when results can be exact. The price of exactness (namely, very lengthy computations) is too great. We settle for the normal distribution approximation for reasons of time and effort. However, availability of suitable computer software will make the direct approach better in some cases.

Thus far we have justified the use of the normal distribution as an approximation to the binomial distribution simply because of a strong resemblance between the two graphs as in Figure 4-7. Actually, this is not a sound justification. There are other distributions (such as the t distribution to be examined later) with the same basic bell shape, yet they cannot be approximated by the normal distribution since unacceptable errors result. The justification that allows us to use the normal distribution as an approximation to the binomial distribution results from more advanced mathematics. Specifically, the central limit theorem (discussed in Section 5-6) tells us that

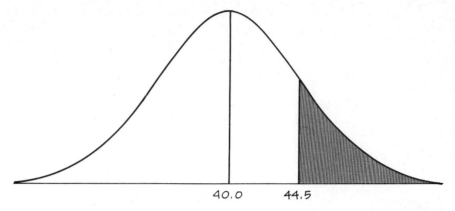

Number of employees staying five or more years

Figure 5-17

if $np \geq 5$ and $nq \geq 5$, then the binomial random variable is approximately normally distributed with the mean and standard deviation given as

$$\mu = np$$
$$\sigma = \sqrt{npq}$$

Results of higher mathematics are used to show that the sampling distribution of sample means from *any* population tends to be normally distributed as long as the sample size is large enough. This will be discussed in more detail later. Relative to binomial experiments, past experience has shown that the normal distribution is a reasonable approximation to the binomial distribution as long as $np \geq 5$ and $nq \geq 5$. In effect, we are stating that there is a formal theorem that justifies our approximation. Unfortunately (or fortunately, depending upon your perspective) it is not practical to outline here the details of the proof for that theorem. For the present, you should accept the intuitive evidence of the strong resemblance between the binomial and normal distributions and take the more rigorous evidence on faith.

Figure 5-17 shows the normal distribution that approximates the binomial experiment involving the 100 new employees. The mean of 40.0 was obtained by applying the formula $\mu = np$, which describes the mean for any binomial experiment. (See Section 4-5.)

$$\mu = np = 100 \cdot 0.40 = 40.0$$

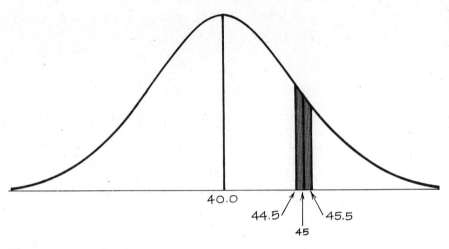

Figure 5-18

Similarly, the standard deviation can be calculated as

$$\sigma = \sqrt{npq} = \sqrt{100 \cdot 0.40 \cdot 0.60} = \sqrt{24} = 4.9$$

Note that in Figure 5-17, most of the values are within 3 standard deviations of the mean (approximately between 25 and 55), which agrees with Figure 5-4. We seek the probability of getting *at least* 45 employees who stay, so we must include the probability of *exactly* 45. But the *discrete* value of 45 is approximated in the *continuous* normal distribution by the interval from 44.5 to 45.5. Such conversions from a discrete to a continuous distribution are called **continuity corrections** (see Figure 5-18). If we ignore or forget the continuity correction, the additional error will be very small as long as n is large.

Reverting to standard procedures associated with normal distributions and Table A-3, we recognize that the probability corresponding to the shaded region of Figure 5-17 cannot be obtained directly. Our strategy will be to find the probability representing the area immediately to the left of the shaded region and subtract that value from 0.5.

$$z = \frac{x - \mu}{\sigma} = \frac{44.5 - 40.0}{4.9} = 0.92$$

Table A-3 indicates that a z score of 0.92 yields a probability of 0.3212, so that the shaded region has a probability of 0.5 − 0.3212, or 0.1788. There is a probability of 0.1788 that the 100 employees will include at least 45 who stay five years or longer. Without the continuity correction, we would have used 45 in place of 44.5, and an answer of 0.1539 would have resulted, but the probability of 0.1788 is the better result.

Listed below is a computer printout from the statistics package STATDISK. One program in that package computes binomial probabilities and the sample display shows one page of the output that results

X	P(X)	Cum.Prob.	X	P(X)	Cum.Prob.
44	0.05763	0.82111	55	0.00083	0.99913
45	0.04781	0.86892	56	0.00044	0.99957
46	0.03811	0.90703	57	0.00023	0.99980
47	0.02919	0.93622	58	0.00011	0.99991
48	0.02149	0.95771	59	0.00005	0.99996
49	0.01520	0.97291	60	0.00002	0.99998
50	0.01034	0.98325	61	0.00001	0.99999
51	0.00676	0.99001	62	0.00000	0.99999
52	0.00424	0.99425	63	0.00000	0.99999
53	0.00256	0.99681	64	0.00000	0.99999
54	0.00149	0.99830	65	0.00000	0.99999

when we request the binomial probabilities corresponding to $n = 100$ and $p = 0.40$. The table shows that the cumulative probability corresponding to $x = 44$ is 0.82111, so that the probability of getting any value of 45 and above is $1 - 0.82111$, or 0.17889, which is very close to our result of 0.1788. For this example, the STATDISK program above is an alternative to the normal approximation method, but every program will have limitations that do not apply to the approximation method.

In this example, the normal distribution does serve as a good approximation to the binomial experiment, but this is not always the case. For large values of n with p not too close to 0 or 1, the normal distribution approximates the binomial fairly accurately, but for *small values of* n *with* p *near 0 or 1, the binomial distribution is approximated very poorly by the normal distribution.* The approximation is suitable only when $np \geq 5$ and $nq \geq 5$; both conditions must be met. Table 5-2 lists some specific values of p along with the corresponding minimum values of n.

Suppose, for example, that a couple plans to have four children and they seek the probability of having three girls and one boy. With $n = 4$, $p = 0.5$, $q = 0.5$, and $x = 3$, we see that $np = 4 \cdot 0.5 = 2$ and $nq = 4 \cdot 0.5 = 2$, so the normal distribution should *not* be used. Even though the distribution is symmetrical and mound shaped, it does not fit a normal distribution well. Table 5-2 indicates that they should plan on at least ten children if they really want to use the normal distribution as an approximation to the binomial. (Admittedly, this is not much of an inducement for a large family.) The binomial probability formula can be applied easily to the given situation.

In Figure 5-19 we summarize the procedure for using the normal distribution as an approximation to the binomial distribution. The following example illustrates that procedure.

Table 5-2

Minimum sample required to approximate a binomial distribution by a normal distribution

p	n must be at least
0.001	5000
0.01	500
0.1	50
0.2	25
0.3	17
0.4	13
0.5	10
0.6	13
0.7	17
0.8	25
0.9	50
0.99	500
0.999	5000

Figure 5-19
Solving binomial
probability problems

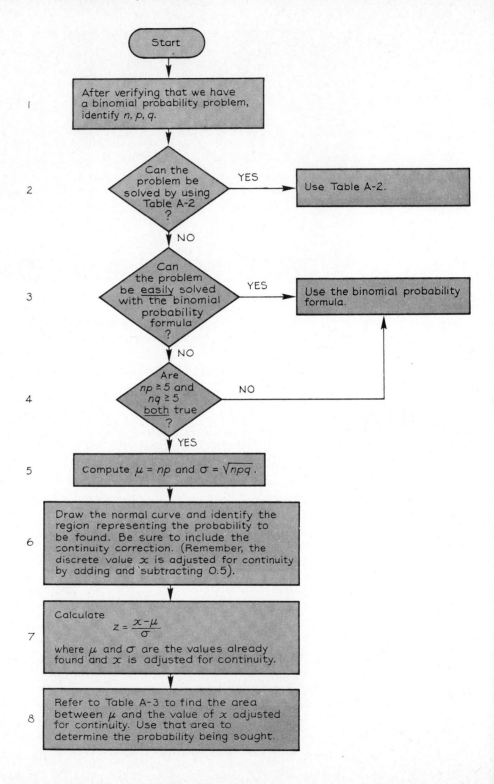

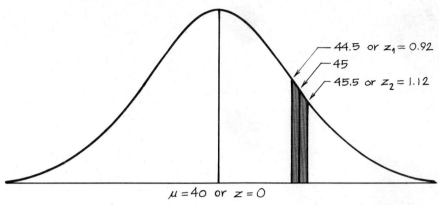

$$\mu = 40 \text{ or } z = 0$$

Number of employees retained

Figure 5-20

E X A M P L E

A firm experiences a 40% retention rate for employees staying at least five years. Find the probability that of the next 100 new employees, *exactly* 45 will stay at least five years.

Solution

Refer to Figure 5-19. In Step 1 we verify that the conditions described do satisfy the criteria for the binomial distribution and $n = 100$, $p = 0.40$, $q = 0.60$, and $x = 45$. Proceeding to Step 2, we see that the table cannot be used because n is too large. In Step 3, the binomial probability formula applies, but

$$P(45) = \frac{100!}{(100 - 45)! \ 45!} \cdot 0.40^{45} \cdot 0.60^{100-45}$$

is too difficult to compute. Most calculators cannot evaluate anything above 70!, but the availability of calculators or a computer would simplify this approach. In Step 4 we get

$$np = 100 \cdot 0.40 = 40 \geq 5$$
$$nq = 100 \cdot 0.60 = 60 \geq 5$$

and since np and nq are both at least 5, we conclude that the normal approximation to the binomial is satisfactory. We now go on to Step 5, where we obtain the values of μ and σ as follows.

$$\mu = np = 100 \cdot 0.40 = 40$$
$$\sigma = \sqrt{npq} = \sqrt{100 \cdot 0.40 \cdot 0.60} = \sqrt{24} = 4.9$$

Now we go to Step 6, where we draw the normal curve shown in Figure 5-20. The shaded region of Figure 5-20 represents the probability we want. Use of the continuity correction results in the representation of 45 by the region extending from 44.5 to 45.5 We now proceed to Step 7.

The format of Table A-3 requires that we first find the probability corresponding to the region bounded on the left by the vertical line through the mean of 40 and on the right by the vertical line through 45.5 so that one calculation required in Step 7 is as follows.

$$z_2 = \frac{x - \mu}{\sigma} = \frac{45.5 - 40}{4.9} = 1.12$$

We also need the probability corresponding to the region bounded by 40 and 44.5, so we calculate

$$z_1 = \frac{x - \mu}{\sigma} = \frac{44.5 - 40}{4.9} = 0.92$$

Finally, in Step 8 we use Table A-3 to find that a probability of 0.3686 corresponds to $z_2 = 1.12$ and 0.3212 corresponds to $z_1 = 0.92$. Consequently, the entire shaded region of Figure 5-20 depicts a probability of $0.3686 - 0.3212 = 0.0474$. Using the software package STATDISK, we find that the probability is 0.04781; the discrepancy of 0.0004 is very small.

Table 5-3

The values of $P(x)$ are computed with the binomial probability formula or from Table A-2

x	$P(x)$
0	0+
1	0+
2	0.002
3	0.009
4	0.028
5	0.067
6	0.122
7	0.175
8	0.196
9	0.175
10	0.122
11	0.067
12	0.028
13	0.009
14	0.002
15	0+
16	0+

As one last example that can be checked with Table A-2, let's assume that an experiment consists of randomly selecting 16 births and recording the number of boys. With $n = 16$, $p = 0.5$, and $q = 0.5$, we could let x assume the values $0, 1, 2, \ldots, 15, 16$ and apply the binomial probability formula to obtain the results summarized in Table 5-3. The graph of Table 5-3 consists of the rectangles shown in Figure 5-21. (Figure 5-21 provides a visual comparison of a binomial distribution and the approximating normal distribution.)

The approximating normal curve is characterized by $\mu = np = 16 \cdot 0.5 = 8$ and $\sigma = \sqrt{npq} = \sqrt{16 \cdot 0.5 \cdot 0.5} = \sqrt{4} = 2$. As a basis for comparison, we find the probability of getting exactly nine boys using the binomial distribution and the normal distribution. Using Table 5-3 we see that $P(9) = 0.175$. Using the normal distribution with the continuity correction, $P(9)$ is computed to be 0.1747 and the two results agree.

We now have three different methods for determining probabilities in binomial experiments, which are summarized in Figure 5-19.

$$P(x) = \frac{n!}{(n - x)! \; x!} \, p^x q^{n-x}$$

$$= \frac{16!}{(16 - x)! \; x!} \cdot 0.5^x \cdot 0.5^{16-x}$$

$$= \frac{16!}{(16 - x)! \; x!} \cdot 0.5^{16}$$

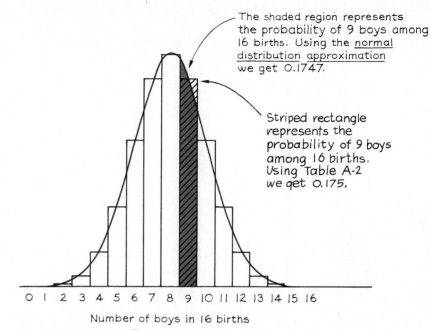

The shaded region represents the probability of 9 boys among 16 births. Using the <u>normal distribution approximation</u> we get 0.1747.

Striped rectangle represents the probability of 9 boys among 16 births. Using Table A-2 we get 0.175.

0 1 2 3 4 5 6 7 8 9 10 11 12 13 14 15 16

Number of boys in 16 births

Figure 5-21   Normal approximation compared to binomial result

5-5   Exercises A
Normal as Approximation to Binomial

In Exercises 5-105 through 5-112, check that np ≥ 5 and nq ≥ 5 in order to determine whether the normal distribution is a suitable approximation. In each case, also find the values of μ and σ.

5-105 $n = 25, p = 0.250$ **5-106** $n = 50, p = 0.333$

5-107 $n = 90, p = 0.048$ **5-108** $n = 63, p = 0.068$

5-109 $n = 250, p = 0.010$ **5-110** $n = 657, p = 0.994$

5-111 $n = 84, p = 0.950$ **5-112** $n = 125, p = 0.961$

In Exercises 5-113 through 5-120, find the indicated binomial probabilities by using (a) Table A-2 in Appendix A and (b) the normal distribution as an approximation to the binomial probability distribution.

5-113 With $n = 12$ and $p = 0.50$, find $P(8)$.

5-114 With $n = 15$ and $p = 0.40$, find $P(7)$.

5-115 With $n = 20$ and $p = 0.70$, find $P(12)$.

5-116   With $n = 25$ and $p = 0.30$, find $P(2)$.

5-117   With $n = 12$ and $p = 0.50$, find P(at least 8).

5-118   With $n = 15$ and $p = 0.40$, find P(at least 7).

5-119   With $n = 20$ and $p = 0.70$, find P(at most 12).

5-120   With $n = 25$ and $p = 0.30$, find P(at most 2).

5-121   Find the probability of getting at least 60 boys in 100 births.

5-122   Find the probability of passing a true-false test of 100 questions if 65% is passing and all responses are random guesses.

5-123   In an eastern college, 60% of the entering freshmen graduate. Find the probability that, of 2000 entering freshmen, there will be at most 1150 graduates.

5-124   The probability of a particular computer component being defective is 0.01. Find the probability of getting at most 60 defects in a random sample of 5000 components.

5-125   A quarterback completes 35% of his passes. What is the probability of his completing fewer than 165 of his next 400 passes?

5-126   The failure rate on a civil service examination is 30%. Find the probability of getting more than 325 failures on the next 1000 exams.

5-127   On a question in an I.Q. test, 25% of the respondents answer correctly. Find the probability that, of 250 responses, fewer than 70 are correct.

5-128   A certain seed has a 75% germination rate. If 50,000 seeds are planted, find the probability that more than 37,400 will germinate.

5-129   An engine on a light aircraft has a 99.9% chance of completing a flight without failure. Find the probability that of 6000 flights, there are more than ten engine failures. (Assume that all aircraft have one engine.)

5-130   A magazine subscription service has found that 8% of its telephone solicitations result in orders. Find the probability that of 120 contacts, there are fewer than 10 orders.

5-131   An experiment in parapsychology involves the guessing of a selected number from 1 through 5. If this test is repeated on 250 subjects, find the probability of getting more than 60 correct responses. Would it be unusual to get more than 70 correct responses?

5-132   The probability of winning anything in the weekly New York State lottery is about 1/500. In a small village 3000 tickets are sold. What is the probability that no prizes are won in that village?

5-133   Experience has shown that 12% of all subjects fail a visual perception test. Find the probability that of 200 randomly selected subjects, there are at least 20 failures.

5-134   An airline company experiences a 7% rate of no-shows on advance reservations. Find the probability that of 250 randomly selected advance reservations, there will be at least 10 no-shows.

5-135   A certain genetic characteristic appears in one-quarter of all offspring. Find the probability that of 40 randomly selected offspring, fewer than 5 exhibit the characteristic in question.

5-136   The IRS finds that of all taxpayers whose returns are audited, 70% end up paying additional taxes. Find the probability that of 500 randomly selected returns, at least 400 end up paying additional taxes.

5-5   Exercises B
Normal as Approximation to Binomial

5-137   In a binomial experiment with $n = 25$ and $p = 0.4$, find P(at least 18) using:
(a)   The table of binomial probabilities (Table A-2).
(b)   The binomial probability formula.
(c)   The normal distribution approximation.

5-138   The cause of death is related to heart disease in 52% of the cases studied. Find the probability that in 500 randomly selected cases, the number of heart-disease–related deaths differs from the mean by more than two standard deviations.

5-139   An airline company works only with advance reservations and experiences a 7% rate of no-shows. How many reservations could be accepted for an airliner with a capacity of 250 if there is at least a 0.95 probability that all reservation holders who show will be accommodated?

5-140   A company manufactures integrated circuit chips with a 23% rate of defects. What is the minimum number of chips that must be manufactured if there must be at least a 90% chance that 5000 good chips can be supplied?

5-6   The Central Limit Theorem

The central limit theorem is one of the most important topics in the study of statistics, but students traditionally have some difficulty with it. Before considering the statement of the central limit theorem, we will first try to develop an intuitive understanding of one of its consequences.

The sampling distribution of sample means tends to be a normal distribution. This implies that if we collect samples all of the same size, compute their means, and then develop a histogram of those means, it will tend to assume the bell shape of a normal distribution. This is true regardless of the shape of the distribution of the original population.

Table 5-4

Thirty collections of samples with ten random numbers between 1 and 9 in each sample. The right column consists of the corresponding sample means.

Sample	Data										Sample mean
1	2	7	5	5	2	1	7	7	9	4	4.9
2	5	8	1	1	5	7	1	4	1	4	3.7
3	7	6	9	8	5	1	6	4	7	9	6.2
4	7	3	1	7	3	6	7	9	4	3	5.0
5	9	7	7	6	1	6	8	3	4	7	5.8
6	5	3	3	4	2	5	9	9	1	9	5.0
7	5	5	3	9	5	3	1	9	1	5	4.6
8	4	3	9	5	5	9	1	7	7	8	5.8
9	2	1	7	8	6	7	7	9	8	3	5.8
10	3	4	5	6	8	4	8	3	4	5	5.0
11	5	3	2	2	6	8	1	5	5	9	4.6
12	7	5	9	6	8	2	2	7	2	1	4.9
13	3	1	4	1	7	9	3	2	3	8	4.1
14	6	2	7	4	4	5	2	6	8	6	5.0
15	9	6	2	9	4	2	6	3	5	5	5.1
16	9	2	2	3	6	2	6	6	8	3	4.7
17	5	4	2	1	9	4	2	9	4	2	4.2
18	8	1	2	1	4	3	2	8	5	4	3.8
19	5	8	9	6	2	7	9	3	8	5	6.2
20	5	6	8	7	5	9	6	4	8	7	6.5
21	7	9	9	8	3	5	5	1	4	6	5.7
22	8	4	7	8	7	8	7	7	1	8	6.5
23	5	5	1	7	5	7	7	2	9	8	5.6
24	9	5	2	5	9	2	5	3	5	8	5.3
25	4	5	8	4	2	9	2	6	6	1	4.7
26	1	7	7	3	4	7	7	2	8	7	5.3
27	8	1	1	7	6	2	2	1	4	9	4.1
28	9	4	3	7	3	7	8	4	3	2	5.0
29	1	2	9	3	8	2	4	6	2	8	4.5
30	2	9	3	3	1	2	6	7	8	7	4.8

The central limit theorem qualifies the preceding remarks and includes additional aspects, but stop and try to understand the thrust of these remarks before continuing.

Let's begin with some concrete numbers. Table 5-4 contains a block of data consisting of 300 sample scores. These scores were generated through a computer simulation, but they could have been extracted from a telephone directory or a book of random numbers. (Now *there's* exciting reading, although the plot is a little thin.) In Figure 5-22 we illustrate the histogram of the 300 sample scores, and we can see that their distribution is essentially uniform. Now consider the 300 scores to be 30 samples, with

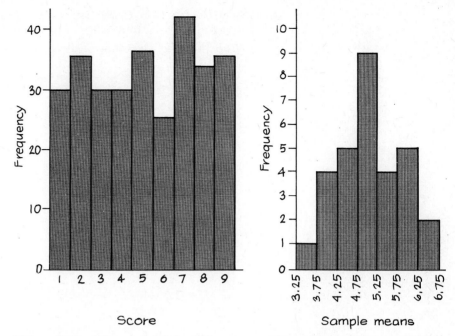

Figure 5-22 Histogram of the 300 *original scores* randomly selected (between 1 and 9)

Figure 5-23 Histogram of the 30 *sample means*. Each sample mean is based on 10 raw scores randomly selected between 1 and 9, inclusive.

10 scores in each sample. The resulting 30 sample means are listed in Table 5-4 and illustrated in the histogram of Figure 5-23. Note that the shape of Figure 5-23 is roughly that of a normal distribution. It is important to observe that even though the original population has a uniform distribution, the sample means seem to have a normal distribution. It was observations exactly like this that led to the formulation of the central limit theorem. If our sample means were based on samples of size larger than ten, Figure 5-23 would more closely resemble a normal distribution. We are now ready to consider the central limit theorem.

Let's assume that the variable x represents scores that may or may not be normally distributed, and that the mean of the x values is μ while the standard deviation is σ. Suppose we collect a sample of size n and calculate the sample mean $\bar{x}$. What do we know about the collection of all sample means that we produce by repeating this experiment, collecting a sample of size n to get the sample mean? The central limit theorem tells us that as the sample size n increases, the sample means will tend to approach a normal distribution with mean μ and standard deviation $\sigma/\sqrt{n}$. This conclusion is not intuitively obvious, and it was arrived at

Reliability and Validity

The reliability of data refers to the consistency with which the same results occur, but the validity of data refers to how well the data measure what they are supposed to measure. The reliability of an I.Q. test can be judged by comparing scores for the test given on one date to the scores for the same test given on another date. Many critics claim that the popular I.Q. tests are reliable but not valid. That is, they produce consistent results, but they do not really measure intelligence levels. It is much easier to use statistics to analyze reliability than validity, but validity is the more important characteristic. To test the validity of an intelligence test, we might compare the test scores to another indicator of intelligence, such as academic performance.

through extensive research and analysis. The formal rigorous proof requires advanced mathematics and is beyond the scope of this text, so we can only illustrate the theorem and give examples of its practical use.

CENTRAL LIMIT THEOREM

If the variable x has some distribution so that the mean value is μ and the standard deviaton is σ, the sample means $\bar{x}$ (based on random samples of size n) will, as n increases, approach a **normal** distribution with mean μ and standard deviation $\sigma/\sqrt{n}$.

NOTATION

In random samples of size n, the mean of the sample means is denoted by $\mu_{\bar{x}}$ so that $\mu_{\bar{x}} = \mu$.

Also, the standard deviation of the sample means is denoted by $\sigma_{\bar{x}}$ so that $\sigma_{\bar{x}} = \sigma/\sqrt{n}$.

$\sigma_{\bar{x}}$ is often called the **standard error of the mean.**

Statisticians have found that for samples of size larger than 30, the sample means can be approximated reasonably well by a normal distribution with mean μ and standard deviation $\sigma/\sqrt{n}$. Sometimes even smaller sample sizes can produce means that approximate a normal distribution. If the original population is itself normally distributed, then samples of size $n = 1$ are large enough.

Comparison of Figures 5-22 and 5-23 should confirm that the original numbers have a nonnormal distribution, while the sample means approximate a normal distribution. The central limit theorem also indicates that the mean of *all* such sample means should be μ (the mean of the original population) and the standard deviation of all such sample means should be $\sigma/\sqrt{n}$ (where σ is the standard deviation of the original population and n is the sample size of 10). We can find μ and σ for the original population of numbers between 1 and 9 by noting that, if those numbers occur with equal frequency as they should, then the population mean μ is given by

$$\mu = \frac{1 + 2 + 3 + 4 + 5 + 6 + 7 + 8 + 9}{9} = 5.0$$

Similarly, we can find σ by again using 1, 2, 3, 4, 5, 6, 7, 8, 9 as an ideal or theoretical representation of the population. Following this course, σ is computed to be 2.58. The mean and standard deviation of the sample means can now be found as follows.

$$\mu_{\bar{x}} = \mu = 5.0$$

$$\sigma_{\bar{x}} = \frac{\sigma}{\sqrt{n}} = \frac{2.58}{\sqrt{10}} = \frac{2.58}{3.16} = 0.82$$

The preceding results represent *all* sample means of size $n = 10$. For the 30 sample means shown in Table 5-4, we have a mean of 5.08 and a standard deviation of 0.75. We can see that our real data conform quite well to the theoretically predicted values for $\mu_{\bar{x}}$ and $\sigma_{\bar{x}}$.

We can use a simple random selection process beginning with a uniform distribution to artificially create normally distributed data. Realistic and practical simulations become possible through this use of the central limit theorem.

EXAMPLE

A traffic study shows that for a certain bridge, the mean number of occupants in a car is 1.850 people, and the standard deviation is 0.310 people. For a random sample of 64 cars, find the probability that the mean number of occupants is between 1.850 and 1.900.

Solution

By the central limit theorem and the large sample size, we know that the sampling distribution of such sample means is the normal distribution. Given the population mean ($\mu = 1.850$) and standard deviation ($\sigma = 0.310$), we proceed to find $\mu_{\bar{x}}$ and $\sigma_{\bar{x}}$ as follows.

$$\mu_{\bar{x}} = \mu = 1.850$$

$$\sigma_{\bar{x}} = \frac{\sigma}{\sqrt{n}} = \frac{0.310}{\sqrt{64}} = 0.03875$$

In Figure 5-24, we show the shaded area corresponding to the probability we seek. We find that area by first determining the value of the z score.

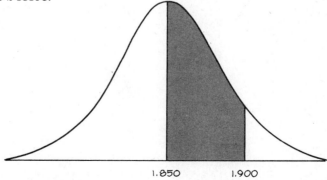

Figure 5-24

$$z = \frac{\bar{x} - \mu_{\bar{x}}}{\sigma_{\bar{x}}} = \frac{\bar{x} - \mu}{\dfrac{\sigma}{\sqrt{n}}} = \frac{1.900 - 1.850}{\dfrac{0.310}{\sqrt{64}}} = 1.29$$

From Table A-3 we find that $z = 1.29$ corresponds to a probability of 0.4015 so that the probability we seek is 0.4015.

EXAMPLE

An elevator in a large office building can safely carry up to 5000 pounds of people. A study shows that the population of elevator riders has its mean and standard deviation given by $\mu = 148.0$ pounds and $\sigma = 15.2$ pounds. If a sign allows up to 32 passengers and the elevator is filled with 32 randomly selected riders, what is the probability that the 5000-pound limit will be exceeded?

Solution

The safe load limit will be exceeded if the 32 riders average more than $5000 \div 32$ or 156.25 pounds. That is, the limit will be exceeded if $\bar{x}$ exceeds 156.25 pounds for the sample of 32 riders. Even though the weights might not be normally distributed, the sample size is large enough (greater than 30) so that the sample means approximate a normal distribution. Using the known values of μ and σ, we can find $\mu_{\bar{x}}$ and $\sigma_{\bar{x}}$ for a sample of 32 as follows.

$$\mu_{\bar{x}} = \mu = 148.0$$

$$\sigma_{\bar{x}} = \frac{\sigma}{\sqrt{n}} = \frac{15.2}{\sqrt{32}} = 2.687$$

In Figure 5-25, we show a working sketch of the normal distribution that applies to this situation.

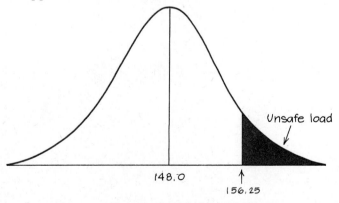

Mean weight (pounds)

Figure 5-25

We find the area of the shaded region by first finding the area of the region to its immediate left.

$$z = \frac{\overline{x} - \mu_{\overline{x}}}{\sigma_{\overline{x}}} = \frac{156.25 - 148.0}{2.687} = 3.07$$

From Table A-3 we see that a z score of 3.07 corresponds to an area of 0.4989, so the shaded region is $0.5 - 0.4989 = 0.0011$. The probability of exceeding 5000 pounds is only 0.0011 and we see that it would be rare for the safe load limit to be exceeded. We now know that the original working sketch of Figure 5-25 shows the shaded region to be much larger than it should be. But at that stage, the graph was helpful in making the correct calculations.

It is interesting to note that, as the sample size increases, the sample means tend to vary less, since $\sigma_{\overline{x}} = \sigma/\sqrt{n}$ gets smaller as n gets larger. For example, assume that I.Q. scores have a mean of 100 and a standard deviation of 15. Samples of 36 will produce means with $\sigma_{\overline{x}} = 15/\sqrt{36} = 2.5$, so that 99% of all such samples will have means between 93.6 and 106.4. If the sample size is increased to 100, $\sigma_{\overline{x}}$ becomes $15/\sqrt{100}$, or 1.5, so that 99% of the samples will have means between 96.1 and 103.9. (These particular computations and concepts are explained in Section 7-2.)

These results are supported by common sense: As the sample size increases, the corresponding sample mean will tend to be closer to the true population mean. The effect of an unusual or outstanding score tends to be dampened as it is averaged in as part of a sample.

Our use of $\sigma_{\overline{x}} = \sigma/\sqrt{n}$ assumes that the population is infinite. When we sample with replacement of selected data, for example, the population is effectively infinite. Yet realistic applications involve sampling without replacement, so successive samples depend on previous outcomes.

NOTATION

Just as n denotes the *sample* size, N denotes the size of a *population*. For finite populations of size N, we should incorporate the **finite population correction factor** $\sqrt{(N - n) \div (N - 1)}$ so that $\sigma_{\overline{x}}$ is found as follows.

$$\sigma_{\overline{x}} = \frac{\sigma}{\sqrt{n}} \sqrt{\frac{N - n}{N - 1}}$$

If the sample size n is small in comparison to the population size N, the finite population correction factor will be close to 1. Consequently its impact will be negligible and it can therefore be ignored.

Statisticians have devised the following **rule of thumb.**

RULE

Use the finite population correction factor when computing $\sigma_{\bar{x}}$ if the population is finite and $n > 0.05N$. That is, use the correction factor only if the sample size is greater than 5% of the population size.

EXAMPLE

For 1000 fuses, the breaking points have a mean of 7.5 amperes and a standard deviation of 1.0 ampere. If a sample of 150 fuses is obtained without replacement, find the probability that the sample yields a mean between 7.5 amperes and 7.6 amperes.

Solution

With $\mu = 7.5$ amperes, $\sigma = 1.0$ ampere, $N = 1000$, and $n = 150$, we compute

$$\mu_{\bar{x}} = \mu = 7.5 \text{ amperes}$$

$$\sigma_{\bar{x}} = \frac{\sigma}{\sqrt{n}}\sqrt{\frac{N-n}{N-1}} = \frac{1.0 \text{ ampere}}{\sqrt{150}}\sqrt{\frac{1000-150}{1000-1}} = 0.075 \text{ ampere}$$

The finite population correction factor was used because $n > 0.05N$, or $150 > 0.05(1000)$. This means that the sample size is large in comparison to the population size. Having determined the values of $\mu_{\bar{x}}$ and $\sigma_{\bar{x}}$, we can apply the central limit theorem by using the methods associated with the normal distribution. Figure 5-26 illustrates the normal distribution and the region representing the desired probability.

$$z = \frac{\bar{x} - \mu_{\bar{x}}}{\sigma_{\bar{x}}} = \frac{7.6 - 7.5}{0.075} = \frac{0.1}{0.075} = 1.33$$

From Table A-3 we see that $z = 1.33$ corresponds to a probability of 0.4082 so that P(mean is between 7.5 and 7.6 amperes) = 0.4082. If we did not use the finite population correction factor, our answer would have been 0.3888 instead of the better result of 0.4082. Figure 5-27 outlines most of the key points presented in this section.

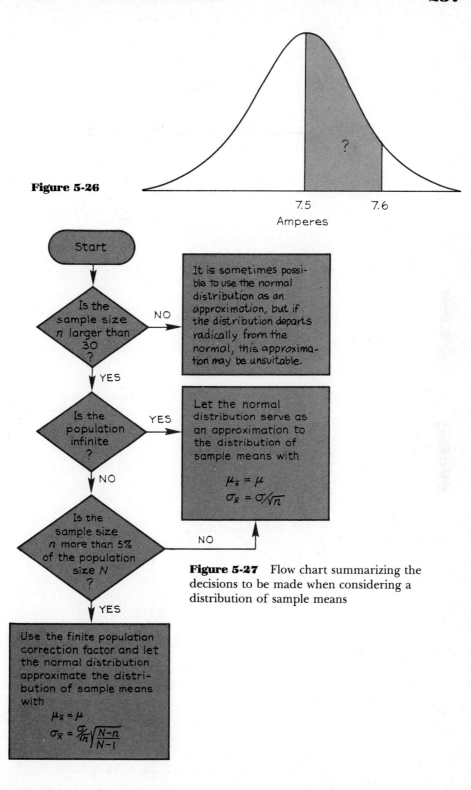

Figure 5-26

Figure 5-27 Flow chart summarizing the decisions to be made when considering a distribution of sample means

One final provocative query: Can it be that normally distributed populations actually reflect a composite total of factors that are not themselves normally distributed? The weights of newborn whales, for example, may be affected by genetic and environmental factors such as the weights and diets of predecessors and the random selection of certain genes. Perhaps the contributing factors were not themselves normally distributed, but when the total net effect is realized, the central limit theorem applies by suggesting that the weight of the newborn represents a mean of the contributing sample factors. The normal distribution of the sample means is then reflected by the normal distribution of weights of newborn whales. This may help explain the frequent occurrence of the normal distribution among various physical, sociological, and psychological measurements.

5-6 Exercises A
The Central Limit Theorem

5-141 A sample of $n = 49$ is taken from a very large population that has a mean of 50 and a standard deviation of 14. Find the probability that the sample mean $\bar{x}$ will be between 50 and 54.

5-142 A sample of $n = 100$ is taken from a very large population that has a mean of 50 and a standard deviation of 14. Find the probability that the sample mean $\bar{x}$ will be between 50 and 54.

5-143 A sample of $n = 64$ is taken from a population of 2000. (Samples are taken without replacement.) The population mean μ is 200 and the standard deviation σ is 40. Find the probability that the sample mean $\bar{x}$ will be between 200 and 209.

5-144 A sample of $n = 70$ is taken from a population of 1573. The population mean μ is 78.4, and the population standard deviation σ is 8.2. Find the probability that the sample mean $\bar{x}$ will be less than 77.1.

5-145 The manager of a clothing store knows that in the past, summer daily gross receipts averaged $900 with a standard deviation of $300. In attempting to evaluate the effect of a new advertising approach, she needs to determine the probability of getting 36 days with a mean gross income between $900 and $1000. Find that probability.

5-146 On one I.Q. test, the mean score is 100 and the standard deviation is 15. A sample of 50 scores is selected from a very large population. Find the probability that the mean of the sample group is more than 103.

5-147 A test in extrasensory perception is given to 500 subjects and the mean and standard deviation are computed to be 48 and 8, respectively. If a random sample of 25 is selected without replacement, find the probability that the mean of the sample group is more than 50.

5-148   A study of 10,000 males of age 25 who smoke at least two packs of cigarettes daily shows that the mean life span is 65.3 years while the standard deviation is 3.4 years. If a sample of 40 is selected, find the probability that the mean life span of the sample is less than the retirement age of 65 years.

5-149   A strobe light is designed so that the mean time between flashes is 10.00 seconds, with a standard deviation of 0.40 second. A sample of 81 lights is selected from the 2000 lights produced in one week. Find the probability that for the sample, the mean time between flashes is between 10.00 seconds and 10.10 seconds.

5-150   A battery is designed to last for 25 hours of operation during normal use. The batteries are produced in batches of 2000 and 50 batteries from each batch are tested. If the mean life of the sample is less than 24 hours, the entire batch is rejected. Assuming that $\mu = 25$ hours and $\sigma = 3$ hours, find the probability that a batch will be rejected.

5-151   A college entrance examination produces a mean score of 500, while the standard deviation is 100. Find the probability that, for a sample of 750 randomly selected subjects, the mean is between 495 and 505.

5-152   Among 800 subjects given a test for reaction times the mean is 0.820 second with a standard deviation of 0.180 second. If a sample of 33 is selected from this population of 800 subjects, find the probability that the sample mean is less than 0.825 second.

5-153   In a county of 230,762 households, the mean annual income per household is $11,387 and the standard deviation is $3419. Find the probability that, in a sample of 3% of the households, the mean annual income is between $10,500 and $11,500.

5-154   In a county of 40,250 households, the mean annual income per household is $11,387 and the standard deviation is $3419. Find the probability that, in a sample of 1500 households, the mean annual income is between $10,500 and 11,500.

5-155   A visual perception test is given to children about to enter kindergarten and the mean score is 4.85 while the standard deviation is 0.45. If a sample of 120 subjects is randomly selected from this population, find the probability that the sample mean is more than 4.90.

5-156   A study of the time required for airplanes to land at a certain airport shows that the mean is 280 seconds with a standard deviation of 20 seconds. (The times are not normally distributed.) For a sample of 36 randomly selected incoming airplanes, find the probability that their mean landing time is greater than 275 seconds.

5-157   A population has a standard deviation of 20. Samples of size n are taken randomly and the means of the samples are computed. What happens to the standard error of the mean if the sample size is increased from 100 to 400?

5-158 A population has a standard deviation of 20. Samples of size n are randomly selected and the means of the samples are computed. What happens to the standard error of the mean if the sample size is decreased from 64 to 16?

5-159 If samples of size n are selected from populations of size N, identify those cases in which the finite population correction factor can be ignored. If it cannot be ignored, find the value of $\sqrt{(N-n)/(N-1)}$. Assume that sampling is done without replacement.

(a) $N = 5000$, $n = 200$ (b) $N = 12,000$, $n = 1000$
(c) $N = 4000$, $n = 500$ (d) $N = 8000$, $n = 3000$
(e) $N = 1500$, $n = 50$ (f) $N = 750$, $n = 50$
(g) $N = 673$, $n = 32$ (h) $N = 866$, $n = 73$
(i) $N = 50,000$,   (j) $N = 8362$, $n = 935$
      $n = 10,000$

5-160 If $\sigma = 15$, find the standard error of the mean $\sigma_{\bar{x}}$ for each part of Exercise 5-159. Incorporate the finite population correction factor whenever necessary.

In Exercises 5-161 through 5-164, be sure to check for the use of the finite population correction factor and include it whenever necessary.

5-161 A population consists of 500 beginning students at a certain college. For this group, the mean I.Q. is 117.0, and the standard deviation is 8.0. If a sample of 200 of these students is randomly selected, find the probability that the mean for this sample group is below 118.0.

5-162 A study involves a population of 300 women who are 5 feet tall and are between 18 and 24 years of age. This population has a mean weight of 121.5 pounds and a standard deviation of 6.5 pounds. If 50 members of this population are randomly selected, find the probability that the mean weight of this sample group is greater than 120.0 pounds.

5-163 A psychologist collects a population of 250 volunteers who will participate in a behavior modification experiment. A pretest of this population produces behavior indices with a mean of 436 and a standard deviation of 24. If 20% of the population is randomly selected for one phase of treatment, find the probability that the mean for this sample group is between the desired limits of 430 and 440.

5-164 In doing an economic impact study, a sociologist identifies a population of 1200 households with a mean annual income of $23,460 and a standard deviation of $3,750. If 10% of these households are randomly selected for a more detailed survey, find the probability that the mean for this sample group will fall between the acceptable limits of $23,000 and $24,000.

5-6   Exercises B
The Central Limit Theorem

5-165   Assume that a population of 10,000 has $\sigma = 15$. If a sample of 56 is selected, find the probability that a random sample mean will differ from the population mean by more than 4.

5-166   Assume that a population is infinite. Find the probability that the mean of a sample of 100 differs from the population mean by more than $\sigma/4$.

5-167   The accompanying frequency table summarizes the number of defective units produced by a machine on 87 different days. Find μ and σ for this population. If 32 of the 87 days are randomly selected, use the central limit theorem to find the probability that the mean for the 32 days is greater than 30.0.

5-168   A sample of size 50 is randomly selected from a population of size N, with the result that the standard error of the mean is one-tenth the value of the population standard deviation. Find the size of the population.

x	f
0– 4	2
5– 9	0
10–14	5
15–19	8
20–24	12
25–29	17
30–34	20
35–39	14
40–44	6
45–49	3

Computer Project
Normal Probability Distributions

Listed below are two BASIC programs, which may require some minor modification in order to run on certain computers. The first program will produce 36 randomly generated numbers. The second program will produce a mean of 36 randomly generated numbers.

(a)   Enter and run the first program and manually construct a histogram of these 36 values.

(b)   Enter the second program and run it 36 times to get 36 sample means. Then manually construct a histogram of these 36 values. (*Hint:* Instead of entering RUN 36 times, we can run the program 36 times by including these two lines:

$$5 \text{ FOR } J = 1 \text{ to } 36$$
$$65 \text{ NEXT } J$$

(c)   Compare the two resulting histograms.

```
10  RANDOMIZE
20  FOR I = 1 to 36
30    PRINT INT(100*RND(X))
40  NEXT I
50  END

10  RANDOMIZE
20  LET T = 0
30  FOR I = 1 to 36
30    LET T = T + INT(100*RND(X))
50  NEXT I
60  PRINT T/36
70  END
```

Review

The main concern of this chapter is the concept of a normal probability distribution, the most important of all continuous probability distributions. Many real and natural occurrences yield data that are normally distributed or can be approximated by a normal distribution. The normal distribution, which appears to be bell-shaped when graphed, can be described algebraically by an equation, but the complexity of that equation usually forces us to use a table of values instead.

Table A-3 represents the **standard normal distribution,** which has a mean of 0 and a standard deviation of 1. This table relates deviations away from the mean with areas under the curve. Since the total area under the curve is 1, those areas correspond to probability values.

In the early sections of this chapter, we worked with the standard procedures used in applying Table A-3 to a variety of different situations. We saw that Table A-3 can be applied indirectly to normal distributions that are nonstandard. (That is, μ and σ are not 0 and 1, respectively.) We were able to find the number of standard deviations that a score x is away from the mean μ by computing $z = (x - \mu)/\sigma$.

In Sections 5-3 and 5-4 we considered real and practical examples as we converted from a nonstandard to a standard normal distribution. In Section 5-5 we saw that we can sometimes approximate a binomial probability distribution by a normal distribution. If $np \geq 5$ and $nq \geq 5$, the binomial random variable x is approximately normally distributed with the mean and standard deviation given as $\mu = np$ and $\sigma = \sqrt{npq}$. Since the binomial probability distribution deals with discrete data while the normal distribution deals with continuous data, we introduced the **continuity correction,** which should be used in normal approximations to binomial distributions if n is small. Finally, in Section 5-6, we considered the distribution of sample means that can come from normal or non-normal populations. The **central limit theorem** asserts that the distribution of sample means $\bar{x}$ (based on random samples of size n) will, as n increases, approach a normal distribution with mean μ and standard deviation $\sigma/\sqrt{n}$. This means that if samples are of size n where $n > 30$, we can approximate the distribution of those sample means by a normal distribution. The **standard error of the mean** is $\sigma/\sqrt{n}$ as long as the population is infinite or the sample size is not more than 5% of the population. But if the sample n exceeds 5% of the population N, then the standard error of the mean must be adjusted by the **finite population correction factor** with $\sigma/\sqrt{n}$ multiplied by $\sqrt{(N - n)/(N - 1)}$. Figure 5-27 summarizes these concepts.

In Chapter 6 we apply many of the concepts introduced in this chapter as we study the extremely important process of testing hypotheses. Since basic concepts of this chapter serve as critical prerequisites for the following material, it would be wise to master these ideas and methods now.

IMPORTANT FORMULAS

Standard normal distribution has $\mu = 0$ and $\sigma = 1$.
Standard score or z score:

$$z = \frac{x - \mu}{\sigma}$$

Prerequisites for approximating binomial by normal:

$$np \geq 5$$
$$nq \geq 5$$

Parameters used when approximating binomial by normal:

$$\mu = np$$
$$\sigma = \sqrt{npq}$$

Parameters used when applying central limit theorem:

$$\mu_{\bar{x}} = \mu$$

$$\sigma_{\bar{x}} = \frac{\sigma}{\sqrt{n}} \qquad \text{(standard error of the mean)}$$

Finite population correction factor used when $n > 0.05N$:

$$\sigma_{\bar{x}} = \frac{\sigma}{\sqrt{n}} \sqrt{\frac{N - n}{N - 1}}$$

Vocabulary List

Define and give an example of each term.

normal distribution
standard normal
  distribution
z score
standard score
continuity correction

central limit
  theorem
standard error of
  the mean
finite population
  correction factor

Review Exercises

5-169   A landfill operation collects daily waste amounts, which are normally distributed with a mean of 26.4 tons and a standard deviation of 2.4 tons.
   (a)   For a randomly selected day, find the probability that the amount of waste is between 26.4 tons and 30.0 tons.

(b) For a randomly selected day, find the probability that the amount of waste is less than 27.0 tons.

(c) For a randomly selected day, find the probability that the amount of waste is between 26.0 tons and 28.0 tons.

(d) If the landfill operation is overburdened for 5% of the days it is open, find the minimum amount that causes an overburden.

(e) For 40 randomly selected days, find the probability that the mean amount of waste is less than 26.0 tons.

5-170 Scores on a hearing test are normally distributed with a mean of 600 and a standard deviation of 100.

(a) If one subject is randomly selected, find the probability that the score is between 600 and 735.

(b) If one subject is randomly selected, find the probability that the score is more than 450.

(c) If one subject is randomly selected, find the probability that the score is between 500 and 800.

(d) If a job requires a score in the top 80%, find the lowest acceptable score.

(e) If 50 subjects are randomly selected, find the probability that their mean score is between 600 and 635.

5-171 A sociologist finds that for a certain segment of the population, the numbers of years of formal education are normally distributed with a mean of 13.20 years and a standard deviation of 2.95 years.

(a) For a person randomly selected from this group, find the probability that he or she has between 13.20 and 13.50 years of education.

(b) For a person randomly selected from this group, find the probability that he or she has at least 12.00 years of education.

(c) Find the first quartile, Q_1. That is, find the value separating the lowest 25% from the highest 75%.

(d) If an employer wants to establish a minimum education requirement, how many years of education would be required if only the top 5% of this group would qualify?

(e) If 35 people are randomly selected from this group, find the probability that their mean years of education is at least 12.00 years.

5-172 Scores on a standard I.Q. test are normally distributed with a mean of 100 and a standard deviation of 15.

(a) Find the probability that a randomly selected subject will achieve a score between 90 and 120.

(b) Find the probability that a randomly selected subject will achieve a score above 105.

(c) If 30 subjects are randomly selected and tested, find the probability that their mean I.Q. score is above 105.

(d) Find P_{95}, the I.Q. score separating the top 5% from the lower 95%.

(e) Find P_{15}, the I.Q. score separating the bottom 15% from the top 85%.

5-173 The probability of a particular computer component being defective is 0.2. If 500 of these components are randomly selected and tested, find the probability of getting at least 120 defective components.

5-174 It has been found that 4% of the respondents fail to answer a particular question on a survey. Find the probability that among 150 respondents, at least five fail to answer the question.

5-175 A study has shown that among people without any preschool education, 32% were employed at age 19. If these figures are correct, find the probability that for a group of 100 people without any preschool education, 40 or fewer are employed at age 19.

5-176 The annual divorce rate is reported as 22.8 divorces per 1000 married women. If 225 married women are randomly selected, find the probability that at least five of them will be divorced in the next year.

5-177 Errors on a scale are normally distributed with a mean of 0 kilograms and a standard deviation of 1 kilogram. One item is randomly selected and weighed. (The errors can be positive or negative.)
(a) Find the probability that the error is between 0 and 0.74 kilograms.
(b) Find the probability that the error is greater than 1.76 kilograms.
(c) Find the probability that the error is greater than -1.08 kilograms.

5-178 Suppose that 800 scores have a very unusual distribution with $\mu = 120.0$ and $\sigma = 4.0$. If 50 of these scores are randomly selected, find the probability that their mean is above 121.0.

Chapter 5
Case Study Activity

In Table 5-4, we listed 30 collections of samples with ten random numbers between 1 and 9 in each sample. Use a table of random digits, or a telephone book, or a computer to observe or generate 50 collections of samples with ten random numbers between 0 and 9. Construct a frequency table and histogram of the combined sample of 500 scores, and find the mean and standard deviation. Then find the mean of each sample and, using the 50 sample means, construct a frequency table and histogram and find the mean and standard deviation. Compare the results obtained from the combined sample of 500 scores to those obtained for the 50 sample means.

CHAPTER 6

6-1 Overview
We define chapter **objectives**.

6-2 Testing a Claim About a Mean
We present the **general procedure** for testing hypotheses. The hypotheses considered in this section relate to claims made about a population mean.

6-3 P-Values (Optional)
We present the **P-value approach** commonly used in computer software packages.

6-4 _t_ Test
We present the **t distribution,** which is used in certain hypothesis tests instead of the normal distribution.

6-5 Tests of Proportions
We describe the method of testing a hypothesis made about a population **proportion** or **percentage.**

6-6 Tests of Variances
We describe the method of testing a hypothesis made about the **standard deviation** or **variance** of a population. In such cases, we use the **chi-square distribution.**

Testing
Hypotheses

A fundamental concept in this chapter is that of significance. In Section 6-2, we consider a manufacturer's claim that a new type of tire has a skid distance less than the old tire's mean of 152 feet. A sample of 36 tires yields a mean skid distance of 148 feet, but is that value *significantly* less than 152 feet? Do these sample results really support the manufacturer's claim of a shorter skid distance? This chapter provides us with the ability to answer such questions so that decisions can be made about a variety of different claims.

6-1 Overview

Our main objective in this chapter is to develop an understanding of the concepts that underlie hypothesis testing. We also want you to develop the skills required to execute successfully the test of a hypothesis relating to a mean, proportion, or variance.

A hypothesis is a statement that something is true. The following statements are examples of hypotheses that can be tested by the procedures that we develop in this chapter:

- A consumer claims that a brewery gives less than 32 ounces of beer per bottle.
- A manufacturer claims that a new type of snow tire has a shorter skid distance.
- A tobacco company claims that its cigarettes contain less than 40 milligrams of nicotine.
- A senator claims that 60% of her constituents favor a gun control bill.
- A sociologist claims that the unemployment rate is 9.2%.

Suppose you take a dime from your pocket and claim that it favors heads when it is flipped. You have formulated a hypothesis. Most reasonable people wishing to test your claim would begin by examining the coin to see if it is two-headed or possesses any other gross irregularities. If the dime seems normal, the next reasonable step is to flip the coin several times to see what happens. Let's suppose that heads occurs 94 times out of 100 tosses. On the basis of those sample data, most people would agree with the claim that the coin favors heads.

Perhaps the dime is actually fair and does not favor heads or tails. Perhaps the occurrence of 94 heads out of 100 tosses is simply a very unusual manifestation for a coin that usually exhibits normal behavior. But since the probability of flipping a fair dime 100 times and getting 94 heads is incredibly small, it is more reasonable to conclude that the coin does favor heads. This is the way statisticians think when they are testing hypotheses.

We operate under the assumption that when events that appear to be very unusual occur, they probably occur not by chance but by the influence of another factor. If a gambler started a friendly dice game and began by rolling ten consecutive 7s, it is not likely that he or she would be congratulated for such a streak of good fortune. A more likely reaction would be a request to examine the dice. The suspicious players would be behaving like statisticians as they attributed very unusual events to causes other than pure chance.

This intuitive discussion should reveal a fundamental concept that underlies the method of testing hypotheses. That method involves a variety of standard terms and conditions in the context of an organized procedure. We suggest that you begin the study of this chapter by first reading Section 6-2 casually to obtain a general idea of its concepts. Then read the material more carefully to gain familiarity with the terminology. Subsequent readings should incorporate the details and refinements into the basic procedure. You are not expected to master the principles of hypothesis testing in one reading. It may take several readings for the material to become understandable.

6-2   Testing a Claim About a Mean

The coin and dice examples mentioned in the introduction are largely intuitive and devoid of some of the components that are required in a formal statistical test of a hypothesis. For example, we stated that the occurrence of 94 heads in 100 tosses is a very unusual event, but we made no attempt to specify the exact criteria used to identify unusual events. We will begin this section with another illustrative example, but we will include more of the necessary details. We will then proceed to define the terms, conditions, and procedures that constitute the formal method of a standard hypothesis test.

Consider the claim that a new and more expensive type of snow tire has a shorter skid distance. Because of the costs involved, the owner of a large fleet of cars will purchase these tires only if strongly convinced that the tires really do skid less. The only way to prove or disprove the claim is to test all such tires, but that is clearly impractical. Instead, sample data must be used to form a conclusion. Let's assume that we have obtained sample test results from an independent testing laboratory. We learn that 36 of these tires were tested and, under standard conditions, the mean skid distance for that sample group is found to be $\bar{x} = 148$ feet. The testing laboratory also informs us that skid distances of the traditional type of snow tire are normally distributed with a mean of 152 feet and a standard deviation of 12 feet. We might be tempted to conclude that the hypothesis of a shorter skid distance is correct simply because the sample mean of 148 feet is less than the population mean of 152 feet. But let's analyze this critically. We know that sample data fluctuate and display errors of various amounts. Tires produced by the same workers and the same machines do not necessarily provide identical performances. Recognizing this, we formulate a key question: Does the sample mean score of 148 represent a statistically **significant** decrease from the population mean of 152, or is the difference more likely due to chance variations in the skid distances? Obviously, if the sample of new tires produced a mean skid distance greater than 152 feet, we would immediately conclude that their added cost is not warranted. Also, if the sample mean were something like 50 feet, the improvement would obviously be significant and not due to chance. However, many sample mean values are not quite so obvious in their implications. Where do we draw the line?

Why Professional Articles Are Rejected

In an editorial published in the *Journal of Applied Psychology,* John P. Campbell cited the major reasons for rejecting submitted articles. After citing meaningless topics, unclear writing, and inappropriate methodology, Campbell stated that "a much less frequent disqualifier was low statistical power. In some instances too small a sample was a reason for rejection." He also stated that a frequent reason for rejection was that "the procedure (not the statistical analysis) used in the study could not answer the question(s) that were asked. For example, there was every reason to believe that the measures used in a study had no reliability or validity or that alternative explanations for the results were much more likely." He went on to say that the acceptance or rejection of a submitted paper depended heavily on expertise, sound measurement, suitable methodology, and clear writing.

What critical value separates significant decreases from those due to chance fluctuations? From now on, whenever we use the term *significant,* we will mean *statistically significant.*

Let's pursue the claim that the new tires have shorter skid distances by examining the data in relation to what we know about sample means. Specifically, from the central limit theorem (Section 5-6) we know that sample means tend to be normally distributed with a mean equal to the population mean and with a standard deviation equal to $\sigma/\sqrt{n}$.

$$\mu_{\bar{x}} = \mu$$

$$\sigma_{\bar{x}} = \frac{\sigma}{\sqrt{n}}$$

We can use that knowledge to determine how unusual a sample mean of 148 is for a sample of 36 tires randomly selected from the population. If 148 could easily occur in a sample of ordinary tires, then we would conclude that the claim of shorter skid distances is unwarranted. But if it is very unlikely that a random sample of 36 tires will produce a mean

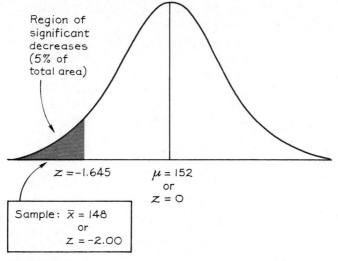

Region of
significant
decreases
(5% of
total area)

$z = -1.645$ $\mu = 152$
or
$z = 0$

Sample: $\bar{x} = 148$
or
$z = -2.00$

Figure 6-1

as low as 148, then we would be inclined to agree with the hypothesis of shorter skid distances.

What exactly do we mean by unusual or very unlikely? Let's arbitrarily select 5% (or a probability of 0.05) as a level that separates a significant difference from a chance fluctuation. (The choice of 0.05 is very common.) That is, we are defining *unusual* to mean that the event has a 5% chance (or less) of occurring. If the probability of getting a sample mean of 148 (or lower) is 0.05 or less, we will conclude that the sample mean represents a significant decrease from the population mean of 152 and we will therefore agree with the claim of a shorter skid distance. Otherwise, we will reject that claim.

The test of the claim now centers on the probability of getting a sample mean of 148. Distributions of sample means can be approximated by normal distributions and since $\mu = 152$ and $\sigma = 12$, we get

$$\mu_{\bar{x}} = \mu = 152$$

$$\sigma_{\bar{x}} = \frac{\sigma}{\sqrt{n}} = \frac{12}{\sqrt{36}} = \frac{12}{6} = 2.00$$

(see Section 5-6). We are applying the central limit theorem and our knowledge about the sample size ($n = 36$) and the population of ordinary tires ($\mu = 152$, $\sigma = 12$) to get these results. Figure 6-1 illustrates these results, as well as our choice of 5% as the level separating chance fluctuations from significant differences.

We find that $\bar{x} = 148$ is equivalent to $z = -2.00$ by computing

$$z = \frac{\bar{x} - \mu_{\bar{x}}}{\sigma_{\bar{x}}} = \frac{148 - 152}{12/\sqrt{36}} = \frac{-4}{2} = -2.00$$

We determine that $z = -1.645$ is the cutoff separating unusual results from chance fluctuations by observing that if the shaded region in the left tail represents 5%, or 0.05, of the total area, then the rightmost limit of that region and $z = 0$ must encompass 45%, or 0.45, of the total. From Table A-3, we see that 0.4500 is halfway between the values for $z = 1.64$ and $z = 1.65$, so we split the difference and make z negative since it is below the mean. This gives $z = -1.645$.

In effect, Figure 6-1 shows that a sample mean of 148 is unusually low. Unusually low sample means have a probability of 0.05 or less, and they fall within the shaded region of Figure 6-1. We therefore support the claim that these new tires have shorter skid distances; that is, the experimental results do warrant such support. Our final conclusion is this: *There is sufficient sample evidence to support the manufacturer's claim that the skid distance is shorter.* Because this kind of procedure is used in decision-making and because the consequences of decisions can often be great, we will examine this method closely.

A **hypothesis test,** or **test of significance,** involves procedures that allow us to make inferences about whole populations by analyzing samples. In this decision-making process, we begin by hypothesizing (sometimes just guessing) about the population. After gathering sample data, we try to determine whether the data support the hypothesis or whether they are statistically significant. If we hypothesize that the new tires have shorter skid distances and then find that the sample mean score is below the mean for ordinary tires, we question the *significance* of that decrease. We have already indicated that such a decrease would be statistically significant and the hypothesis is supported.

We need to formalize this decision-making process. The following are some of the standard terms and components required to do this (see also Table 6-1).

- *Null hypothesis* (denoted by H_0): The statement of a zero or null difference that is directly tested. This will correspond to the original claim if that claim includes the condition of no change or difference. Otherwise, the null hypothesis is the negation of the original claim.

Table 6-1	The null hypothesis is true.	The null hypothesis is false.
We decide to reject the null hypothesis.	Type I error	Correct decision
We fail to reject the null hypothesis.	Correct decision	Type II error

Drug Approval Requires Strict Procedure

The Pharmaceutical Manufacturing Association has reported that the development and approval of a new drug will cost around $87 million and will take around eight years. Extensive laboratory and animal testing is followed by FDA approval for human testing, which is done in three phases. Phase I human testing involves about 80 people, phase II involves about 250 people, and phase III involves between 1000 and 3000 volunteers. Overseeing such an expensive, time-consuming, and complex process would be enough to give anyone a headache, but the process does protect us from dangerous or worthless drugs.

We test the null hypothesis directly in the sense that the final conclusion will be either rejection of H_0 or failure to reject H_0.

- *Alternative hypothesis* (denoted by H_1): The statement that must be true if the null hypothesis is false.
- *Type I error:* The mistake of rejecting the null hypothesis when it is true.
- *Type II error:* The mistake of failing to reject the null hypothesis when it is false.
- *α (alpha):* Symbol used to represent the probability of a type I error.
- *β (beta):* Symbol used to represent the probability of a type II error.
- *Test statistic:* A sample statistic or a value based on the sample data. It is used in making the decision about the rejection of the null hypothesis.
- *Critical region:* The set of all values of the test statistic that would cause us to reject the null hypothesis.
- *Critical value(s):* The value(s) that separates the critical region from the values of the test statistic that would not lead to rejection of the null hypothesis. The critical value(s) depends on the nature of the null hypothesis, the relevant sampling distribution, and the level of significance $α$.
- *Significance level:* The probability of rejecting the null hypothesis when it is true. Typical values selected are 0.05 and 0.01. That is, the values of $α = 0.05$ and $α = 0.01$ are typically used. (We use the symbol $α$ to represent the significance level.)
- *Elation:* The feeling experienced when the techniques of hypothesis testing are mastered.

Determination of the null and alternative hypotheses can be confusing and somewhat difficult. Yet that determination is a critical beginning for the very important process of hypothesis testing. If we are making our own claims, we should arrange the null and alternative hypotheses so that the most serious error is a type I error. In the discussion that follows, we will assume that we are testing a claim made by someone else. Part of the difficulty arises from what might superficially appear to be an inconsistency: In some cases the null hypothesis corresponds to the original claim, while in other cases the null hypothesis is just the opposite. Ideally, all claims would be made in such a way that they would correspond directly to the null hypothesis. Unfortunately, our real world is not ideal, and many claims correspond to what becomes the alternative hypothesis. This text was written with that real consideration in mind. As a result, some examples and exercises involve claims that correspond to the null hypothesis, while others involve claims that correspond to the alternative hypothesis.

Follow these three steps to determine the null and alternative hypotheses:

1. Identify the claim that was made and translate it into symbolic form.

2. Write the symbolic form that must be true when the original claim is false.

3. Of the two symbolic statements, only one of them should contain the condition of equality. Call the statement containing the condition of equality the null hypothesis. The alternative hypothesis is the statement that does not contain the condition of equality.

In conducting a formal statistical hypothesis test, we are *always* testing the *null hypothesis*, whether it corresponds to the original claim or not. Sometimes the null hypothesis corresponds to the original claim and sometimes it corresponds to the opposite of the original claim. Since we always test the null hypothesis, we will be testing the original claim in some cases and the opposite of the original claim in other cases. Carefully examine the following examples.

	Original Claim			
	The mean grade is 75.	The mean grade is not 75.	The mean grade is at least 75.	The mean grade is above 75.
Step 1: Symbolic form of original claim.	$\mu = 75$	$\mu \neq 75$	$\mu \geq 75$	$\mu > 75$
Step 2: Symbolic form that is true when original claim is false.	$\mu \neq 75$	$\mu = 75$	$\mu < 75$	$\mu \leq 75$
Step 3: Null hypothesis H_0 (must contain equality).	H_0: $\mu = 75$	H_0: $\mu = 75$	H_0: $\mu \geq 75$	H_0: $\mu \leq 75$
Alternative hypothesis H_1 (cannot contain equality).	H_1: $\mu \neq 75$	H_1: $\mu \neq 75$	H_1: $\mu < 75$	H_1: $\mu > 75$

We always test the null hypothesis and our initial conclusion will always be one of the following:

1. Fail to reject the null hypothesis, H_0.

2. Reject the null hypothesis, H_0.

Some texts say that we "accept the null hypothesis" instead of "fail to reject the null hypothesis." Whether we use the term *accept* or the term *fail to reject*, we should recognize that *we are not proving the null hypothesis;*

we are merely saying that the sample evidence is not strong enough to warrant rejection of the null hypothesis. It's like a jury saying that there is not enough evidence to convict a suspect. The term *accept* is somewhat misleading since it seems incorrectly to imply that the null hypothesis has been proved. The phrase *fail to reject* says, more correctly, "let's withhold judgment because the available evidence isn't strong enough." In this text, we will use *fail to reject the null hypothesis* instead of *accept the null hypothesis*.

We either fail to reject the null hypothesis or we reject the null hypothesis. That type of conclusion is fine for those of us with the wisdom to take a statistics course, but it's usually necessary to use simple nontechnical terms in stating what the conclusion suggests. The following example illustrates how this can be accomplished.

Original claim	Null hypothesis	Initial conclusion	Nontechnical translation
The mean grade is 75.	$H_0: \mu = 75$ (Original claim is H_0.)	Reject H_0 →	There is sufficient evidence to warrant rejection of the claim that the mean grade is 75.
		Fail to reject H_0 →	There is not sufficient evidence to warrant rejection of the claim that the mean grade is 75.
The mean grade is greater than 75.	$H_0: \mu \leq 75$ (Original claim is H_1.)	Reject H_0 →	The sample data support the claim that the mean grade is greater than 75.
		Fail to reject H_0 →	There is not sufficient evidence to support the claim that the mean grade is greater than 75.

Most of our hypothesis testing will be concerned with values of *parameters*—numerical characteristics of a population such as the mean, variance, or proportion. The following rules will be helpful in these tests:

RULES

1. For the sake of simplicity, always arrange the original claim and its alternative so that the null hypothesis contains the condition of equality. The alternative hypothesis will then involve exactly one of the three signs >, <, or ≠. (If *we* are making the claim, we should arrange the null and alternative hypotheses so that the type I error is the more serious error.)

2. H_0 is presumed true until significant sample evidence suggests that it be rejected. As a result of the test, we either reject H_0 or we fail to reject H_0. These methods of statistically testing H_0 *never prove* that H_0 is true.
3. The value of α should be selected *before* the test is conducted. If a type I error is very serious and has dire consequences, α should be a small value such as 0.01. Less serious type I errors usually involve larger values of α. For the significance level α, the values of 0.05 and 0.01 are very common.

The essential steps for testing hypotheses are as follows:

1. Identify the specific claim or hypothesis to be tested, and express it in symbolic form.

2. Express in symbolic form the statement that would be true if the original claim is false.

3. Of the two expressions found in the first two steps, only one should contain the condition of equality, and that is the null hypothesis H_0. The other expression is the alternative hypothesis H_1.

4. Based on the seriousness of a type I error, select α (the probability of making a type I error). The probability of a type II error (β) will be determined when the sample size is fixed. (We will not deal with β.)

5. Determine which sample statistic is appropriate. Also determine the appropriate sampling distribution.

6. Using the sample data, compute the test statistic.

7. Using the computed test statistic and the corresponding critical value, either reject H_0 or assert that the sample data do not warrant a rejection of H_0. (At this point, it is usually wise to graph the appropriate sampling distribution with the test statistic, critical region, and critical value(s) identified.)

8. In simple nontechnical terms, state what the results suggest. Figure 6-2 summarizes these key steps. To the right of the diagram are the actual results of the steps applied to the example at the beginning of this section.

It is easy to become entangled in a web of cookbook-type steps without ever understanding the underlying rationale for the procedure. The key to that understanding lies with recognition of this concept: **If an event can easily occur, we attribute it to chance, but if the event appears to be unusual, we attribute that significant departure to the presence of different characteristics.** If we keep this idea in mind as we examine various examples, the method of hypothesis testing will become meaningful instead of a rote mechanical process.

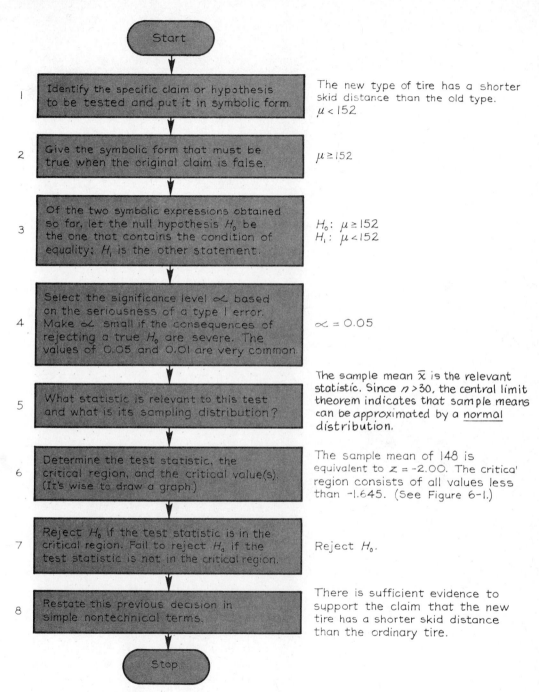

The new type of tire has a shorter skid distance than the old type. $\mu < 152$

$\mu \geq 152$

$H_0: \mu \geq 152$
$H_1: \mu < 152$

$\alpha = 0.05$

The sample mean $\bar{x}$ is the relevant statistic. Since $n > 30$, the central limit theorem indicates that sample means can be approximated by a <u>normal</u> distribution.

The sample mean of 148 is equivalent to $z = -2.00$. The critical region consists of all values less than -1.645. (See Figure 6-1.)

Reject H_0.

There is sufficient evidence to support the claim that the new tire has a shorter skid distance than the ordinary tire.

Figure 6-2   Procedure to be used in testing hypotheses. Adjacent to the diagram, we have summarized the key steps for the example discussed at the beginning of this section.

Two-tailed test

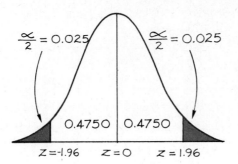

Claim: Prisoners have a mean I.Q. equal to 100.

$H_0: \mu = 100$
$H_1: \mu \neq 100$
$\alpha = 0.05$

Reject H_0 if the sample mean $\bar{x}$ is significantly above 100 (right tail) or below 100 (left tail).

Left-tailed test

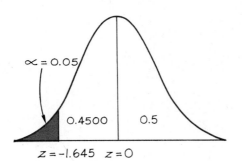

Claim: Boxers have a mean I.Q. less than 100.

$H_0: \mu \geq 100$
$H_1: \mu < 100$
$\alpha = 0.05$

Reject H_0 if the sample mean $\bar{x}$ is significantly below 100 (left tail)

Right-tailed test

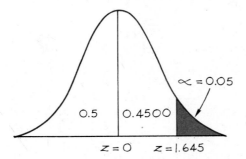

Claim: Teachers have a mean I.Q. greater than 100.

$H_0: \mu \leq 100$
$H_1: \mu > 100$
$\alpha = 0.05$

Reject H_0 if the sample mean $\bar{x}$ is significantly greater than 100 (right tail).

Figure 6-3   Critical regions are shaded.

Product Testing Is Big Business

The United States Testing Company in Hoboken, New Jersey, is the world's largest independent product-testing laboratory. Senior Vice President Noel Schwartz says that they are often hired to verify advertising claims. "We have to produce the data to show that nine out of ten customers really did prefer X brand to Y brand. The television networks and the Federal Trade Commission demand it." Schwartz goes on to say that the most difficult part of his job is "telling a client when his product stinks. But if we didn't do that, we'd have no credibility." He says that there have been a few cases when a client wanted to have results fabricated, but most clients want honest results. Some clients sought other laboratories when they received unfavorable reports.

United States Testing evaluates laundry detergents, cosmetics, insulation materials, zippers, pantyhose, football helmets, toothpaste, fertilizer, and a variety of other products.

Our first example of hypothesis testing involves the claim that the new type of tire has a shorter skid distance. The test used is called a **left-tailed** test because the critical region of Figure 6-1 is in the extreme left region under the curve. We reject the null hypothesis H_0 if our test statistic is in the critical region, because that indicates a significant conflict between the null hypothesis and the sample data. Some tests will be **right-tailed,** with the critical region located in the extreme right region under the curve. Other tests may be **two-tailed,** because the critical region is comprised of two components located in the two extreme regions under the curve. Figure 6-3 illustrates these possibilities. In the two-tailed case, α is divided equally between the two components comprising the critical region. In each case, we convert the claim into symbolic form and then determine the symbolic alternative. The null hypothesis H_0 becomes the symbolic statement containing the condition of equality. The alternative hypothesis becomes the other symbolic statement. We reject H_0 if there is significant evidence supporting H_1. For this reason, critical regions correspond to the extremes indicated by H_1. In Figure 6-3, the inequality sign points to the critical region.

Sign used in H_1	Type of test
$>$	Right-tailed
$<$	Left-tailed
$\neq$	Two-tailed

EXAMPLE

The engineering department of a car manufacturer claims that the fuel consumption rate of one model is equal to 35 miles per gallon. The advertising department wants to test this claim to see if the announced figure should be higher or lower than 35 miles per gallon. The quality-control group suggests that $\sigma = 4$ miles per gallon and a sample of 50 cars yields $\bar{x} = 33.6$ miles per gallon. Test the claim of the engineering department by using a 0.05 level of significance.

Solution

We outline our test of the manufacturer's claim by following the scheme in Figure 6-2. The result is shown in Figures 6-4 and 6-5. The test is two-tailed because a sample mean significantly greater than 35 miles per gallon (right tail) or less than 35 miles per gallon (left tail) is strong evidence against the null hypothesis that $\mu = 35$ miles per gallon. Our sample mean of 33.6 miles per gallon is found to be equivalent to $z = -2.47$ through the following computation:

$$z = \frac{\bar{x} - \mu_{\bar{x}}}{\sigma_{\bar{x}}} = \frac{33.6 - 35}{4/\sqrt{50}} = -2.47$$

The critical z values are found by distributing $\alpha = 0.05$ equally between the two tails to get 0.025 in each tail. We then refer to Table A-3 (since we are assuming a normal distribution) to find the z value corresponding to 0.5—0.025 or 0.4750. After finding $z = 1.96$, we use the property of symmetry to conclude that the left critical value is -1.96.

Figure 6-4

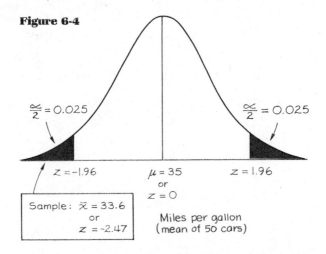

$\frac{\alpha}{2} = 0.025$ $\frac{\alpha}{2} = 0.025$

$z = -1.96$ $\mu = 35$ $z = 1.96$
             or
             $z = 0$

Sample: $\bar{x} = 33.6$
        or
        $z = -2.47$

Miles per gallon
(mean of 50 cars)

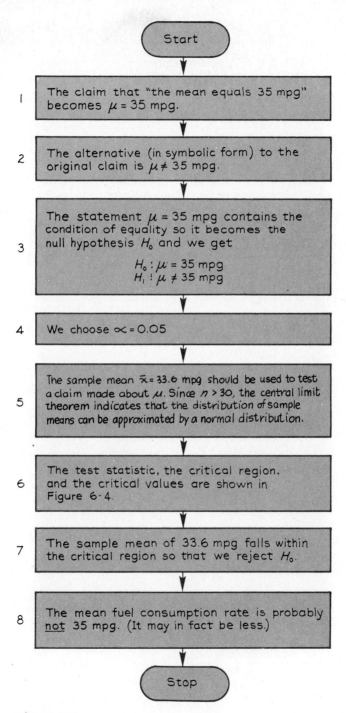

Figure 6-5

EXAMPLE

A brewery distributes beer in bottles labeled 32 ounces. The local Bureau of Weights and Measures randomly selects 50 of these bottles, measures their contents, and obtains a sample mean of 31.0 ounces. Assuming that σ is known to be 0.75 ounces, is it valid at the 0.01 significance level to conclude that the brewery is cheating the consumer?

Solution

The brewery is cheating the consumer if they give significantly less than 32 ounces of beer. We outline the test of the claim that the mean is less than 32 ounces by again following the model of Figure 6-2. The results are presented in Figures 6-6 and 6-7. The z value of −9.43 is computed as follows:

$$z = \frac{\bar{x} - \mu_{\bar{x}}}{\sigma_{\bar{x}}} = \frac{31 - 32}{0.75/\sqrt{50}} = -9.43$$

The critical z value is found in Table A-3 as the z value corresponding to an area of 0.4900.

In presenting the results of a hypothesis test, it is not necessary to show all of the steps included in Figure 6-7. However, the results should include the null hypothesis, the alternative hypothesis, the calculation of the test statistic, a graph such as Figure 6-6, the initial conclusion (reject

Figure 6-6

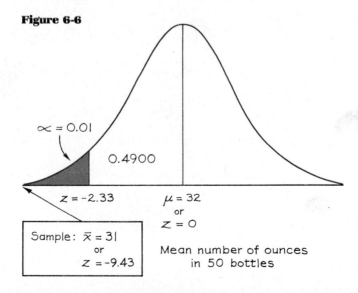

$\propto$ = 0.01

0.4900

z = −2.33

μ = 32
or
z = 0

Sample: $\bar{x}$ = 31
or
z = −9.43

Mean number of ounces
in 50 bottles

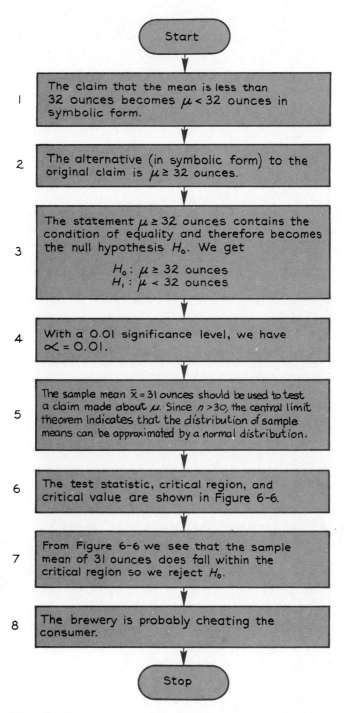

Figure 6-7

H_0 or fail to reject H_0) and the final conclusion stated in nontechnical terms. The graph should show the test statistic, critical value(s), critical region, and significance level.

A potentially unrealistic feature of the preceding example is the assumption that $\sigma = 0.75$. Realistic tests of hypotheses must often be made without knowledge of the population standard deviation. **If the sample is large enough (31 or larger), we can compute the sample standard deviation, and we may be able to use that value of s as an estimate for σ.** For smaller samples with unknown standard deviations, we may be able to use a t statistic (discussed in Section 6-4).

One of the steps in our procedure for testing hypotheses involves the selection of $P(\text{type I error}) = \alpha$, but we do not select $P(\text{type II error}) = \beta$. We could select both α and β—the required sample size would then be determined—but the usual procedure used in research and industry is to determine in advance the values of α and n, so that the value of β is determined. The following practical considerations may be relevant to some experiments:

1. To decrease both α and β, increase the sample size n.
2. For any fixed sample size n, a decrease in α will cause an increase in β. Conversely, an increase in α will cause a decrease in β.

A more detailed discussion of β and computations involving β are normally beyond the scope of an introductory statistics text, so we will not pursue the role of β any further.

6-2 Exercises A
Testing a Claim About a Mean

6-1 The null hypothesis should include the condition of equality, while the alternative hypothesis should involve one of the three signs $>$, $<$, or $\neq$. For each of the following claims, identify H_0 and H_1 as in the following example: The mean I.Q. of doctors is greater than 110.

$$H_0: \mu \le 110$$
$$H_1: \mu > 110$$

(a) The mean age of professors is more than 30 years.
(b) The mean I.Q. of criminals is above 100.
(c) The mean I.Q. of college students is at least 100.
(d) The mean annual household income is at least $12,300.
(e) The mean monthly maintenance cost of an aircraft is $3,271.
(f) The mean height of females is 1.6 meters.
(g) The mean annual salary of air traffic controllers is more than $26,000.
(h) The mean annual salary of college presidents is under $50,000.

(i)   The mean weight of girls at birth is at most 3.2 kilograms.
(j)   The mean life of a car battery is not more than 46 months.

6-2   Identify the type I error and the type II error corresponding to each claim in Exercise 6-1.

6-3   For each claim in Exercise 6-1, categorize the hypothesis test as a right-tailed test, a left-tailed test, or a two-tailed test.

6-4   Find the critical z value for the following conditions. In each case, assume that the normal distribution applies so that Table A-3 can be used.
(a)   Right-tailed test; $\alpha = 0.05$.
(b)   Right-tailed test; $\alpha = 0.01$.
(c)   Two-tailed test; $\alpha = 0.05$.
(d)   Two-tailed test; $\alpha = 0.01$.
(e)   Left-tailed test; $\alpha = 0.05$.
(f)   Left-tailed test; $\alpha = 0.02$.
(g)   Two-tailed test; $\alpha = 0.10$.
(h)   Right-tailed test; $\alpha = 0.005$.
(i)   Right-tailed test; $\alpha = 0.025$.
(j)   Left-tailed test; $\alpha = 0.025$.

In each of the following exercises, test the given hypotheses by following the procedure suggested by Figure 6-2. Draw the appropriate graph, as in Figure 6-1.

6-5   Test the claim that $\mu \leq 100$ given a sample of $n = 81$ for which $\bar{x} = 100.8$. Assume that $\sigma = 5$, and test at the $\alpha = 0.01$ significance level.

6-6   Test the claim that $\mu \leq 40$ given a sample of $n = 150$ for which $\bar{x} = 41.6$. Assume that $\sigma = 9$, and test at the $\alpha = 0.01$ significance level.

6-7   Test the claim that $\mu \geq 20$ given a sample of $n = 100$ for which $\bar{x} = 18.7$. Assume that $\sigma = 3$, and test at the $\alpha = 0.05$ significance level.

6-8   Test the claim that $\mu \geq 15.5$ given a sample of $n = 45$ for which $\bar{x} = 14.3$. Assume that $\sigma = 5.5$, and test at the $\alpha = 0.05$ significance level.

6-9   Test the claim that $\mu = 500$ given a sample of $n = 300$ for which $\bar{x} = 510$. Assume that $\sigma = 100$, and test at the $\alpha = 0.10$ significance level.

6-10   Test the claim that $\mu = 65$ given a sample of $n = 50$ for which $\bar{x} = 66.1$. Assume that $\sigma = 4$, and test at the $\alpha = 0.05$ significance level.

6-11   Test the claim that $\mu > 40$ given a sample of $n = 50$ for which $\bar{x} = 42$. Assume that $\sigma = 8$, and test at the $\alpha = 0.05$ significance level.

6-12   Test the claim that $\mu < 75.0$ given a sample of $n = 32$ for which $\bar{x} = 73.8$. Assume that $\sigma = 4.2$, and test at the $\alpha = 0.10$ significance level.

6-13   A sociologist finds that for a certain population, the mean number of years of education is 13.20, while the standard deviation is 2.95. In one region, a random sample of 60 people is drawn from this population, and the sample mean is 13.87 years. At the 0.05 level of significance, test the claim that the mean for this region is the same as the mean of the population.

6-14   Triple X sugar is packed in boxes labeled 5 pounds. When 100 packages are randomly selected and measured, the sample mean and standard deviation are found to be 4.95 pounds and 0.15 pound, respectively. At the 0.01 significance level, test the claim that the sample comes from a population in which the mean weight is less than 5 pounds. (Assume that the sample standard deviation can be used for σ.)

6-15   On a college entrance examination, the mean score is 500, while the standard deviation is 100. A high school principal boasts better than average scores for her graduates. To support this claim, she randomly selects 75 graduates and determines that the mean score for this sample is 515. Is her claim actually supported by the sample data? Let $\alpha = 0.05$.

6-16   The mean time between failures for a certain type of radio used in light aircraft is 420 hours. Suppose 35 new radios have been modified for more reliability, and tests show that the mean time between failures for this sample is 385 hours. Assume that σ is known to be 24 hours and let $\alpha = 0.05$. Test the claim that the modifications improved reliability. (Note that improved reliability should correspond to a *longer* mean time between failures.)

6-17   A certain nighttime cold medicine bears a label indicating the presence of 600 milligrams of acetaminophen in each fluid ounce of the drug. The Food and Drug Administration randomly selects 65 1-ounce samples and finds that the mean acetaminophen content is 589 milligrams, while the standard deviation is 21 milligrams. With $\alpha = 0.01$, test the claim that the population mean is equal to 600 milligrams. (Assume that the sample standard deviation can be used for σ.)

6-18   A paint is applied to tin panels and baked for 1 hour so that the mean index of hardness is 35.2. Suppose 38 test panels are painted and baked for 3 hours producing a sample mean index of hardness equal to 35.9. Assuming that $\sigma = 2.7$, test (at the $\alpha = 0.05$ significance level) the claim that longer baking does not affect hardness of the paint.

6-19   Excluding Monday mornings, a secretary can type an average of 50 words per minute (with a standard deviation of 4 words). The secretary's boss will invest in a new typewriter if the new model can be shown to increase productivity. Suppose 40 tests are conducted on the new model with a resulting mean of 53 words per minute. At the $\alpha = 0.01$ significance level, test the claim that the new model increases productivity.

6-20   A psychological test of motor skills requires the subject to sort cards into different boxes. The mean time for this task is 125.3 seconds, with a standard deviation of 15.0 seconds. For 50 pianists given the test, the mean time was 120.5 seconds. At the $\alpha = 0.01$ significance level, test the claim that pianists require at least as much time as everyone else.

6-21   Tests on automobile braking reaction times for normal young men have produced a mean and standard deviation of 0.610 second and 0.123 second, respectively. When 40 young male graduates of a driving school were randomly selected and tested for their braking reaction times, a mean of 0.587 second resulted. At the $\alpha = 0.10$ significance level, test the claim of the driving instructor that his graduates had faster reaction times.

6-22   A dairy produces milk with an average of 18.24 grams of fat in each quart. ($\sigma = 3.88$ grams.) The dairy supervisor hypothesizes that skinny cows would produce milk with less fat. After 50 quarts of milk are randomly selected from 50 different skinny cows, the fat contents are measured with a resulting mean of 17.98 grams. At the $\alpha = 0.05$ significance level, test the supervisor's hypothesis.

6-23   The mean and standard deviation for fuel consumption in miles per gallon for a given car are 22.6 and 3.4, respectively. A revolutionary new spark plug is used and a sample of 33 tests yields a mean of 20.1 miles per gallon. At the $\alpha = 0.05$ significance level, test the claim that the new spark plug does not change fuel consumption as measured in miles per gallon.

6-24   An economist knows that in one region, the mean annual household income is $23,460 while the standard deviation is $3750. A random sample of 200 households is drawn from that region in a way that includes only households with public employees. The mean for this sample group is $19,870. At the 0.02 level of significance, test the claim that households with public employees have a mean income at least as high as the others in the same region. (Assume that the same standard deviation applies.)

6-2   Exercises B
Testing a Claim About a Mean

6-25   At the $\alpha = 0.03$ significance level, test the claim that the following I.Q. scores come from a special group in which the mean is above 100. Use the sample standard deviation as an estimate for σ.

101	110	114	105	79	144	111	99	101	107	103	82
107	90	91	99	95	117	93	103	120	82	123	112
107	89	105	130	106	103	100	118	98	101		

6-26   A brewery claims that consumers are getting a mean volume equal to 32 ounces of beer in their quart bottles. The Bureau of Weights and Measures randomly selects 36 bottles and obtains the following measures in ounces:

32.09	31.89	31.06	32.03	31.42	31.39	31.75	31.53	32.42	31.56
31.95	32.00	31.39	32.09	31.67	31.47	32.45	32.14	31.86	32.09
32.34	32.00	30.95	33.53	32.17	31.81	31.78	32.64	31.06	32.64
32.20	32.11	31.42	32.09	33.00	32.06				

Using the sample standard deviation as an estimate for σ, test the claim of the brewery at the 0.05 significance level.

6-27   The accompanying frequency table summarizes the numbers of defective units produced by a machine for a sample of randomly selected days. Find $\bar{x}$ and s for this sample and use s as an estimate of σ. Then test the claim that the mean number of defective parts per day is equal to 30.0. Use a 0.04 level of significance.

x	f
0– 4	2
5– 9	0
10–14	5
15–19	8
20–24	12
25–29	17
30–34	20
35–39	14
40–44	6
45–49	3

6-28   The accompanying stem-and-leaf plot represents randomly selected test scores of applicants for a civil service job. Use the sample data to find $\bar{x}$ and s, and use s as an estimate of σ. Test the claim that these sample scores come from a population with a mean of at least 75 (the passing grade). Use a 0.10 level of significance.

5	0 6 7
6	0 2 2 4 5
7	1 1 1 3 4 8 8
8	0 0 0 0 1 1 5 5 7 9 9
9	0 4 8 9 9
10	0

6-3 P-Values (Optional)

In Sections 6-1 and 6-2 we introduced the basic procedure used to test a hypothesis or claim made about a population mean. We saw that the conclusion involved a decision either to reject or to fail to reject the null hypothesis, and that decision was determined by a comparison of the test statistic and the critical value. The decision criterion is essentially this: If the test statistic falls in the critical region bounded by the critical value, we reject the null hypothesis. Otherwise, we fail to reject the null hypothesis.

In addition to many professional articles, many software packages refer to another decision criterion, which involves a **probability-value** or **P-value.** Listed below are excerpts from the displayed results of three typical software packages.

$$P\text{-value} = 0.0228 \quad \text{(from STATDISK)}$$

THE TEST IS STATISTICALLY
SIGNIFICANT AT ALPHA $= 0.0228$ $\quad$ (from Minitab)

2-TAIL PROB. 0.0456 $\quad$ (from SPSS)

All of these displays express, in different formats, results using the P-value approach.

While the basic procedure for testing hypotheses remains the same, the decision criterion appears to be different from the one presented in Section 6-2. In this section we will examine this alternative approach.

Let's begin by recalling the essential components of the snow tire example from Section 6-2.

Claim: The new tires have a mean skid distance less than 152 ft.

Null hypothesis: H_0: $\mu \geq 152$

Alternative hypothesis: H_1: $\mu < 152$

Significance level: $\alpha = 0.05$

Test statistic: $z = -2.00$ (see Figure 6-8)

Critical value: $z = -1.645$ (see Figure 6-8)

Conclusion: Reject the null hypothesis; that is, there is sufficient evidence to support the claim that the new tire has a shorter skid distance than 152 ft.

With the P-value approach, we find the probability of observing a value of $\bar{x}$ that is at least as extreme as the $\bar{x}$ found from the sample data. Examine Figure 6-9 and observe that the area to the left of $\bar{x} = 148$ (or $z = -2.00$) can be found by using Table A-3. With $z = 2.00$ in Table A-3, we get the area of 0.4772 as shown in Figure 6-9. That area is subtracted from 0.5 to yield the left-tail area of 0.0228. Now that we have a P-value of 0.0228, what do we do with it?

Figure 6-8

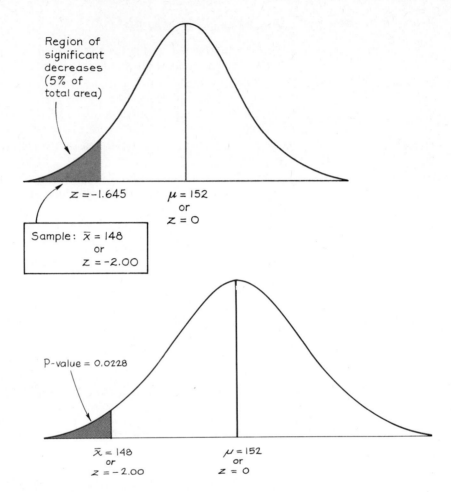

Figure 6-9

Some statisticians prefer simply to report the P-value and leave the conclusion to the reader. Others prefer to use this decision criterion in deciding to reject or fail to reject the null hypothesis:

1. *Reject* the null hypothesis if the P-value is less than the significance level α.
2. *Fail to reject* the null hypothesis if the P-value is equal to or greater than the significance level α.

In this example, the P-value of 0.0228 is less than the significance level of 0.05, so we reject the null hypothesis as we did in Section 6-2. The procedure used in Section 6-2 is sometimes called the **classical** or **traditional approach,** while the procedure of this section is called the probability-value or P-value approach. The classical approach involves a comparison of the test statistic and critical value, while the P-value approach involves either a comparison of the P-value and significance level or a conclusion based on the P-value alone.

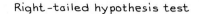

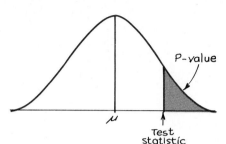

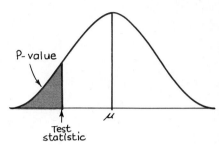

Beware of P-Value Misuse

John P. Campbell, editor of the *Journal of Applied Psychology,* wrote that "books have been written to dissuade people from the notion that smaller P-values mean more important results or that statistical significance has anything to do with substantive significance. It is almost impossible to drag authors away from their P-values, and the more zeros after the decimal point, the harder people cling to them." While it might be necessary to provide a statistical analysis of the results of a study, we should place strong emphasis on the significance of the results themselves.

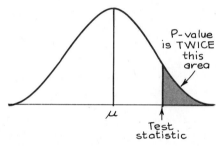

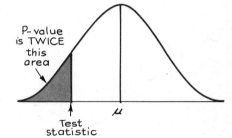

Figure 6-10   P-values

In a right-tailed test, the P-value is obtained by finding the area to the right of the test statistic. However, we must be careful to note that in a two-tailed test, the P-value is *twice* the area of the extreme region bounded by the test statistic. This makes sense when we recognize that the P-value gives us the probability of getting a sample mean that is *at least as extreme* as the sample mean actually obtained, and the two-tailed case has critical or extreme regions in *both* tails. See Figure 6-10.

In Section 6-2 we included an example of a two-tailed hypothesis test, and that example used the classical approach to hypothesis testing. We have extracted the essential components of that example so that they may be compared to the P-value approach. (We use the decision criterion that involves a comparison of the significance level α and the P-value.) Note that the only real difference is the decision criterion, which leads to the same conclusion in both cases.

In Section 6-2, we stated that the significance level α should be selected *before* a hypothesis test is conducted. Many statisticians consider this to be a good practice since it helps to prevent us from using the data

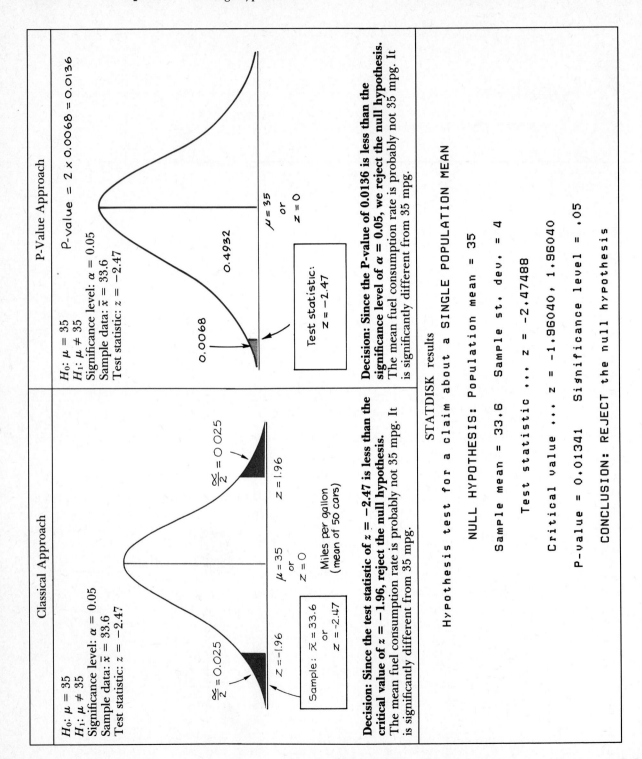

Classical Approach

H_0: $\mu = 35$
H_1: $\mu \neq 35$
Significance level: $\alpha = 0.05$
Sample data: $\bar{x} = 33.6$
Test statistic: $z = -2.47$

$\frac{\alpha}{2} = 0.025$

$z = -1.96$

$\mu = 35$
or
$z = 0$

$z = 1.96$

$\frac{\alpha}{2} = 0.025$

Miles per gallon
(mean of 50 cars)

Sample: $\bar{x} = 33.6$
or
$z = -2.47$

Decision: Since the test statistic of z = −2.47 is less than the critical value of z = −1.96, reject the null hypothesis. The mean fuel consumption rate is probably not 35 mpg. It is significantly different from 35 mpg.

P-Value Approach

H_0: $\mu = 35$
H_1: $\mu \neq 35$
Significance level: $\alpha = 0.05$
Sample data: $\bar{x} = 33.6$
Test statistic: $z = -2.47$

P-value $= 2 \times 0.0068 = 0.0136$

0.0068

0.4932

$\mu = 35$
or
$z = 0$

Test statistic:
$z = -2.47$

Decision: Since the P-value of 0.0136 is less than the significance level of $\alpha = 0.05$, we reject the null hypothesis. The mean fuel consumption rate is probably not 35 mpg. It is significantly different from 35 mpg.

STATDISK results

Hypothesis test for a claim about a SINGLE POPULATION MEAN

NULL HYPOTHESIS: Population mean = 35

Sample mean = 33.6   Sample st. dev. = 4

Test statistic ... z = -2.47488

Critical value ... z = -1.96040, 1.96040

P-value = 0.01341   Significance level = .05

CONCLUSION: REJECT the null hypothesis

to support subjective conclusions or beliefs. They feel that this practice becomes especially important with the P-value approach because we may be tempted to adjust the significance level based on the resulting P-value. With a 0.05 level of significance and a P-value of 0.06, we should fail to reject the null hypothesis, but it is sometimes tempting to say that a probability of 0.06 is small enough to warrant rejection of the null hypothesis. Consequently, we should always select the significance level first. Other statisticians feel that prior selection of a significance level reduces the usefulness of P-values. They contend that no significance level should be specified, and the conclusion should be left to the reader. We shall use the decision criterion that involves a comparison of a significance level and the P-value.

6-3   Exercises A
P-Values

In Exercises 6-29 through 6-32, use the P-value and significance level to choose between rejecting the null hypothesis or failing to reject the null hypothesis.

6-29   P-value: 0.03; significance level: $\alpha = 0.05$

6-30   P-value: 0.04; significance level: $\alpha = 0.01$

6-31   P-value: 0.405; significance level: $\alpha = 0.10$

6-32   P-value: 0.09; significance level: $\alpha = 0.10$

In Exercises 6-33 through 6-40, first find the P-value. Then either reject or fail to reject the null hypothesis by assuming a significance level of $\alpha = 0.05$.

6-33 $H_0: \mu \geq 152$
$H_1: \mu < 152$
Test statistic: $z = -1.40$

6-34 $H_0: \mu \geq 100$
$H_1: \mu < 100$
Test statistic: $z = -0.46$

6-35 $H_0: \mu \leq 15.7$
$H_1: \mu > 15.7$
Test statistic: $z = 1.94$

6-36 $H_0: \mu \leq 428$
$H_1: \mu > 428$
Test statistic: $z = 1.66$

6-37 $H_0: \mu = 75.0$
$H_1: \mu \neq 75.0$
Test statistic: $z = 1.66$

6-38 $H_0: \mu = 1365$
$H_1: \mu \neq 1365$
Test statistic: $z = -1.30$

6-39 $H_0: \mu = 2.53$
$H_1: \mu \neq 2.53$
Test statistic: $z = -1.94$

6-40 $H_0: \mu = 12.8$
$H_1: \mu \neq 12.8$
Test statistic: $z = 3.00$

In Exercises 6-41 through 6-48, use the P-value approach to test the given hypotheses.

6-41 Test the claim that $\mu \geq 100$, given a sample of $n = 45$ for which $\bar{x} = 95$. Assume that $\sigma = 15$, and test at the $\alpha = 0.05$ significance level.

6-42 Test the claim that $\mu \leq 500$, given a sample of $n = 35$ for which $\bar{x} = 508$. Assume that $\sigma = 90$, and test at the $\alpha = 0.01$ significance level.

6-43 Test the claim that $\mu = 75.6$, given a sample of $n = 81$ for which $\bar{x} = 78.8$. Assume that $\sigma = 12.0$, and test at the $\alpha = 0.01$ significance level.

6-44 Test the claim that $\mu = 98.6$, given a sample of $n = 200$ for which $\bar{x} = 97.1$. Assume that $\sigma = 2.35$, and test at the $\alpha = 0.10$ significance level.

6-45 At the 0.05 level of significance, test the claim that the mean lengths of long distance telephone calls is at least 10.00 minutes. Sample data consist of 60 randomly selected long distance calls, and the mean for this sample group is 8.94 minutes. Assume that the standard deviation for all such calls is 8.53 minutes.

6-46 The manager of a moving company claims that the average (mean) load weighs more than 10,000 pounds. Test the claim at the 0.01 level of significance. A sample of 540 loads has a mean of 10,410 pounds, and assume that $\sigma = 4227$ pounds.

6-47 In a survey of 950 patrons attending a certain movie, the mean age was found to be 26.1 years. Test the claim that the mean age of people seeing this movie is 25 years. Use a 0.01 level of significance and assume that $\sigma = 13.08$ years.

6-48 At the 0.05 level of significance, test the claim that the mean asking price of a used car is under $4000. Use sample data consisting of 80 cars with a mean price of $3780. Assume that $\sigma = \$3922$.

6-3 Exercises B
P-Values

6-49 What do you know about the null hypothesis if a P-value is found to be greater than 0.5?

6-50 For a random sample of 40 people, what is the mean I.Q. score if a P-value of 0.04 is obtained when testing the claim that $\mu \geq 100$? Assume that $\sigma = 15$.

6-51 For a random sample of 60 adults, what is the mean number of years of education if a P-value of 0.2005 is obtained when testing the claim that $\mu \leq 13.20$ years? Assume that $\sigma = 2.95$ years.

6-52   Find the smallest mean above $23,460 that leads to a rejection of the claim that the mean annual household income is $23,460. Assume a 0.02 significance level, and assume that $\sigma = \$3750$. The sample consists of 50 randomly selected households.

6-4   *t* Test

In Sections 6-2 and 6-3 we introduced the general method for testing hypotheses, but all the examples and exercises involved situations in which the normal distribution can be used. The population standard deviation σ is given, the samples are large, and each hypothesis tested relates to a population mean. In those cases, we can apply the central limit theorem and use the normal distribution as an approximation to the distribution of sample means. A very unrealistic feature of those examples and exercises is the assumption that σ is known. If σ is unknown and the sample is large, we can treat s as if it were σ and proceed as in Section 6-2 or 6-3. This estimation of σ by s is reasonable because large random samples tend to be representative of the population. But small random samples may exhibit unusual behavior, and they cannot be so trusted. Here we consider tests of hypotheses about a population mean, where the samples are small and σ is unknown. We begin by referring to Figure 6-11, which outlines the theory we are describing (see page 296).

Starting at the top of Figure 6-11, we see that our immediate concerns lie only with hypotheses made about one population mean. (In following sections we will consider hypotheses made about population parameters other than the mean.) Figure 6-11 summarizes the following observations:

1. In *any* population, the distribution of sample means can be approximated by the normal distribution as long as the random samples are large. This is justified by the central limit theorem.

2. In populations with distributions that are essentially normal, samples of *any* size will yield means having a distribution that is approximately normal. The value of μ would correspond to the null hypothesis, and the value of σ must be known. If σ is unknown and the samples are large, we can use s as a substitute for σ, since large random samples tend to be representative of the populations from which they come.

3. In populations with distributions that are essentially normal, assume that we randomly select small samples and we do not know the value of σ. The distribution of these sample means is approximately a *student t distribution*, which is described in this section.

4. If our random samples are small, σ is unknown, and the population is grossly nonnormal, then we can use nonparametric methods, some of which are discussed in Chapter 11.

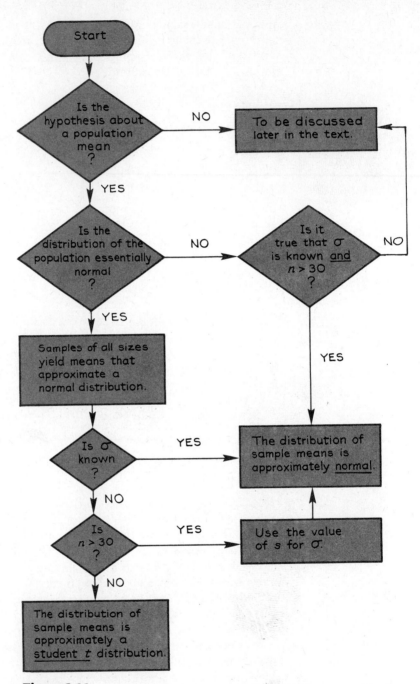

Figure 6-11

We refer to large and small samples, but the number that separates large and small samples is not derived with absolute exactness. It is somewhat arbitrary. However, there is widespread agreement that samples of size $n > 30$ are large enough so that the distribution of their means can be approximated by a normal distribution.

Around 1908, William S. Gosset (1876–1937) developed specific distributions for small samples in response to certain applied problems he encountered while working for a brewery. The Irish brewery where he worked did not allow the publication of research results, so Gosset published under the pseudonym "Student." The small sample results are extremely valuable because there are many real and practical cases for which large samples are impossible or impractical and for which σ is unknown. Gosset, in his original research that led to small sample techniques, found that temperature and ingredient changes associated with brewing often allowed experimentation that provided only small samples. Factors such as cost and time often severely limited the size of a sample, so the normal distribution could not be an appropriate approximation to the distribution of these small sample means. As a result of those earlier experiments and studies of small samples, we can now use the student *t* distributions instead.

TEST STATISTIC

If a population is essentially normal, then the distribution of

$$t = \frac{\bar{x} - \mu}{s/\sqrt{n}}$$

is essentially a **student *t* distribution** for all samples of size n.

We will not discuss the complicated mathematical equations that correspond to this student *t* distribution. Instead, we list critical values for this distribution in Table A-4. (Note that the student *t* distribution of Table A-4 involves only the critical values of *t* corresponding to common choices of α.) The critical *t* value is obtained by locating the proper value for degrees of freedom in the left column and then proceeding across that corresponding row until reaching the number directly below the applicable value of α. Roughly stated, degrees of freedom correspond to the number of values that may vary after certain restrictions have been imposed on all values. For example, if 10 scores must total 50, we can freely assign values to the first 9 scores, but the tenth score would then be determined so that there would be 9 degrees of freedom. In tests on a mean, the number of degrees of freedom is simply the sample size minus 1.

degrees of freedom $= n - 1$

For example, suppose we are testing at the $\alpha = 0.05$ significance level the hypothesis that $\mu = 100$ for some population, and we have only a sample of 20 for which $\bar{x} = 102$ and $s = 5$. (We assume that the sample is random and that the population is essentially normal.) The critical t value is extracted from Table A-4 by noting the following:

1. The test involves two tails since H_0 is $\mu = 100$ and H_1 becomes $\mu \neq 100$.
2. $\alpha = 0.05$ for the two-tailed case.
3. The sample size is $n = 20$, so there are $20 - 1$, or 19, degrees of freedom.

Locating 19 at the left column and 0.05 (two tails) at the top row, we determine that the critical t value is 2.093. Actually, $t = \pm 2.093$ since the test is two-tailed. The test statistic is obtained by computing

$$t = \frac{\bar{x} - \mu}{s/\sqrt{n}} = \frac{102 - 100}{5/\sqrt{20}} = 1.789$$

We would therefore fail to reject H_0, since the test statistic would not be in the critical region. (We will soon present another example in much more detail and with the critical region, critical values, and test statistic depicted in an appropriate graph.)

In order to use the student t distribution, we require that the parent population be essentially normal. The population may not be exactly normal, but if it has only one mode and is basically symmetric, we will generally get good results if we use the student t distribution. If there is strong evidence that the population has a very nonnormal distribution, then nonparametric methods (see Chapter 11) may apply.

The student t distribution has the same general shape and symmetry of the normal distribution, but it reflects the greater variability that is expected with small samples. The standard normal and student t distribution both have a mean of zero, but the standard deviation for the standard normal distribution is one, while that of the student t distribution is greater than one. For a visual comparison, see Figure 6-12. In Figure 6-12 we can also see that, as n gets larger, the student t distribution gets closer to the normal distribution. For values of $n > 30$, the differences between the student t and normal distributions are so small that we can use the critical z values instead of developing a much larger table of critical t values. Consequently, the values in the bottom row of Table A-4 are equal to the corresponding critical z values found from the normal distribution.

Let's use a few examples to illustrate the use of the student t distribution. Remember that the student t distribution applies when we test a claim about a population mean and all the following conditions are met:

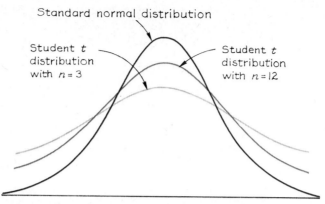

Figure 6-12 The student *t* distribution has the same general shape and symmetry as the normal distribution, but it reflects the greater variability that is expected with small samples.

Conditions for Student *t* Distribution

1. The sample is small ($n \leq 30$); and
2. σ is unknown; and
3. the parent population is essentially normal.

E X A M P L E

A pilot training program usually takes an average of 57.2 hours, but new teaching methods were used on the last class of 25 students. Computations reveal that for this experimental class, the completion times had a mean of 54.8 hours and a standard deviation of 4.3 hours. At the $\alpha = 0.05$ significance level, test the claim that the new teaching techniques reduce the instruction time.

Solution

Let μ represent the mean completion time for the new teaching method. The claim that it reduces instruction time is equivalent to the claim that $\mu < 57.2$ hours. We use the format of Figure 6-2 to outline the test of this claim, and the results are shown in Figures 6-13 and 6-14. We compute the test statistic as follows:

$$t = \frac{\overline{x} - \mu}{s/\sqrt{n}} = \frac{54.8 - 57.2}{4.3/\sqrt{25}} = -2.791$$

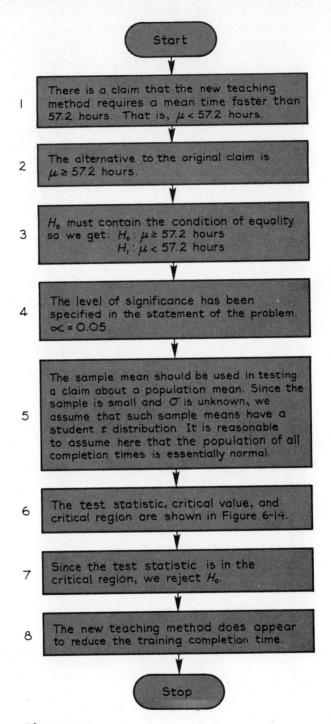

Start

1. There is a claim that the new teaching method requires a mean time faster than 57.2 hours. That is, $\mu < 57.2$ hours.

2. The alternative to the original claim is $\mu \geq 57.2$ hours.

3. H_0 must contain the condition of equality so we get: $H_0 : \mu \geq 57.2$ hours
$H_1 : \mu < 57.2$ hours

4. The level of significance has been specified in the statement of the problem. $\alpha = 0.05$.

5. The sample mean should be used in testing a claim about a population mean. Since the sample is small and σ is unknown, we assume that such sample means have a student t distribution. It is reasonable to assume here that the population of all completion times is essentially normal.

6. The test statistic, critical value, and critical region are shown in Figure 6-14.

7. Since the test statistic is in the critical region, we reject H_0.

8. The new teaching method does appear to reduce the training completion time.

Stop

Figure 6-13

We find the critical *t* value from Table A-4, where we locate 25 − 1, or 24, degrees of freedom at the left column and $\alpha = 0.05$ (one tail) across the top. The critical *t* value of 1.711 is obtained; but since small values of $\bar{x}$ will cause the rejection of H_0, we recognize that $t = -1.711$ is the actual *t* value that is the boundary for the critical region.

It is easy to lose sight of the underlying rationale as we go through this hypothesis-testing procedure, so let's review the essence of the test. We set out to determine whether the sample mean of 54.8 hours is *significantly* below the value of 57.2 hours. Knowing the distribution of sample means (of which 54.8 is one) and choosing a level of significance (5% or $\alpha = 0.05$), we are able to determine the cutoff for what is a significant difference and what is not. Any sample mean equivalent to a *t* score below −1.711 represents a significant difference. The mean of 54.8 hours is significantly below 57.2 hours, so it appears that the new teaching method does reduce the instruction time.

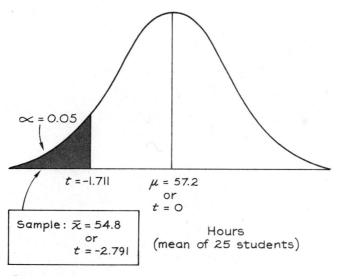

Figure 6-14

EXAMPLE

A tobacco company claims that its best-selling cigarettes contain at most 40 milligrams of nicotine. Test this claim at the 1% significance level by using the results of 15 randomly selected cigarettes for which $\bar{x} = 42.6$ milligrams and $s = 3.7$ milligrams.

Solution

In this example we will list only certain key elements of the solution. The null and alternative hypotheses are as follows.

$$H_0: \mu \leq 40 \text{ milligrams}$$
$$H_1: \mu > 40 \text{ milligrams}$$

The significance level of 1% corresponds to $\alpha = 0.01$. The sample mean should be used in testing a claim about the population mean, and we assume that such sample means have a student t distribution because of the following:

1. It is reasonable to expect that the nicotine contents of all cigarettes of a certain brand are essentially a normal distribution.
2. σ is unknown.
3. The sample size ($n = 15$) is small. It is less than or equal to 30.

The test statistic is

$$t = \frac{\overline{x} - \mu}{s/\sqrt{n}} = \frac{42.6 - 40}{3.7/\sqrt{15}} = 2.722$$

In this right-tailed test, the critical value of 2.625 is found from Table A-4 by noting that $\alpha = 0.01$ (one tail) and degrees of freedom = $15 - 1$, or 14. Since the test statistic of $t = 2.722$ does fall in the critical region, we reject H_0 and conclude that there is sufficient evidence to warrant rejection of the tobacco company's claim. These cigarettes appear to contain significantly more than 40 mg of nicotine.

Commercials, Commercials, Commercials

As we are all exposed to a multitude of commercials and advertisements, we may wonder about the validity of the claims being made. Who regulates or tests these claims? Television networks have their own clearance departments, which require either verification or changes in commercials. The National Advertising Division is a branch of the Council of Better Business Bureaus, which also investigates advertising claims. The Federal Trade Commission and local district attorneys also get into the act. As a result of these efforts, Firestone had to drop a claim that its tires resulted in 25% faster stops, Warner Lambert had to spend $10 million informing customers that Listerine doesn't prevent or cure colds, and General Motors had to stop saying that the Vega handles better. But not all advertising claims are so scrutinized. Many deceptive ads are voluntarily discontinued as soon as any inquiry is started. Others escape rigorous testing simply because the regulatory mechanisms cannot keep up with the thousands of claims that come and go.

P-Values

In the examples presented in this section, we used the classical approach to hypothesis testing. However, much of the literature and many computer packages will use the P-value approach. Shown below, for example, is a sample of the computer display for the last example.

```
Hypothesis test for a claim about a SINGLE POPULATION MEAN

   NULL HYPOTHESIS: Population mean <= 40

   Sample mean = 42.6

   Sample st. dev. = 3.7

   Test statistic ... t = 2.72155

   Critical value ... t = 2.62610

P-value = 8.27152E-03          Significance level = .01

   CONCLUSION: REJECT the null hypothesis
```

Since the *t* distribution table (Table A-4) includes only selected values of α, we cannot usually find the specific P-value from Table A-4. Instead, we can use that table to identify limits that contain the P-value. In the last example we found the test statistic to be $t = 2.722$, and we know that the test is one-tailed with 14 degrees of freedom. By examining the row of Table A-4 corresponding to 14 degrees of freedom, we see that the test statistic of 2.722 falls between the table values of 2.977 and 2.625, which, in a one-tailed test, correspond to $\alpha = 0.005$ and $\alpha = 0.01$. While we cannot determine the exact value of the P-value, we do know that it must fall between 0.005 and 0.01 so that

$$0.005 < \text{P-value} < 0.01.$$

With a significance level of 0.01 and a P-value less than 0.01, we would reject the null hypothesis as we did in the classical approach.

In the next section we test hypotheses relating to claims made about population proportions or percentages.

6-4   Exercises A
t Test

In Exercises 6-53 and 6-54, find the critical t value suggested by the given data.

6-53 (a) H_0: $\mu = 12$ (b) H_0: $\mu \leq 50$
          $n = 27$ $n = 17$
          $\alpha = 0.05$ $\alpha = 0.10$
       (c) H_0: $\mu \geq 1.36$ (d) H_0: $\mu = 1.36$
          $n = 6$ $n = 6$
          $\alpha = 0.01$ $\alpha = 0.01$
       (e) H_0: $\mu \geq 10.75$
          $n = 29$
          $\alpha = 0.01$

6-54 (a) H_0: $\mu \leq 100$ (b) H_1: $\mu \neq 500$
          $n = 27$ $n = 16$
          $\alpha = 0.10$ $\alpha = 0.05$
       (c) H_1: $\mu < 67.5$ (d) H_1: $\mu > 98.4$
          $n = 12$ $n = 7$
          $\alpha = 0.05$ $\alpha = 0.05$
       (e) H_1: $\mu \neq 75$
          $n = 24$
          $\alpha = 0.05$

In Exercises 6-55 through 6-60, assume that the population is normally distributed.

6-55 Test the claim that $\mu \leq 10$, given a sample of 9 for which $\bar{x} = 11$ and $s = 2$. Use a significance level of $\alpha = 0.05$.

6-56 Test the claim that $\mu \leq 32$, given a sample of 27 for which $\bar{x} = 33.5$ and $s = 3$. Use a significance level of $\alpha = 0.10$.

6-57 Test the claim that $\mu \geq 98.6$, given a sample of 18 for which $\bar{x} = 98.2$ and $s = 0.8$. Use a significance level of $\alpha = 0.025$.

6-58 Test the claim that $\mu \geq 100$, given a sample of 22 for which $\bar{x} = 95$ and $s = 18$. Use a 5% level of significance.

6-59 Test the claim that $\mu = 75$, given a sample of 15 for which $\bar{x} = 77.6$ and $s = 5$. Use a significance level of $\alpha = 0.05$.

6-60 Test the claim that $\mu = 500$, given a sample of 20 for which $\bar{x} = 541$ and $s = 115$. Use a significance level of $\alpha = 0.10$.

In Exercises 6-61 through 6-76, test the given hypothesis by following the procedure suggested by Figure 6-2. Draw the appropriate graph. In each case, assume that the population is approximately normal.

6-61   A manufacturer of chains claims that her product has a mean breaking level of at least 3000 pounds. Test this claim at the $\alpha = 0.005$ significance level if a random sample of 10 chains produces a mean and standard deviation of 2892 pounds and 480 pounds, respectively.

6-62   The skid properties of a snow tire have been tested and the mean skid distance of 154 feet has been established for standardized conditions. A new, more expensive tire is developed, but tests on a sample of 20 new tires yield a mean skid distance of 141 feet with a standard deviation of 12 feet. Because of the cost involved, the new tires will be purchased only if they skid less at the $\alpha = 0.005$ significance level. Based on the sample, will the new tires be purchased?

6-63   Teaching method A produces a mean score of 77 on a standard test. Teaching method B has been introduced, and the first sample of 19 students achieved a mean score of 79 with a standard deviation of 8. At the 5% level of significance, test the claim that method B is better.

6-64   An aircraft manufacturer randomly selects 12 planes of the same model and tests them to determine the distance they require for takeoff. The sample mean and standard deviation are found to be 524 meters and 23 meters, respectively. At the 5% level of significance, test the claim that the mean for all such planes is more than 500 meters.

6-65   A battery indicates that it supplies 1.50 volts, but a sample of 19 such batteries is tested and yields a mean of 1.56 volts and a standard deviation of 0.10 volt. At the $\alpha = 0.02$ significance level, can we conclude that these batteries have a mean voltage level of 1.50 volts?

6-66   A pill is supposed to contain 20.0 mg of phenobarbitol. A random sample of 30 pills yields a mean and standard deviation of 20.5 mg and 1.5 mg, respectively. Are these sample pills acceptable at the $\alpha = 0.02$ significance level?

6-67   The Federal Aviation Administration randomly selects five light aircraft of the same type and tests the left wings for their loading capacities. The sample mean and standard deviation are 16,735 pounds and 978 pounds, respectively. At the 5% level of significance, test the claim that the mean loading capacity for all such aircraft is equal to 17,850 pounds.

6-68   A standard test for braking reaction times has produced an average of 0.75 second for young males. A driving instructor claims that his class of young males exhibits an overall reaction time that is below the average. Test his claim if it is known that 13 of his students (randomly selected) produced a mean of 0.71 second and a standard deviation of 0.06 second. Test at the 1% level of significance.

6-69   Ten randomly selected power lawnmowers are tested for noise. The mean and standard deviation are found to be 95 decibels and 5 decibels, respectively. At the $\alpha = 0.05$ significance level, test the claim that the mean level for all such lawnmowers is below 100 decibels.

6-70 In a study of consumer credit, 25 randomly selected credit-card holders were surveyed, and the mean amount they charged in the past 12 months was found to be $1756, while the standard deviation was $843. Use a 0.025 level of significance to test the claim that the mean amount charged by all credit-card holders was greater than $1500.

6-71 A sociologist designs a test to measure prejudicial attitudes and claims that the mean population score is 60. The test is then administered to 28 randomly selected subjects and the results produce a mean and standard deviation of 69 and 12, respectively. At the 5% level of significance, test the sociologist's claim.

6-72 A biofeedback experiment involves measurement of muscle tension by using an instrument attached to a person's forehead. For a sample group of 16 randomly selected subjects, the mean and standard deviation are found to be 6.3 microvolts and 1.8 microvolts, respectively. At the 5% level of significance, test the claim that the mean of all subjects equals 5.4 microvolts.

6-73 A transportation study involved sampling the number of people who ride a particular bus. For 20 randomly selected working days, the mean number of passengers was found to be 23.6, while the standard deviation was 9.3. At the 0.10 level of significance, test the claim that the mean number of passengers was below 30.

6-74 The mean down time for a computer has been 6.34 hours per week. A new supervisor initiates new procedures; the down times (based on a sample of 23 different weeks) now have a mean of 5.77 hours and a standard deviation of 1.82 hours. At the 5% level of significance, test the new supervisor's claim that the mean down time has been reduced.

6-75 A long-range missile misses its target by an average of 0.88 mile. A new steering device is supposed to increase accuracy, and a random sample of eight missiles is equipped with this new mechanism and tested. These eight missiles miss by distances with a mean of 0.76 mile and a standard deviation of 0.04 mile. At $\alpha = 0.01$, does the new steering mechanism lower the miss distance?

6-76 At the 0.10 level of significance, test the claim that a brewery fills bottles with amounts having a mean greater than 32 ounces. A sample of 27 bottles produces a mean of 32.2 ounces and a standard deviation of 0.4 ounce.

6-4   Exercises B
t Test

6-77   A standard final examination in an elementary statistics course produces a mean score of 75. At the 5% level of significance, test the claim that the following sample scores reflect an above-average class:

> 79 79 78 74 82 89 74 75 78 73
> 74 84 82 66 84 82 82 71 72 83

6-78   A sample of beer cans labeled 16 ounces is randomly selected and the actual contents accurately measured. The results (in ounces) are as follows. Is the consumer being cheated?

> 15.8 16.2 16.3 15.9 15.5
> 15.9 16.0 15.6 15.8

6-79   What do you know about the P-value in each of the following cases?
(a) H_0: $\mu \leq 5.00$
  $n = 10$
  Test statistic: $t = 2.205$
(b) H_0: $\mu = 5.00$
  $n = 20$
  Test statistic: $t = 2.678$
(c) H_0: $\mu \geq 5.00$
  $n = 16$
  Test statistic: $t = -1.234$

6-80   Some computer programs approximate critical t values by

$$t = \sqrt{DF \cdot (e^{A^2/DF} - 1)}$$

$$\text{where } DF = n - 1,$$

$$e = 2.718,$$

$$A = z\left(\frac{8\,DF + 3}{8\,DF + 1}\right)$$

and z is the critical z score.

Use this approximation to find the critical t score corresponding to $n = 10$ and a significance level of 0.05 in a right-tailed case. Compare the result to the critical t found in Table A-4.

6-5   Tests of Proportions

In Section 1-4 we presented the four different levels of measurement (nominal, ordinal, interval, ratio). We saw that data at the nominal level of measurement lacked any real numerical significance and was essentially qualitative in nature. One way to make a quantitative analysis using such qualitative data is to represent that data in the form

Misleading Statistics

Money reported that among airlines, Air North had the second highest complaint rate in one particular year. But reporter George Bernstein investigated and found that Air North's high complaint rate of 38.6 per 100,000 passengers really represented only 22 complaints. (Air North flew 57,000 passengers that year.) "That many complaints could have come from a single delayed flight," wrote Bernstein. He went on to state that "if nothing else, this proves how dangerous statistics can be, and how easy it is for them to be blown out of proportion if you don't know exactly what's behind them." The following year, Air North had only three complaints.

of a proportion or percentage. For example, we cannot average a sample of 390 Democrats and 330 Republicans, but we can analyze the fact that 54.2% of this sample of 720 voters is comprised of Democrats. In this section we present the method used for testing hypotheses made about population proportions or percentages, a method that can be used with sample data at the nominal level of measurement. This method is very useful in á variety of applications, including surveys, polls, and quality control considerations involving the proportions of defective parts.

Throughout this section we assume that individual trials are independent so that the relevant probability remains constant for each trial. Recall that with a fixed number of independent trials having constant probabilities, as long as each trial has two outcomes, we have a binomial experiment (see Section 4-4). When we deal with proportions or percentages, we can usually make the assumptions that enable us to use a binomial distribution.

In Section 5-5 we saw that under suitable circumstances (namely, $np \geq 5$ and $nq \geq 5$) the binomial distribution can be approximated by a normal distribution with the mean and standard deviation given by $\mu = np$ and $\sigma = \sqrt{n \cdot p \cdot q}$. If $np \geq 5$ and $nq \geq 5$ are not both true, we may be able to use Table A-2 or the binomial probability formula described in Section 4-4, but this section deals only with situations in which the normal distribution is a suitable approximation for the distribution of sample proportions. Replacing μ and σ by their binomial counterparts, we get

$$z = \frac{x - \mu}{\sigma} = \frac{x - np}{\sqrt{n \cdot p \cdot q}}$$

where x is the number of successes in n trials.

To test a hypothesis made about a population proportion or percentage, we will therefore follow the standard procedure described in Section 6-2, but the value of the test statistic will be found by computing z, as follows.

TEST STATISTIC

$$z = \frac{x - np}{\sqrt{n \cdot p \cdot q}}$$

where
n = number of trials
p = population proportion (given in the null hypothesis)
$q = 1 - p$

Let's assume that your county Republican party leader has claimed that the Democratic presidential candidate will receive no more than 48% of all votes cast in the county. Let us also assume that this claim is to be tested on the basis of the survey of 720 randomly selected registered voters of which 390 have indicated a preference for the Democratic candidate. The sample proportion of 390/720 (or 54.2%) is more than the claimed value of 48%, but we must now determine whether it is *significantly* more. We will assume a significance level of 5%. The sample data can now be summarized:

$$n = 720$$
$$x = 390$$

The population parameters p and q are suggested by the claim that we are testing:

$$p = 0.48$$
$$q = 1 - 0.48 = 0.52$$

The significance level of 5% indicates that $\alpha = 0.05$. Following the pattern developed in Section 6-2 (see Figure 6-2), we get the following null and alternative hypotheses:

$$H_0: p \leq 0.48$$
$$H_1: p > 0.48$$

The statistic relevant to this test is found from

$$z = \frac{x - np}{\sqrt{n \cdot p \cdot q}}$$

and it is normally distributed. Specifically, we get

$$z = \frac{x - np}{\sqrt{n \cdot p \cdot q}} = \frac{390 - (720)(0.48)}{\sqrt{(720)(0.48)(0.52)}} = 3.31$$

This test statistic, the critical region, and critical value are shown in Figure 6-15. The test statistic does fall in the critical region, and we

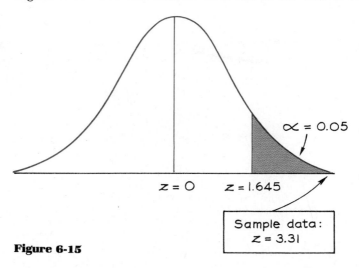

Figure 6-15

Quality Control in a Nuclear Power Plant

The construction of several nuclear power plants has been plagued by quality control problems. Ohio's Zimmer Nuclear Power Station, for example, was cited for 15,000 violations of quality assurance regulations, and construction was eventually stopped because of inability to maintain proper levels of quality. It was estimated that 70% of the welds on structural beams were unacceptable. For each welder, inspectors examined random welds and, based on those sample results, either accepted or rejected all the welds made by that individual. Since there were hundreds of welders, this project was monumental.

therefore reject the null hypothesis H_0. This indicates that the sample result of a 54.2% Democratic preference among the 720 voters does represent a proportion of voters *significantly* greater than the 48% value claimed by the county Republican leader. If the 48% value is correct, there is less than a 5% chance of getting the sample results used in this example. Rather than concluding that an unusual sample has been obtained, we conclude that the true population proportion is more likely to be greater than 0.48 or 48%. This conclusion may in reality be wrong, even though it follows from accepted standard statistical techniques. But that is the nature of statistics. We are led to likely, but not certain, conclusions.

If the original problem is stated with a percentage, convert the percentage to the equivalent decimal form. For example, the claim that at least 45% of all married people are women would suggest the null hypothesis H_0: $p \geq 0.45$. In our computations, we should assume that $p = 0.45$ and $q = 0.55$. In addition to accommodating proportions and percentages, the methods described here also apply to tests of hypotheses made about probabilities. Whether we have a proportion, a percentage, or probability, the value of p must be between 0 and 1, and the sum of p and q must be exactly 1.

E X A M P L E

A senator claims that 60% or more of his constituents favor a gun-control bill. An independent pollster contacts 500 constituents selected at random and finds 273 that favor the bill. What can we conclude about the senator's claim? Assume a 5% level of significance.

Solution

We summarize the solution to this problem by following the pattern suggested by Figure 6-2. The results are shown in Figures 6-16 and 6-17 (page 312) where the test statistic is computed as follows:

$$z = \frac{x - np}{\sqrt{n \cdot p \cdot q}} = \frac{273 - (500)(0.6)}{\sqrt{(500)(0.6)(0.4)}} = -2.465$$

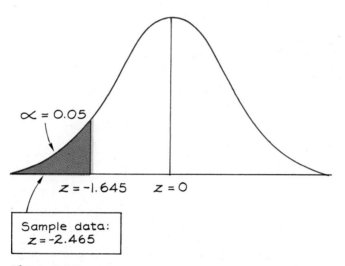

Figure 6-16

P-Values

The examples of this section followed the traditional approach to hypothesis testing, but it would be easy to use the P-value approach since the test statistic is a z score. The P-value is obtained by using the same procedure described in Section 6-3. In a right-tailed test, the P-value is the area to the right of the test statistic. In a left-tailed test, the P-value is the area to the left of the test statistic. In a two-tailed test, the P-value is twice the area of the extreme region bounded by the test statistic. We reject the null hypothesis if the P-value is less than the significance level.

The last example was left-tailed, so the P-value is the area to the left of the test statistic $z = -2.465$. Table A-3 indicates that the area between $z = 0$ and $z = -2.47$ is 0.4932, so the P-value is $0.5 - 0.4932$ or 0.0068. Since the P-value of 0.0068 is less than the significance level of 0.05, we reject the null hypothesis that 60% or more of the constituents favor the gun-control bill.

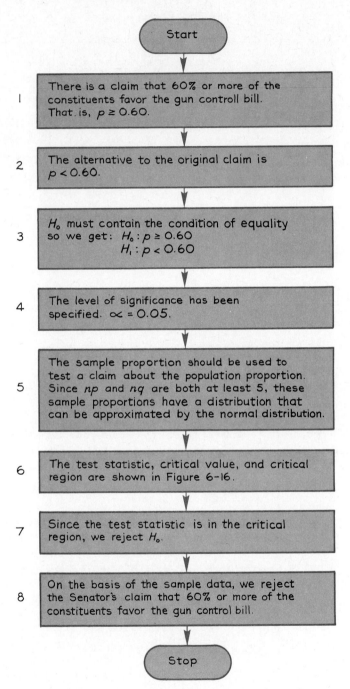

Start

1 | There is a claim that 60% or more of the constituents favor the gun control bill. That is, $p \geq 0.60$.

2 | The alternative to the original claim is $p < 0.60$.

3 | H_0 must contain the condition of equality so we get: $H_0 : p \geq 0.60$ $H_1 : p < 0.60$

4 | The level of significance has been specified. $\alpha = 0.05$.

5 | The sample proportion should be used to test a claim about the population proportion. Since np and nq are both at least 5, these sample proportions have a distribution that can be approximated by the normal distribution.

6 | The test statistic, critical value, and critical region are shown in Figure 6-16.

7 | Since the test statistic is in the critical region, we reject H_0.

8 | On the basis of the sample data, we reject the Senator's claim that 60% or more of the constituents favor the gun control bill.

Stop

Figure 6-17

6-5 Exercises A
Tests of Proportions

In Exercises 6-81 through 6-100, test the given hypotheses. Include the steps listed in Figure 6-2, and draw the appropriate graph.

6-81 At the 0.05 significance level, test the claim that the proportion of defects p for a certain product equals 0.3. Sample data consist of $n = 100$ randomly selected products, of which 45 are defective.

6-82 At the 0.05 significance level, test the claim that the proportion of males p at a given college equals 0.6. Sample data consist of $n = 80$ randomly selected students, of which 54 are males.

6-83 At the 0.01 level of significance, test the claim that the proportion of voters p who favor nuclear disarmament is less than 0.7. Sample data consist of 50 randomly selected voters, of whom 31 favor nuclear disarmament.

6-84 Test the claim that the proportion of adults who have credit cards is greater than 0.75. Sample data consist of 400 randomly selected adults, of whom 325 have credit cards. Use a 0.01 level of significance.

6-85 A manufacturer considers his production process to be out of control when defects exceed 3%. A random sample of 500 items includes exactly 22 defects, but the manager claims that this represents a chance fluctuation and that production is not really out of control. Test the manager's claim at the 5% level of significance.

6-86 An obstetrician develops a drug that is claimed to increase the probability of a baby being male. To test the effectiveness of the drug, it is given to 150 volunteers before conception, and the 150 births result in 84 boys and 66 girls. Is the claim supported by the experimental data? Use a 1% level of significance.

6-87 The mathematics department has had a consistent failure rate of 12%. In one experimental semester, more active counseling plus rigorous enforcement of prerequisites results in 338 failures among 3217 mathematics students. At the 0.05 significance level, test the department's claim that the failure rate has been lowered.

6-88 An airline reservations system suffers from a 7% rate of no-shows. A new procedure is instituted whereby reservations are confirmed on the day preceding the actual flight, and a study is then made of 5218 randomly selected reservations made under the new system. If 333 no-shows are recorded, test the claim that the no-show rate is lower with the new system. Use a 5% level of significance.

6-89 In a genetics experiment, the Mendelian law is followed as expected if one-eighth of the offspring exhibit a certain recessive trait. Analysis of

500 randomly selected offspring indicates that 83 exhibited the necessary recessive trait. Is the Mendelian law being followed as expected? Use a 2% level of significance.

6-90 In testing depressed patients, a doctor observes that a certain drug causes an unfavorable reaction in 15% of the cases. The doctor then alters the drug to determine whether there is any significant change in the unfavorable reaction rate. The new drug is used on 100 depressed patients and only 7 unfavorable reactions occur. At the 2% level of significance, test the claim that there is no change in the 15% rate.

6-91 At the 0.01 level of significance, test the claim that more than 1/3 of all adults smoke. A random sample of 324 adults shows that 120 smoke.

6-92 At the 0.01 level of significance, test the claim that among those who dine at a restaurant, more than 40% order steak. A random sample of 165 dinner orders shows that 80 of them were for steak.

6-93 In a poll of 1553 randomly selected adult Americans, 92% recognize the name Christopher Columbus. At the 0.01 significance level, test the claim that 95% of all adult Americans recognize the name Columbus.

6-94 A poll is conducted to test the claim of the Republican presidential candidate that he is favored by 56% of the voters. When 1500 randomly selected subjects are polled, 54% of them indicate a preference for the Republican. Test this candidate's claim at the 0.02 significance level.

6-95 The unemployment rate was 8.2% in February. In March the Bureau of the Census polled 50,000 randomly selected employable subjects. Of these, 3785 subjects were classified as unemployed. At the 5% level of significance, test the assertion of the labor leader who claimed that the March unemployment rate was not below the February rate.

6-96 At the 0.05 significance level, test the claim that at least 80% of the students who take a statistics course receive a passing grade. Sample data consist of 90 randomly selected statistics students of whom 64 received a passing grade.

6-97 In a study of 500 aircraft accidents involving spatial disorientation of the pilot, it was found that 91% of those accidents resulted in fatalities. At the 0.05 level of significance, test the claim that three-fourths of all such accidents will result in fatalities.

6-98 Test the claim that among those with preschool educational training, 2/3 went on to graduate from high school. Sample data consist of 295 randomly selected subjects with preschool training, and 71% of them graduated from high school. Use a 0.05 level of significance.

6-99 Test the claim that fewer than 1/2 of San Francisco residential telephones have unlisted numbers. A random sample of 400 such phones results in an unlisted rate of 39%. Use a 0.01 level of significance.

6-100 The Kennedy-Nixon presidential race was extremely close. Kennedy won by 34,227,000 votes to Nixon's 34,108,000 votes. At the 0.01 level of significance, test the claim that the true population proportion for Kennedy exceeded 0.5. Assume that the voters represent a random sampling of those eligible.

6-5 Exercises B
Tests of Proportions

6-101 Show that $z = (x - np)/\sqrt{n \cdot p \cdot q}$ is equivalent to

$$z = \frac{\hat{p} - p}{\sqrt{p(1 - p)/n}}$$

where $\hat{p} = x/n$.

6-102 A reporter claims that 10% of the residents of her city feel that the mayor is doing a good job. Test her claim if it is known that, in a random sample of 15 residents, there are none who feel that the mayor is doing a good job. Use a 5% level of significance. Since $np = 1.5$ and is not at least 5, the normal distribution is not a suitable approximation to the distribution of sample proportions.

6-103 A supplier of chemical waste containers finds that 3% of a sample of 500 units are defective. Being somewhat devious, he wants to make a claim that the defective rate is no more than some specified percentage, and he doesn't want that claim rejected at the 0.05 level of significance if the sample data are used. What is the *lowest* defective rate he can claim under these conditions?

6-104 In a study of 500 aircraft accidents involving spatial disorientation of the pilot, it was found that 91% of those accidents resulted in fatalities. Someone with a vested interest wants to claim that the percentage is at least some particular value, and they don't want that claim rejected at the 0.01 level of significance if the sample data are used. What is the *highest* percentage that can be claimed under these conditions?

6-6 Tests of Variances

The preceding sections of this chapter deal with tests of hypotheses made about means and proportions. In testing these hypotheses, we use the normal and student t distributions in ways that are very similar. They have the same basic bell shapes, they are both symmetric about zero, and the test statistics involve comparable computations. Here, we encounter a very different distribution as we test hypotheses made about a population variance or standard deviation. Recall that the standard deviation is simply the square root of the variance, so if we know the value of one, we also know the value of the other.

NOTATION

σ **Population standard deviation**
σ^2 **Population variance**
s **Sample standard deviation**
s^2 **Sample variance**

For this reason, the comments we make about variances will also apply to standard deviations.

Many real and practical situations demand decisions or inferences about variances. In manufacturing, quality-control engineers want to ensure that a product is, on the average, acceptable. But the engineers also want to produce items of *consistent* quality so there are only a few defective products. This consistency is measured by the variance.

As a specific example, let us consider aircraft altimeters. Due to mass-production techniques and a variety of other factors, these altimeters do not all give exact readings. (Some errors are built in.) It would be easy to change the overall average reading by simply shifting the scale on the altimeters, but it would be very difficult to change the variance of the readings. Even if the overall average were perfect, a very large variance would indicate that some altimeters give excessively high or low readings and seriously jeopardize safety. In this case, quality control engineers want to keep the variance below some tolerable level. When the variance exceeds that level, production is considered to be out of control, and corrective action must be taken.

In this section, **we assume that the population in question has normally distributed values.** We made this same assumption earlier in this chapter, but it is a more critical assumption here. In using the student t distribution of Section 6-4, for example, we require that the population of values be approximately normal, and we can accept deviations away from normality that are not too severe. However, when we deal with variances by using the distribution to be introduced shortly, departures away from normality will lead to gross errors. Consequently the assumption of a normally distributed population must be followed much more strictly.

In a normally distributed population with variance σ^2, we randomly select independent samples of size n and compute the variance s^2 for each sample. The quantity $(n - 1)s^2/\sigma^2$ has a distribution called the **chi-square distribution.** We denote chi-square by χ^2 and we pronounce it "kigh square." (The specific mathematical equations used to define this distribution are not given here since they are confusing at this stage. Instead, you can refer to Table A-5 for the critical values of the chi-square distribution.)

The test statistic used in tests of hypotheses about variances is χ^2 (chi-square):

TEST STATISTIC

$$\chi^2 = \frac{(n-1)s^2}{\sigma^2}$$

where
n = sample size
s^2 = sample variance
σ^2 = population variance (given in the null hypothesis)

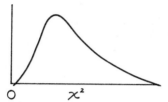

Figure 6-18 Chi-square distribution

The chi-square distribution resembles the student t distribution in that there is actually a different distribution for each sample size n. In Figure 6-18 we see that the chi-square distribution does not have the same symmetric bell shape of the normal and student t distributions. The chi-square distribution has a longer right tail. Unlike the normal and student t distributions, the chi-square distribution does not include negative numbers.

In using Table A-5 to determine critical values of the chi-square distribution, we must first determine the degrees of freedom, which are $n - 1$ as in the student t distribution.

degrees of freedom = $n - 1$

We can use this expression for Section 6-4 and this section, but in later chapters we will encounter situations in which the degrees of freedom are not $n - 1$. For that reason, we should not universally equate degrees of freedom with $n - 1$.

Once we have determined degrees of freedom, the significance level α, and the type of test (left-tailed, right-tailed, or two-tailed), we can use Table A-5 to find the critical chi-square values. An important feature of this table is that each critical value separates an area to the *right* that corresponds to the value given in the top row.

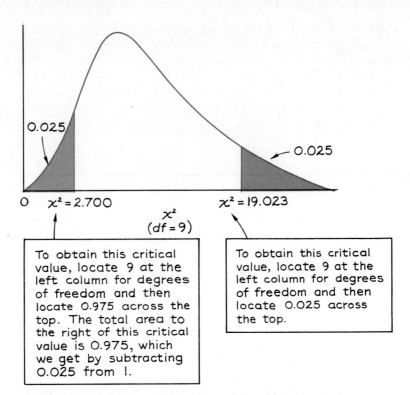

Figure 6-19 Finding critical values of the chi-square distribution using Table A-5

E X A M P L E

Find the critical values of χ^2 that determine critical regions containing areas of 0.025 in each tail. Assume that the relevant sample size is 10 so that the degrees of freedom are $10 - 1$, or 9.

Solution

See Figure 6-19 and refer to Table A-5. The critical value to the right (19.023) is obtained in a straightforward manner by locating 9 in the degrees-of-freedom column at the left and 0.025 across the top. The left critical value of 2.700 once again corresponds to 9 in the degrees-of-freedom column, but we must locate 0.975 across the top since the values in the top row are always *areas to the right* of the critical value.

In a right-tailed test, the value of α will correspond exactly to the areas given in the top row of Table A-5. In a left-tailed test, the value of $1 - \alpha$ will correspond exactly to the areas given in the top row of Table

The Power of Your Vote

In the electoral college system, the power of a voter in a large state exceeds that of a voter in a small state. When voting power is measured as the ability to affect the outcome of an election, we see that a New Yorker has 3.312 times the voting power of a resident of the District of Columbia. This result is included in John Banzhof's article, "One Man, 3.312 Votes."

As an example, the outcome of the 1916 Presidential election could have been changed by shifting only 1,983 votes in California. If the same number of votes were changed in a much smaller state, the resulting change in electoral votes would not have been sufficient to alter the outcome.

A-5. In a two-tailed test, the values of $\alpha/2$ and $1 - \alpha/2$ will correspond exactly to the areas given in the top row of Table A-5.

Having discussed the underlying theory and the mechanics of using Table A-5, we now give an example of a hypothesis test.

EXAMPLE

Test the claim that scores on a standard I.Q. test have a variance equal to 225 if a sample of 41 randomly selected subjects achieve scores with a variance of 258. Use a significance level of $\alpha = 0.05$.

Solution

We again use the basic method for testing hypotheses outlined in Figure 6-2. The claim that the variance equals 225 is $\sigma^2 = 225$ in symbolic form. The alternative is $\sigma^2 \neq 225$ and we get the following null and alternative hypotheses.

$$H_0: \sigma^2 = 225$$
$$H_1: \sigma^2 \neq 225$$

We can see that this is a two-tailed test since a sample variance significantly above or below 225 will be a basis for rejecting H_0. The significance level of $\alpha = 0.05$ has already been stipulated. The sample variance should obviously be used in testing a claim about the population variance, and the χ^2 distribution is therefore appropriate since I.Q. scores tend to be normally distributed. We compute the test statistic as follows.

$$\chi^2 = \frac{(n-1)s^2}{\sigma^2} = \frac{(41-1) \cdot 258}{225} = 45.867$$

We find the right critical χ^2 value of 59.342 in Table A-5 by locating 40 degrees of freedom and an area of 0.025. We find the left critical χ^2 value of 24.433 in Table A-5 by locating 40 degrees of freedom and an area of 0.975. (Since $\alpha = 0.05$, there is an area of 0.025 in each tail, but the left critical value is found by locating the area to its right. See Figure 6-20 on the following page.) The test statistic based on the sample data is not in the critical region, so we fail to reject H_0. On the basis of the available evidence, we cannot refute the claim that the variance equals 225.

One example of variance occurs in the waiting lines of banks. In the past, customers traditionally entered a bank and selected one of several lines formed at different windows. A different system growing in popularity involves one main waiting line, which feeds the various windows as

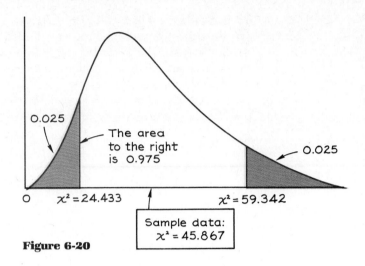

0.025

The area
to the right
is 0.975

0.025

O $x^2 = 24.433$ $x^2 = 59.342$

Sample data:
$x^2 = 45.867$

Figure 6-20

vacancies occur. The mean waiting time isn't reduced, but the variability among waiting times is decreased and the irritation of being caught in a slow line is also diminished.

EXAMPLE

With individual lines at its various windows, a bank finds that the standard deviation for normally distributed waiting times on Friday afternoons is 6.2 minutes. The bank experiments with a single main waiting line and finds that for a random sample of 25 customers, the waiting times have a standard deviation of 3.8 minutes. At the $\alpha = 0.05$ significance level, test the claim that a single line causes lower variation among the waiting times.

Solution

We wish to test $\sigma < 6.2$ based on a sample of $n = 25$ for which $s = 3.8$. We begin by identifying the null and alternative hypotheses. (We will use σ^2 instead of σ since the χ^2 distribution involves variances directly.)

$$H_0: \sigma^2 \geq (6.2)^2$$
$$H_1: \sigma^2 < (6.2)^2$$

The significance level of $\alpha = 0.05$ has already been selected so we proceed to compute the value of χ^2 based on the given data:

$$\chi^2 = \frac{(n-1)s^2}{\sigma^2} = \frac{(25-1)(3.8)^2}{(6.2)^2} = 9.016$$

This test is left-tailed since H_0 will be rejected only for small values of χ^2; with $\alpha = 0.05$ and $n = 25$, we go to Table A-5 and align 24 degrees of freedom with an area of 0.95 to obtain the critical χ^2 value of 13.848 (see Figure 6-21). Since the test statistic falls within the critical region, we reject H_0 and conclude that the 3.8-minute standard deviation is significantly less than the 6.2-minute standard deviation that corresponds to multiple waiting lines. That is, the single main line does appear to lower the variation among waiting times.

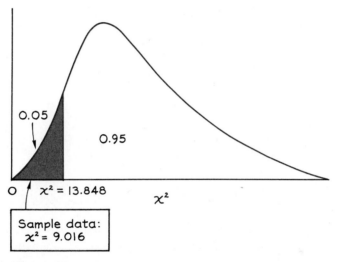

Figure 6-21

We have already illustrated some of the important uses of the chi-square distribution. Other valuable uses, such as situations in which we wish to analyze the significance of differences between expected frequencies and the frequencies that actually occur, are considered in Chapter 10.

6-6 Exercises A
Tests of Variances

In Exercises 6-105 and 6-106, use Table A-5 to find the critical values of χ^2 based on the given data.

6-105

(a) $\alpha = 0.05$
  $n = 20$
  H_0: $\sigma^2 = 256$

(b) $\alpha = 0.05$
  $n = 20$
  H_0: $\sigma^2 \geq 256$

(c) $\alpha = 0.01$
$n = 23$
$H_0: \sigma^2 = 10$

(d) $\alpha = 0.01$
$n = 23$
$H_0: \sigma^2 \le 10$

(e) $\alpha = 0.005$
$n = 15$
$H_1: \sigma^2 < 23.4$

6-106 (a) $\alpha = 0.10$
$n = 6$
$H_1: \sigma^2 < 100$

(b) $\alpha = 0.05$
$n = 40$
$H_1: \sigma^2 > 500$

(c) $\alpha = 0.025$
$n = 81$
$H_0: \sigma^2 \ge 144$

(d) $\alpha = 0.01$
$n = 50$
$H_0: \sigma^2 = 225$

(e) $\alpha = 0.05$
$n = 75$
$H_1: \sigma^2 \ne 31.5$

In Exercises 6-107 through 6-126, test the given hypotheses. Follow the pattern outlined in Figure 6-2 and draw the appropriate graph. In all cases, assume that the population is normally distributed.

6-107 At the $\alpha = 0.05$ significance level, test the claim that $\sigma^2 > 100$ if a random sample of 27 yields a variance of 194.

6-108 At the $\alpha = 0.05$ significance level, test the claim that $\sigma^2 > 225$ if a random sample of 30 yields a variance of 380.

6-109 At the $\alpha = 0.05$ significance level, test the claim that $\sigma^2 \ge 416$ if a random sample of 17 yields a variance of 247.

6-110 At the $\alpha = 0.10$ significance level, test the claim that $\sigma^2 \ge 90$ if a random sample of 21 yields a variance of 53.

6-111 At the $\alpha = 0.01$ significance level, test the claim that $\sigma^2 = 2.38$ if a random sample of 12 yields a variance of 5.00.

6-112 At the $\alpha = 0.05$ significance level, test the claim that $\sigma^2 = 100$ if a random sample of 27 yields a variance of 57.

6-113 At the $\alpha = 0.01$ significance level, test the claim that $\sigma = 10.0$ if a random sample of 18 yields a standard deviation of 14.5.

6-114 At the $\alpha = 0.05$ significance level, test the claim that $\sigma = 52.0$ if a random sample of 18 yields a standard deviation of 71.2.

6-115 At the $\alpha = 0.05$ significance level, test the claim that $\sigma^2 \le 9.00$ if a random sample of 81 yields a variance of 12.25.

6-116 At the $\alpha = 0.025$ level of significance, test the claim that $\sigma < 98.6$ if a random sample of 51 yields a standard deviation of 79.000.

6-117 If the standard deviation for the weekly down times of a computer is low, then availability of the computer is predictable and planning is facilitated. If 12 weekly down times for a computer are randomly selected and the standard deviation is computed to be 2.85 hours, at the 0.025 significance level, test the claim that $\sigma > 2.00$ hours.

6-118 The reaction times for adults on a standard test have a standard deviation of 0.21 second. We randomly select 24 subjects and give each of them two glasses of beer before the test. Their scores produce a standard deviation of 0.29 second. At the 0.05 significance level, test the claim that the standard deviation for the reaction times of all adults who drink two glasses of beer is greater than 0.21 second.

6-119 A machine pours medicine into a bottle in such a way that the standard deviation of the weights is 0.15 ounce. A new machine is tested on 71 bottles, and the standard deviation for this group is 0.12 ounce. At the 0.05 significance level, test the claim that the new machine produces less variance.

6-120 The weights of 20 newborn girls are randomly selected and the standard deviation for this group is computed to be 1.18 pounds. Test the claim that the weights of newborn girls have a greater standard deviation than the weights of all newborn boys. (Assume that $\sigma = 0.99$ pound for boys and let $\alpha = 0.05$.)

6-121 Test the claim that scores on a personality test have a variance equal to 225 if a sample of 41 randomly selected subjects achieve scores with a variance of 135. Use a significance level of $\alpha = 0.05$.

6-122 Test the claim that scores on a standard I.Q. test have a standard deviation equal to 15 if a sample of 24 randomly selected subjects yields a standard deviation of 10. Use a significance level of $\alpha = 0.01$.

6-123 Repeat Exercise 6-122 after changing the significance level to $\alpha = 0.05$.

6-124 A software firm finds that the times required to run one particular computer program have a standard deviation of 52 hours. A sample of 30 new computer programs produces a standard deviation of 60 hours. At the 0.05 significance level, test the claim that the standard deviation for the new computer programs is equal to 52 hours.

6-125 When 22 bolts are tested for hardness, their indices have a standard deviation of 65.0. Test the claim that the standard deviation of the hardness indices for all such bolts is greater than 50.0. Test at the 0.025 level of significance.

6-126 A small-engine repair shop has found that repair times have a standard deviation of 28.0 minutes. A new repair system is being tested, and a sample of 16 repair times produces a standard deviation of 17.3 minutes. At the 0.01 level of significance, test the claim that the new system produces a lower standard deviation than the old system.

6-6 Exercises B
Tests of Variances

6-127 For large numbers of degrees of freedom, we can approximate values of χ^2 as follows.

$$\chi^2 = \tfrac{1}{2}[z + \sqrt{2k - 1}]^2$$

where k = number of degrees of freedom and z = the critical score, found from Table A-3.

For example, if we want to approximate the two critical values of χ^2 in a two-tailed hypothesis test with $\alpha = 0.05$ and a sample size of 150, we let $k = 149$ with $z = -1.96$, followed by $k = 149$ and $z = 1.96$.

(a) Use this approximation to estimate the critical values of χ^2 in a two-tailed hypothesis test when $n = 101$ and $\alpha = 0.05$. Compare the results to those found in Table A-5.

(b) Use this approximation to estimate the critical values of χ^2 in a two-tailed hypothesis test when $n = 150$ and $\alpha = 0.05$.

6-128 Do Exercise 6-127 using the approximation

$$\chi^2 = k\left[1 - \frac{2}{9k} + z\sqrt{\frac{2}{9k}}\right]^3.$$

where k and z are as described in that exercise.

6-129 The caffeine contents (in mg) for a dozen randomly selected cans of a soft drink are given below. At the 0.025 level of significance, test the claim that the standard deviation for all such cans is less than 2.0 mg.

34.2	33.7	31.9	34.3	31.6	32.7
33.1	35.2	31.6	32.9	33.0	32.4

6-130 A consumer's testing company tests 60 different randomly selected coffee makers and records the estimated annual energy consumption as measured in kilowatt hours. The results are summarized in the accompanying frequency table. At the 0.025 level of significance, test the claim that the standard deviation for all such coffee makers is greater than 20 kwh.

x	f
100–119	5
120–139	12
140–159	20
160–179	15
180–199	8

Computer Project
Testing Hypotheses

(a) Most statistics software packages include a procedure for testing a claim made about a single mean. Such a procedure is often referred to as a "*t* test". Use the *t* test option in a software package to solve Exercise 6-77.

(b) Construct a program that takes as input the following items:

1. A set of sample data.
2. The value of the population mean μ.

(Assume that the sample data are randomly selected from a population with an essentially normal distribution.) Output should consist of the sample mean $\bar{x}$, the sample standard deviation s, the value of $s/\sqrt{n}$, and the value of the test statistic z (if the sample is large, with $n > 30$) or t (if the sample is small, with $n \leq 30$). The test statistic should be identified as being a z score or a t score.

Review

In this chapter, we studied standard methods used in statistical tests of hypotheses made about the values of population means, proportions, percentages, variances, and standard deviations. (A hypothesis is simply a statement that something is true.) When sample data conflict with the given hypothesis, we want to decide whether the differences are due to chance fluctuations or whether the differences are so significant that they are not likely to occur by chance. We are able to select exact **levels of significance;** 0.05 and 0.01 are common values. Sample results are said to reflect significant differences when their occurrences have probabilities less than the chosen level of significance.

Section 6-2 presented in detail the procedure for testing hypotheses. The essential steps are summarized in Figure 6-2. We defined **null hypothesis, alternative hypothesis, type I error, type II error, test statistic, critical region, critical value** and **significance level.** All these standard terms are commonly used in discussing tests of hypotheses. We also identified the three basic types of tests: **right-tailed, left-tailed,** and **two-tailed** (see Figure 6-3).

We introduced the method of testing hypotheses in Section 6-2 by using examples in which only the normal distribution applies, and we introduced other distributions in subsequent sections. The brief table that follows summarizes the hypothesis tests covered in this chapter.

Section 6-2 outlined the classical approach to testing hypotheses, while the P-value approach was presented in Section 6-3. In the classical approach we make a decision about the null hypothesis by comparing the test statistic and critical value. With the P-value approach we base that

IMPORTANT FORMULAS

Parameter to which hypothesis refers	Applicable distribution	Assumption	Test statistic	Table of critical values
μ (population mean)	Normal	σ is known and population is normally distributed	$z = \dfrac{\bar{x} - \mu}{\sigma/\sqrt{n}}$	Table A-3
	Normal	$n > 30$	$z = \dfrac{\bar{x} - \mu}{\sigma/\sqrt{n}}$	Table A-3
	Student t	σ is unknown and $n \leq 30$ and population is normally distributed	$t = \dfrac{\bar{x} - \mu}{s/\sqrt{n}}$	Table A-4
p (population proportion)	Normal	$np \geq 5$ and $nq \geq 5$	$z = \dfrac{x - np}{\sqrt{n \cdot p \cdot q}}$	Table A-3
σ^2 (population variance), σ (population standard deviation)	Chi-square	Population is normally distributed	$\chi^2 = \dfrac{(n-1)s^2}{\sigma^2}$	Table A-5

decision on a comparison of the significance level and the P-value, which represents the probability of getting a sample that is at least as extreme as the one obtained.

In addition to presenting the method for testing hypotheses, we also introduced the student t and chi-square distributions. In Chapter 7 we will refer to these distributions again as we investigate estimations of parameters and ways to determine how large certain samples should be.

Vocabulary List

Define and give an example of each term.

hypothesis	type II error	P-value
hypothesis test	test statistic	t distribution
test of significance	critical region	student t
null hypothesis	critical value	distribution
alternative	significance level	degrees of freedom
hypothesis	left-tailed test	chi-square
type I error	right-tailed test	distribution
	two-tailed test	

Review Exercises

In Exercises 6-131 and 6-132, find the appropriate critical values.

6-131 (a) $\alpha = 0.05$
$n = 160$
$H_0: p \geq 0.5$

(b) $\alpha = 0.01$
$n = 35$
$H_0: \mu \leq 16.5$

(c) $\alpha = 0.01$
$n = 12$
$H_0: \mu = 38.4$

(d) $\alpha = 0.01$
$n = 25$
$H_0: \sigma^2 \geq 225$

(e) $\alpha = 0.05$
$n = 30$
$H_0: \sigma^2 = 84.3$

6-132 (a) $\alpha = 0.10$
$n = 15$
$H_0: \mu = 1.23$

(b) $\alpha = 0.10$
$n = 15$
$H_0: \sigma^2 = 123$

(c) $\alpha = 0.06$
$n = 100$
$H_0: \mu = 72.3$

(d) $\alpha = 0.05$
$n = 10$
$H_0: \sigma = 15$

(e) $\alpha = 0.01$
$n = 30$
$H_1: \sigma < 5.8$

In Exercises 6-133 and 6-134, respond to each of the following:
(a) Give the null hypothesis in symbolic form.
(b) Is this test left-tailed, right-tailed, or two-tailed?
(c) In simple terms devoid of symbolism and technical language, describe the type I error.
(d) In simple terms devoid of symbolism and technical language, describe the type II error.
(e) What is the probability of making a type I error?

6-133   At the 0.01 level of significance, the claim is that the mean treatment time for a dentist is at least 20.0 minutes.

6-134   At the 5% level of significance, the claim is that the mean reading in a biofeedback experiment is 6.2 microvolts.

6-135   At the 0.05 level of significance, test the claim that $\mu = 10.0$ seconds. Sample data consist of 100 observations, for which $\bar{x} = 8.2$ seconds. Assume that $\sigma = 6.0$ seconds.

6-136   Test the claim that hypertense patients have a mean systolic blood pressure level of 174.0 mm. Use a 0.05 level of significance. Sample data consist of 60 observations for which $\bar{x} = 177.8$ mm. Assume that $\sigma = 20.0$ mm.

6-137   Test the claim that a certain x-ray machine gives radiation dosages with a mean below 5.00 milliroentgens. Sample data consist of 36 observations with a mean of 4.13 milliroentgens and a standard deviation of 1.91 milliroentgens. Use a 0.01 level of significance.

6-138   Use a 0.025 level of significance to test the claim that vehicle speeds at a certain location have a mean above 55.0 mph. A random sample of 50 vehicles produces a mean of 61.3 mph and a standard deviation of 3.3 mph.

6-139   At the 0.01 level of significance, test the claim that $\mu \leq 25.5$ ft. Sample data consist of 25 observations for which $\bar{x} = 27.3$ ft and $s = 3.7$ ft.

6-140   Test the claim that the mean female reaction time to a highway signal is less than 0.700 second. When 18 females are randomly selected and tested, their mean is 0.668 second. Assume that $\sigma = 0.100$ second, and use a 5% level of significance.

6-141   The following sample scores have been randomly selected from a normally distributed population. At the 0.01 significance level, test the claim that the population has a mean of 100.

101	106	98	92	97	80	89	88
110	112	100	100	103	97	97	

6-142   Test the claim that the mean time required for new employees to learn a certain task is 18.4 minutes. A random sample of 18 new employees produces a mean and standard deviation of 19.6 minutes and 2.7 minutes, respectively. Use a 0.10 significance level.

6-143   At the 0.05 level of significance, test the claim that 20% of adult Americans are unable to read. A random sample of 600 adult Americans included exactly 132 who could not read.

6-144   Test the claim that 25% of those earning a bachelor's degree in business are women. A random sample of 200 business graduates earning the bachelor's degree includes 80 women and 120 men.

6-145   Of 200 females randomly selected and interviewed in the Midwest, 27% believed (incorrectly) that birth control pills prevent venereal disease. Use these sample results to test (at the 5% level of significance) the claim that more than one-fifth of all midwestern women believe that birth control pills prevent venereal disease.

6-146   Of 80 workers randomly selected and interviewed, 55 were opposed to an increase in social security taxes. At the 5% level of significance, test the claim that the majority (more than 50%) of such workers are opposed to the increase in taxes.

6-147   At the 0.05 level of significance, test the claim that $\sigma = 4.0$ grams. A random sample includes 20 observations for which $s = 2.8$ grams.

6-148   One type of pump is designed to remove 600 gallons of water per hour. It has been determined that the actual amounts removed have a standard deviation of 10.5 gallons. A new regulator valve is installed and a sample of 20 modified pumps results in a standard deviation of 16.4 gallons. At the 0.05 level of significance, test the claim that the modified pump has the same standard deviation.

6-149   One large high school has found that students taking a standard college aptitude test earn scores with a variance of 6410. A counselor claims that the current group of test subjects includes a group with more varied aptitudes. Test the claim that the variance is larger than 6410 if a random sample of 60 students produces a variance of 8464. Use a 0.10 level of significance.

6-150   Using the sample data in Exercise 6-141, test the claim that $\sigma = 12$. Use a 1% level of significance.

Chapter 6
Case Study Activity

Conduct a survey by asking the question "Do you favor or oppose the death penalty for people convicted of murder?" Survey at least 50 people and test the claim that the proportion in favor is less than or equal to 0.5. Identify the population from which you are sampling, and identify any factors that might suggest that your sample is not representative of the population. (If you record the response along with the sex of the respondent, you may be able to use the same data in Chapter 10.)

CHAPTER 7

7-1   Overview
We identify **objectives.**

7-2   Estimates and Sample Sizes of Means
We estimate the value of a population **mean** by the **point estimate** and **confidence interval** and present the method of determining the sample size.

7-3   Estimates and Sample Sizes of Proportions
We estimate the value of a population **proportion** by the **point estimate** and **confidence interval** and present the method of determining the sample size.

7-4   Estimates and Sample Sizes of Variances
We estimate the value of a population **variance** by the **point estimate** and **confidence interval** and present the method of determining sample size.

Estimates and Sample Sizes

CHAPTER PROBLEM

Suppose you are assigned a research project that requires you to determine the percentage of medical school seniors who plan to specialize in internal medicine. You do some preliminary research and find that there are currently about 17,000 medical school seniors. It is obvious that you cannot survey all of them. How many should you sample? 500? 2000? 10%? You're worried that if you don't arrive at a proper sample size, you might need a doctor of internal medicine.

Planning ahead, you know that the sample you obtain is likely to be imperfect in the sense that the sample percentage will differ from the true population percentage by some "error due to sampling." If you sample 500 senior medical students and find that 13.0% of them plan to specialize in internal medicine, just how accurate is that sample result? How can you express the results in a way that reveals their credibility?

We have raised two fundamental questions and the contents of this chapter will enable us to answer both of them. While the preceding situation involves a percentage (or proportion), we will also consider the same concepts and apply them to means, standard deviations, and variances.

7-1   Overview

Chapter 6 introduced one aspect of inferential statistics, hypothesis testing. We used sample data to make **decisions** about claims or hypotheses. In this chapter we use sample data to make **estimates** of the values of population parameters. Section 7-2 begins by using the statistic $\bar{x}$ in estimating the value of the parameter μ. In subsequent sections we use sample proportions and sample variances to estimate the values of population proportions and population variances. We also identify some ways of determining how large samples should be.

We apply these methods of estimating and determining sample size to population means, proportions, percentages, variances, and standard deviations. The same distributions whose parameters you studied in Chapter 6 are included in the methods we develop in this chapter. Like hypothesis testing, the fundamental concepts of this chapter are very important and basic to the subject of inferential statistics. Practical applications will become apparent through the examples and exercises.

7-2   Estimates and Sample Sizes of Means

Let's assume that the following 50 I.Q. scores represent a randomly selected sample of the seniors in a large suburban high school.

110	122	119	95	98	100	100	105	112	111
111	116	110	109	108	112	115	123	145	102
85	90	126	127	135	112	113	127	129	99
144	140	97	98	88	117	77	83	119	121
103	100	133	130	109	112	105	106	107	115

$\bar{x} = 111.4$
$s = 15.2$

Using only these sample results, we want to estimate the value of the mean I.Q. of *all* the seniors in that high school.

We could use statistics such as the sample median, midrange, or mode as estimates of μ, but the sample mean $\bar{x}$ usually provides the best estimate of μ. This is not simply an intuitive conclusion. It is based on careful study and an analysis of the distributions of various estimators. For many populations, the distribution of sample means $\bar{x}$ has a smaller variance than the distribution of the other possible estimators, so $\bar{x}$ tends to be more consistent. For all populations we say that $\bar{x}$ is an **unbiased** estimator. This means that the distribution of $\bar{x}$ values tends to center about the value of μ. For these reasons, we will use $\bar{x}$ as the best estimate of μ. Because $\bar{x}$ is a single number that corresponds to a point on the number scale, we call it a point estimate.

> **DEFINITION**
>
> The sample mean $\bar{x}$ is the best **point estimate of the population mean μ.**

Technology Clouds Television Ratings

In the past it was a relatively simple task to provide television advertisers with accurate data about the numbers of viewers who saw their commercials. The simple task, which consisted of rating three major television networks, is now complicated by the presence of local television stations, cable TV, satellite receiving dishes, video-cassette recorders, and remote control devices. One Nielsen study found that when shows are recorded on home VCR's for future viewing, only 80% of those shows are actually played back at a later time. In addition, viewers avoid most commercials when playing back shows. Large amounts of money are at stake since one ratings point equates to about $55 million in annual advertising per year for each network.

Computing the sample mean of the preceding 50 I.Q. scores, we get $\bar{x} = 111.4$, which becomes our point estimate of the mean I.Q. for all the seniors in the high school. Even though 111.4 is our *best* estimate of μ, we really have no indication of just how good that estimate is. (Sometimes even the best is very poor.) Suppose that we have only the first two sample I.Q. scores of 110 and 122. Their mean of 116 is the best point estimate of μ, but we cannot expect this best estimate to be very good since it is based on such a small sample. Mathematicians have developed an estimator which does reveal how good it is. We will first present this estimator, illustrate its use, and then explain the underlying rationale.

NOTATION

$z(\alpha/2)$ is the positive standard z value that separates an area of $\alpha/2$ in the right tail of the standard normal distribution (see Figure 7-1). Note that $z(\alpha/2)$ is a single number and is *not* the product of z and $\alpha/2$.

EXAMPLE

If $\alpha = 0.05$, then $z(\alpha/2) = 1.96$; 1.96 is the standard z value that separates a right-tail region with an area of 0.05/2 or 0.025. We find $z(\alpha/2)$ by noting that the region to its left must be $0.5 - 0.025$, or 0.475. That is, the area bounded by the centerline and $z(\alpha/2)$ is 0.475. In Table A-3, an area of 0.4750 corresponds exactly to a z score of 1.96.

Figure 7-1

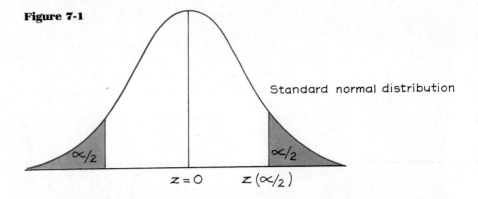

Standard normal distribution

$z = 0$ $z(\alpha/2)$

DEFINITION

The **maximum error of the estimate of** μ is given by

$$E = z(\alpha/2) \frac{\sigma}{\sqrt{n}}$$

and there is a probability of $1 - \alpha$ that $\bar{x}$ differs from μ by less than E.

DEFINITION

The **confidence interval** (or **interval estimate**) for the population mean is given by $\bar{x} - E < \mu < \bar{x} + E$.

We will use the above form of the confidence interval, but another equivalent form is $\mu = \bar{x} \pm E$.

DEFINITION

The **degree of confidence** is the probability $1 - \alpha$ that the parameter μ is contained in the confidence interval. (The probability is often expressed as the equivalent percentage value.) The degree of confidence is also referred to as the **level of confidence** or the **confidence coefficient.**

EXAMPLE

Referring to the sample I.Q. data given at the beginning of this section, we see that $\bar{x} = 111.4$, $n = 50$, and $s = 15.2$. For the standard I.Q. test we will assume that $\sigma = 15$, and the sample standard

The Nielsen Television Rating System

Television shows are canceled and renewed on the basis of their ratings. The A. C. Nielsen Company monitors the viewing habits of about 1700 families by attaching meters to their television sets. Computers automatically telephone these meters and collect the data at the Nielsen center in Florida. The families are very carefully selected so that practically all homes in the United States have an equal chance of being chosen. Many people criticize this system because they don't believe that the small number of Nielsen families can accurately reflect the viewing habits of millions of Americans. However, the concepts of statistics developed in this text show that valid results can generally be obtained with a sample size of 1700, if the sample is random.

deviation of 15.2 lends credence to that assumption. With a 0.95 degree of confidence, use this data to find each value.

(a)   The maximum error of estimate E.

(b)   The confidence interval.

Solution

(a)   The 0.95 degree of confidence implies that $\alpha = 0.05$, so that

$$E = z(\alpha/2) \frac{\sigma}{\sqrt{n}} = 1.96 \cdot \frac{15}{\sqrt{50}} = 4.16$$

(Recall from the last example that when $\alpha = 0.05$, $z(\alpha/2) = 1.96$.)

(b) $\bar{x} - E < \mu < \bar{x} + E$ becomes $111.4 - 4.16 < \mu < 111.4 + 4.16$, or $107.24 < \mu < 115.56$. We interpret the results of the confidence intervals as follows: If we were to select different samples from the given population and use the above method for finding the corresponding confidence intervals, in the long run 95% of those intervals would contain μ.

Assume that in the preceding example the I.Q. data actually comes from a population with a mean of 110. Then the confidence interval obtained from the given sample data does contain the population mean since 110 is between 107.24 and 115.56. This is illustrated by Figure 7-2. The confidence interval obtained in the preceding example is represented along with a few other typical confidence intervals that might evolve from other samples (see Figure 7-2, page 336).

Note the wording of the interpretation of the confidence interval given in part (b) of the preceding example. We must be careful to interpret correctly the meaning of a confidence interval. Let's assume for the present discussion that we are using a 0.95 degree of confidence. Given the appropriate data, we can calculate the values of $\bar{x} - E$ and $\bar{x} + E$, which we will call the *confidence interval limits*. In general, these limits will have a 95% chance of enclosing μ. That is, based on sample data to be collected, there will be a 0.95 probability that the confidence interval will contain μ. If we recognize that different samples produce different confidence intervals, in the long run we will be correct 95% of the time when we say that μ is between the confidence interval limits. But once we use actual sample data to find specific limits, those limits either enclose μ or they do not, and we cannot determine if they do or don't without knowing the whole population. It is incorrect to state that μ has a 95% chance of falling within specific limits since μ is a constant, not a random variable, and it will either fall within the interval or it won't—and there's no probability involved in that.

Excerpts from a Department of Transportation Circular

The following excerpts from a Department of Transportation circular concern some of the accuracy requirements for navigation equipment used in aircraft. Note the use of the confidence interval.

"The total of the error contributions of the airborne equipment, when combined with the appropriate flight technical errors listed, should not exceed the following with a 95% confidence (2-sigma) over a period of time equal to the update cycle."

"The system of airways and routes in the United States has widths of route protection used on a VOR system use accuracy of ± 4.5 degrees on a 95% probability basis."

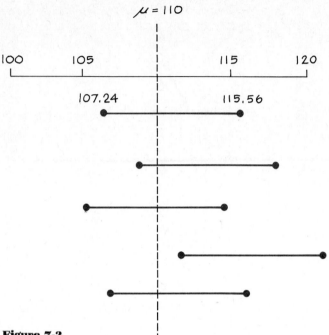

Figure 7-2

Thus far we have defined the confidence interval for the mean and illustrated its use. We now explain why the confidence interval has the form given in the definition.

The basic underlying idea relates to the central limit theorem, which indicates that the distribution of sample means is approximately normal as long as the samples are large ($n > 30$). The central limit theorem was also used to determine that sample means have a mean of μ while the standard deviation of means from samples of size n is $\sigma/\sqrt{n}$. That is,

$$\mu_{\bar{x}} = \mu$$

$$\sigma_{\bar{x}} = \frac{\sigma}{\sqrt{n}}$$

Recall that $\sigma_{\bar{x}}$ is called the standard error of the mean. It is the standard deviation of means computed from samples of size n. Since a z score is the number of standard deviations a value is away from the mean, we conclude that $z(\alpha/2)\sigma/\sqrt{n}$ represents a number of standard deviations away from μ. **There is a probability of $1 - \alpha$ that a sample mean will differ from μ by less than $z(\alpha/2)\sigma/\sqrt{n}$.** See Figure 7-3 and note that the unshaded inner regions total $1 - \alpha$ and correspond to sample means that differ from μ by less than $z(\alpha/2)\sigma/\sqrt{n}$. In other words, a sample mean error of $\mu - \bar{x}$ will be between $-z(\alpha/2)\sigma/\sqrt{n}$ and $z(\alpha/2)\sigma/\sqrt{n}$. This can be expressed as one inequality:

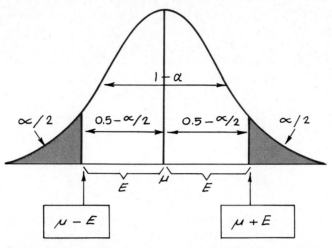

Figure 7-3 (i) There is a $1 - \alpha$ probability that a sample mean will be in error by less than E or $z(\alpha/2)\sigma/\sqrt{n}$.
(ii) There is a probability of α that a sample mean will be in error by more than E (in one of the shaded tails).

$$-z(\alpha/2)\frac{\sigma}{\sqrt{n}} < \mu - \bar{x} < z(\alpha/2)\frac{\sigma}{\sqrt{n}}$$

or

$$-E < \mu - \bar{x} < E$$

Adding $\bar{x}$ to each component of the inequality, we get the confidence interval $\bar{x} - E < \mu < \bar{x} + E$.

Applying these general concepts to the specific I.Q. data given at the beginning of this section, we can see that there is a 95% chance that a sample mean will be in error by less than

$$z(\alpha/2)\frac{\sigma}{\sqrt{n}} = 1.96 \cdot \frac{15}{\sqrt{50}} = 4.16$$

Of the confidence intervals we construct by this method, in the long run an average of 95% of them will involve sample means that are in error by less than 4.16. If $\bar{x} = 111.4$ is really in error by less than 4.16, then μ would be between 107.24 and 115.56. Another way of stating that is $107.24 < \mu < 115.56$, which corresponds to our confidence interval.

Determining Sample Size

If we begin with the expression for E and solve for n, we get

Formula 7-1
$$n = \left[\frac{z(\alpha/2)\sigma}{E}\right]^2$$

which may be used in determining the sample size necessary to produce results accurate to a desired degree of confidence. This equation should be used when we know the value of σ and we want to determine the sample size necessary to establish, with a probability of $1 - \alpha$, the value of μ to within $\pm E$. The existence of such an equation is somewhat remarkable, since it implies that the sample size does not depend on the size of the population.

EXAMPLE

We wish to be 99% sure that a random sample of I.Q. scores yields a mean that is within two units of the true mean. How large should the sample be? Assume that σ is 15.

Solution

We seek n given that $\alpha = 0.01$ (from 99% confidence) and the maximum allowable error is 2.0. Applying the equation for sample size n, we get

$$n = \left[\frac{z(\alpha/2)\sigma}{E} \right]^2 = \left[\frac{2.575 \times 15}{2.0} \right]^2 = (19.3125)^2 = 373$$

Therefore we should obtain at least 373 randomly selected I.Q. scores if we require 99% confidence that our sample mean is within 2.0 units of the true mean.

If we can settle for less accurate results and accept a maximum error of four instead of two, we can see that the required sample size is reduced from 373 to 94. Direct application of the equation for n produces a value of 93.24, which is *rounded up* to 94. (We always round up in sample size computations so that the required number is at least adequate instead of being slightly inadequate.)

Doubling the maximum error from two to four caused the required sample size to decrease to one-fourth of its original value. Conversely, if we want to halve the maximum error, we must quadruple the sample size. In the equation, n is inversely proportional to the square of E and directly proportional to the square of $z(\alpha/2)$. All of this implies that we can obtain more accurate results with greater confidence, but the sample size will be substantially increased. Since larger samples generally require more time and money, there may be a need for a tradeoff between the sample size and the confidence level.

Small Sample Cases

Unfortunately, the application of the equation for sample size (Formula 7-1) requires prior knowledge of σ. Realistically, σ is usually unknown unless previous research results are available. When we do sample from a population in which σ is unknown, a preliminary study must be conducted so that σ can be reasonably estimated. Only then can we determine the sample size required to meet our error tolerance and confidence demands. However, if we intend to construct a confidence interval, do not know σ, and do not plan a preliminary study, we can use the student t distribution as long as the population is essentially normal. In a normally distributed population with unknown σ, we can replace σ by s if $n > 30$. Otherwise we can replace $z(\alpha/2)$ by $t(\alpha/2)$ in the computation of E to get the maximum error E in the following expression.

MAXIMUM ERROR

$$E = t(\alpha/2) \frac{s}{\sqrt{n}}$$

when *all* the following conditions are met:

1. $n \leq 30$; and
2. σ is unknown; and
3. the population is normal.

EXAMPLE

Twenty randomly selected high school seniors are given I.Q. tests with the following sample results:

$$\bar{x} = 111.4 \quad \text{and} \quad s = 15.2$$

Assuming that σ is unknown, construct the 95% confidence interval for the mean I.Q. of all seniors.

Solution

Assume here that the population is essentially normal. With σ unknown and a sample of 20 scores, we know from Section 6-4 that the distribution of such sample means is a student t distribution. We therefore calculate E using the student t distribution and s.

$$E = t(\alpha/2) \frac{s}{\sqrt{n}} = 2.093 \cdot \frac{15.2}{\sqrt{20}} = 7.11$$

and $\bar{x} - E < \mu < \bar{x} + E$ becomes

$$111.4 - 7.11 < \mu < 111.4 + 7.11 \quad \text{or} \quad 104.29 < \mu < 118.51$$

A Professional Speaks About Sampling Error

Daniel Yankelovich, in an essay for *Time*, commented on the sampling error often reported along with poll results. He stated that sampling error refers only to the inaccuracy created by using random sample data to make an inference about a population. Yankelovich noted that the sampling error does not address issues of poorly stated, biased, or emotional questions. He says that "most important of all, warning labels about sampling error say nothing about whether or not the public is conflict-ridden or has given a subject much thought. This is the most serious source of opinion poll misinterpretation."

Shown below is the STATDISK computer display of the results from the data in the preceding example. The user enters the level of confidence, the sample size, sample mean, and sample standard deviation. (STATDISK allows entry of a population standard deviation if it is known, but this is not shown on the display below.)

```
Confidence level.......... = .095

Sample size............... = 20

Sample mean............... = 111.4

Sample standard deviation = 15.2


       The Confidence Interval is:
---------------------------------------
104.2839 < Population mean < 118.5161
```

7-2 Exercises A
Estimates and Sample Sizes of Means

7-1 (a) If $\alpha = 0.05$, find $z(\alpha/2)$.
(b) If $\alpha = 0.02$, find $z(\alpha/2)$.
(c) Find $z(\alpha/2)$ for the value of α corresponding to a confidence level of 96%.
(d) If $\alpha = 0.05$, find $t(\alpha/2)$ for a sample of 20 scores.
(e) If $\alpha = 0.01$, find $t(\alpha/2)$ for a sample of 15 scores.

7-2 (a) If $\alpha = 0.10$, find $z(\alpha/2)$.
(b) Find $z(\alpha/2)$ for the value of α corresponding to a confidence level of 95%.
(c) Find $z(\alpha/2)$ for the value of α corresponding to a confidence level of 80%.
(d) If $\alpha = 0.10$, find $t(\alpha/2)$ for a sample of 10 scores.
(e) If $\alpha = 0.02$, find $t(\alpha/2)$ for a sample of 25 scores.

In Exercises 7-3 through 7-10, use the given data to find the maximum error of estimate E. Be sure to use the correct expression for E.

7-3 $\alpha = 0.05$, $\sigma = 15$, $n = 100$ **7-4** $\alpha = 0.05$, $\sigma = 15$, $n = 64$

7-5 $\alpha = 0.01$, $\sigma = 40$, $n = 25$ **7-6** $\alpha = 0.01$, $\sigma = 80$, $n = 16$

7-7 $\alpha = 0.05$, $s = 15$, $n = 25$ **7-8** $\alpha = 0.05$, $s = 30$, $n = 25$

7-9 $\alpha = 0.01$, $s = 15$, $n = 100$ **7-10** $\alpha = 0.01$, $s = 30$, $n = 64$

7-11 Find the 95% confidence interval for μ if $\sigma = 5$, $\bar{x} = 70.4$, and $n = 36$.

7-12 Find the 95% confidence interval for μ if $\sigma = 7.3$, $\bar{x} = 84.2$, and $n = 40$.

7-13 Find the 99% confidence interval for μ if $\sigma = 2$, $\bar{x} = 98.6$, and $n = 100$.

7-14 Find the 90% confidence interval for μ if $\sigma = 5.5$, $\bar{x} = 123.6$, and $n = 75$.

7-15 A sample of 40 randomly selected adults is tested for pulse rates. The resulting mean is 75.8 beats per minute. Assuming that $\sigma = 10$ beats per minute, find the 95% confidence interval for the mean pulse rate of all adults.

7-16 A sample of 40 randomly selected I.Q. scores produces a mean of 96.8. Assuming that $\sigma = 15$, find the 99% confidence interval of μ based on these sample results.

7-17 A psychologist administers a test to determine the braking reaction time of drivers. Two hundred subjects produce a mean time of 0.61 second and a standard deviation of 0.12 second. Find the 95% confidence interval for the mean reaction time of all drivers.

7-18 A random sample of 40 newborn boys produces a mean weight at birth of 7.61 pounds and a standard deviation of 1.00 pound. Find the 99% confidence interval for the mean weight of all newborn boys.

7-19 An electronics company manufactures radios and conducts extensive tests on 25 randomly selected models of a type used in airplanes. The mean time between failures is found to be 322.4 hours for the sample group while $s = 45.0$ hours. Find the 99% confidence interval for the mean time between failures for all such radios.

7-20 The birth process of a newly discovered mammal is being studied, and the lengths of 18 observed pregnancies have been recorded. The mean and standard deviation for these 18 times is 97.3 days and 2.2 days, respectively. Find the 95% confidence interval for the mean time of pregnancy.

7-21 A country mints a coin that is supposed to weigh 1 ounce. A random sample of 25 such coins produces a mean of 0.996 ounce with a standard deviation of 0.004 ounce. Construct the 95% confidence interval for the mean weight of all such coins.

7-22 A botanist measures the heights of 16 seedlings and obtains a mean and standard deviation of 72.5 cm and 4.5 cm, respectively. Find the 90% confidence interval for the mean height of seedlings in the population from which the sample was selected.

7-23 Find the sample size necessary to estimate a population mean to within three units if $\sigma = 16$ and we want 95% confidence in our results.

7-24 Find the sample size necessary to estimate a population mean. Assume that $\sigma = 20$, the maximum allowable error is 1.5, and we want 95% confidence in our results.

7-25 On a standard I.Q. test, σ is 15. How many random I.Q. scores must be obtained if we want to find the true population mean (with an allowable error of 0.5) and we want 99% confidence in the results?

7-26 Do Exercise 7-25 assuming that the maximum error is 0.25 unit instead of 0.5 unit.

7-27 An assembly line supervisor observes 50 employees assigned to a given task, and in each case she records the time required for the employee to learn the operation. A mean of 36.2 minutes results from these times. Assuming that $\sigma = 5.0$ minutes, find the 95% confidence interval for the mean learning time for all such employees.

7-28 A magazine reporter is conducting independent tests to determine the distance a certain car will travel while consuming only 1 gallon of gas. A sample of five cars is tested and a mean of 28.2 miles is obtained. Assuming that $\sigma = 2.7$ miles, find the 99% confidence interval for the mean distance traveled by all such cars using 1 gallon of gas.

7-29 A sample of seven monkeys is studied. Among the results we find that the times required for the monkeys to learn a task had a mean of 14.7 minutes and a standard deviation of 2.5 minutes. If previous testing has shown that $\sigma = 2.5$ minutes, find the 95% confidence interval for the mean time required to learn the task.

7-30 A sample of 25 employees exposed to loud noises is randomly selected for a hearing test and the resulting mean score is found to be 68.5. Assuming that σ is known to be 5.5, construct the 95% confidence interval for the mean of all such employees.

7-31 Do Exercise 7-30 assuming that σ is unknown and $s = 5.5$ is computed from the sample.

7-32 In sampling the caloric content of 24 bottles of "light" beer, the mean and standard deviation are found to be 107.3 calories and 3.9 calories, respectively. Find the 90% confidence interval for the mean caloric content of all such bottles.

7-33 A sample of computer connect times (in hours) is obtained for 18 randomly selected students. For this sample, the mean is 16.2 while the standard deviation is 3.4. Construct the 99% confidence interval for the mean connect time of all such students.

7-34 Ten randomly selected cars are tested for fuel consumption and the mean and standard deviation are found to be 25.1 mpg and 1.9 mpg, respectively. Construct the 95% confidence interval for the mean of all such cars.

7-35 Assume that $\sigma = 2.64$ inches for heights of adult males. Find the sample size necessary to estimate the mean height of all adult males to within 0.5 inch if we want 99% confidence in our results.

7-36 We want to estimate the mean weight of one type of coin minted by a certain country. How many coins must we sample if we want to be 99% confident that the sample mean is within 0.001 ounce of the true mean? Assume that a pilot study has shown that $\sigma = 0.004$ ounce can be used.

7-37 We want to determine the mean weight of all boxes of cereal labeled 400 grams. We need to be 98% confident that our sample mean is within 3 grams of the population mean, and a pilot study suggests that $\sigma = 10$ grams. How large must our sample be?

7-38 A psychologist has developed a new test of spatial perception, and she wants to estimate the mean score achieved by adult male pilots. How many people must she test if she wants the sample mean to be in error by no more than 2.0 points, with 95% confidence? A pilot study suggests that $\sigma = 21.2$.

7-2 Exercises B
Estimates and Sample Sizes of Means

7-39 Solve for n and show all work:

$$E = \frac{z(\alpha/2)\sigma}{\sqrt{n}}$$

7-40 Why *don't* we take

$$^{\bullet}E = \frac{t(\alpha/2)s}{\sqrt{n}}$$

and solve for n to get an equation that can be used to determine the sample size required for cases in which the student t distribution applies?

7-41 The development of Formula 7-1 assumes that the population is infinite, or we are sampling with replacement, or the population is very large. If we have a relatively small population and we sample without replacement, we should modify E to include the finite population correction factor as follows:

$$E = z(\alpha/2)\frac{\sigma}{\sqrt{n}}\sqrt{\frac{N-n}{N-1}} \qquad \text{where } N \text{ is the population size}$$

(a) Show that the above expression can be solved for n to yield

$$n = \frac{N\sigma^2[z(\alpha/2)]^2}{(N-1)E^2 + \sigma^2[z(\alpha/2)]^2}$$

(b) Do Exercise 7-36 assuming that the coins are selected without replacement from a population of $N = 100$ coins.

7-42 The standard error of the mean is $\sigma/\sqrt{n}$ provided that the population size is infinite. If the population size is finite and is denoted by N, then the correction factor

$$\sqrt{\frac{N-n}{N-1}}$$

should be used whenever $n > 0.05N$. This correction factor multiplies the standard error of the mean, as shown in Exercise 7-41. Find the 95% confidence interval for the mean of 100 I.Q. scores if a sample of 30 scores produces a mean and standard deviation of 132 and 10, respectively.

7-43 The Internal Revenue Service conducts a study of estates valued at more than \$300,000 and determines the value of bonds for a randomly selected sample with the results (in dollars) given below. Find an interval estimate of the mean value of bonds for all such estates. Assume that the sample standard deviation can be used as an estimate of the population standard deviation.

45,300	36,200	72,500	50,500	15,300	58,500
26,200	97,100	74,200	83,700	72,000	10,000
63,000	15,000	49,200	37,500	81,000	24,000
145,000	27,900	53,100	27,500	94,000	23,800
74,600	36,800	65,900	29,400	86,300	25,600
53,200	47,200	61,800	33,200	18,200	75,000

7-44 In one region of a city, a random survey of households includes a question about the number of people in the household. The results are given in the accompanying frequency table. Construct the 90% confidence interval for the mean size of all such households. Assume that the sample standard deviation can be used as an estimate of the population standard deviation.

Household size	f
1	15
2	20
3	37
4	23
5	14
6	4
7	2

7-45 Using the data from Exercise 7-43 as a pilot study, how many random estates must be surveyed in order to estimate the mean bond value? Assume that we want to be 95% confident that the sample mean is in error by at most $1000.

7-46 Using the data from Exercise 7-44 as a pilot study, how many random households must be surveyed if we want to estimate the mean household size? Assume that we want 99% confidence that the sample mean is in error by at most 0.1.

7-3 Estimates and Sample Sizes of Proportions

In this section we consider the same concepts of estimating and sample size determination that were in Section 7-2, but we apply those concepts to proportions instead of means. We already studied proportions in Section 6-5, where we presented a method of conducting tests of hypotheses about population proportions. As in Section 6-5, we again assume that the conditions for the binomial distribution are essentially satisfied. We consider binomial experiments for which $np \geq 5$ and $nq \geq 5$. These assumptions enable us to use the normal distribution as an approximation to the binomial distribution.

Although we make repeated references to proportions, you should remember that the theory and procedures also apply to probabilities and percents. Proportions and probabilities are both expressed in decimal or fraction form. If we intend to deal with percents we can easily convert them to proportions by deleting the percent sign and dividing by 100. The symbol p may therefore represent a proportion, a probability, or the decimal equivalent of a percent. We continue to use p as the population proportion in the same way that we use μ to represent the population mean. In Section 6-5, we represented a sample proportion by x/n where x was the number of successes in n trials. Continued use of x/n here would lead to some awkward expressions, so we introduce the following new notation.

NOTATION

$$\hat{p} = \frac{x}{n}$$

In this way, p represents the population proportion while $\hat{p}$ represents the sample proportion. In previous chapters we stipulated that $q = 1 - p$, so it now becomes natural to stipulate that $\hat{q} = 1 - \hat{p}$.

The term $\hat{p}$ denotes the sample proportion that is analogous to the relative frequency definition of a probability. As an example, suppose that a pollster is hired to determine the proportion of adult Americans who favor socialized medicine. Let's assume that 2000 adult Americans are surveyed with 1347 favorable reactions. The pollster seeks the value of p, the true proportion of all adult Americans favoring socialized medicine. Sample results indicate that $x = 1347$ and $n = 2000$, so that

$$\hat{p} = \frac{x}{n} = \frac{1347}{2000} = 0.6735$$

Just as $\bar{x}$ was selected as the point estimate of μ, we now select $\hat{p}$ as the best point estimate of p.

DEFINITION

The sample proportion $\hat{p}$ is the best **point estimate of the population proportion p.**

Of the various estimators that could be used for p, $\hat{p}$ is deemed best because it is unbiased and the most consistent. (It is unbiased in the sense that the distribution of sample proportions tends to center about the value of p. It is most consistent in the sense that the variance of sample proportions tends to be smaller than the variance of the other unbiased estimators.)

We assume in this section that the binomial conditions are essentially satisfied and that the normal distribution can be used as an approximation to the distribution of sample proportions. This allows us to draw from results established in Section 5-5 and to conclude that the mean number of successes μ and the standard deviation of the number of successes σ are given by

$$\mu = np$$
$$\sigma = \sqrt{npq}$$

where p is the probability of a success. Both of these parameters pertain to n trials, and we now convert them to a "per trial" basis simply by dividing by n:

$$\text{mean of sample proportions} = \frac{np}{n} = p$$

$$\text{standard deviation of sample proportions} = \frac{\sqrt{npq}}{n} = \sqrt{\frac{npq}{n^2}} = \sqrt{\frac{pq}{n}}$$

The first result may seem trivial since we have already stipulated that the true population proportion is p. The second result is nontrivial and very useful. In the last section, we saw that the sample mean $\bar{x}$ has a probability of $1 - \alpha$ of being within $z(\alpha/2)\sigma/\sqrt{n}$ of μ. Similar reasoning leads us to conclude that $\hat{p}$ has a probability of $1 - \alpha$ of being within

$z(\alpha/2)\sqrt{pq/n}$ of p. But if we already know the value of p or q, we have no need for estimates or sample size determinations. Consequently, we must replace p and q by their point estimates of $\hat{p}$ and $\hat{q}$ so that an error factor can be computed in real situations. This leads to the following results.

DEFINITION

The **maximum error of the estimate of p** is given by

$$E = z(\alpha/2)\sqrt{\frac{\hat{p}\hat{q}}{n}}$$

and the probability that $\hat{p}$ differs from p by less than E is $1 - \alpha$.

DEFINITION

The **confidence interval** (or interval estimate) for the population proportion p is given by

$$\hat{p} - E < p < \hat{p} + E$$

The following example illustrates the construction of a confidence interval for a proportion.

EXAMPLE

Of 856 Americans polled, 360 smoked at least one cigarette in the last week. Find the 95% confidence interval for the true population proportion of Americans who smoked at least one cigarette in the last week.

Solution

The sample results show that $n = 856$ and $x = 360$, so $\hat{p} = 360/856 = 0.421$ and $\hat{q} = 1 - \hat{p} = 1 - 0.421 = 0.579$. A confidence level of 95% requires that $\alpha = 0.05$ so that $z(\alpha/2) = 1.96$. We first calculate E to get

$$E = z(\alpha/2)\sqrt{\frac{\hat{p}\hat{q}}{n}} = 1.96\sqrt{\frac{(0.421)(0.579)}{856}} = 0.033$$

We now determine the confidence interval.

$$\hat{p} - E < p < \hat{p} + E \quad \text{or} \quad 0.421 - 0.033 < p < 0.421 + 0.033$$

which becomes $0.388 < p < 0.454$.

The next example involves the situation described in the problem posed in the beginning of this chapter. In this example, note that the sample size is given as 500, but the value of x is not given directly. Instead, the sample percent corresponds to $\hat{p}$.

EXAMPLE

In a poll of 500 randomly selected senior medical students, 13.0% indicated that they plan to specialize in internal medicine. Construct the 99% confidence interval for the true proportion of all senior medical students who plan to specialize in internal medicine.

Solution

The sample results directly reveal that $n = 500$ and $\hat{p} = 0.130$ so that $\hat{q} = 1 - 0.130 = 0.870$. With a confidence level of 99%, we have $\alpha = 0.01$ so that $z(\alpha/2) = 2.575$. We first find

$$E = z(\alpha/2)\sqrt{\frac{\hat{p}\hat{q}}{n}} = 2.575\sqrt{\frac{(0.130)(0.870)}{500}} = 0.039$$

so that

$$\hat{p} - E < p < \hat{p} + E$$

becomes

$$0.130 - 0.039 < p < 0.130 + 0.039$$

or

$$0.091 < p < 0.169$$

or, using percents,

$$9.1\% < p < 16.9\%$$

That is, there is a 99% chance that the confidence interval will contain the true population percent. We now have our sample results expressed with information about the credibility of those results.

Sample Size

Having discussed point estimates and confidence intervals for p, we now consider the problem of determining how large a sample should be when we want to find the approximate value of a population proportion.

Large Sample Size Is Not Good Enough

Better Homes and Gardens conducted a survey by including questionnaires in two of its issues. The number of respondents was an impressively large 302,602. One result showed that 76% of the respondents felt that "family life in America is in trouble." Along with an analysis of the survey results, the magazine presented data indicating that the poll results did not really represent a cross-section of our nation's population. They pointed out, for example, that a disproportionately high number of respondents had attended college. The median income level of respondents was also well above the national median. They also noted that only readers with "a particularly strong interest in the subject" would be likely to complete the questionnaire. They concluded that the survey results represented the views of only the 302,602 respondents and could not be treated as sample data from which to draw broader inferences. Unlike *Better Homes and Gardens,* many people often make the mistake of concluding that sample data can be used for inferences simply because the sample is large. Biased sample data cannot be used for inferences no matter how large the sample is.

In the previous section we started with the expression for the error E and solved for n. Following that reasonable precedent, we begin with

$$E = z(\alpha/2)\sqrt{\frac{\hat{p}\hat{q}}{n}}$$

and we solve for n to get

SAMPLE SIZE

$$n = \frac{[z(\alpha/2)]^2\,\hat{p}\hat{q}}{E^2}$$

$\hat{p}$	$\hat{q}$	$\hat{p}\hat{q}$
0.1	0.9	0.09
0.2	0.8	0.16
0.3	0.7	0.21
0.4	0.6	0.24
0.5	0.5	0.25
0.6	0.4	0.24
0.7	0.3	0.21
0.8	0.2	0.16
0.9	0.1	0.09

But if we are going to determine the necessary sample size, we can assume that the sampling has not yet taken place, so $\hat{p}$ and $\hat{q}$ are not known. Mathematicians have cleverly circumvented this problem by showing that, in the absence of $\hat{p}$ and $\hat{q}$ we can assign the value of 0.5 to each of those statistics and the resulting sample size will be at least sufficient. The underlying reason for the assignment of 0.5 is found in the conclusion that the product $\hat{p} \cdot \hat{q}$ achieves a maximum possible value of 0.25 when $\hat{p} = 0.5$ and $\hat{q} = 0.5$. See the accompanying table, which lists some values of $\hat{p}$ and $\hat{q}$. In practice, this means that no knowledge of $\hat{p}$ or $\hat{q}$ requires that the preceding expression for n evolves into

SAMPLE SIZE

$$n = \frac{[z(\alpha/2)]^2 \cdot 0.25}{E^2}$$

where the ocurrence of 0.25 reflects the substitution of 0.5 for each of $\hat{p}$ and $\hat{q}$. If we have evidence supporting specific known values of $\hat{p}$ or $\hat{q}$ we can substitute those values and thereby reduce the sample size accordingly. For example, if $\hat{p} = 0.6$, then $\hat{q} = 0.4$ and $\hat{p}\hat{q} = 0.24$, which is less than 0.25, so that the resulting value of n will be smaller.

The next two examples are intended to illustrate these points. Again we are faced with the useful and fascinating result that the size of a sample depends, not on the total population size, but on the levels of accuracy and confidence we desire. It is the absolute size of the sample that is important, not the size of the sample relative to the population. It is the sample size number itself that determines its credibility, not the percent of the population. The result of the next example shows that under the stated conditions, a sample size of 752 is appropriate because the magnitude of that number itself is sufficient. Whether 752 is 2%, 1%, or 0.001% of the population size is irrelevant.

In the beginning of this chapter, we stated the problem of a research project in which we sought the percent of senior medical students who plan to specialize in internal medicine. This next example shows how to determine the number of such students that should be polled.

EXAMPLE

We want to estimate, with a maximum error of 0.03, the true proportion of all senior medical students who plan to specialize in internal medicine, and we want 90% confidence in our results. Also, we assume we have no prior information suggesting a possible value of p. How large should our sample be?

Solution

With a confidence level of 90%, we have $\alpha = 0.10$ so that $z(\alpha/2) = 1.645$. We are given $E = 0.03$, but in the absence of $\hat{p}$ or $\hat{q}$ we use the last expression for n.

$$n = \frac{[z(\alpha/2)]^2 \cdot 0.25}{E^2} = \frac{[1.645]^2 \cdot 0.25}{0.03^2} = 751.67 = 752 \text{ (rounded up)}$$

To be 90% confident that we come within 0.03 of the true proportion of those who plan to specialize in internal medicine, we should poll 752 randomly selected senior students.

How One Telephone Survey Was Conducted

A *New York Times*/CBS News survey was based on 1417 interviews of adult men and women in the United States. The sponsors reported that there is 95% certainty that the sample results differ by no more than 3 percentage points from the percentage that would have been obtained if every adult American had been interviewed. A computer was used to select telephone exchanges in proportion to the population distribution. After selecting an exchange number, the computer generated random numbers to develop a complete phone number so that both listed and unlisted numbers could be included.

EXAMPLE

We want to estimate the proportion of those who believe that the extended use of marijuana weakens self-discipline. We want an error of no more than 0.02 and a confidence level of 96%. A previous survey indicates that p should be close to 0.85. How large should our sample be?

Solution

With a 96% confidence level, we have $\alpha = 0.04$ and $z(\alpha/2) = 2.05$. We are given $E = 0.02$, and since $\hat{p}$ is around 0.85, we conclude that $\hat{q} = 1 - \hat{p} = 1 - 0.85 = 0.15$. We can now use our original expression for n to get

$$n = \frac{[z(\alpha/2)]^2\, \hat{p}\hat{q}}{E^2} = \frac{[2.05]^2(0.85)(0.15)}{(0.02)^2}$$

$$= 1339.55 = 1340 \text{ (rounded up)}$$

Rounding *up*, we find that the sample size should be 1340. If we had no prior knowledge of the value of p, we would have used 0.25 for $\hat{p}\hat{q}$ and our required sample size would have been 2627, almost twice as large!

Newspaper, television, and radio reports often give poll results. A good reporter supplies information that will reveal the quality of the results. The following is one statement from a national news magazine.

The study is drawn from personal interviews conducted during the month of June with a representative sample of 1016 men and women of voting age. There is a 95% certainty that the figures have an error factor of plus or minus 3%.

If we let $E = 0.03$ and choose a confidence level of 95%, we will find that $n = 1068$, so we can see that the data are approximately correct. We could verify the statement another way by letting $E = 0.03$ and $n = 1016$. Solving for $z(\alpha/2)$ we get 1.91, indicating a confidence level of 94.4%, which is reasonably close to the claimed value of 95%.

Polling is an important and common practice in the United States. The concepts you have just studied should remove much of the mystery and misunderstanding often created by polls.

7-3 Exercises A
Estimates and Sample Sizes of Proportions

7-47   In a binomial experiment, a trial is repeated n times with x successes. For the given data, find $\hat{p}$, $\hat{q}$, and the best point estimate for the value of p.

   (a) $n = 1000$, $x = 450$
   (b) $n = 283$, $x = 172$
   (c) $n = 17519$, $x = 873$
   (d) $n = 1879$, $x = 1653$
   (e) $n = 2366$, $x = 1103$

7-48   Using the sample data from Exercise 7-47, find the maximum error of estimate E. In each case, assume that $\alpha = 0.05$.

In Exercises 7-49 through 7-52, use the given data to find the appropriate confidence interval for the population proportion p.

7-49 $n = 400$, $x = 100$, 95% confidence

7-50 $n = 900$, $x = 400$, 95% confidence

7-51 $n = 512$, $x = 309$, 98% confidence

7-52 $n = 12{,}485$, $x = 3456$, 99% confidence

7-53   You want to determine the precentage of individual tax returns that include capital gains deductions. How many such returns must be randomly selected and checked? We want to be 90% confident that our sample percent is in error by no more than four percentage points.

7-54   A pollster is hired to determine the percentage of voters favoring the Republican presidential nominee. If we require 99% confidence that the estimated value is within two percentage points of the true value, how large should the random sample be?

7-55   A quality-control engineer wants to determine the proportion of defective flashbulbs that her employer produces. How many bulbs should be tested if she needs to have 95% confidence that the estimated proportion is within 0.01 of the true proportion?

7-56   A political candidate is planning his campaign strategy and needs to know the extent to which his name is recognized. How many voters should be polled if the candidate wants to be 98% sure that the sample percentage of those who recognize his name is within three percentage points of the true percentage?

7-57   A pollster surveys 1500 randomly selected eligible voters and finds that 855 favored the Republican presidential nominee. Construct the 95% confidence interval for the true proportion of all voters favoring the Republican presidential nominee.

7-58 A quality-control analyst tests 500 flashbulbs and finds 13 that are defective. Construct the 98% confidence interval for the true proportion of defective flashbulbs.

7-59 Of 150 interviews, 119 registered opposition to busing students to achieve racial balance. Estimate the true proportion of all who are opposed to busing students.

7-60 Of 3000 random subjects tested, 27 were illiterate. Find the 99% confidence interval for the true proportion of illiterates.

7-61 Do Exercise 7-55 assuming there is strong additional evidence that the proportion of defective bulbs is about 0.012.

7-62 Do Exercise 7-56 assuming there is additional evidence that the candidate is known by 70% of all voters.

7-63 An educator intends to verify reliable information that the illiteracy rate in the United States is about 1%. How many randomly selected subjects should be tested if we want 95% confidence that the sample is in error by not more than one-half of one percentage point?

7-64 An advertising agent has sound evidence that 15% of all radio listeners prefer country music and wants to verify that the given precent is correct. How many randomly selected radio listeners must be polled? The agent wants 90% confidence that the sample is in error by no more than four precentage points.

7-65 A multiple-choice test question is considered easy if at least 80% of the responses are correct. A sample of 6503 responses to one question indicates that 5463 of those responses were correct. Construct the 99% confidence interval for the true proportion of correct responses. Is it likely that this question is really easy?

7-66 Of 281 aviation accidents, 95 resulted in fatalities. Based on this sample, find the 98% confidence interval for the true proportion of all aviation accidents that result in fatalities.

7-67 A survey of 777 audited tax returns shows that 536 resulted in additional tax payments. Construct the 99% confidence interval for the true proportion of all audited returns that result in additional payments.

7-68 Of 200 adult Americans surveyed, 154 indicated that they had at least one credit card. Find the 90% confidence interval for the true proportion of all adult Americans who have at least one credit card.

7-69 If a sample of 80 cottage renters includes 27.5% who are new, find the 95% confidence interval for the true proportion of new renters.

7-70 In a random sample of 250 high school students, it was found that 85% of them later graduated. Find the 98% confidence interval for the true proportion of all high school students who graduate.

7-71   Of 600 people who completed the first item on a questionnaire, 24% responded "always," 60% responded "sometimes," and the others responded "never." Construct the 99% confidence interval for the true proportion of people who respond "never."

7-72   A consumer research team finds that of 520 people surveyed, 25% prefer brand X, 40% prefer brand Y, and the rest prefer brand Z. Construct the 95% confidence interval for the true proportion of people who prefer brand Z.

7-3   Exercises B
Estimates and Sample Sizes of Proportions

7-73   A poll conducted for *Time* reported that its results were based on a representative sample of 1016 people of voting age. An error factor of plus or minus 3% was also reported. Verify that the 3% error is approximately correct at the 95% confidence level by finding E for the given sample size and the 95% level of confidence.

7-74   A newspaper article indicates that an estimate of the unemployment rate involves a sample of 47,000 people. If the reported unemployment rate must have an error no larger than 0.2 percentage point and the rate is known to be about 8%, find the corresponding confidence level.

7-75   For any values of $\hat{p}$ and $\hat{q}$, the product $\hat{p} \cdot \hat{q}$ can be at most 0.25. Find an expression for the largest possible maximum error of the estimate, assuming a confidence level of 95%.

7-76   Solve for n showing all steps:

$$E = z(\alpha/2)\sqrt{\frac{\hat{p}\hat{q}}{n}}$$

7-77   A rectangle has sides of length p and q, and the perimeter is two units. What values of p and q will cause the rectangle to have the largest possible area? To what concept of Section 7-3 does this problem relate?

7-78   In this section we developed two formulas used for determining sample size, and in both cases we assume that the population is infinite, or we are sampling with replacement, or the population is very large. If we have a relatively small population and we sample without replacement, we should modify E to include the finite population correction factor as follows:

$$E = z(\alpha/2)\sqrt{\frac{\hat{p}\hat{q}}{n}}\sqrt{\frac{N-n}{N-1}}$$

where N is the size of the population

(a) Show that the above expression can be solved for n to yield

$$n = \frac{N\hat{p}\hat{q}[z(\alpha/2)]^2}{\hat{p}\hat{q}[z(\alpha/2)]^2 + (N-1)E^2}$$

(b) Do Exercise 7-53 assuming that there is a finite population of size $N = 500$ tax returns and sampling is done without replacement.

7-4   Estimates and Sample Sizes of Variances (Optional)

In Section 6-6 we considered tests of hypotheses made about population variances or standard deviations. We noted that many real and practical situations, such as quality control in a manufacturing process, require inferences about variances or standard deviations. In addition to making products with good average quality, the manufacturer must make products of *consistent* quality that do not run the gamut from extremely poor to extremely good. This consistency can be measured by variance and standard deviation, so these statistics become important in maintaining the quality of products. There are many other situations in which variance and standard deviation are critically important.

As in Section 6-6 we assume here that the population in question has normally distributed values. This assumption is again a strict requirement since the chi-square distribution is so sensitive to departures from normality that gross errors can easily arise. We describe this sensitivity by saying that inferences about σ^2 (or σ) made on the basis of the chi-square distribution are not *robust* against departures from normality. In contrast, inferences made about μ based on the student t distribution are reasonably robust since departures from normality that are not too extreme will not lead to gross errors.

In this section we extend the concepts of the previous sections to variances. These concepts are point estimates, confidence intervals, and sample size determination. (Because of the nature of the chi-square distribution, the techniques of this section will not closely parallel those of the preceding two sections.) Since the sample mean $\bar{x}$ was given as the best point estimate of μ and the sample proportion $\hat{p} = x/n$ was given as the best estimate of p, you should not be too surprised by the following definition.

DEFINITION

The sample variance s^2 is the best **point estimate of the population variance** σ^2.

Since sample variances tend to center about the value of the population variance, we say that s^2 is an unbiased estimator of σ^2. Also, the variance of s^2 values tends to be smaller than the variance of the other unbiased estimators. For these reasons we decree that, among the various possible statistics we could use to estimate σ^2, the best is s^2. Since s^2 is the best point estimate of σ^2, it would be natural to expect that s is the best point estimate of σ, but this is not the case. For reasons we will not pursue, s is a biased estimator of σ; if the sample size is large, however, the bias is small so that we can use s as a reasonably good estimate of σ.

Although s^2 is the best point estimate of σ^2, there is no indication of just how good this best estimate is. To compensate for that deficiency, we need to develop a more informative interval estimate (or confidence interval).

Recall that in Section 6-6 we tested hypotheses about σ^2 by using the test statistic $(n - 1)s^2/\sigma^2$, which has a chi-square distribution. In Figure 7-4 we illustrate the chi-square distribution of $(n - 1)s^2/\sigma^2$ for samples of size $n = 20$. We also show the two critical values of χ^2 corresponding to a 95% level of significance. In a normal or t distribution, the left and right critical values are the same numbers with opposite signs, but from Figure 7-4 we can see that this is not the case with the chi-square distribution. Because we will use those values in developing confidence intervals for standard deviations and variances, we introduce the following notation.

NOTATION

With a total area of α divided equally between the two tails of a chi-square distribution, χ_L^2 denotes the left-tailed critical value and χ_R^2 denotes the right-tailed critical value.

A sample of 20 scores corresponds to 19 degrees of freedom, and if we refer to Table A-5 we find that the critical values shown in Figure 7-4 separate areas to the right of 0.025 and 0.975.

Figure 7-4 shows that, for a sample of 20 scores taken from a normally distributed population, the statistic $(n - 1)s^2/\sigma^2$ has a 0.95 probability of falling between 8.907 and 32.852. In general, there is a probability of $1 - \alpha$ that the statistic $(n - 1)s^2/\sigma^2$ will fall between χ_L^2 and χ_R^2. In other words (and symbols), there is a $1 - \alpha$ probability that both of the following are true:

$$\frac{(n - 1)s^2}{\sigma^2} < \chi_R^2$$

and

$$\frac{(n - 1)s^2}{\sigma^2} > \chi_L^2$$

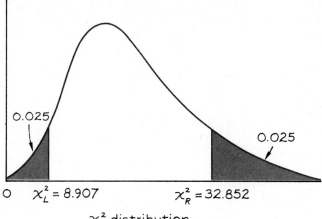

Figure 7-4 χ^2 distribution $(n = 20)$

If we multiply both of the preceding inequalities by σ^2 and divide each inequality by the appropriate critical value of χ^2, we see that the two inequalities can be expressed in the equivalent forms

$$\frac{(n-1)s^2}{\chi_R^2} < \sigma^2$$

and

$$\frac{(n-1)s^2}{\chi_L^2} > \sigma^2$$

These last two inequalities can be combined into one inequality,

$$\frac{(n-1)s^2}{\chi_R^2} < \sigma^2 < \frac{(n-1)s^2}{\chi_L^2}$$

There is a probability of $1 - \alpha$ that the population variance σ^2 is contained in the above interval.

These results provide the foundation for the following definition, which applies to normally distributed populations.

DEFINITION

The **confidence interval** (or **interval estimate**) for the population variance σ^2 is given by

$$\frac{(n-1)s^2}{\chi_R^2} < \sigma^2 < \frac{(n-1)s^2}{\chi_L^2}$$

TV Watching Sets Records

The A. C. Nielsen Company recently found that the average television viewing per household has reached a new peak: 7 hours and 2 minutes per day! Among the explanations for this new record are the increase in cable subscriptions and the fact that more televisions are left on while people are doing other things. While a lengthy report about these statistics referred to the "average" viewing time, it did not explicitly state that the value is a mean, and no measure of dispersion (such as the standard deviation) was given.

The confidence interval (or interval estimate) for σ can be found by simply taking the square root of each component of the preceding inequality:

$$\sqrt{\frac{(n-1)s^2}{\chi_R^2}} < \sigma < \sqrt{\frac{(n-1)s^2}{\chi_L^2}}$$

EXAMPLE

A bank experiments with a single waiting line that feeds all windows as openings occur. An employee observes 30 randomly selected customers, and their waiting times produce a standard deviation of 3.8 minutes. Construct the 95% confidence interval for the true value of σ^2.

Solution

With a sample size $n = 30$, we get $n - 1 = 29$ degrees of freedom. Since we seek 95% confidence, we divide the 5% chance of error between the two tails so that each tail contains a proportion of 0.025. In the 29th row of Table A–5, we find that 0.025 in the left and right tails indicates that $\chi_L^2 = 16.047$ and $\chi_R^2 = 45.722$. With these values of χ_L^2 and χ_R^2, with $n = 30$, and with $s = 3.8$, we use the preceding definition to obtain

$$\frac{(30-1)(3.8)^2}{45.722} < \sigma^2 < \frac{(30-1)(3.8)^2}{16.047}$$

which becomes $9.159 < \sigma^2 < 26.096$.

If the previous problem had required the 95% confidence interval for σ instead of σ^2, we could have taken the square root of each component to get $\sqrt{9.159} < \sigma < \sqrt{26.096}$, which can then be expressed as $3.026 < \sigma < 5.108$.

The problem of determining the sample size necessary to estimate σ^2 to within given tolerances and confidence levels becomes much more complex than it was in similar problems that dealt with means and proportions. Instead of developing very complicated procedures, we supply Table 7–1, which lists approximate sample sizes.

EXAMPLE

You wish to estimate σ^2 to within 10% and you need 99% confidence in your results. How large should your sample be? Assume that the population is normally distributed.

Solution

From Table 7–1, 99% confidence and an error of 10% for σ^2 correspond to a sample of size 1400. You should randomly select 1400 values from the population.

Table 7-1

To be 95% confident that s^2 is within	of the value of σ^2, the sample size n should be at least
1%	77,210
5%	3,150
10%	806
20%	210
30%	97
40%	57
50%	38

To be 99% confident that s^2 is within	of the value of σ^2, the sample size n should be at least
1%	133,362
5%	5,454
10%	1,400
20%	368
30%	172
40%	101
50%	67

To be 95% confident that s is within	of the value of σ, the sample size n should be at least
1%	19,205
5%	767
10%	192
20%	47
30%	21
40%	12
50%	8

To be 99% confident that s is within	of the value of σ, the sample size n should be at least
1%	33,196
5%	1,335
10%	336
20%	85
30%	38
40%	22
50%	14

7-4 Exercises A
Estimates and Sample Sizes of Variances

7-79   Use the given data to find the best point estimate of σ^2.
(a) $n = 100, \bar{x} = 103, s^2 = 12.5$
(b) $n = 500, \bar{x} = 320, s = 15.3$
(c) $n = 10, \bar{x} = 98.6, s = 1.2$
(d) $n = 3, \bar{x} = 74, s = 1.9$

7-80   Use Table 7-1 to find the approximate minimum sample size necessary to estimate the parameter with the given tolerances and levels of confidence.
(a) σ^2 with a 30% maximum error and 95% confidence
(b) σ with a 10% maximum error and 95% confidence
(c) σ with a 5% maximum error and 99% confidence
(d) σ with a 10% maximum error and 99% confidence
(e) σ with a 1% maximum error and 99% confidence

7-81   Find the χ_L^2 and χ_R^2 values for a sample of 25 scores and a confidence level of 99%.

7-82   Find the χ_L^2 and χ_R^2 values for a sample of 15 scores and a confidence level of 95%.

7-83   Find the χ_L^2 and χ_R^2 values for a sample of 11 scores and a confidence level of 95%.

7-84   Find the χ_L^2 and χ_R^2 values for a sample of 27 scores and a confidence level of 90%.

7-85   Construct a 95% confidence interval about σ^2 if a random sample of 21 scores is selected from a normally distributed population and the sample variance is 100.

7-86   Construct a 95% confidence interval about σ^2 if a random sample of ten scores is selected from a normally distributed population and the sample variance is 225.

7-87   A standard test for motor skills is being developed by a manufacturer, and a sample of 27 randomly selected adults produces scores with a variance of 22. Construct the 95% confidence interval for the true variance of all scores on this test.

7-88   The weekly downtimes of a computer are being studied for planning and replacement purposes. If the downtimes of 15 randomly selected weeks produce a standard deviation of 2.85 hours, find the 90% confidence interval for the true value of the variance of all weekly downtimes.

7-89   The weights of 20 randomly selected newborn girls have a standard deviation of 535 grams. Find the 95% confidence interval for the true standard deviation of the weights of all newborn girls.

7-90 A vending machine pours soup into cups. The standard deviation of 20 randomly selected cups is 4.4 milliliters. Find the 95% confidence interval for the true standard deviation of all cups filled by this machine.

7-91 We randomly select 22 normal adults and measure their pulse rates. The standard deviation for these pulse rates is 9.25 beats per minute. Find the 99% confidence interval for the true variance of the pulse rates of all normal adults.

7-92 A fuse manufacturer randomly selects 30 blown fuses for detailed analysis. The standard deviation for the 30 amperage values that caused the fuses to blow is 0.874 ampere. Find the 95% confidence interval for the variance of the critical amperage levels.

7-93 A drug is intended to lower blood pressure by an amount equivalent to the pressure of 10 millimeters of mercury. To test the success of this drug, a clinic administers it to ten randomly selected normal adults. The resulting decreases produce a standard deviation of 3.1 millimeters of mercury. Find the 90% confidence interval for the true variance of all such decreases.

7-94 A study involves 40 randomly selected cases of drivers arrested for intoxication, and their blood alcohol concentrations produce a standard deviation of 0.0306. Construct the 99% confidence interval for the standard deviation of all such arrested drivers.

7-95 A researcher finds that 40 randomly selected business students have computer connect times (in hours) that yield a variance of 77.6. Find the 99% confidence interval for the true standard deviation of all such business students at the same college.

7-96 An environmentalist randomly selects the pollution indices for 30 different days in the past five years and computes the variance as 91.3. Find the 98% confidence interval for the standard deviation of indices for all days within the past five years.

7-4 Exercises B
Estimates and Sample Sizes of Variances

7-97 The following reaction times (in seconds) are randomly obtained from a normally distributed population: 0.60, 0.61, 0.63, 0.72, 0.91, 0.72. Find the 99% confidence interval for σ.

7-98 The following weights (in grams) are randomly obtained from a normally distributed population: 201, 203, 212, 222, 213, 215, 217, 230, 205, 208, 217, 225. Find the 95% confidence interval for σ.

7-99 The accompanying frequency table summarizes the times (in seconds) required for a sample of 50 students to memorize a poem. Find the 98% confidence interval for the standard deviation of the times of all students.

x	f
120–149	3
150–179	7
180–209	12
210–239	15
240–269	9
270–299	4

7-100 Under what conditions, if any, can the numerical value of σ exceed that of σ^2?

Computer Project
Estimates and Sample Sizes

(a) We know from this chapter that the formula

$$n = \frac{[z(\alpha/2)]^2 \cdot 0.25}{E^2} = \frac{[1.96]^2 \cdot 0.25}{E^2} = \frac{0.9604}{E^2}$$

can be used to determine the sample size necessary to estimate a population proportion to within a maximum error E, and that result will correspond to a 95% confidence level. Develop a computer program that takes the value of E as input. Output should consist of the sample size corresponding to a 95% confidence level. Run the program to determine the sample sizes corresponding to a variety of different values of E ranging from 0.001 to 0.150.
(b) Computer software packages designed for statistics commonly provide programs for generating confidence intervals. Use such a program to find the confidence intervals referred to in Exercises 7-15, 7-43, 7-57, and 7-98.

Review

In this chapter we continued our study of inferential statistics by introducing the concepts of **point estimate, confidence interval** (or **interval estimate**), and ways of determining the **sample size** necessary to estimate parameters to within given error factors. In Chapter 6 we used sample data to make **decisions** about hypotheses, but the central concern of this chapter is the **estimate** of parameter values. The parameters are population means, proportions, and variances.

The table to the right summarizes some of the key results of this chapter.

I M P O R T A N T   F O R M U L A S

Parameter	Point estimate	Confidence interval	Sample size
μ	$\bar{x}$	$\bar{x} - E < \mu < \bar{x} + E$ where $E = z(\alpha/2) \dfrac{\sigma}{\sqrt{n}}$ (if σ is known or if $n > 30$) or $E = t(\alpha/2) \dfrac{s}{\sqrt{n}}$ (if σ is unknown and $n \leq 30$)	$n = \left[\dfrac{z(\alpha/2)\sigma}{E} \right]^2$
p	$\hat{p} = \dfrac{x}{n}$	$\hat{p} - E < p < \hat{p} + E$ where $E = z(\alpha/2) \sqrt{\dfrac{\hat{p}\hat{q}}{n}}$	$n = \dfrac{[z(\alpha/2)]^2 \hat{p}\hat{q}}{E^2}$ or $n = \dfrac{[z(\alpha/2)]^2 \cdot 0.25}{E^2}$
σ^2	s^2	$\dfrac{(n-1)s^2}{\chi_R^2} < \sigma^2 < \dfrac{(n-1)s^2}{\chi_L^2}$	See Table 7-1

Vocabulary List

Define and give an example of each term.

point estimate
maximum error of
   estimate
confidence interval
interval estimate

degree of
   confidence
confidence interval
   limits

Review Exercises

7-101 (a) Evaluate $z(\alpha/2)$ for $\alpha = 0.10$.
      (b) Evaluate χ^2_L and χ^2_R for $\alpha = 0.05$ and a sample of 10 scores.
      (c) Evaluate $t(\alpha/2)$ for $\alpha = 0.05$ and a sample of 10.
      (d) What is the largest possible value of $p \cdot q$?

7-102 (a) Evaluate $z(\alpha/2)$ for a confidence level of 96%.
      (b) Evaluate $t(\alpha/2)$ for a confidence level of 99% and a sample size of 16.
      (c) Evaluate χ^2_L and χ^2_R for a confidence level of 99% and a sample size of 20.
      (d) For the same set of data, confidence intervals are constructed for the 95% and 99% confidence levels. Which interval has limits that are farther apart?

7-103 A sociologist develops a test designed to measure a person's attitudes about disabled people, and 16 randomly selected subjects are given the test. Their mean is 71.2 while their standard deviation is 10.5. Construct the 99% confidence interval for the mean score of all subjects.

7-104 Using the sample data in Exercise 7-103, construct the 99% confidence interval for the standard deviation of the scores of all subjects.

7-105 Given $n = 60$, $\bar{x} = 83.2$ kilograms, $s = 4.1$ kilograms. Assume that the given statistics represent sample data randomly selected from a normally distributed population.
      (a) What is the best point estimate of μ?
      (b) Construct the 95% confidence interval about μ.

7-106 Use the sample data given in Exercise 7-105.
      (a) What is the best point estimate of σ^2?
      (b) Construct the 95% confidence interval about σ.

7-107 Given $n = 16$, $\bar{x} = 83.2$ kilograms, $s = 4.1$ kilograms. Assume that the given statistics represent sample data randomly selected from a normally distributed population.
      (a) What is the best point estimate of μ?
      (b) Construct the 95% confidence interval about μ.

7-108   Use the sample data given in Exercise 7-107.
(a) What is the best point estimate of σ^2?
(b) Construct the 95% confidence interval about σ.

7-109   A psychologist is collecting data on the time it takes to learn a certain task. For 50 randomly selected adult male subjects, the sample mean and standard deviation are computed to be 16.40 minutes and 4.00 minutes, respectively. Construct the 98% interval estimate for the mean time required by all adult males.

7-110   Using the sample data from Exercise 7-109, construct the 99% confidence interval for σ.

7-111   A pollster is expected to determine the percentage of voters favoring some form of capital punishment. If there must be 95% confidence that the estimated value is within two percentage points of the true value, how large should the random sample be?

7-112   A newspaper editor plans to supervise a survey of regional opinions. How many people must be surveyed if, for the first question, she wants to be 94% confident that the percent of respondents saying "no" is in error by no more than five percentage points.

7-113   The president of the student body at a very large university wants to determine the percent of students who are registered voters. How many students must be surveyed if we want 90% confidence that the sample is in error by no more than five percentage points?

7-114   A psychologist wants to determine the proportion of students in a large school district who have divorced parents. How many students must be surveyed if he wants 96% confidence that the sample proportion is in error by no more than 0.06?

7-115   A medical researcher wishes to estimate the cholesterol level of all 6-year-old boys, and there is strong evidence suggesting that $\sigma = 50$. If the researcher wants to be 95% confident of obtaining a sample mean that is off by no more than four units, how large must the sample be?

7-116   A sociologist wants to determine the mean value of cars owned by retired people. If the sociologist wants to be 96% confident that the mean of the sample group is off by no more than $250, how many retired people must be sampled? A pilot study suggests that the standard deviation is $3050.

7-117   An advertising firm wants to estimate the mean time spent by preschool children watching television on Saturday morning. A pilot study suggests that $\sigma = 0.8$ hours. How many subjects must be surveyed for 98% confidence that the sample mean is off by no more than 0.02 hours?

7-118 A botanist wants to determine the mean diameter of pine trees in a forest. She conducts a preliminary study to establish that the standard deviation of all such trees is about 6.35 cm. How many randomly selected trees must be measured if she wants 98% confidence that the sample mean is in error by no more than 0.5 cm?

7-119 A medical treatment is applied in 300 cases and 18 cures result. Construct the 95% confidence interval for the proportion of cures.

7-120 Of 240 bicycles inspected on an assembly line, 14 were found to have at least one missing part. Construct the 98% confidence interval for the proportion of all such bicycles with at least one missing part.

7-121 Of 1533 adults surveyed in a Gallup poll, 47% correctly identified Freud. Construct a 99% confidence interval for the true proportion of adults who can correctly identify Freud.

7-122 Of 520 software packages sold to computer owners, 37.9% were for educational purposes. If this is a random sample, construct a 95% confidence interval for the true proportion of software in the educational category.

Chapter 7
Case Study Activity

Determine the sample size needed to estimate the mean weight of a passenger car in your region. The calculation of that number requires the value of the population standard deviation, which is unknown. Conduct a preliminary pilot study by finding the weights of at least 30 different passenger cars. Those weights can usually be found on vehicle registration documents. List the year and make of each vehicle and identify any factors suggesting that your sample is not representative of the population of all passenger cars in your region. Calculate the standard deviation of the sample weights and use the result as an estimate of the population standard deviation so that the required sample size can be determined. Assume that you want to be 98% confident that the sample mean is within 25 pounds of the true population mean.

CHAPTER

8-1   Overview

We identify chapter **objectives.** This chapter deals with methods of testing hypotheses made about two population parameters.

8-2   Tests Comparing Two Variances

We present the method for testing hypotheses made about **two population variances** or standard deviations.

8-3   Tests Comparing Two Means

We present methods for testing hypotheses made about **two dependent means** or **two independent means** with equal or unequal variances.

8-4   Tests Comparing Two Proportions

We use the pooled estimate of p_1 and p_2 to test hypotheses made about the **population proportions** p_1 and p_2.

Tests Comparing Two Parameters

CHAPTER PROBLEM

A private educational service offers a course designed to increase scores on standard I.Q. tests. To test the effectiveness of this course, ten randomly selected subjects take I.Q. tests before and after the course. How do we analyze these results to test the claim that the course is effective? For the course to be effective, the mean test score after the course should be higher than the mean score of the tests taken before the course. We want to test the claim that the "after" mean is greater than the "before" mean. In this chapter, we will learn how to conduct this type of test as well as other types of tests involving two sets of data.

8-1   Overview

In Chapter 6 we introduced the method of hypothesis testing, but that chapter involved only tests of claims made about a single population parameter. In this chapter we extend the methods of hypothesis testing to cases involving the comparison of two variances, two means, or two proportions. In reality, there are many cases where the main objective is a comparison of *two* groups of data instead of a comparison of *one* group to some known value. For example, a manufacturer may want to know which of two different production techniques yields better results. A college dean may be interested in the differences between the grades given at two different schools. A psychologist may need to know whether two different I.Q. tests produce similar or different results. A doctor may be interested in the comparative effectiveness of two different cold medicines. A sociologist may want to compare last year's college entrance examination scores to those of 20 years ago. A psychologist may wish to compare variability in reaction times between a group of nondrinkers and a group of people who have each consumed a double martini. In all of these cases, we want to compare two population parameters.

Chapters 6 and 7 started with means and then moved on to proportions, followed by variances. Here we depart from this order because we will sometimes use results of tests on two variances as a *prerequisite* for a test involving two means. Consequently, we begin this chapter with a discussion of the method for testing hypotheses involving a comparison of two variances.

8-2   Tests Comparing Two Variances

In many cases where we want to compare parameters from each of two separate populations, the **variances** are the most relevant parameters. For example, a manufacturer of automobile batteries may want to compare two different production methods of batteries that last, on the average, approximately four years. Consider the case in Table 8-1.

We assume throughout this section that we have two independent populations that are approximately normally distributed. A sample of size n_1 is drawn from the first population, while a sample of size n_2 is drawn from the second. Since we want to compare variances, we compute the respective sample variances s_1^2 and s_2^2 and use those statistics in testing for equality of σ_1^2 and σ_2^2.

From Table 8-1 we see that both methods seem to produce batteries that last the same length of time, but the batteries produced by method A exhibit much less consistency. The two production methods appear to differ radically in the variability of the battery lives, and this is reflected in the large difference between the two sample variances. Thus it is through a comparison of the two *variances* that we find the crucial differ-

Table 8-1	
Life (in years) of car batteries	
Production method A	Production method B
2.0	3.7
2.1	3.9
2.5	3.9
3.0	3.9
3.3	4.0
4.2	4.0
4.2	4.0
4.3	4.1
6.8	4.2
7.6	4.3
$\overline{x}_1 = 4.00$	$\overline{x}_2 = 4.00$
$s_1^2 = 3.59$	$s_2^2 = 0.03$

Cheating Success

Recently, 18 students from a high school in a depressed area of Los Angeles took an advanced-placement exam for college calculus. After all the students passed (seven with the highest possible scores), representatives of the Educational Testing Service challenged the validity of 14 students' results. There was a significant discrepancy between the expected scores and the actual scores. However, a retest of 12 of the students confirmed the original scores; the students knew their calculus! It turned out that an interested teacher had recruited good students, and they had all worked diligently in special classes before, during, and after regular school hours. What appeared to be cheating was, in fact, the result of hard work by a group of dedicated students and an exceptional teacher.

ence between the two production methods. Since not all comparisons of variances involve such obvious differences, we need more standardized and objective procedures. Even for the data in Table 8-1, the difference between the variances of 3.59 and 0.03 must be weighed against the sample sizes to determine whether this "obvious" difference is statistically significant.

For simplicity of notation, let's stipulate that s_1^2 **always represents the larger of the two sample variances** and s_2^2 represents the smaller of the two sample variances. We can do this because identification of the samples through subscript notation is arbitrary. Extensive analyses have shown that **for two normally distributed populations with equal variances (that is, $\sigma_1^2 = \sigma_2^2$), the sampling distribution of**

$$F = \frac{s_1^2}{s_2^2}$$

is the F distribution shown in Figure 8-1 and described in Table A-6.

If the two populations really do have equal variances, then $F = s_1^2/s_2^2$ tends to be close to 1 since s_1^2 and s_2^2 tend to be close in value. But if the two populations have radically different variances, s_1^2 and s_2^2 tend to be very different numbers. Denoting the larger of the sample variances by s_1^2, we see that the ratio s_1^2/s_2^2 will be a large number whenever s_1^2 and s_2^2 are far apart in value. Consequently, a value of F near 1 will be evidence in favor of the conclusion that $\sigma_1^2 = \sigma_2^2$. A large value of F will be evidence against the conclusion of equality of the population variances. The critical F values are summarized in Table A-6.

When we use Table A-6 we obtain critical F values that are determined by the following three values:

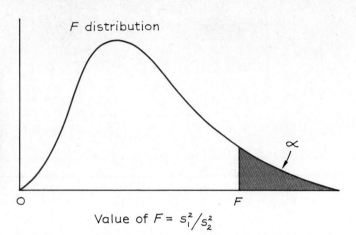

Figure 8-1   Properties of the F distribution. (1) All values of F are nonnegative ($F \geq 0$). (2) Instead of being symmetric, the F distribution is skewed to the right. (3) There is a different F distribution for each different pair of degrees of freedom for numerator and denominator.

Exit Polls on Their Way Out?

The major television networks have been using exit polls to make early projections of winners in races for senate and gubernatorial seats as well as the presidency. In an exit poll, voters are randomly selected in various precincts as they "exit" the polling area after voting. These voters are surveyed on their election choices, age, sex, and race. Such polls have no way to verify if the data supplied by the voters is correct. Coupled with other statistical flaws, exit polls can easily produce substantial errors. This lack of reliability, combined with pressure to discontinue the practice of making early projections, may eventually cause the exit of exit polls.

1. The significance level α.
2. The degrees of freedom for the numerator, $(n_1 - 1)$.
3. The degrees of freedom for the denominator, $(n_2 - 1)$.

We should be careful to ensure that n_1 corresponds to the sample yielding a variance of s_1^2 while n_2 corresponds to the sample having variance s_2^2. In Table A-6, we choose the appropriate value of α and then intersect the column representing the degrees of freedom for s_1^2 with the row representing the degrees of freedom for s_2^2. The appropriateness of α is determined by the level of significance desired and whether the test is one-tailed or two-tailed. In a one-tailed test we use the F distribution corresponding to the significance level α, but in a two-tailed test we divide α equally between the two tails and refer to the section of Table A-6 identified as $\alpha/2$. Since we stipulate that the larger sample variance is s_1^2, our tests will be either two-tailed or right-tailed. The following example illustrates the method of testing hypotheses about two population variances.

E X A M P L E

A department store manager experiments with two methods for checking out customers, and the following sample data are obtained. At the 0.02 significance level, test the claim that $\sigma_1^2 = \sigma_2^2$. Assume that the sample data came from normally distributed populations.

Sample A	Sample B
$n_1 = 16$	$n_2 = 10$
$s_1^2 = 225$	$s_2^2 = 100$

Solution

The solution is summarized in Figures 8-2 and 8-3. With the claim of $\sigma_1^2 = \sigma_2^2$ and the alternative of $\sigma_1^2 \neq \sigma_2^2$, we have a two-tailed test, so the significance level of $\alpha = 0.02$ leads to an area of 0.01 in the left-tailed critical region and an area of 0.01 in the right-tailed critical region. But as long as we stipulate that the larger of the two sample variances be placed in the numerator, we need concern ourselves only with the right-tailed critical value. With an area of 0.01 in the right tail, with $n_1 = 16$ and $n_2 = 10$, we locate the critical F value of 4.9621, which corresponds to 15 degrees of freedom in the numerator and 9 degrees of freedom in the denominator. The test statistic is computed as

$$F = \frac{s_1^2}{s_2^2} = \frac{225}{100} = 2.25$$

From Figure 8-3 we see that the test statistic does not fall within the critical region, so we fail to reject the null hypothesis. The two methods are not significantly different. Refer to Figures 8-2 and 8-3 for the complete hypothesis test (see pages 374–375).

So far in this section we have discussed only tests comparing two population variances, but the same methods and theory can be used to compare two population standard deviations. Any claim about two population standard deviations can be easily restated in terms of the corresponding population variances. For example, suppose we want to test the claim that $\sigma_1 = \sigma_2$, and we are given the following sample data taken from normal populations.

Sample 1	Sample 2
$n_1 = 25$	$n_2 = 17$
$s_1 = 12.72$	$s_2 = 8.43$

Restating the claim as $\sigma_1^2 = \sigma_2^2$, we get $s_1^2 = (12.72)^2$ and $s_2^2 = (8.43)^2$, and we can proceed with the F test in the usual way. Rejection of $\sigma_1^2 = \sigma_2^2$ is equivalent to rejection of $\sigma_1 = \sigma_2$. Failure to reject $\sigma_1^2 = \sigma_2^2$ implies failure to reject $\sigma_1 = \sigma_2$.

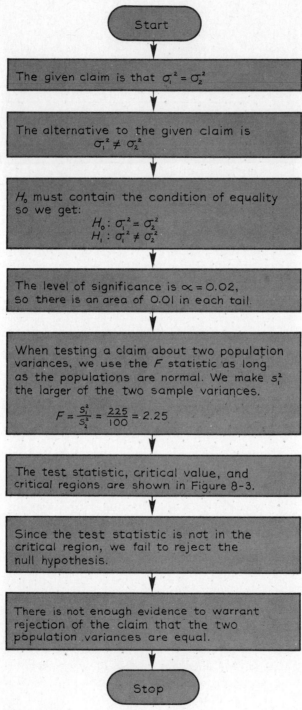

Figure 8-2

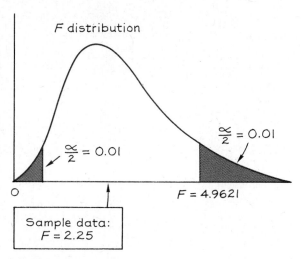

Figure 8-3

More Police, Fewer Crimes?

Does an increase in the number of police officers result in lower crime rates? The question was studied in a New York City experiment that involved a 40% increase in police officers in one precinct while adjacent precincts maintained a constant level of officers. Statistical analysis of the crime records showed that crimes visible from the street (such as auto thefts) did decrease, but crimes not visible from the street (such as burglaries) were not significantly affected.

Note that in all tests of hypotheses made about population variances and standard deviations, the values of the means are irrelevant. In the next section we consider tests comparing two population means, and we will see that those tests sometimes depend upon the results of a hypothesis test about population variances (F test).

8-2   Exercises A
Tests Comparing Two Variances

In Exercises 8-1 through 8-8, test the claim that the two samples come from populations having equal variances. Use a significance level of $\alpha = 0.05$ and assume that all populations are normally distributed. Follow the pattern suggested by Figure 6-2 and draw the appropriate graphs.

8-1   Sample A: $n = 10$, $s^2 = 50$
Sample B: $n = 10$, $s^2 = 25$

8-2   Sample A: $n = 10$, $s^2 = 50$
Sample B: $n = 15$, $s^2 = 25$

8-3   Sample A: $n = 20$, $s^2 = 45.2$
Sample B: $n = 5$, $s^2 = 5.8$

8-4   Sample A: $n = 8$, $s^2 = 8.47$
Sample B: $n = 12$, $s^2 = 1.33$

8-5   Sample A: $n = 20$, $s^2 = 110$
Sample B: $n = 25$, $s^2 = 265$

8-6   Sample A: $n = 41$, $s^2 = 15.1$
Sample B: $n = 16$, $s^2 = 33.2$

8-7   Sample A: $n = 5$, $s = 14.3$
Sample B: $n = 15$, $s = 1.1$

8-8   Sample A: $n = 25$, $s = 3.9$
Sample B: $n = 10$, $s = 6.3$

In Exercises 8-9 through 8-12, test the claim that the variance of population A exceeds that of population B. Use a 5% level of significance and assume that all populations are normally distributed. Follow the pattern outlined in Figure 6-2 and draw the appropriate graphs.

8-9   Sample A: $n = 10$, $s^2 = 48$ **8-10**   Sample A: $n = 50$, $s^2 = 18.2$
           Sample B: $n = 10$, $s^2 = 12$ Sample B: $n = 20$, $s^2 = 8.7$

8-11   Sample A: $n = 16$, $s^2 = 225$ **8-12**   Sample A: $n = 35$, $s^2 = 42.3$
           Sample B: $n = 200$, $s^2 = 160$ Sample B: $n = 25$, $s^2 = 16.2$

8-13   On a test of manual strength, a sample of 25 randomly selected men earn scores that have a variance of 130. When 27 randomly selected women take the same test, the variance of their scores is 75. At the 0.02 significance level, test the claim that the population variances are not equal.

8-14   When 15 randomly selected adult males are given a test on reaction times, their scores produce a variance of 1.04. When 17 other randomly selected adult males are given a double martini before taking the same test, their scores produce a variance of 3.26. At the 0.05 significance level, test the claim that the population of all drinkers will have a variance larger than the population of all nondrinkers.

8-15   A scientist wants to compare two delicate weighing instruments by repeated weighings of the same object. The first scale is used 30 times and the weights have a standard deviation of 72 milligrams. The second scale is used 41 times and the weights have a standard deviation of 98 milligrams. At the 0.05 significance level, test the claim that the second scale produces greater variance.

8-16   An experiment is devised to study the variability of grading procedures among college professors. Two different professors are asked to grade the same set of 25 exam solutions, and their grades have variances of 103.4 and 39.7, respectively. At the 0.05 significance level, test the claim that the first professor's grading exhibits greater variance.

8-2 Exercises B
Tests Comparing Two Variances

8-17   Given the following samples, test the claim that they come from populations with equal variances. Assume that both populations are normally distributed and use a 5% level of significance.

Sample A:	47	52	54	49	38	68	52
	47	48	51				
Sample B:	101	86	105	91	93	99	96
	112	94	102	115	86	117	108

8-18 Given the following sample I.Q. scores, test the claim that they come from populations with equal standard deviations. Assume that both populations are normally distributed.

Sample A: 100 101 110 115 95 88 120
109 116 102 107 93 99 95
Sample B: 112 113 123 127 107 100 132 118
121 128 114 119 105 111 120 135

8-19 Given the accompanying frequency tables, use a 0.05 level of significance to test the claim that both samples come from populations with equal standard deviations.

x	f	x	f
50–59	5	50–59	5
60–69	12	60–69	7
70–79	32	70–79	8
80–89	10	80–89	17
90–99	2	90–99	4

8-20 Two samples of equal size produce variances of 37 and 57. At the 0.05 significance level, we test the claim that the variance of the second population exceeds that of the first, and that claim is upheld by the data. What is the approximate minimum size of each sample?

8-21 Do Exercise 8-20 after changing the significance level to 0.01.

8-22 A sample of 21 scores produces a variance of 67.2 and another sample of 25 produces a variance that causes rejection of the claim that the two populations have equal variances. If this test is conducted at the 0.02 level of significance, find the maximum variance of the second sample if you know that it is smaller than that of the first sample.

8-3 Tests Comparing Two Means

In this section we consider tests of hypotheses made about two population means. Many real and practical situations use such tests successfully. For example, an educator may want to compare mean test scores produced by two teaching methods. A manager may want to test for a difference in the mean weight of cereal loaded into boxes by two machines. A car manufacturer may want to test for a difference in the mean longevity of batteries produced by two suppliers. A psychologist may want to test for a difference in mean reaction times between men and women. A farmer may want to test for a difference in mean crop production for two irrigation methods. A medical researcher may want to test for a difference between a new drug and one currently in use.

The Gender Gap in Wages

Many articles note that on average, full-time female workers earn only about 66¢ for each $1 earned by full-time male workers. This discrepancy is attributed to such factors as the following: Women tend to have less seniority; they tend to enter and leave the labor market more often; they tend to choose lower paying jobs more often; and they tend to have less education. But do these factors account for the wage discrepancy, or is sex discrimination another real factor? Researchers at the Institute for Social Research at the University of Michigan analyzed the effects of various key factors and found that of the 34¢ gap, 11.9¢ is due to differences in education and seniority, work interruptions and job choices. The other 22.1¢ remains "unexplained" by such labor factors.

The way in which we compare means using sample data taken from two populations is affected by the presence or absence of a relationship between those samples.

DEFINITION

Two samples are **dependent** if the values in one are related to the values in the other in some way.

Consider the three sets of sample data tabulated below. In the first example, we would expect the sample of wives' ages and the sample of husbands' ages to be two dependent samples, since there is a tendency for people to marry partners in their own age group. In the second example, we can again expect the sample of x scores and the sample of y scores to be dependent samples, since each xy pair represents two blood pressure readings for the same person. In the third example, the sample of x scores and the sample of y scores are independent samples because the sample of five females is completely unrelated to the sample of seven males.

	Smith	Jones	Brown	Carter	Mayer
x (age of wife)	22	48	27	29	32
y (age of husband)	23	51	25	29	32

	Ann	Bob	Carol	Don	Eve
x (blood pressure before treatment)	120	150	136	166	192
y (blood pressure after treatment)	118	130	136	152	170

x (reaction times of females)	0.70	0.68	0.59	0.72	0.74	
y (reaction times of males)	0.63	0.55	0.58	0.59	0.57	0.60	0.59

When dealing with two dependent samples, it is very wasteful to reduce the sample data to $\bar{x}_1$, s_1, n_1, $\bar{x}_2$, s_2, and n_2 since the relationship between pairs of values would be completely lost. Instead, we compute the *differences (d)* between the pairs of data as follows:

x	22	48	27	29	32
y	23	51	25	29	32
$d = x - y$	-1	-3	2	0	0

NOTATION

Let $\bar{d}$ denote the mean value of d or $x - y$ for the paired sample data.

EXAMPLE

For the d values of -1, -3, 2, 0, 0 taken from the preceding table, we get

$$\bar{d} = \frac{\Sigma d}{n} = \frac{(-1) + (-3) + 2 + 0 + 0}{5} = \frac{-2}{5} = -0.4$$

NOTATION

Let s_d denote the standard deviation of the d values for the paired sample data.

EXAMPLE

For the d values of -1, -3, 2, 0, 0, we get $s_d = 1.8$ as follows.

$$s_d = \sqrt{\frac{n(\Sigma d^2) - (\Sigma d)^2}{n(n-1)}} = \sqrt{\frac{5(14) - (-2)^2}{5(5-1)}} = \sqrt{3.3} = 1.8$$

NOTATION

Let n denote the number of *pairs* of data.

EXAMPLE

For the data of the last table, $n = 5$.

In repeated random sampling from two normal and dependent populations in which the mean of the paired differences is μ_d, the following test statistic possesses a student t distribution with $n - 1$ degrees of freedom:

TEST STATISTIC

$$t = \frac{\bar{d} - \mu_d}{s_d/\sqrt{n}}$$

Note that the involved populations must be normally distributed. If the populations depart radically from normal distributions, we should not use the methods of this section. Instead, we may be able to apply the sign test (Section 11-2) or the Wilcoxon signed-ranks test (Section 11-3).

If we claim that there is no difference between the two population means, then we are claiming that $\mu_d = 0$. This makes sense if we recognize that $\bar{d}$ should be around zero if there is no difference between the two population means.

In the following example we illustrate a complete hypothesis test for a situation involving dependent samples.

EXAMPLE

A private educational service offers a course designed to increase scores on I.Q. tests. To test the effectiveness of this course, ten randomly selected subjects agreed to take I.Q. tests before and after the course. The results follow. At the 5% level of significance, test the claim that the course increases I.Q. scores.

Subjects	I.Q. Test Scores									
	A	*B*	*C*	*D*	*E*	*F*	*G*	*H*	*I*	*J*
Before course	96	110	98	113	88	92	106	119	100	97
After course	99	112	107	110	88	101	107	123	91	99
Difference	−3	−2	−9	3	0	−9	−1	−4	9	−2

Solution

Since each pair of scores is for the same person, we can conclude that the values are dependent. Each difference represents the "before score" − "after score." If the course is effective, we would expect the before scores to be less than the after scores, so the differences would tend to be negative and $\bar{d}$ would be negative and *significantly* below zero. Thus the claim of improved scores is equivalent to the claim that $\mu_d < 0$. We therefore have $H_0: \mu_d \geq 0$ and $H_1: \mu_d < 0$. Figures 8-4 and 8-5 summarize the key features of this hypothesis test (see pages 382–383).

The mean of the differences found in the sample data is $\bar{d}$, where

$$\bar{d} = \frac{(-3) + (-2) + (-9) + 3 + 0 + (-9) + (-1) + (-4) + 9 + (-2)}{10}$$

$$= -1.8$$

The standard deviation of the differences s_d is computed to be 5.3. There are 10 pairs of data, so $n = 10$. These components are used in finding the value of the test statistic.

$$t = \frac{\bar{d} - \mu_d}{s_d/\sqrt{n}} = \frac{-1.8 - 0}{5.3/\sqrt{10}} = -1.074$$

The critical t value of -1.833 can be found in Table A-4 after you note that this is a left-tailed test, $\alpha = 0.05$, and there are $10 - 1$, or 9, degrees of freedom. (The test is left-tailed since only negative values of $\bar{d}$ significantly less than zero will cause rejection of the null hypothesis; the alternate hypothesis is $\mu_d < 0$.)

This test could also be conducted by using the STATDISK software package or any other software package designed to treat the case of two dependent samples. (This test could also be run as a t test for a single population mean; enter the sample data as $-3, -2, -9, 3, 0, -9, -1, -4, 9, -2$ and use zero as the value of the claimed population mean.) After selecting the appropriate case, STATDISK prompts the user to enter the necessary data and the results are shown below.

```
          Hypothesis test for a claim
        about two DEPENDENT populations
  NULL HYPOTHESIS: First Mean >= Second Mean
        Mean of differences = -1.8000
       St. dev. of differences = 5.3083
          Test statistic ... t = -1.0723
          Critical value ... t = -1.8337
              P-value = 0.1557
          Significance level = .05
  CONCLUSION: FAIL TO REJECT the null hypothesis
```

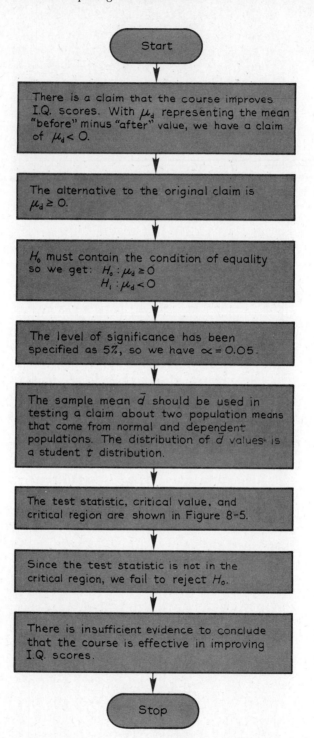

Figure 8-4

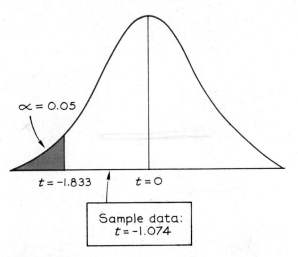

$\alpha = 0.05$

$t = -1.833$ $t = 0$

Sample data:
$t = -1.074$

Figure 8-5

Note that the P-value of 0.1557 is greater than the significance level of 0.05. This indicates that the sample results could easily occur by chance, assuming that the null hypothesis is true.

As we consider other tests of hypotheses made about two population means, we begin to encounter a maze, which can easily lead to confusion. Questions must be answered regarding the independence of samples, knowledge of σ_1 and σ_2, and sample size before the correct procedure can be selected. Most of the confusion can be avoided by referring to Figure 8-6, which summarizes the procedures discussed in this section. We illustrate the use of Figure 8-6 through specific examples and then present the underlying theory that led to its development. Since we have already presented an example involving dependent populations, our next examples will involve independent populations.

EXAMPLE

Two machines fill packages, and samples selected from each machine produce the following results:

Machine A	Machine B
$n_1 = 50$	$n_2 = 100$
$\bar{x}_1 = 4.53$ kilograms	$\bar{x}_2 = 4.01$ kilograms

If the standard deviations of the contents filled by machine A and machine B are 0.80 kilogram and 0.60 kilogram, respectively, test the claim that the mean contents produced by machine A equal the mean for machine B. Assume a 0.05 significance level.

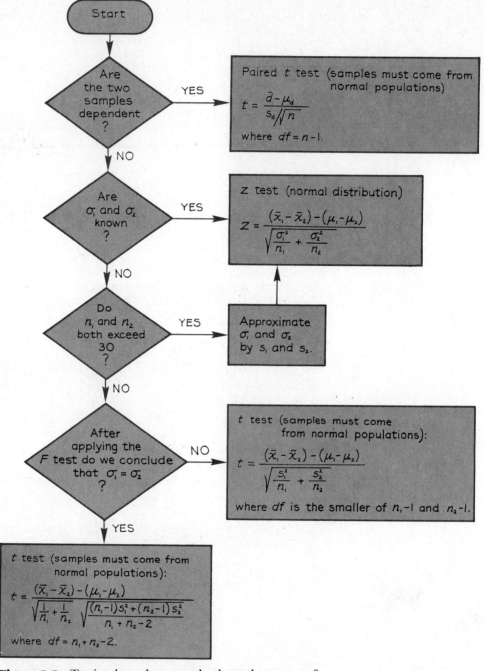

Figure 8-6 Testing hypotheses made about the means of two populations.

Solution

The two means are independent and σ_1 and σ_2 are known. Referring to Figure 8-6, we see that we should use a normal distribution test with

$$z = \frac{(\bar{x}_1 - \bar{x}_2) - (\mu_1 - \mu_2)}{\sqrt{\dfrac{\sigma_1^2}{n_1} + \dfrac{\sigma_2^2}{n_2}}} = \frac{(4.53 - 4.01) - 0}{\sqrt{\dfrac{0.80^2}{50} + \dfrac{0.60^2}{100}}} = 4.06$$

With the null and alternative hypotheses described as

$$H_0: \mu_1 = \mu_2 \quad (\text{or } \mu_1 - \mu_2 = 0)$$
$$H_1: \mu_1 \neq \mu_2 \quad (\text{or } \mu_1 - \mu_2 \neq 0)$$

and $\alpha = 0.05$, we conclude that the test involves two tails. From Table A-3 we extract the critical z values of 1.96 and -1.96. The test statistic of 4.06 is well into the critical region, and we therefore reject H_0 and conclude that the population means corresponding to the two machines are not equal. It appears that machine A fills with amounts that are significantly greater than those of machine B.

The preceding example is somewhat contrived because we seldom know the values of σ_1 and σ_2. It is rare to sample from two populations with unknown means but known standard deviations. To cover more realistic cases involving independent samples with unknown standard deviations or variances, we next examine the sizes of the two samples, as suggested by Figure 8-6. In this next example and the remaining examples of this section, it is a requirement that the samples come from normally distributed populations. If this condition is not satisfied, we may be able to use the Wilcoxon rank-sum test described in Section 11-4.

If both samples are large (greater than 30), we can estimate σ_1 and σ_2 by s_1 and s_2. We can then proceed as in the last example. However, if either sample is small, we must apply the F test to determine whether the two sample variances are indicative of equal population variances. The next example illustrates these points since the populations are independent, both population standard deviations are unknown, and both samples are small (less than or equal to 30). We see that the F test suggests that the two population standard deviations are not equal and, from Figure 8-6, we see that the circumstances of this next example cause us to turn right at the last diamond.

How Valid Are Crime Statistics?

Police departments are often judged by the number of arrests because that statistic is readily available and clearly understood. However, this encourages police to concentrate on easily solved minor crimes (such as marijuana smoking) at the expense of more serious crimes, which are difficult to solve. A study of Washington, D.C., police records revealed that police can also manipulate statistics another way. The study showed that more than 1000 thefts in excess of $50 were intentionally valued at less than $50 so that they would be classified as petty larceny and not major crime.

EXAMPLE

Random samples of similar calculators are obtained from two competing manufacturers. Analyses of the lengths of their lives are as follows:

Brand X	Brand Y
$n_1 = 11$	$n_2 = 15$
$\bar{x}_1 = 4.3$ years	$\bar{x}_2 = 4.6$ years
$s_1 = 1.1$ years	$s_2 = 0.4$ year

At the 0.05 significance level, test the claim that the mean lives of both brands are equal.

Solution

Since the samples are independent, neither standard deviation is known, and both samples are small, Figure 8-6 indicates that we should begin by applying the F test discussed in Section 8-2. We want to decide whether $\sigma_1 = \sigma_2$, so we formulate the following null and alternative hypotheses:

$$H_0: \sigma_1 = \sigma_2$$
$$H_1: \sigma_1 \neq \sigma_2$$

With $\alpha = 0.05$, we do a complete hypothesis test to decide whether or not $\sigma_1 = \sigma_2$. We then do another complete hypothesis test of the claim that $\mu_1 = \mu_2$ using the appropriate student t distribution. For the preliminary test we get

$$F = \frac{s_1^2}{s_2^2} = \frac{1.1^2}{0.4^2} = 7.5625$$

The critical F value obtained from Table A-6 is 3.1469. (The test involves two tails with $\alpha = 0.05$, and the degrees of freedom for the numerator and the denominator are 10 and 14, respectively.) These results cause us to reject the null hypothesis of equal variances and we decide that $\sigma_1 \neq \sigma_2$. In Figure 8-6, we answer no to the question contained in the fourth diamond, and we test the claim that $\mu_1 = \mu_2$ by using the student t distribution and the test statistic given in the box to the right of the fourth diamond. With

$$H_0: \mu_1 = \mu_2 \quad (\text{or } \mu_1 - \mu_2 = 0)$$
$$H_1: \mu_1 \neq \mu_2 \quad (\text{or } \mu_1 - \mu_2 \neq 0)$$
$$\alpha = 0.05$$

we compute the test statistic based on the sample data.

$$t = \frac{(\bar{x}_1 - \bar{x}_2) - (\mu_1 - \mu_2)}{\sqrt{\dfrac{s_1^2}{n_1} + \dfrac{s_2^2}{n_2}}} = \frac{(4.3 - 4.6) - 0}{\sqrt{\dfrac{1.1^2}{11} + \dfrac{0.4^2}{15}}}$$

$$= -0.864$$

This is a two-tailed test with $\alpha = 0.05$ and 10 degrees of freedom, so the critical t values obtained from Table A-4 are $t = 2.228$ and $t = -2.228$. The computed t value of -0.864 does not fall within the critical region and we fail to reject the null hypothesis of equal means. Based on the available sample data, we cannot reject the claim that the two brands of calculators have equal mean lives.

   The next example illustrates a hypothesis test comparing two means for a situation featuring the following characteristics:

- The two samples come from independent and normal populations.
- σ_1 and σ_2 are unknown.
- Both sample sizes are small (≤ 30).
- The sample variances suggest, through the F test, that $\sigma_1 = \sigma_2$.

The conditions inherent in this next example cause us to follow the path leading to the bottom of the flow chart in Figure 8-6.

E X A M P L E

A principal wants to compare the effectiveness of two algebra teachers and obtains sample data from their students' exams.

Teacher A	Teacher B
$n_1 = 16$	$n_2 = 12$
$\bar{x}_1 = 73.2$	$\bar{x}_2 = 80.7$
$s_1 = 12.8$	$s_2 = 8.5$

At the 0.05 significance level, test the claim that both teachers are equally effective in the sense that their students produce the same mean exam score.

Solution

Referring to Figure 8-6 we begin by questioning the independence of the two populations and conclude that they are independent because separate groups of subjects are used. We continue with the flow chart by noting that σ_1 and σ_2 are not known and that neither n_1 nor n_2 exceeds 30. At this stage, the flow chart brings us to the last diamond, which requires application of the F test. With $H_0: \sigma_1^2 = \sigma_2^2$, $H_1: \sigma_1^2 \neq \sigma_2^2$, and $\alpha = 0.05$, we compute

$$F = \frac{s_1^2}{s_2^2} = \frac{12.8^2}{8.5^2} = 2.2677$$

With $\alpha = 0.05$ in a two-tailed F test and with 15 and 11 as the degrees of freedom for the numerator and the denominator, respectively, we use Table A-6 to obtain the critical F value of 3.3299. Since the computed test statistic of 2.2677 is not within the critical region, we fail to reject the null hypothesis of equal variances (or standard deviations). Leaving the last diamond in Figure 8-6, we follow the yes path and apply the required t test as follows:

$$H_0: \mu_1 = \mu_2 \quad (\text{or } \mu_1 - \mu_2 = 0)$$
$$H_1: \mu_1 \neq \mu_2 \quad (\text{or } \mu_1 - \mu_2 \neq 0)$$
$$\alpha = 0.05$$

$$t = \frac{(\bar{x}_1 - \bar{x}_2) - (\mu_1 - \mu_2)}{\sqrt{\dfrac{1}{n_1} + \dfrac{1}{n_2}} \sqrt{\dfrac{(n_1 - 1)s_1^2 + (n_2 - 1)s_2^2}{n_1 + n_2 - 2}}}$$

$$= \frac{(73.2 - 80.7) - 0}{\sqrt{\dfrac{1}{16} + \dfrac{1}{12}} \sqrt{\dfrac{(16 - 1)(12.8)^2 + (12 - 1)(8.5)^2}{16 + 12 - 2}}} = -1.756$$

With $\alpha = 0.05$ in this two-tailed t test, and with $n_1 + n_2 - 2 = 16 + 12 - 2 = 26$ degrees of freedom, we obtain critical t values of 2.056 and -2.056. The computed test statistic of -1.756 does not fall in the critical region, so we fail to reject the null hypothesis of equal means. Based on the given sample data, there appears to be no significant difference between the effectiveness of the two teachers.

The calculations in this last example might seem somewhat complex, but we should recognize that calculators and computers can be used to ease that burden. STATDISK, for example, can be used to conduct this test. The user selects the option indicating a hypothesis test involving the means of two independent samples, and the user is then prompted for the necessary data. The resulting display is shown below. Note that the results of the prerequisite F test are included in the display. Also note the inclusion of the P-values in both tests. Using the given sample statistics, it takes about 60 seconds to run this test on STATDISK! (*above right*)

One of the most difficult aspects of tests comparing two means is the determination of the correct test to be used. Careful and consistent use of Figure 8-6 should help us avoid the common error of using the wrong procedures for a situation involving a hypothesis test. There is a danger of being overwhelmed by the overall complexity of the work when dealing with the five different cases considered here. However, we can use

```
       Hypothesis test for a claim
     about two INDEPENDENT populations
      NULL HYPOTHESIS: Mean 1 = Mean 2
              F TEST RESULTS
        Test Statistic F = 2.26768
             P-Value = 0.08769
     FAIL TO REJECT equality of variances
        Test Statistic ... t = -1.75599
          Degrees of Freedom = 26
    Critical Value ... t = -2.05609 , 2.05609
          Significance Level = .05
             P-Value = 0.09087
  CONCLUSION: FAIL TO REJECT the null hypothesis
  ------------------------------------------------
```

Figure 8-6 to decompose a complex problem into simpler components that can be treated individually.

It is not practical to outline in full detail the derivations leading to the general test statistics given in Figure 8-6, but we can give some reasons for their existence. We have already discussed the case for dependent populations having normal distributions. Actual experiments and mathematical derivations show that, in repeated random samplings from two normal and dependent populations, the values of $\bar{d}$ possess a student t distribution with mean μ_d and standard deviation $\sigma_d/\sqrt{n}$.

Two cases of Figure 8-6 lead to the normal distribution with the test statistic given by

$$z = \frac{(\bar{x}_1 - \bar{x}_2) - (\mu_1 - \mu_2)}{\sqrt{\sigma_1^2/n_1 + \sigma_2^2/n_2}}$$

This expression is essentially an application of the central limit theorem, which tells us that sample means $\bar{x}$ are normally distributed with mean μ and standard deviation $\sigma/\sqrt{n}$.

In Section 5-6 we saw that, when samples are size 31 or larger, the normal distribution serves as a reasonable approximation to the distribution of sample means. By similar reasoning, the values of $\bar{x}_1 - \bar{x}_2$ also tend to approach a normal distribution with mean $\mu_1 - \mu_2$. When both samples are large, we conclude that the values of $\bar{x}_1 - \bar{x}_2$ will have a standard deviation of

$$\sqrt{\frac{\sigma_1^2}{n_1} + \frac{\sigma_2^2}{n_2}}$$

by using a property of variances: **the variance of the differences between two independent random variables equals the variance of the**

	Table 8-2	
x	y	$x - y$
1	7	−6
3	0	3
1	0	1
3	4	−1
2	6	−4
8	3	5
3	9	−6
3	7	−4
9	0	9
1	1	0
4	2	2
9	4	5
4	9	−5
9	6	3
5	2	3
6	7	−1
5	6	−1
5	8	−3
3	3	0
3	6	−3
1	3	−2
0	2	−2
4	4	0
3	1	2
3	9	−6
6	4	2
9	5	4
5	1	4
2	2	0
9	6	3
2	2	0
6	1	5
1	9	−8
7	0	7
9	1	8
6	5	1

first random variable plus the variance of the second random variable. That is, the variance of values $\bar{x}_1 - \bar{x}_2$ will tend to equal $\sigma_{\bar{x}_1}^2 + \sigma_{\bar{x}_2}^2$ provided that $\bar{x}_1$ and $\bar{x}_2$ are independent. This is a difficult concept, and we therefore illustrate it by the specific data given in Table 8-2. The x and y scores were independently and randomly selected as the last digits of numbers in a telephone book. The variance of the x values is 7.68, the variance of the y values is 8.48, and the variance of the $x - y$ values is 17.22.

$$\left. \begin{array}{l} s_x^2 = 7.68 \\ s_y^2 = 8.48 \\ s_{x-y}^2 = 17.22 \end{array} \right\} \quad s_{x-y}^2 \approx s_x^2 + s_y^2$$

We can see that $s_x^2 + s_y^2$ is roughly equal to s_{x-y}^2. By comparing the values of x, y, and $x - y$, we see that the variance is largest for the $x - y$ values. The x values range from 0 to 9, the y values range from 0 to 9, but the $x - y$ values exhibit greater variation by ranging from −8 to 9. If our x, y, and $x - y$ sample sizes were much larger than the sample sizes of 36 for Table 8-2, we would approach these theoretical values: $\sigma_x^2 = 8.25$, $\sigma_y^2 = 8.25$, and $\sigma_{x-y}^2 = 16.50$.

These theoretical values illustrate that $\sigma_{x-y}^2 = \sigma_x^2 + \sigma_y^2$ when x and y are independent random variables. When we deal with means of large random sample sizes, the standard deviation of those sample means is $\sigma/\sqrt{n}$, so the variance is σ^2/n.

We can now combine our additive property of variances with the central limit theorem's expression for variance of sample means to obtain the following result:

$$\sigma_{\bar{x}_1 - \bar{x}_2}^2 = \sigma_{\bar{x}_1}^2 + \sigma_{\bar{x}_2}^2 = \frac{\sigma_1^2}{n_1} + \frac{\sigma_2^2}{n_2}$$

In the preceding expression, we assume that we have population 1 with variance σ_1^2 and population 2 with variance σ_2^2. Samples of size n_1 are randomly drawn from population 1 and the mean $\bar{x}$ is computed. The same is done for population 2. Here, $\sigma_{\bar{x}_1 - \bar{x}_2}^2$ denotes the variance of $\bar{x}_1 - \bar{x}_2$ values. This result shows that the standard deviation of $\bar{x}_1 - \bar{x}_2$ values is

$$\sqrt{\frac{\sigma_1^2}{n_1} + \frac{\sigma_2^2}{n_2}}$$

Since z is a standard score that corresponds in general to

$$z = \frac{\text{(sample statistic)} - \text{(population mean)}}{\text{(population standard deviation)}}$$

we get

$$z = \frac{(\bar{x}_1 - \bar{x}_2) - (\mu_1 - \mu_2)}{\sqrt{\sigma_1^2/n_1 + \sigma_2^2/n_2}}$$

by noting that the sample values of $\bar{x}_1 - \bar{x}_2$ will have a mean of $\mu_1 - \mu_2$ and the standard deviation just given.

The test statistic located at the bottom of Figure 8-6 is appropriate for tests of hypotheses about two means from independent and normal populations when the samples are small and the population variances appear to be equal. The numerator of $(\bar{x}_1 - \bar{x}_2) - (\mu_1 - \mu_2)$ again describes the difference between the sample statistic $(\bar{x}_1 - \bar{x}_2)$ and the mean of all such values $(\mu_1 - \mu_2)$. The denominator represents the standard deviation of $\bar{x}_1 - \bar{x}_2$ values, which come from repeated sampling when the stated assumptions are satisfied. Since these assumptions include equality of population variances, the denominator should be an estimate of

$$\sqrt{\frac{\sigma^2}{n_1} + \frac{\sigma^2}{n_2}} = \sigma\sqrt{\frac{1}{n_2} + \frac{1}{n_2}}$$

Both n_1 and n_2 will be known, so we need to estimate only σ, and we pool (combine) the sample variances to get the best possible estimate. The pooled estimate of σ is

$$\sqrt{\frac{(n_1 - 1)s_1^2 + (n_2 - 1)s_2^2}{n_1 + n_2 - 2}}$$

where the expression under the square root sign is just a weighted average of s_1^2 and s_2^2. (Weights of $n_1 - 1$ and $n_2 - 1$ are used.) See Exercise 8-56.

We use the test statistic

$$t = \frac{(\bar{x}_1 - \bar{x}_2) - (\mu_1 - \mu_2)}{\sqrt{\dfrac{s_1^2}{n_1} + \dfrac{s_2^2}{n_2}}}$$

in tests of hypotheses about two means coming from independent and normal populations when the samples are small and the two population variances appear to be different. The form of that test statistic seems to follow from similar reasoning used in the other situations, but that is not the case. In fact, no exact test has been found for testing the equality of two means when all the following conditions are met:

1. The two population variances are unknown.
2. The two population variances are unequal.
3. At least one of the samples is small ($n \leq 30$).
4. The two populations are independent and normal.

The approach given in Figure 8-6 for this case is only an approximate test, which is widely used. Also, in Figure 8-6 we indicate that the number of degrees of freedom for this case is found by selecting the smaller of $n_1 - 1$ and $n_2 - 1$. This is a more conservative and simplified alternative to computing the number of degrees of freedom, as follows:

$$df = \frac{(A + B)^2}{\dfrac{A^2}{n_1 - 1} + \dfrac{B^2}{n_2 - 1}} \quad \text{where } A = \frac{s_1^2}{n_1} \text{ and } B = \frac{s_2^2}{n_2}$$

The two bottom cases of Figure 8-6 begin with a preliminary F test. If we apply the F test with a certain level of significance and then do a t test at that same level of significance, the overall result will not be at that same level of significance. Also, in addition to being sensitive to differences in population variances, the F statistic is also sensitive to departures from normal distributions so that it is possible to reject a null hypothesis for the wrong reason. Because of these factors, some statisticians do not recommend a preliminary F test, while others consider this approach to be better. In any event, there is no universal agreement.

8-3   Exercises A
Tests Comparing Two Means

8-23   At the 5% level of significance, test the claim that the sample x and y values come from populations having equal means. Assume that the populations are independent.

x	1	2	2	3	5	6	8	4	6	7	2	5	3	2
y	2	0	1	4	4	4	9	3	5	7	2	6	2	1

8-24   Do Exercise 8-23 assuming that the two populations are dependent and normal.

8-25   Two machines pour beer into liter containers in such a way that the standard deviations of the contents are 0.10 liter for the first machine and 0.15 liter for the second machine. A sample of 20 containers from the first machine is found to have a mean volume of 1.00 liter, while a sample of 15 containers from the second machine has a mean volume of 0.95 liter. At the 0.05 significance level, test the claim that the mean volumes put out by both machines are equal.

8-26   At the 0.05 significance level, test the claim that two ambulance services have the same mean response time. A sample of 50 responses from the first firm produces a mean of 12.2 minutes and a standard deviation of 1.5 minutes. A sample of 50 responses from the second firm produces a mean of 14.0 minutes with a standard deviation of 2.1 minutes.

8-27   A small commuter airline owns two jets and has subcontracted its maintenance operations to two separate firms. Samples of the monthly downtimes of these subcontractors follow. At the 0.05 significance level, test the claim that the mean monthly downtimes of both firms are equal.

	Firm A	Firm B
$n =$	12	14
$\bar{x} =$	14.4 hours	16.3 hours
$s =$	3.1 hours	1.6 hours

8-28   A number of pregnant women agree to participate in an experiment to see if vitamin pills will affect the weights of their newborn children. Some women are given the vitamins, while others are given a placebo of no medicinal value. To eliminate variations due to sex, only the male births are included in the sample data given below. At the 0.05 level of significance, test the claim that there is no difference between the mean weights of the two groups.

	Vitamin	Placebo
$n_1 = 40$		$n_2 = 35$
$\bar{x}_1 =$	3.39 kilograms	$\bar{x}_2 =$ 3.18 kilograms
$s_1 =$	0.44 kilogram	$s_2 =$ 0.53 kilogram

8-29   A dose of the drug captopril, designed to lower systolic blood pressure, is administered to ten randomly selected volunteers. The results follow. At the $\alpha = 0.05$ significance level, test the claim that systolic blood pressure is not affected by the pill.

Before pill	120	136	160	98	115	110	180	190	138	128
After pill	118	122	143	105	98	98	180	175	105	112

8-30   Samples of two competing cold medicines are tested for the amount of acetaminophen, and the results follow. At the 0.05 significance level, test the claim that the mean amount of acetaminophen is the same in each brand.

	Brand X	Brand Y
$n =$	25	35
$\bar{x} =$	503 milligrams	520 milligrams
$s =$	14 milligrams	18 milligrams

8-31   A student hears that fish is a "brain food" that helps make people more intelligent. She participates in an experiment involving 12 randomly selected volunteers who take an I.Q. test and then begin a diet consisting solely of fish. A second I.Q. test is given at the end of the experiment and the results are listed in the following table. At the 0.05 significance level, test the claim that the fish diet has no effect on I.Q. scores.

Before diet	98	110	105	121	100	88	112	92	99	109	103	104
After diet	98	112	106	118	102	97	115	90	99	110	105	109

8-32   A manufacturer produces two models of car batteries. Sample results of each type follow. Assuming that model A and model B have standard deviations of 1 month and 2 months, respectively, test the claim that the mean lives of each type are equal. Use a 0.05 level of significance.

Model A	Model B
$n = 15$	$n = 18$
$\bar{x} = 38$ months	$\bar{x} = 35$ months

8-33   Samples of two different car models are tested for fuel economy by determining the miles traveled using 1 gallon of gas. The results follow. At the 0.05 significance level, test the claim that the means of the numbers of miles traveled by the two models are equal.

Car A	Car B
$n_1 = 10$	$n_2 = 12$
$\bar{x}_1 = 27.2$ miles	$\bar{x}_2 = 31.6$ miles
$s_1 = 4.1$ miles	$s_2 = 2.1$ miles

8-34   A large firm collects sample data on the lengths of telephone calls (in minutes) made by employees in two different divisions, and the results are given below. At the 0.02 level of significance, test the claim that there is no difference between the mean times of all long distance calls made in the two divisions.

Sales division	Customer service division
$n_1 = 40$	$n_2 = 20$
$\bar{x}_1 = 10.26$	$\bar{x}_2 = 6.93$
$s_1 = 8.65$	$s_2 = 4.93$

8-35   An investor is considering two possible locations for a new restaurant and commissions a study of the pedestrian traffic at both sites. At each location, the pedestrians are observed in 1-hour units and, for each hour, an index of desirable characteristics is compiled. The sample results are given below. At the 0.05 level of significance, test the claim that both sites have the same mean.

East	West
$n = 35$	$n = 50$
$\bar{x} = 421$	$\bar{x} = 347$
$s = 122$	$s = 85$

8-36   Researchers study commercial air-filtering systems for noise pollution with sample results as follows. At the 5% level of significance, test the claim that there is no difference in the mean noise levels.

Unit A	Unit B
$n = 8$	$n = 6$
$\bar{x} = 87.5$	$\bar{x} = 91.3$
$s = 0.8$	$s = 1.1$

8-37 Two separate counties use different procedures for selecting jurors. We want to test the claim that the mean waiting time of prospective jurors is the same for both counties. The 40 randomly selected subjects from one county produce a mean of 183.0 minutes and a standard deviation of 21.0 minutes. The 50 randomly selected subjects from the other county produce a mean of 253.1 minutes and a standard deviation of 29.2 minutes. Test the claim at the 0.05 significance level.

8-38 A test of driving ability is given to a random sample of ten student drivers before and after they completed a formal driver education course. The results follow. At the $\alpha = 0.05$ significance level, test the claim that the mean score is not affected by the course.

Before course	100	121	93	146	101	109	149	130	127	120
After course	136	129	125	150	110	138	136	130	125	129

8-39 Cigarettes randomly selected from two brands are analyzed to determine their nicotine content. The results follow. At the $\alpha = 0.05$ significance level, test the claim that both brands have equal nicotine content.

Brand X	Brand Y
$n_1 = 20$	$n_2 = 40$
$\bar{x}_1 = 35.1$ milligrams	$\bar{x}_2 = 36.5$ milligrams
$s_1 = 4.2$ milligrams	$s_2 = 3.3$ milligrams

8-40 A course is designed to increase readers' speed and comprehension. To evaluate the effectiveness of this course, a test is given both before and after the course, and sample results follow. At the 0.05 significance level, test the claim that the scores are higher after the course.

Before	100	110	135	167	200	118	127	95	112	116
After	136	160	120	169	200	140	163	101	138	129

8-41 Two types of string are tested for strength, and the sample data follow. Test the claim that there is no difference between the two types of string.

Breaking load of strings A and B in kilograms											
A	23	25	25	28	19	31	35	30	26			
B	18	17	16	24	20	21	25	15	15	16	18	21

8-42 Ten randomly selected volunteers test a new diet with the following results. At the 5% level of significance, test the claim that the diet is effective. All the weights are given in kilograms.

Subject	A	B	C	D	E	F	G	H	I	J
Before diet	68	54	59	60	57	62	62	65	88	76
After diet	65	52	52	60	58	59	60	63	78	75

8-43   Random samples of two brands of snow tires are tested for stopping distances under standardized ice conditions; the results follow. At the 5% level of significance, test the claim that there is no difference between the two brands.

Badmonth	Brimstone
$n = 36$	$n = 20$
$\bar{x} = 42.7$ meters	$\bar{x} = 46.3$ meters
$s = 6.1$ meters	$s = 6.2$ meters

8-44   In writing an antiunion article, a management consultant presents statistics that purportedly show nonunion masons are more productive than their union counterparts. At the 0.05 significance level, test the consultant's claim that the mean number of bricks set in 1 hour by nonunion workers exceeds the corresponding mean for union masons. The following data are based on random samples consisting of bricks set in 1 hour.

Nonunion	Union
$n = 15$	$n = 10$
$\bar{x} = 24.3$	$\bar{x} = 23.3$
$s = 3.6$	$s = 1.8$

8-45   Two different firms design their own I.Q. tests, and a psychologist administers both versions to randomly selected subjects. The results are given below. At the 0.02 level of significance, test the claim that both versions produce the same mean score.

Subject	A	B	C	D	E	F	G	H	I	J
Test I	98	94	111	102	108	105	92	88	100	99
Test II	105	103	113	98	112	109	97	95	107	103

8-46   A hearing sensitivity test is given to two groups of employees. Group A consists of clerical personnel working in a quiet environment, while Group B consists of assembly workers constantly exposed to loud noises. The sample results follow. At the 0.05 significance level, test the claim that there is no difference between the two population means.

Group A	Group B
$n = 50$	$n = 40$
$\bar{x} = 76$	$\bar{x} = 64$
$s = 12$	$s = 16$

8-47   Stores and theaters are randomly selected and their inside temperatures are measured in degrees Celsius. At the 5% level of significance, test the claim that theaters are warmer than stores. Use the sample data that follow.

Stores	Theaters
$n = 40$	$n = 32$
$\bar{x} = 18.3$	$\bar{x} = 22.2$
$s = 0.8$	$s = 0.9$

8-48   The following chart lists a random sampling of the ages of married couples. Each husband's age is listed above the age of his wife. At the 0.05 significance level, test the claim that there is no difference between the mean ages of husbands and wives.

Ages of married couples									
28.1	33.0	29.8	53.1	56.7	41.6	50.6	21.4	62.0	19.7
28.4	27.6	32.7	52.0	58.1	41.2	50.7	20.6	61.1	18.1

8-49   Two procedures are used for controlling the aircraft traffic at an airport. Sample results based on each of the two procedures follow. At the 0.05 significance level, test the claim that the use of system 2 results in a mean number of operations per hour exceeding the mean for system 1.

System 1	System 2
$n_1 = 24$ hours	$n_2 = 24$ hours
$\bar{x}_1 = 63.0$ operations per hour	$\bar{x}_2 = 60.1$ operations per hour
$s_1 = 5.2$ operations per hour	$s_2 = 3.2$ operations per hour

8-50   An elementary school principal is confronted by an irate group of parents, who charge that the sixth-grade students in their school are not reading as well as the sixth-graders in a nearby school. From each of the two schools, the principal randomly selects 15 scores on a standard reading test and computes the sample data for those two groups of sixth-grade students given below. At the 0.05 level of significance, test the principal's claim that both groups have the same mean.

A	B
$\bar{x} = 96.8$	$\bar{x} = 103.5$
$s = 14.2$	$s = 16.3$

8-3 Exercises B
Tests Comparing Two Means

8-51 A confidence interval for the difference between (independent) population means μ_1 and μ_2 can be constructed for the case involving known values of σ_1 and σ_2 as follows.

$$(\bar{x}_1 - \bar{x}_2) - E < (\mu_1 - \mu_2) < (\bar{x}_1 - \bar{x}_2) + E$$

where $E = z(\alpha/2) \sqrt{\dfrac{\sigma_1^2}{n_1} + \dfrac{\sigma_2^2}{n_2}}$

Construct the 95% confidence interval to estimate the difference between the mean temperature of stores and the mean temperature of theaters. Use the sample data given in Exercise 8-47 and assume that the sample standard deviations can be used as estimates of the corresponding population standard deviations.

8-52 In Exercise 8-51, we presented the confidence interval for the difference between two independent population means. That confidence interval applies to the case where both σ_1 and σ_2 are known.

(a) Construct the general form of the confidence interval for the difference between two independent population means, where the following conditions are satisfied:
1. σ_1 and σ_2 are unknown.
2. $n_1 \leq 30$ and $n_2 \leq 30$.
3. The values of s_1 and s_2 suggest that $\sigma_1 = \sigma_2$.

(b) Use the results of part (a) to construct the 95% confidence interval for $\mu_1 - \mu_2$, where the sample data is found in Exercise 8-36.

8-53 The accompanying frequency tables summarize the ages of adults surveyed in two different regions. At the 0.05 level of significance, test the claim that both populations have the same mean age.

Region A

Age	f
21–30	24
31–40	37
41–50	58
51–60	19
61–70	12

Region B

Age	f
21–30	23
31–40	28
41–50	35
51–60	39
61–70	42

8-54 When testing hypotheses about two population means, under what conditions do we require the populations to have normal distributions, and when can we ignore that requirement? (See Figure 8-6.)

8-55 Use the x and y data of Table 8-2 to find the variance for the values of $x + y$. How does that variance compare to s_x^2 and s_y^2?

8-56 If

$$s_1^2 = \frac{\Sigma(x_1 - \overline{x}_1)^2}{n_1 - 1}$$

and

$$s_2^2 = \frac{\Sigma(x_2 - \overline{x}_2)^2}{n_2 - 1}$$

show that

$$s^2 = \frac{\Sigma(x_1 - \overline{x}_1)^2 + \Sigma(x_2 - \overline{x}_2)^2}{n_1 + n_2 - 2}$$

is equivalent to

$$s = \sqrt{\frac{(n_1 - 1)s_1^2 + (n_2 - 1)s_2^2}{n_1 + n_2 - 2}}$$

This result is used in developing the pooled estimate of σ for the test statistic found at the bottom of Figure 8-6.

8-4 Tests Comparing Two Proportions

In this section we consider tests of hypotheses made about two population proportions. The concepts and procedures we develop can be used to answer questions such as the following:

- Is there a difference between the proportion of homicides among 17-year-olds and the proportion of homicides in the general population?
- Is there a difference between the proportion of men and women in management positions?
- Is there a difference between the percentage of students who passed mathematics courses and the percentage of students who passed history courses?
- Is there a difference between the proportion of IBM computers sold in California and the proportion sold in New York?

Throughout this section we assume that our sample data come from two independent populations and that both samples are large (greater than 30).

The Golden Gate to Heaven

The Golden Gate Bridge and the Bay Bridge both lead out of San Francisco, they were both built in the 1930's, and they are both about 200 feet above water level. Yet 700 suicides have jumped from the Golden Gate, whereas only about 150 have jumped from the Bay Bridge. One obviously relevant factor to be considered in analyzing this discrepancy is the fact that pedestrians are allowed on the Golden Gate, but not on the Bay Bridge. But psychologist Richard Seider claims that even when the pedestrian factor is taken into account, the Golden Gate still has three times as many suicides. He attributes much of the preference for the Golden Gate to the fact that it leads to more fashionable regions, and its suicides tend to receive more notoriety.

N O T A T I O N

For population 1 we let:

p_1 denote the population proportion
n_1 denote the size of the sample
x_1 denote the number of successes

$$\hat{p}_1 = \frac{x_1}{n_1}$$

The corresponding meanings are attached to p_2, n_2, x_2, and $\hat{p}_2$ which come from population 2.

Just as $\hat{p}_1$ represents the sample estimate of p_1, the sample estimate of p_2 is represented by $\hat{p}_2$. We know from Chapter 5 that such proportions have a distribution that is approximately normal, and the differences $\hat{p}_1 - \hat{p}_2$ also have a distribution that is approximately normal. Since the means of $\hat{p}_1$ and $\hat{p}_2$ are p_1 and p_2, respectively, it follows that the mean of the differences $\hat{p}_1 - \hat{p}_2$ will be $p_1 - p_2$. In Section 7-3, we established the fact that sample proportions $\hat{p}$ have a standard deviation of $\sqrt{pq/n}$, and this implies that the variance of the sample proportion $\hat{p}_1$ is p_1q_1/n_1. By similar reasoning, the variance of the $\hat{p}_2$ sample values is p_2q_2/n_2. In Section 8-3, we established the fact that the variance of the differences between two independent random variables is the sum of their individual variances, and we use this property to get

$$\sigma^2_{(\hat{p}_1 - \hat{p}_2)} = \sigma^2_{\hat{p}_1} + \sigma^2_{\hat{p}_2} = \frac{p_1q_1}{n_1} + \frac{p_2q_2}{n_2}$$

However, we are often unable to use this expression, since we usually do not know the values of p_1, p_2, q_1, and q_2. But if we assume that $p_1 = p_2$, we can estimate their common value by pooling the sample data.

D E F I N I T I O N

The **pooled estimate of p_1 and p_2** is denoted by $\bar{p}$ and is given by

$$\bar{p} = \frac{x_1 + x_2}{n_1 + n_2}$$

D E F I N I T I O N

$$\bar{q} = 1 - \bar{p}$$

Using this pooled estimate in place of p_1 and p_2, we find that the standard deviation of the differences between the sample proportions becomes

$$\sigma_{(\hat{p}_1 - \hat{p}_2)} = \sqrt{\frac{\overline{p}\,\overline{q}}{n_1} + \frac{\overline{p}\,\overline{q}}{n_2}} = \sqrt{\overline{p}\,\overline{q}\left(\frac{1}{n_1} + \frac{1}{n_2}\right)}$$

We now know that the sample proportion differences $\hat{p}_1 - \hat{p}_2$ have a distribution that is approximately normal with mean $p_1 - p_2$ and standard deviation as given above. We now summarize these results: As long as n_1 and n_2 are both large, the test statistic

TEST STATISTIC

(For H_0: $p_1 = p_2$, H_0: $p_1 \geq p_2$, or H_0: $p_1 \leq p_2$)

$$z = \frac{(\hat{p}_1 - \hat{p}_2) - (p_1 - p_2)}{\sqrt{\overline{p}\,\overline{q}\left(\frac{1}{n_1} + \frac{1}{n_2}\right)}}$$

where

$$\hat{p}_1 = \frac{x_1}{n_1} \qquad \hat{p}_2 = \frac{x_2}{n_2}$$

$$\overline{p} = \frac{x_1 + x_2}{n_1 + n_2}$$

$$\overline{q} = 1 - \overline{p}$$

has a sampling distribution that is approximately the standard normal distribution. This test statistic applies only to cases where the null hypothesis is $p_1 = p_2$. For testing claims that the difference $p_1 - p_2$ is equal to a nonzero constant, a different test statistic is used. (See Exercise 8-79.) We consider only hypotheses that lead to a null hypothesis of $p_1 = p_2$, or $p_1 \geq p_2$, or $p_1 \leq p_2$, so our tests will be conducted under the assumption that $p_1 = p_2$ or $p_1 - p_2 = 0$. We can then use the preceding test statistic with $p_1 - p_2$ replaced by 0.

EXAMPLE

A sample of 200 New York State voters included 88 Republicans, while a sample of 300 California voters produced 143 Republicans. At the $\alpha = 0.05$ significance level, test the claim that there is no difference between the proportions of New York Republicans and California Republicans.

The given data can be summarized as follows:

New York	California
$n_1 = 200$	$n_2 = 300$
$x_1 = 88$	$x_2 = 143$

Solution

With H_0: $p_1 = p_2$ and H_1: $p_1 \neq p_2$ and the significance level set at 0.05, we proceed with the calculations necessary to determine the value of the test statistic.

From the sample data we compute

$$\hat{p}_1 = \frac{x_1}{n_1} = \frac{88}{200} = 0.440$$

$$\hat{p}_2 = \frac{x_2}{n_2} = \frac{143}{300} = 0.477$$

$$\overline{p} = \frac{x_1 + x_2}{n_1 + n_2} = \frac{88 + 143}{200 + 300} = \frac{231}{500} = 0.462$$

$$\overline{q} = 1 - \overline{p} = 1 - 0.462 = 0.538$$

$$z = \frac{(\hat{p}_1 - \hat{p}_2) - 0}{\sqrt{\overline{p}\,\overline{q}\left(\dfrac{1}{n_1} + \dfrac{1}{n_2}\right)}} = \frac{(0.440 - 0.477) - 0}{\sqrt{(0.462)(0.538)\left(\dfrac{1}{200} + \dfrac{1}{300}\right)}} = -0.81$$

But with $\alpha = 0.05$ in this two-tailed test, the critical z values of 1.96 and -1.96 indicate that the test statistic is not in the critical region. As a result, we fail to reject the null hypothesis of equal proportions. The proportions of Republicans in both states are not significantly different. The P-value can be determined by first noting that $z = -0.81$ corresponds to an area of 0.2910 (See Table A-3). We then get P-value = $2 \times (0.5000 - 0.2910) = 0.4180$. The more precise STATDISK results are displayed below.

```
Hypothesis test for a claim about two Population
                    Proportions
            NULL HYPOTHESIS: P1 = P2
 First Sample:  X1 = 88   N1 = 200   X1/N1 = 0.4400
 Second Sample: X2 = 143  N2 = 300   X2/N2 = 0.4767
            Test Statistic ... Z = -0.8057
       Critical Value ... Z = -1.9604 , 1.9604
Significance Level = .05          P-Value = 0.4204
    CONCLUSION: FAIL TO REJECT the null hypothesis
```

The symbols x_1, x_2, n_1, n_2, $\hat{p}_1$, $\hat{p}_2$, $\bar{p}$, and $\bar{q}$ should have become more meaningful through this example. In particular, you should recognize that, under the assumption of equal proportions, the best estimate of the common proportion is obtained by pooling both samples into one larger sample. Then

$$\bar{p} = \frac{x_1 + x_2}{n_1 + n_2}$$

becomes a more obvious estimate of the common population proportion.

So far we have discussed only proportions in this section, but probabilities are already in decimal or fractional form, so they can directly replace proportions in the preceding discussion. Percentages can also be accommodated by using the corresponding decimal equivalents as illustrated in the following example.

EXAMPLE

A study of smoking habits is conducted in a certain country. In a sample of 150 rural and suburban residents, 44% smoke. In a sample of 400 city residents, 52% smoke. At the 0.01 level of significance, test the claim that city residents comprise a larger proportion of smokers than rural and suburban residents.

Solution

The claim of a greater proportion of smokers in the city suggests that H_1 be $p_1 < p_2$ so that H_0 becomes $p_1 \geq p_2$.

From the given data we extract the following.

Rural-Suburban	City
$n_1 = 150$	$n_2 = 400$
$x_1 = $ 44% of 150 = 66	$x_2 = $ 52% of 400 = 208

With $\hat{p}_1 = x_1/n_1 = 0.44$ and $\hat{p}_2 = x_2/n_2 = 0.52$, we compute $\bar{p}$ and $\bar{q}$.

$$\bar{p} = \frac{x_1 + x_2}{n_1 + n_2} = \frac{66 + 208}{150 + 400} = \frac{274}{550} = 0.498$$

and

$$\bar{q} = 1 - \bar{p} = 0.502$$

With α already set at 0.01, we continue by computing the test statistic based on the sample data.

$$z = \frac{(\hat{p}_1 - \hat{p}_2) - 0}{\sqrt{\bar{p}\,\bar{q}\left(\frac{1}{n_1} + \frac{1}{n_2}\right)}} = \frac{(0.44 - 0.52) - 0}{\sqrt{(0.498)(0.502)\left(\frac{1}{150} + \frac{1}{400}\right)}} = -1.67$$

With $\alpha = 0.01$ in this left-tailed test, we use Table A-3 to obtain the critical z value of -2.33. Since the test statistic does not fall in the critical region, we fail to reject the null hypothesis and conclude that there is insufficient evidence to support the claim that the percentage of city smokers exceeds the percentage of rural and suburban smokers.

The P-value can be determined as follows. From Table A-3 we find that $z = -1.67$ corresponds to 0.4525. The P-value is therefore $0.5000 - 0.4525 = 0.0475$. This P-value shows us that with a larger significance level, such as 0.05, we would have rejected the null hypothesis and concluded that there is a significantly higher percentage of smokers in the city.

Polio Experiment

In 1954 a vast medical experiment was conducted to test the effectiveness of the Salk vaccine as a protection against the devastating effects of polio. Previously developed polio vaccines had been used, but it was discovered that some of those earlier vaccinations actually *caused* paralytic polio. Researchers justifiably developed a cautious and conservative approach to approving new vaccines for general use, and they decided to conduct a large-scale experiment of the Salk vaccine with volunteers.

Because of a variety of physiological factors, the vaccine could not be 100% effective, so its effectiveness had to be proved by a lowered incidence of polio among inoculated children. It was hoped that this experiment would result in a lower incidence of polio among vaccinated children that would be so significant as to be overwhelmingly convincing.

The number of children involved in this experiment was necessarily large because a small sample would not provide the conclusive evidence required by these cautious researchers. Approximately 200,000 children were injected with an ineffective salt solution, while 200,000 other children were injected with the Salk vaccine. Assignments of the real vaccine and the useless salt solution were made on a random basis. The children being injected did not know whether they were given the real vaccine or the salt solution. Even the doctors giving the injections and evaluating subsequent results did not know which injections contained the real Salk vaccine. Only 33 of the 200,000 vaccinated children later developed paralytic polio, while 115 of the 200,000 injected with the salt solution later developed paralytic polio. Statistical analysis of these and other results led to the conclusion that the Salk vaccine was indeed effective against paralytic polio. (See Exercise 8-71.)

8-4   Exercises A
Tests Comparing Two Proportions

8-57   In each of the following, use the given sample data to determine the values of $\hat{p}_1$, $\hat{p}_2$, $\bar{p}$, and $\bar{q}$.

(a)   Sample A: Of 200 voters polled, 67 are Democrats.
Sample B: Of 400 voters polled, 148 are Democrats.

(b)   Sample A: In a sample of 250 adults, 38% smoked during the last week.
Sample B: In a sample of 300 adults, 46% smoked during the last week.

(c)   When 500 randomly selected subjects are divided into two equal groups, 20% of the first group passes a physical fitness test, while 24% of the second group passes the same test.

(d)   In a random sample of 300 men and 400 women, exactly 53% of each group believe in life on other planets.

(e)   Among United States governors 42% favor the passage of a certain bill, while 57% of all United States senators favor passage of the same bill.

8-58   For each part of Exercise 8-57, compute the z test statistic.

8-59   Let samples from two populations be such that $x_1 = 45$, $n_1 = 100$, $x_2 = 115$, and $n_2 = 200$.

(a)   Compute the z test statistic based on the given data.

(b)   If the significance level is 0.05 and the test is two-tailed, find the critical z values.

(c)   Test the claim that the two populations have equal proportions using the significance level of $\alpha = 0.05$.

(d)   Find the P-value.

8-60   Samples taken from two populations yield the data $x_1 = 30$, $n_1 = 250$, $x_2 = 44$, $n_2 = 800$. The hypothesis of equal proportions is to be tested at the $\alpha = 0.02$ significance level.

(a)   Compute the z test statistic based on the data.

(b)   Find the critical z values.

(c)   What conclusion do you reach?

8-61   Test the claim that the sample proportion of 0.26 does not differ significantly from the sample proportion of 0.20 if both sample sizes are 50.

8-62   Test the claim that the sample proportion 0.25 is significantly greater than the sample proportion of 0.20 if both sample sizes are 100. Use a 5% level of significance.

8-63   A manufacturer experiments with two production methods. The first method produces 18 defects in 275 sample items, while the second method produces 27 defects in 320 samples. At the $\alpha = 0.05$ significance level, test the claim that there is no difference between the two proportions of defects.

8-64 At a university, 40% of the 95 faculty members surveyed feel that the library facilities are adequate, while 60% of the 210 students surveyed rate the library facilities as adequate. At the $\alpha = 0.01$ significance level, test the claim that the percentage of students who feel that the library facilities are adequate exceeds the percentage for faculty members. Assume that the student and faculty populations are both large.

8-65 In a random sample of 9584 licensed drivers between the ages of 18 and 24, there were ten fatalities due to car accidents in 1 year. In another random sample of 11,316 licensed drivers in the 25-to-34 age group, there were eight auto fatalities. At the $\alpha = 0.01$ significance level, test the claim that drivers in the 18-to-24 age bracket have a greater proportion of fatal accidents.

8-66 A study in genetics involves two sample groups. Of the 80 people in one group, 23 have blue eyes. Of the 120 people in the second group, 24 have blue eyes. At the $\alpha = 0.05$ significance level, test the claim that the proportion of blue-eyed people is the same in each group.

8-67 In the past year, two professors each taught four sections of a statistics course with 25 students in each section. If 12 students of the first professor failed and 18 students of the second professor failed, test the claim that the rate of failure is the same for the two professors. Assume a significance level of $\alpha = 0.05$.

8-68 A drug is tested on 150 randomly selected volunteers. Among the 80 men receiving this drug, there are 12 noticeable reactions. Among the 70 women who take the drug, there are 20 noticeable reactions. At the 0.05 significance level, test the claim that the proportion of noticeable reactions among women is equal to the proportion of noticeable reactions among men.

8-69 Of 5000 income tax returns reviewed from low-income families, 52 were selected for audit. Of 1000 returns reviewed from high-income families, 72 were selected for audit. At the 0.01 level of significance, test the claim that the proportions of audits from the two groups are equal.

8-70 A survey was conducted of randomly selected television viewers to determine the composition of the audience for a certain show. Of 200 people with college educations, 30 indicated that they watched the show. Of 600 people without college educations, 210 indicated that they watched the show. At the 0.01 significance level, test the claim that a greater proportion of viewers did not have a college education.

8-71 In initial tests of the Salk vaccine, 33 of 200,000 vaccinated children later developed polio. Of 200,000 children vaccinated with a placebo, 115 later developed polio. At the 1% level of significance, test the claim that the Salk vaccine is effective.

8-72 An accreditation committee compares the dropout rates of two similar colleges. Of 250 of the entering freshmen in the first college, the dropout rate is found to be 22%. A random sample of 275 freshmen beginning studies at the second college produces a dropout rate of 18%. At the $\alpha = 0.05$ significance level, test the claim that both colleges have the same dropout rate.

8-73 A manufacturer produces a new and inexpensive fire detection device. Advertising claims include the assertion that the percentage of defective units is not higher than the corresponding percentage for models made by competitors. Five defective units are found in a sample of 150 randomly selected new units, while the competition produces a 2% defective rate in 400 randomly selected units. At the $\alpha = 0.05$ significance level, test the claim made by the manufacturer of the new units.

8-74 As the election approaches, a candidate for a state office recognizes that she has time to campaign effectively in only one of two important counties. A poll of 600 voters in the first county reveals that 57% favor our candidate, while a poll of 800 voters in the second county shows that 49% favor her. At the $\alpha = 0.05$ significance level, test the claim that the first county has a higher percentage of voters who favor this candidate.

8-75 A test question is answered correctly by 80% of the 200 male respondents and 72% of the 300 female respondents. At the 5% level of significance, test the claim that the proportions of correct responses are equal.

8-76 An advertiser studies the proportion of radio listeners who prefer country music. In region A, 38% of the 250 listeners surveyed indicated a preference for country music. In region B, country music was preferred by 14% of the 400 listeners surveyed. At the 0.02 level of significance, test the claim that region A has a greater proportion of listeners who prefer country music.

8-4 Exercises B
Tests Comparing Two Proportions

8-77 Refer to the data summarized in the accompanying table, and use a 0.01 level of significance to test the claim that there is a difference between the proportion of males who are Democrats and the proportion of females who are Democrats.

	Democrat	Republican
Male	100	150
Female	200	50

8-78   A confidence interval for the difference between population proportions p_1 and p_2 can be constructed by evaluating

$$(\hat{p}_1 - \hat{p}_2) - E < (p_1 - p_2) < (\hat{p}_1 - \hat{p}_2) + E$$

where $E = z(\alpha/2)\sqrt{\dfrac{\hat{p}_1\hat{q}_1}{n_1} + \dfrac{\hat{p}_2\hat{q}_2}{n_2}}$

Construct the 95% confidence interval to estimate the difference between the two proportions of radio listeners who prefer country music (see Exercise 8-76).

8-79   To test the null hypothesis that the difference between two population proportions is equal to a nonzero constant c, use

$$z = \frac{(\hat{p}_1 - \hat{p}_2) - c}{\sqrt{\dfrac{\hat{p}_1(1 - \hat{p}_1)}{n_1} + \dfrac{\hat{p}_2(1 - \hat{p}_2)}{n_2}}}$$

As long as n_1 and n_2 are both large, the sampling distribution of the above test statistic z will be approximately the standard normal distribution. Suppose a winery is conducting market research in New York and California. In a sample of 500 New Yorkers, 120 like the wine, while a sample of 500 Californians shows that 210 like the wine. Use a 0.05 level of significance to test the claim that the percentage of Californians who like the wine is 25% more than the percentage of New Yorkers who like it.

8-80   Sample data are randomly drawn from three independent populations. The sample sizes and the numbers of successes follow.

Population 1	Population 2	Population 3
$n = 100$	$n = 100$	$n = 100$
$x = 40$	$x = 30$	$x = 20$

(a)   At the 0.05 significance level, test the claim that $p_1 = p_2$.
(b)   At the 0.05 significance level, test the claim that $p_2 = p_3$.
(c)   At the 0.05 significance level, test the claim that $p_1 = p_3$.
(d)   In general, if hypothesis tests lead to the decisions that $p_1 = p_2$ and $p_2 = p_3$, does it follow that the decision $p_1 = p_3$ will be reached under the same conditions?

Computer Project
Tests Comparing Two Parameters

(a)   Select any one of the tests listed below and develop a computer program that takes the two sets of data as input and gives as output the value of the appropriate test statistic.

- Test comparing two variances.
- Test comparing two proportions.
- Test comparing two means for two sets of dependent data.
- Test comparing two means when either σ_1 and σ_2 are both known or n_1 and n_2 are both larger than 30.

(b) Use existing software that can be run with two sets of data and use it to solve Exercise 8-29.

Review

In this chapter we extended to two populations the method of testing hypotheses introduced in Chapter 6. We began by developing a test for comparing two population variances (or standard deviations) that come from two independent populations having normal distributions. We began with a test for comparing two variances or standard deviations since such a test is sometimes used as part of an overall test for comparing two population means. In Section 8-2 we saw that the sampling distribution of $F = s_1^2/s_2^2$ is the F distribution for which Table A-6 was computed.

In Section 8-3 we considered various situations that can occur when we want to use a hypothesis test for comparing two means. We should begin such a test by determining whether or not the two populations are **dependent** in the sense that they are related in some way. When comparing population means that come from two dependent and normal populations, we compute the differences between corresponding pairs of values. Those differences have a mean and standard deviation denoted by $\overline{d}$ and s_d, respectively. In repeated random samplings, the values of $\overline{d}$ possess a student t distribution with mean μ_d and standard deviation $\sigma_d/\sqrt{n}$.

When using hypothesis tests to compare two population means from independent populations, we encounter four situations that can be summarized best by Figure 8-6. These cases incorporate standard deviations reflecting the property that, if one random variable x has variance σ_x^2 and another independent random variable y has variance σ_y^2, the random variable $x - y$ will have variance $\sigma_x^2 + \sigma_y^2$.

In Section 8-4 we considered hypothesis tests that can be used to compare proportions, probabilities, or percentages that come from two independent populations. We limited our discussion to cases involving large samples only. We saw that the sample proportions have differences

$$\frac{x_1}{n_1} - \frac{x_2}{n_2} \quad \text{or} \quad \hat{p}_1 - \hat{p}_2$$

that tend to have a distribution that is approximately normal with mean $p_1 - p_2$ and a standard deviation estimated by

$$\sqrt{\bar{p}\,\bar{q}\left(\frac{1}{n_1} + \frac{1}{n_2}\right)}$$

when $p_1 = p_2$.

Also, $\bar{q} = 1 - \bar{p}$ and $\bar{p}$ is the pooled proportion $(x_1 + x_2)/(n_1 + n_2)$.

Figure 8-7 provides a reference chart for locating the appropriate test. One of the most difficult aspects of hypothesis testing involves the identification of the most appropriate distribution and the selection of the proper test statistic, and Figure 8-7 should help in that determination.

IMPORTANT FORMULAS

Parameters to which hypothesis refers	Applicable distribution	Test statistic	Table of critical values
σ_1, σ_2 (two standard deviations) or σ_1^2, σ_2^2 (two variances)	F	$F = \dfrac{s_1^2}{s_2^2}$ where $s_1^2 \geq s_2^2$	Table A-6
μ_1, μ_2 (two means): dependent samples	Student t	$t = \dfrac{\bar{d} - \mu_d}{s_d/\sqrt{n}}$	Table A-4
independent samples (use Figure 8-6 to determine the correct case)	Normal or student t	$z = \dfrac{(\bar{x}_1 - \bar{x}_2) - (\mu_1 - \mu_2)}{\sqrt{\dfrac{\sigma_1^2}{n_1} + \dfrac{\sigma_2^2}{n_2}}}$	Table A-3
		$t = \dfrac{(\bar{x}_1 - \bar{x}_2) - (\mu_1 - \mu_2)}{\sqrt{\dfrac{s_1^2}{n_1} + \dfrac{s_2^2}{n_2}}}$	Table A-4
		$t = \dfrac{(\bar{x}_1 - \bar{x}_2) - (\mu_1 - \mu_2)}{\sqrt{\dfrac{1}{n_1} + \dfrac{1}{n_2}}\sqrt{\dfrac{(n_1 - 1)s_1^2 + (n_2 - 1)s_2^2}{n_1 + n_2 - 2}}}$	Table A-4
p_1, p_2 (two proportions)	Normal	$z = \dfrac{(\hat{p}_1 - \hat{p}_2) - (p_1 - p_2)}{\sqrt{\bar{p}\,\bar{q}\left(\dfrac{1}{n_1} + \dfrac{1}{n_2}\right)}}$	Table A-3

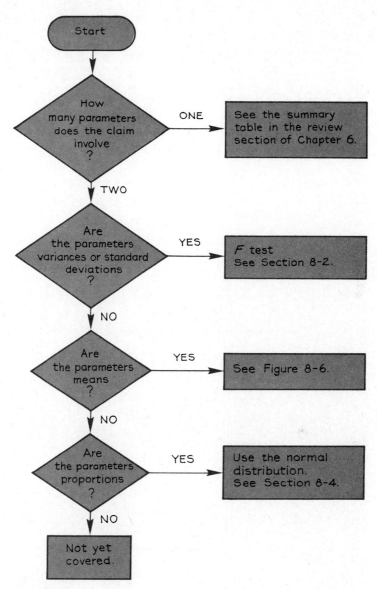

Figure 8—7

Review Exercises

8-81 A manufacturer of shock absorbers operates one plant in the West and one in the East. A random selection of 400 shock absorbers is tested in each plant and 6 defective units are found in the western plant while 11 defective units are found in the eastern plant. At the 0.02 level of significance, test the claim that both plants have the same rate of defects.

8-82 In order to test the effectiveness of a lesson, a teacher gives randomly selected students a pretest and a follow-up test. The results are given below. At the 0.025 level of significance, test the claim that the lesson was effective.

Student	A	B	C	D	E	F	G	H
Before	6	8	5	4	3	5	4	7
After	9	10	8	7	6	8	7	10

8-83 Samples of similar wires are randomly selected from two different manufacturers and tested for their breaking strengths. The results are as follows:

Company A
$n = 10$
$\bar{x} = 82.0$ kilograms
$s = 6.0$ kilograms

Company B
$n = 12$
$\bar{x} = 77.6$ kilograms
$s = 8.1$ kilograms

At the 0.05 significance level, test the claim that the standard deviations of the breaking strengths of both brands are equal.

8-84 Use the sample data from Exercise 8-83 to test the claim that the two companies produce wire having the same breaking strength. Use a 0.05 level of significance.

8-85 The same aptitude test is given to students randomly selected from schools in two different states and the results are as follows:

New York
$n = 208$
$\bar{x} = 121.0$
$s = 9.1$

Alabama
$n = 54$
$\bar{x} = 103.2$
$s = 11.9$

At the 0.05 significance level, test the claim that there is no difference between the mean scores of students from both states.

8-86 A test question is considered good if it discriminates between good and poor students. The first question on a test is answered correctly by 62 of 80 good students, while 23 of 50 poor students give correct answers. At the 5% level of significance, test the claim that this question is answered correctly by a greater proportion of good students.

8-87 Automobiles are selected at random and tested for fuel economy with each of two different carburetors. The following results show the distance traveled on 1 gallon of gas. At the 5% level of significance, test the claim that both carburetors produce the same mean mileage.

	Distance with carburetors A and B								
Car	1	2	3	4	5	6	7	8	9
A	16.1	21.3	19.2	14.8	29.3	20.2	18.6	19.7	16.4
B	18.2	23.4	19.7	14.7	28.7	23.4	19.0	21.2	18.2

8-88 Two different firms manufacture garage door springs that are designed to produce a tension of 68 kilograms. Random samples are selected from each of these two suppliers and tension test results are as follows:

Firm A	Firm B
$n = 20$	$n = 32$
$\bar{x} = 66.0$ kilograms	$\bar{x} = 68.3$ kilograms
$s = 2.1$ kilograms	$s = 0.4$ kilogram

At the 5% level of significance, test the claim that both firms produce the same standard deviation.

8-89 Use the sample data from Exercise 8-88 to test the claim that both firms' springs produce the same mean tension. Use a 0.05 level of significance.

8-90 Do Exercise 8-88 after changing the sample size for Firm A to $n = 40$.

8-91 At the 10% level of significance, test the claim that the sample proportion of 0.4 differs significantly from the sample proportion of 0.6. Assume that both sample sizes are 75.

8-92 In a study of the effects of alcohol on driving, randomly selected subjects are given a test of coordination before and after having consumed several drinks. The results are given below. At the 0.05 level of significance, test the claim that alcohol has no effect.

Subject	A	B	C	D	E	F	G	H
Before	16	13	12	15	14	14	17	16
After	9	6	5	9	7	8	11	6

8-93 Two different production methods are used to make batteries for hearing aids. Batteries produced by both methods are randomly selected and tested for longevity. Eighteen batteries produced by the first method have a standard deviation of 42 hours, while 12 batteries produced by the second method have a standard deviation of 78 hours. At the 5% level of significance, test the claim that the two production methods yield batteries whose lives have equal standard deviations.

8-94 A bank has branches in two different cities, and it uses a standard credit-rating system for all loan applicants. Randomly selected applicants are chosen from each branch and the results are summarized below. At the 0.05 level of significance, test the claim that both populations have the same mean.

City A	City B
$n = 40$	$n = 60$
$\bar{x} = 43.7$	$\bar{x} = 48.2$
$s = 16.2$	$s = 16.5$

8-95   Do Exercise 8-94 after changing the sample size of City B to $n = 20$.

8-96   A poll reveals that 47.0% of 1500 randomly selected voters in Ohio favor a certain candidate, while 48.0% of 500 randomly selected voters in Maryland favor that same candidate. At the 1% level of significance, test the claim that the candidate is favored by the same percentage in both states.

8-97   To test the effectiveness of a physical training program, researchers asked randomly selected participants to run as far as possible in 5 minutes. This test was conducted before and after the training program and the results follow. The numbers represent distances in meters. At the 5% level of significance, test the claim that the training program was effective.

Before course	510	620	705	590	800	1450	790	830	1220	680
After course	1130	680	810	780	1275	1410	970	1050	1380	1050

8-98   A farmer experiments with two different diets for chickens, and sample weights are given below. At the 0.05 level of significance, test the claim that both diets result in the same variance.

Diet A	Diet B
$n = 35$	$n = 25$
$\bar{x} = 7080$ g	$\bar{x} = 6670$ g
$s = 30$ g	$s = 50$ g

8-99   Using the data from Exercise 8-98, test the claim that both diets result in the same mean weight. Use a 0.05 level of significance.

8-100   The manager of a movie theater conducts a study of the ages of those who view two different movies and sample results are given below. At the 0.025 level of significance, test the claim that there is no difference between the two population means.

Movie X	Movie Y
$n = 45$	$n = 65$
$\bar{x} = 22.6$ years	$\bar{x} = 31.0$ years
$s = 5.8$ years	$s = 4.7$ years

Vocabulary List

Define and give an example of each term.

F distribution	dependent samples
numerator degrees of freedom	independent samples
denominator degrees of freedom	pooled estimate of p_1 and p_2

Chapter 8
Case Study Activity

Select two sample groups as follows. Let one group consist of 15 adults who do not attend college, and let the other group consist of 20 college students. Ask everyone in both groups to estimate the present age of the president. At the 0.05 level of significance, test the claim that both populations have the same mean.

CHAPTER 9

9-1 Overview

We identify chapter **objectives.** This chapter presents methods for analyzing the relationship between two variables.

9-2 Correlation

We use the **scatter diagram** and **linear correlation coefficient** to determine whether a linear relationship exists between two variables.

9-3 Regression

We describe linear relationships between two variables by the equation and graph of the **regression line.**

9-4 Variation

We analyze the **variation** between **predicted** and **observed** values.

Correlation and Regression

An environmentalist wants to lobby for reduced speed limits because he has read that slower cars consume less gasoline. Reduced gasoline consumption would help conserve a natural resource while causing less air pollution. A test is conducted with a car traveling at different speeds, and a special gauge records the gasoline consumption rates (in mpg) included in Table 9-1.

Table 9-1

Speed	20	25	30	35	40	45	50	55	60
MPG	22	21	20	23	19	18	16	14	11

Based on the given sample data, can we conclude that there is a relationship between speed and gasoline consumption? If so, what is the relationship?

9-1 Overview

In this chapter, we will learn how to answer questions like the ones asked in the chapter problem. Note that we are analyzing a collection of data that is arranged in pairs. Such paired data is sometimes referred to as **bivariate data.** We have considered paired data in Chapter 8, but the main concern in that chapter was fundamentally different from this chapter. In Chapter 8, for example, we considered collections of paired data such as the one given in Table 9-2.

Table 9-2										
Before	96	110	98	113	88	92	106	119	100	97
After	99	112	107	110	88	101	107	123	91	97

The numbers in Table 9-2 represent the scores of ten subjects on an I.Q. test before and after a special preparatory course. In Chapter 8 we were concerned with the effectiveness of the course. An effective course should result in higher scores after the course has been completed. However, the numbers in Table 9-1 involve the variables of speed and mpg rating, and we are concerned with the relationship between those two variables. Clearly, the mpg values of Table 9-1 will be generally less than the corresponding speeds, but we are not at all interested in that characteristic. For the data of Table 9-1, we are interested in the presence or absence of an association between the two variables; for the data of Table 9-2, we are interested in the differences between the numbers in each pair. These are basically different issues and the statistical methods used to consider them will be different. In short, the key questions are these:

1. Is there some statistical relationship between the speed and mpg values of Table 9-1 and, if so, what is it?
2. Are the "after" values of Table 9-2 significantly greater than the "before" values?

Chapter 8 considered the second question; this chapter will consider the first.

In this chapter we investigate ways of analyzing the relationship between two variables. We begin Section 9-2 by describing the **scatter diagram,** which serves as a graph of the sample data. We then investigate the concept of **correlation,** which is used to decide whether there is a statistically significant relationship between two variables. Section 9-3 investigates **regression analysis** as we attempt to identify the exact nature of the relationship between two variables. Specifically, we show how to determine an equation that relates the two variables. In Section 9-4 we analyze the variation between predicted and observed values.

Throughout this chapter we deal only with **linear** (straight-line) relationships between two variables. (Advanced texts consider more variables and nonlinear relationships.)

As a very simple example, consider Table 9-3.

Table 9-3							
x	1	2	4	5	7	8	10
y	3	5	9	11	15	17	21

Note that each y value is one more than twice the corresponding x value. This strongly suggests that there exists a very definite relationship between the two variables. In deciding that there is a relationship, we are dealing with the concept of correlation. We can identify exactly what the relationship is. The relationship between x and y can be described by the equation $y = 2x + 1$. We deal with the concept of regression when we determine what the relationship is.

9-2   Correlation

As we mentioned in the overview, this section deals with the concept of correlation and scatter diagrams as tools that help us decide whether a linear relationship exists between two variables. Each of the two variables should be normally distributed. That is, for any given value of one variable, the distribution of values of the other variable should be normal. Since these tools are designed to analyze relationships between two variables, the sample data must be collected as paired data. We start with an example.

Is there a linear relationship between the speed at which a car travels and the rate at which it consumes fuel? We will use the sample data presented in Table 9-1.

We can often form intuitive and qualitative conclusions about paired data by constructing a **scatter diagram** similar to the one in Figure 9-1, which represents the data in Table 9-1. The points in the figure seem to follow a downward pattern, so we might conclude that there is a relationship between speed and the gasoline consumption rate measured in miles per gallon (mpg) (see Figure 9-1, page 420).

Using other collections of paired data we may get scatter diagrams similar to the examples illustrated in Figure 9-2. The scatter diagram is easily plotted and doesn't require complex computations. In addition, a large collection of paired data may exhibit a pattern and become meaningful when displayed in this form. However, these advantages are often offset by the subjective and qualitative nature of the conclusions that may be drawn from a scatter diagram.

More precise and objective analyses accompany the computation of

Correlation Shows Trend

We can use the concepts of correlation to establish that there is a relationship between the age and the height of a child. Children obviously grow taller as they age, but that relationship is not perfect. The taller of the two brothers shown is one year younger than his shorter brother. The concepts of correlation accommodate exceptions of this type while enabling us to identify the dominant trend.

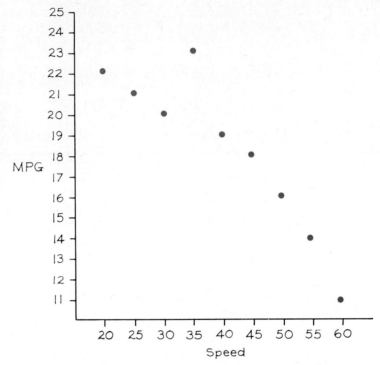

Figure 9-1

the **linear correlation coefficient,** which is denoted by r and is given in Formula 9-1.

Formula 9-1 $$r = \frac{n\Sigma xy - (\Sigma x)(\Sigma y)}{\sqrt{n(\Sigma x^2) - (\Sigma x)^2}\ \sqrt{n(\Sigma y^2) - (\Sigma y)^2}}$$

Since r is calculated using sample data, it is a sample statistic. We might think of r as a point estimate of the population parameter ρ, which is the linear correlation coefficient for all pairs of data in a population.

We now describe the way to compute and interpret the linear correlation coefficient r given a list of paired data. Later in this section we present the underlying theory that led to the development of this formula. Before computing the correlation coefficient r for the data of Table 9-1, we make the following notes relevant to Formula 9-1.

NOTATION

n	denotes the **number of pairs** of data present. In Table 9-1, for example, $n = 9$.
Σ	denotes the addition of the items indicated.

(continued on page 422)

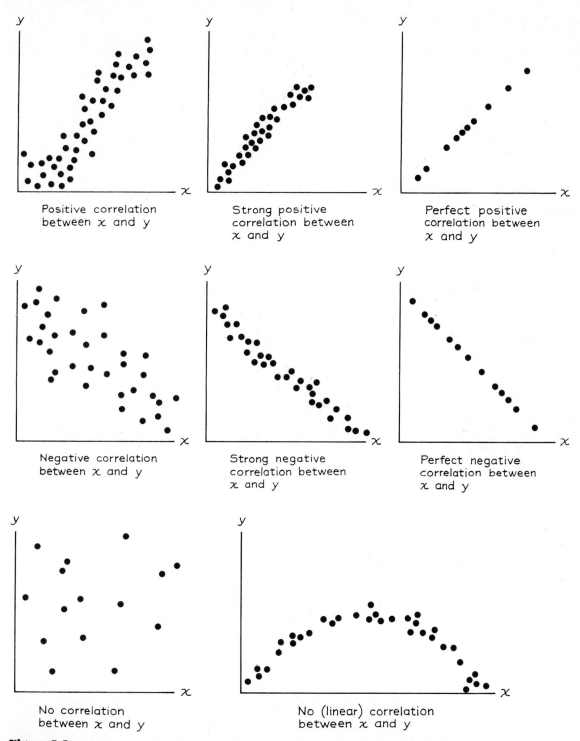

Positive correlation between x and y

Strong positive correlation between x and y

Perfect positive correlation between x and y

Negative correlation between x and y

Strong negative correlation between x and y

Perfect negative correlation between x and y

No correlation between x and y

No (linear) correlation between x and y

Figure 9-2

Student Ratings of Teachers

Many colleges equate high student ratings with good teaching, and this equation is often fostered by the fact that student evaluations are easy to administer and easy to measure.

However, in one study that compared student evaluations of teachers with the amount of material learned by the students, the researchers found a strong *negative* correlation between these two factors. Teachers rated highly by students seemed to induce less learning.

In a related study, an audience gave a high rating to a lecturer who conveyed very little information but was interesting and entertaining.

Σx denotes the sum of all x scores.

Σx^2 indicates that each x score should be squared, and then those squares are added.

$(\Sigma x)^2$ indicates that the x scores should be added and then the total should be squared. It is extremely important to avoid confusion between Σx^2 and $(\Sigma x)^2$.

Σxy indicates that each x score should be multiplied by its corresponding y score. After obtaining all such products, find their sum.

r is the linear correlation coefficient, which measures the strength of the relationship between the paired x and y values in a *sample*. r is a sample statistic.

ρ (rho) is the linear correlation coefficient, which measures the strength of the relationship between all paired x and y values in a *population*. ρ is a population parameter.

For the data of Table 9-1, we compute the individual components and then use the results to determine the value of r.

EXAMPLE

For the sample paired data of Table 9-1 we get $n = 9$, since there are nine pairs of data. The other components required in Formula 9-1 are found from the calculations that follow.

	Speed x	MPG y	xy	x^2	y^2
	20	22	440	400	484
	25	21	525	625	441
	30	20	600	900	400
	35	23	805	1225	529
	40	19	760	1600	361
	45	18	810	2025	324
	50	16	800	2500	256
	55	14	770	3025	196
	60	11	660	3600	121
Totals:	360	164	6170	15,900	3112
	↑ Σx	↑ Σy	↑ Σxy	↑ Σx^2	↑ Σy^2

Solution

Using these values, we compute r as follows:

$$r = \frac{n(\Sigma xy) - (\Sigma x)(\Sigma y)}{\sqrt{n(\Sigma x^2) - (\Sigma x)^2} \sqrt{n(\Sigma y^2) - (\Sigma y)^2}}$$

$$= \frac{9(6170) - (360)(164)}{\sqrt{9(15,900) - 129,600} \sqrt{9(3112) - 26,896}}$$

$$= \frac{-3510}{\sqrt{13,500}\sqrt{1112}}$$

$$= -0.906$$

Terrific! Now that we have computed the value of the linear correlation coefficient r, what does it mean? The way Formula 9-1 was derived, the computed value of r must always be between -1 and $+1$, inclusive. If a computed linear correlation coefficient has a value greater than $+1$ or less than -1, an error must be present in the computations. A strong positive linear correlation between x and y is indicated by a value of r near $+1$. A strong negative linear correlation is indicated by a value of r near -1. If the linear correlation coefficient r is close to 0, we conclude that there is no significant linear correlation between x and y. These vague descriptions of close to 0 and near $+1$ or -1 are made precise by Table A-7, which lists critical values of r for various sample sizes. If the magnitude of the computed linear correlation coefficient r exceeds the critical r value found in Table A-7, we can conclude that there is a statistically significant linear relationship between the variables x and y. If the relationship is found to be significant and the computed r is positive, we say that there is a positive linear correlation. If the relationship is found to be significant and the computed r is negative, we say that there is a negative linear correlation. Table A-7 includes the two significance levels of $\alpha = 0.05$ and $\alpha = 0.01$.

If we have a sample of paired data, we could develop a hypothesis test following the same general pattern of hypothesis testing introduced in Chapter 6. With a null hypothesis of $H_0: \rho = 0$, we have the alternative hypothesis $H_1: \rho \neq 0$. The test statistic is the linear correlation coefficient r and critical values are found in Table A-7.

EXAMPLE

Find the critical value of the linear correlation coefficient if we have nine pairs of data (as in Table 9-1) and the significance level is $\alpha = 0.05$.

Solution

Refer to Table A-7 and locate the critical r value of 0.666 corresponding to $n = 9$ and $\alpha = 0.05$.

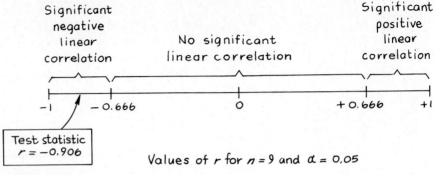

Values of r for $n = 9$ and $\alpha = 0.05$

Figure 9-3

EXAMPLE

From nine pairs of data, r is computed to be -0.906. What can you conclude at the significance level of $\alpha = 0.05$?

Solution

With H_0: $\rho = 0$ and H_1: $\rho \neq 0$ and $\alpha = 0.05$, we proceed to obtain the value of the test statistic and critical value. The *magnitude* of the computed linear correlation coefficient is 0.906, and since it exceeds the critical r value of 0.666, we reject H_0 and conclude that there is a significant linear relationship between the two variables. See Figure 9-3.

We must be careful to avoid the common error of concluding that a significant linear correlation between two variables is proof that there is a cause-and-effect relationship between them. One study, for example, showed that there is a significant positive linear correlation between teachers' salaries and per capita beer consumption. The significance of the correlation implies that teachers are using their raises to buy more beer, right? Wrong. Perhaps increases in teachers' salaries precipitate higher taxes, which in turn cause taxpayers to drown their sorrows and forget their financial difficulties by drinking more beer. Or perhaps higher teachers' salaries and greater beer consumption are both manifestations of some other factor, such as general improvement in the standard of living. In any event, the techniques in this chapter can be used only to establish a **statistical** linear relationship. *We cannot establish the existence or absence of any inherent cause-and-effect relationship* between the two variables. The cause-and-effect issue is considered by the professionals in the different fields, such as psychologists, sociologists, biologists, and so on.

Have Atomic Tests Caused Cancer?

Conquerors was a 1954 movie made in Utah. A few years ago a team of investigative reporters attempted to locate members of the cast and crew. Of the 79 people they found (some living and some dead), 27 had developed cancer, including John Wayne, Susan Hayward, and Dick Powell. Two key questions arise:

1. Is a cancer rate of 27 out of 79 significant, or could it be a coincidence?

2. Was the cancer caused by fallout from the nuclear tests previously conducted in the area?

We can use statistics to answer the first question, but not the second.

A medical researcher may establish a significant correlation between the unhealthy habit of smoking and the unhealthy habit of dying. Yet such a correlation does not prove that smoking causes or hastens deaths. Perhaps people become nervous about the prospect of dying and turn to cigarettes as a way of relieving that tension. Maybe dancing puts a strain on the heart that ultimately leads to death and, in the process, creates a biological urge to smoke. Statisticians cannot determine the inherent cause-and-effect relationship, but they can assist and guide the medical researcher in the analysis of the relevant physiological and biological processes.

Another source of potential error arises with data based on rates or averages. When we use rates or averages for data, we suppress the variation among the individuals, which may easily lead to an inflated correlation coefficient. As an example, one study produced a 0.4 linear correlation coefficient for paired data relating income and education among *individuals,* but the correlation coefficient became 0.7 when regional *averages* were used.

One misuse of the correlation coefficient involves the concept of linearity. The linear correlation coefficient r, as discussed in this section, is significant only if the paired data follow a linear or straight-line pattern. Consider the data of Table 9-4 along with the corresponding scatter diagram of Figure 9-4. Table 9-4 and Figure 9-4 represent nine pairs of data obtained from a physical experiment that consists of shooting an object upward and recording the height of the object at different times after its release. For example, the pair of "2 seconds: 192 feet" indicates that the object is at a height of 192 feet exactly 2 seconds after it is shot upward. While Figure 9-4 exhibits a clearly recognizable pattern, the relationship is not linear. The application of Formula 9-1 to the data of Table 9-4 reveals that $r = 0$, an indication that there is no *linear* relationship between the two variables. A nonlinear relationship becomes very obvious when we examine Figure 9-4. (The techniques for dealing with these nonlinear cases are beyond the scope of this text.)

Table 9-4									
Time (seconds)	0	1	2	3	4	5	6	7	8
Distance (feet) above ground	0	112	192	240	256	240	192	112	0

So far we have presented the formula for computing the linear correlation coefficient r, but we have given no justification for it. Formula 9-1 is actually a simplified form of the equivalent formula

$$r = \frac{\Sigma(x - \bar{x})(y - \bar{y})}{(n - 1)s_x s_y}$$

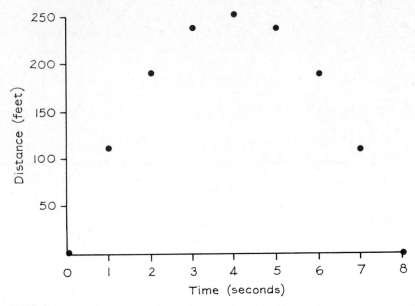

Figure 9-4

While this formula is equivalent to Formula 9-1, we find that Formula 9-1 is generally easier to work with, especially if we use a calculator. (There are several inexpensive calculators that are designed to compute the linear correlation coefficient r directly. The user simply enters the sample data in pairs and then presses the appropriate key to obtain r.)

Formula 9-1 is a short-cut form of the preceding formula for the linear correlation coefficient, but the following discussion refers to the formula given here since its form relates more directly to underlying theory. We will consider the paired data

x	1	1	2	4	7
y	4	5	8	15	23

depicted in the scatter diagram of Figure 9-5. Figure 9-5 includes the point $(\bar{x}, \bar{y})$, which is called the **centroid** of the sample points.

Sometimes r is called **Pearson's product moment,** and that title reflects both the fact that it was first developed by Karl Pearson (1857–1936) and that it is based on the product of the moments $(x - \bar{x})$ and $(y - \bar{y})$. That is, Pearson based the measure of scattering on the statistic $\Sigma(x - \bar{x})(y - \bar{y})$. In any scatter diagram, vertical and horizontal lines through the centroid $(\bar{x}, \bar{y})$ divide the diagram into four quadrants (see Figure 9-5). If the points of the scatter diagram tend to approximate an uphill line (as in the figure), then individual values of $(x - \bar{x})(y - \bar{y})$

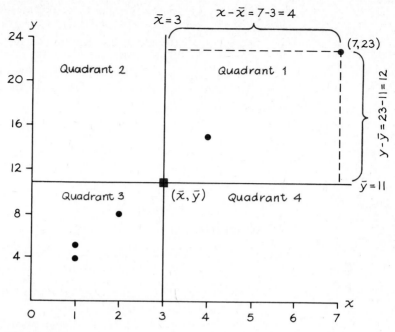

Figure 9-5

tend to be positive, since the points are predominantly found in the first and third quadrants, where the products of $(x - \bar{x})$ and $(y - \bar{y})$ are positive. If the points of the scatter diagram approximate a downhill line, the points are predominantly in the second and fourth quadrants where $(x - \bar{x})$ and $(y - \bar{y})$ are opposite in sign, so $\Sigma(x - \bar{x})(y - \bar{y})$ tends to be negative. If the points follow no linear pattern, they tend to be scattered among the four quadrants, so $\Sigma(x - \bar{x})(y - \bar{y})$ tends to be close to zero.

The sum $\Sigma(x - \bar{x})(y - \bar{y})$ depends on the magnitude of the numbers used, yet r should not be affected by the particular scale used. For example, r should not change whether heights are measured in meters or centimeters. We therefore standardize the moments $(x - \bar{x})$ and $(y - \bar{y})$ by dividing them by s_x and s_y, respectively. The sum

$$\Sigma \frac{(x - \bar{x})(y - \bar{y})}{s_x s_y}$$

is not affected by the scales used, but it is affected by the number of pairs of data. We can average out the preceding sum by simply dividing by n, but we divide by $n - 1$ for theoretical reasons similar to those that caused us to use $n - 1$ in sample standard deviation computations. We get

$$r = \frac{\Sigma(x - \bar{x})(y - \bar{y})}{(n - 1)s_x s_y}$$

which can be algebraically manipulated into the equivalent form of Formula 9-1 (see Exercise 9-32).

We can use the linear correlation coefficent to decide whether there is a linear relationship between two variables. After deciding that a relationship exists, we can then determine what it is. In the next section we will see how the relationship can be described.

9-2 Exercises A
Correlation

For each part of Exercises 9-1 and 9-2, would you expect positive correlation, negative correlation, or no correlation for each of the given sets of paired data?

9-1 (a) The weights and lengths of newborn babies.
(b) People's ages and blood pressures.
(c) The amounts of rainfall and vegetation growth.
(d) The weights of cars and fuel consumption rates measured in miles per gallon.
(e) The hat sizes of adults and their I.Q. scores.

9-2 (a) The diameters and circumferences of circles.
(b) Hours spent studying for tests and the resulting test scores.
(c) The numbers of absences in a course and the grades in that course.
(d) The weights and the heights of children.
(e) Annual per capita income for different nations and infant mortality rates for those nations.

For each part of Exercises 9-3 and 9-4, a sample of paired data produces a linear correlation coefficient r. What do you conclude in each case? Assume a significance level of $\alpha = 0.05$.

9-3 (a) $n = 20, r = 0.5$ (b) $n = 20, r = -0.5$
(c) $n = 50, r = 0.2$ (d) $n = 50, r = -0.2$
(e) $n = 37, r = 0.25$

9-4 (a) $n = 77, r = 0.35$ (b) $n = 22, r = 0.37$
(c) $n = 22, r = 0.40$ (d) $n = 22, r = -0.5$
(e) $n = 6, r = -0.8$

In Exercises 9-5 through 9-8, use the given list of paired data.
(a) Construct the scatter diagram.   (b) Determine n.
(c) Find Σx.   (d) Find Σx^2.
(e) Find $(\Sigma x)^2$.   (f) Find Σxy.
(g) Find r.

9-5

x	1	1	2	3
y	1	5	4	2

9-6

x	1	2	2	3
y	5	4	3	1

9-7

x	0	1	1	2	5
y	3	3	4	5	6

9-8

x	1	3	3	4	5	5
y	5	3	2	2	0	1

In Exercises 9-9 through 9-24:

(a) Construct the scatter diagram.

(b) Compute the linear correlation coefficient r.

(c) Assume that $\alpha = 0.05$ and find the critical value of r from Table A-7.

(d) Based on the results of parts (b) and (c), decide whether there is a significant positive linear correlation, a significant negative linear correlation, or no significant linear correlation. In each case, assume a significance level of $\alpha = 0.05$.

(e) Save all your work. The same data will be used in the exercises of the next section.

9-9 The number of drinks consumed and the corresponding blood alcohol concentrations are listed for various subjects with the same body weight.

Number of drinks	2	2	4	5	8
Blood alcohol concentration	0.05	0.06	0.11	0.13	0.22

9-10 The loads (in pounds) on a spring are listed along with the corresponding lengths (in inches) of the spring.

Load	0	1	3	2	6
Length	5	6	8	7	12

9-11 The assessed values (in thousands of dollars) and the appraised values (in thousands of dollars) are listed for various homes.

Assessed value	33	34	37	42	40
Appraised value	90	85	95	110	110

9-12   The outside air temperature (in degrees Fahrenheit) and altitudes (in thousands of feet) of an aircraft are listed for various times during a short flight.

Temperature	80	70	65	50	60
Altitude	0	4	4	10	6

9-13   The heights (in inches) are listed for boys who were measured at 5 years of age and again at 18 years of age.

Age 5	43	41	42	45	46
Age 18	68	66	66	70	71

9-14   The numbers of pushups and the endurance times (in minutes) are recorded for several participants in a fitness program.

Pushups	24	4	40	30	10
Endurance time	4	2	8	12	5

9-15   The values of exports and incomes on foreign investments are listed (in billions of dollars) for various years.

Exports	16	20	27	39	56	63	66
Incomes on foreign investments	2	3	4	7	11	11	13

9-16   The following table lists the ages and diastolic blood pressures (in millimeters of mercury) of six randomly selected subjects.

Ages	35	38	40	69	29	42
Blood pressures	89	81	100	90	59	82

9-17   Two different tests are designed to measure one's understanding of a certain topic. Two tests are given to ten different subjects and the results are listed in the following table.

Test X	75	78	88	92	95	67	55	73	74	80
Test Y	81	73	85	85	89	73	66	81	81	81

9-18   The researchers in a laboratory experiment with a car at different speeds in an attempt to study the fuel consumption rates as measured in miles per gallon (MPG). The accompanying data are obtained.

Speed	15	23	30	35	42	45	50	54	60	65
MPG	14	17	20	24	26	23	18	15	11	10

9-19 A farmer notices that crickets seem to chirp faster on warm days. The accompanying table lists the number of chirps made by a cricket in one minute, along with the corresponding temperature (degrees Fahrenheit).

Chirps per minute	66	58	94	120	83	119	121	65	55	50
Temperature	55	53	62	70	59	69	69	55	52	51

9-20 A researcher obtains the current unemployment and suicide rates for a sample of eight urban areas, and the results are summarized in the accompanying table. (The unemployment rate is the number of unemployed per 100, while the suicide rate is the number per 100,000.)

Unemployment	8.6	6.1	6.3	8.4	14.9	12.1	12.7	9.3
Suicide	19.1	12.0	10.2	27.5	29.0	31.4	40.4	23.7

9-21 A manager in a factory randomly selects 15 assembly line workers and develops scales to measure their dexterity and productivity levels. The results are listed in the following table.

Productivity	63	67	88	44	52	106	99	110	75	58	77	91	101	51	86
Dexterity	2	9	4	5	8	6	9	8	9	7	4	10	7	4	6

9-22 Randomly selected subjects are given a standard I.Q. test and then tested for their receptivity to hypnosis. The results are listed in the following table.

I.Q.	103	113	119	107	78	153	114	101	103	111	105	82	110	90	92
Receptivity to hypnosis	55	55	59	64	45	72	42	63	62	46	41	49	57	52	41

9-23 The following table lists per capita cigarette consumption in the United States for various years, along with the percentage of the population admitted to mental institutions as psychiatric cases.

Cigarette consumption	3522	3597	4171	4258	3993	3971	4042	4053
Percentage of psychiatric admissions (in percentage points)	0.20	0.22	0.23	0.29	0.31	0.33	0.33	0.32

9-24   In the following table, all figures are in kilograms.

	Rat	Man	Dolphin	Elephant	Chimp	Lion	Bat	Crow
Body mass	0.2	60	180	8000	60	200	0.02	0.3
Brain mass	0.0025	1.7	1.8	5	0.4	0.25	0.001	0.01

9-2   Exercises B
Correlation

9-25   Attempt to compute the linear correlation coefficient r for the data in the table and comment on the results. Also, plot the scatter diagram.

x	0	3	5	5	6
y	2	2	2	2	2

9-26   Do Exercise 9-10 after interchanging each x value with the corresponding y value. How is the value of the linear correlation coefficient affected?

9-27   Do Exercise 9-10 after changing the length of 12 inches to 36 inches. How much effect does an extreme value have on the value of the linear correlation coefficient?

9-28   Do Exercise 9-10 after converting all of the lengths from inches to centimeters (1 in. = 2.54 cm). What effect does this conversion have on the value of the linear correlation coefficient?

9-29   Compute the linear correlation coefficient for the paired data in the following table by using

$$r = \frac{\Sigma(x - \bar{x})(y - \bar{y})}{(n - 1)s_x s_y}$$

and compare your result to Exercise 9-6.

x	1	2	2	3
y	5	4	3	1

9-30   The graph of $y = x^2$ is a parabola, not a straight line, so we might expect that the value of r would not reflect a linear correlation between x and y. Using $y = x^2$, make a table of x and y values for $x = 0, 1, 2, \ldots, 10$ and calculate the value of r. What do you conclude? How do you explain the result?

9-31   An approximation sometimes used to find the critical values of the linear correlation coefficient r is given by

$$r = \frac{t}{\sqrt{t^2 + n - 2}}$$

where the t value is found from Table A-4 by assuming a two-tailed case with $n - 2$ degrees of freedom. Assuming that $\alpha = 0.05$, use the above approximation to find all of the critical values for the values of n listed in Table A-7. Of the 28 critical values listed for $\alpha = 0.05$, how many differ from the approximated values rounded to three decimal places?

9-32   Show that

$$\frac{\Sigma(x - \bar{x})(y - \bar{y})}{(n - 1)s_x s_y} = \frac{n\Sigma xy - (\Sigma x)(\Sigma y)}{\sqrt{n(\Sigma x^2) - (\Sigma x)^2} \; \sqrt{n(\Sigma y^2) - (\Sigma y)^2}}$$

9-3 Regression

In Section 9-2 we tested paired data for the presence or absence of the statistical relationship of a linear correlation. In this section we identify that relationship. The relationship is expressed in the form of a linear, or straight-line, equation. Such an equation is often useful in predicting the likely value of one variable given a value of another. The straight line that concisely summarizes the relationship between the two variables is the **regression line.**

Sir Francis Galton (1822–1911), a cousin of Charles Darwin, studied the phenomenon of heredity in which certain characteristics regress or revert to more typical values. Galton noted, for example, that children of tall parents tend to be shorter than their parents, while short parents tend to have children taller than themselves (when fully grown, of course). These original studies of regression evolved into a fairly sophisticated branch of mathematics called *regression analysis,* which includes the consideration of linear and nonlinear relationships.

This section is confined to linear relationships for two basic reasons. First, the real relationship between two variables is often a linear relationship, or it can be effectively approximated by a linear relationship. Second, nonlinear or curvilinear regression problems introduce complexities beyond the scope of this introductory text.

Let's reconsider the paired data of Table 9-1, which lists speeds of a car along with the corresponding gasoline consumption rates. We saw in Section 9-2 that the value of the linear correlation coefficient is computed to be $r = -0.906$. We also saw that at the 0.05 level of significance, there is a negative linear-correlation between speed and gasoline consumption. Knowing that there is a linear correlation between those two variables, we now seek to find the equation of the straight line that relates them. We again stipulate that the variable x represents speed while y represents fuel consumption. We want an equation of the form $y = mx + b$, where the values of m and b can be found from the paired data by using Formulas 9-2 and 9-3 given below.

Formula 9-2

$$m = \frac{n(\Sigma xy) - (\Sigma x)(\Sigma y)}{n(\Sigma x^2) - (\Sigma x)^2}$$

Formula 9-3

$$b = \frac{(\Sigma y)(\Sigma x^2) - (\Sigma x)(\Sigma xy)}{n(\Sigma x^2) - (\Sigma x)^2}$$

Some inexpensive calculators accept entries of paired data and provide the m and b values directly. These formulas appear formidable, but three observations make the required computations easier. First, if the correlation coefficient r has been computed by Formula 9-1, the values of Σx, Σy, Σx^2, $(\Sigma x)^2$, and Σxy have already been computed. These values can now be used again in Formulas 9-2 and 9-3. Second, examine the denominators of the formulas for m and b and note that they are identical. This means that the computation of $n(\Sigma x^2) - (\Sigma x)^2$ need be done only once, and the resulting value can be used in both formulas. Third, the regression line always passes through the centroid $(\bar{x}, \bar{y})$, so the equation $\bar{y} = m\bar{x} + b$ must be true. This implies that $b = \bar{y} - m\bar{x}$, and it is usually easier to evaluate b by computing $\bar{y} - m\bar{x}$ than by using Formula 9-3.

Formula 9-4

$$b = \bar{y} - m\bar{x}$$

We now use the speed–gasoline consumption data to find the equation of the regression line. We will use the following statistics already found in the first example of Section 9-2.

$n = 9$ $\Sigma x^2 = 15{,}900$
$\Sigma x = 360$ $\Sigma y^2 = 3112$
$\Sigma y = 164$ $\Sigma xy = 6170$

Having determined the values of the individual components, we can now compute m by using Formula 9-2.

$$m = \frac{n(\Sigma xy) - (\Sigma x)(\Sigma y)}{n(\Sigma x^2) - (\Sigma x)^2} = \frac{9(6170) - (360)(164)}{9(15{,}900) - (360)^2}$$

$$= \frac{-3510}{13{,}500}$$

$$= -0.26$$

We now use Formula 9-4 to find the value of b.

$$b = \bar{y} - m\bar{x} = \frac{164}{9} - (-0.26)\left(\frac{360}{9}\right) = 28.62$$

Using these results, the general straight-line equation of $y = mx + b$ becomes $y = -0.26x + 28.62$. The graph of that straight line is shown in

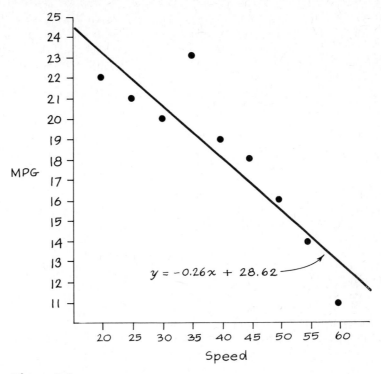

Figure 9-6

the scatter diagram of Figure 9-6. From that graph, we can see that the regression line is the line that best approximates the original points. We should realize that the regression equation $y = -0.26x + 28.62$ is based on one particular set of sample data. Another sample drawn from the same population will probably lead to a slightly different equation. We might think of these equations as estimates of the equation $y = Mx + B$, which represents the total population of paired data. We will now review some of the basic principles related to straight-line equations and graphs.

The equation $y = mx + b$ is called the **slope-intercept** form of a straight line, since the constants m and b represent slope and y-intercept, respectively. The y-intercept b is simply the value along the vertical y-axis where the straight line intercepts or crosses that axis. For example, the straight line represented by the equation $y = mx + 2$ crosses the y-axis at $y = 2$. The slope m represents the steepness of the straight line. Going from left to right, an uphill line has positive slope, while a downhill line has a negative slope. Given any two points (x_1, y_1) and (x_2, y_2) on a straight line, the slope may be computed by $m = (y_2 - y_1)/(x_2 - x_1)$. In effect, the slope measures the changes in y values relative to the changes in x values. Here we are concerned with two basic procedures:

1. Obtaining the equation of a straight line given the coordinates of different points.
2. Graphing the straight line after we have determined its equation.

Unusual Economic Indicators

Forecasting and predicting are important goals of statistics. Investors constantly seek indicators that can be used to forecast stock market trends. Among the more colorful indicators are the hemline index, the Super-Bowl omen, and the aspi-

rin count. Developed in 1967, the hemline index is based on the length of women's skirts. According to one theory, rising hemlines precede a rise in the Dow Jones industrial average, while falling hemlines are followed by a drop in that average. According to the Super-Bowl omen, a Super-Bowl victory by a team with NFL origins is indicative of a year in which the New York Stock Exchange index will rise. A victory by a team with AFL origins means that the market will fall. As of this writing, the Super-Bowl indicator has been correct in 16 of the past 18 years. The aspirin count theory also seems

to work fairly well. It suggests that bad times cause an increase in aspirin sales, and such an increase is followed by a falling market. A drop in aspirin sales should be followed by rising stock market prices. But consider this: In a 10-year sequence of stock market increases and decreases, there are only 1024 different possible strings of rises and drops. Among the millions of different possible indicators, surely some of them must correspond exactly to the stock market fluctuations, but that does not mean that they will be reliable predictors.

Formulas 9-2, 9-3, and 9-4 can be used to determine the equation of the regression line, but it is often helpful (or necessary) to determine that equation by other means. One alternative that will yield approximate results begins with a rough sketch of the straight line that appears to fit the data best. Then you need to identify the x and y coordinates of two separate points on that line and compute m by evaluating $(y_2 - y_1)/(x_2 - x_1)$. After inserting the resulting value of m in the equation $y = mx + b$, continue by substituting the x and y values of either known point into the equation and solve for the only remaining unknown, b. This is all illustrated in the next example.

EXAMPLE

Without using Formulas 9-2, 9-3, or 9-4, estimate the equation of the straight line that best fits the paired data given in the following table.

x	1	2	3	6	8	10
y	1800	1400	1300	1000	600	500

Solution

After constructing the scatter diagram, draw the straight line that appears to best fit the given data (see Figure 9-7, below). The points (3, 1400) and (10, 400) appear to be on the estimated regression line, even though those pairs of data are not present in the original list given in the table. We use the coordinates of those points on the line to find the slope m.

$$m = \frac{y_2 - y_1}{x_2 - x_1} = \frac{400 - 1400}{10 - 3} = \frac{-1000}{7} = -142.9$$

The general equation $y = mx + b$ now becomes $y = -142.9x + b$. We substitute the x and y coordinates of either point in this last equation. Choosing (3, 1400), we get

$$1400 = -142.9(3) + b$$

or

$$1400 = -428.7 + b$$

or

$$1828.7 = b$$

The equation suggested by the line in Figure 9-7 is therefore $y = -142.9x + 1828.7$. For comparison, the corresponding equation obtained through application of Formulas 9-2 and 9-3 is $y = -135.9x + 1779.7$. Thus the estimated line and the resulting equation served as a fairly good approximation to the true regression line.

Figure 9-7

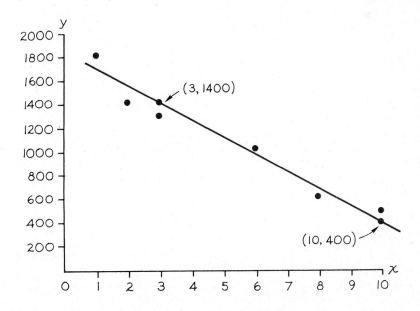

The second major concern of this brief review on straight lines deals with the graphing of the regression line once we have obtained its equation. Using the equation, we can substitute a value for x and then compute the corresponding y value. These values are the x and y coordinates of a single point on the line. Repeat the same process and plot a second point. Both points can be graphed and, as we all well know, two points determine a straight line. See the following example.

E X A M P L E

A list of paired data results in the regression values of $m = 3$ and $b = 2$. Graph the equation $y = 3x + 2$.

Solution

Before substituting for x blindly, we should examine the original values of x so that we can select convenient numbers near the smallest and largest x values. Let's assume that the regression equation came from paired data in which the minimum and maximum x values were 1.8 and 6.3. Then 2 and 6 would make excellent choices for x. If $x = 2$, then $y = 3(2) + 2 = 8$. If $x = 6$, then $y = 3(6) + 2 = 20$. We have generated the pairs of coordinates $(2, 8)$ and $(6, 20)$ that are plotted and connected, as in Figure 9-8.

Figure 9-8

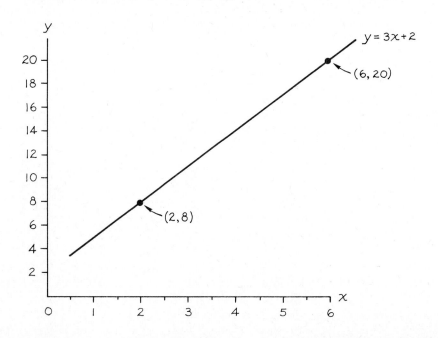

The next example incorporates concepts from this section, along with the linear correlation coefficient from the preceding section.

EXAMPLE

Two different tests are designed to measure the understanding of a certain topic and are administered to ten subjects. The results follow. Find the linear correlation coefficient r, the equation of the regression line, and plot the scatter diagram and regression line on the same graph. Determine whether there is a significant linear correlation and then predict the likely score on test Y for someone receiving a score of 70 on test X. Use a significance level of 0.05.

Test X	75	78	88	92	95	67	55	73	74	80
Test Y	81	73	85	85	89	73	66	81	81	81

Solution

With H_0: $\rho = 0$ and H_1: $\rho \neq 0$, and $\alpha = 0.05$, we proceed to evaluate the test statistic. For the sample data, we determine the following:

$$n = 10 \qquad \Sigma y^2 = 63,629$$
$$\Sigma x = 777 \qquad (\Sigma x)^2 = 603,729$$
$$\Sigma y = 795 \qquad (\Sigma y)^2 = 632,025$$
$$\Sigma x^2 = 61,661 \qquad \Sigma xy = 62,432$$

We now compute the value of the linear correlation coefficient r.

$$r = \frac{n\Sigma xy - (\Sigma x)(\Sigma y)}{\sqrt{n(\Sigma x^2) - (\Sigma x)^2} \sqrt{n(\Sigma y^2) - (\Sigma y)^2}}$$

$$= \frac{10(62,432) - (777)(795)}{\sqrt{10(61,661) - 603,729} \sqrt{10(63,629) - 632,025}}$$

$$= \frac{6605}{\sqrt{12,881} \sqrt{4265}} = 0.891$$

With $n = 10$, the test statistic of $r = 0.891$ indicates a significant positive linear correlation since it exceeds the critical r value of 0.632 (for $\alpha = 0.05$). That is, at the 0.05 level of significance, we conclude that there is a significant positive linear correlation. Having determined that r is significant, we could proceed to obtain the regression values of m and b.

$$m = \frac{n\Sigma xy - (\Sigma x)(\Sigma y)}{n(\Sigma x^2) - (\Sigma x)^2} = \frac{10(62,432) - (777)(795)}{10(61,661) - 603,729}$$

$$= \frac{6605}{12,881} = 0.51$$

$$b = \bar{y} - m\bar{x} = 79.5 - 0.51(77.7) = 39.87$$

The general equation of the regression line $y = mx + b$ becomes $y = 0.51x + 39.87$. The scatter diagram appears in Figure 9-9, along with the graph of the regression line. The line $y = 0.51x + 39.87$ is plotted by generating the coordinates of two different points on that line. Specifically, when $x = 60$, the regression equation becomes $y = 0.51(60) + 39.87 = 70.47$, so $(60, 70.47)$ is on the regression line and appears as the leftmost asterisk in Figure 9-9. Letting $x = 90$ implies that $y = 0.51(90) + 39.87 = 85.77$, so that $(90, 85.77)$ represents a second point on the regression line that corresponds to the right asterisk in the figure. Finally, we use the equation of the regression line to find the last item requested. To predict the score on test Y given a score of 70 on test X, we simply substitute 70 for x in the equation of the regression line. If $x = 70$, $y = 0.51x + 39.87$ becomes $y = 0.51(70) + 39.87 = 75.57$, which is our predicted point estimate of the score.

By examining Figure 9-9 and the computed value of the linear correlation coefficient for the given data, we conclude that the regression line fits the data reasonably well. It is therefore useful in making projections or predictions that do not go far beyond the scope of the

Figure 9-9

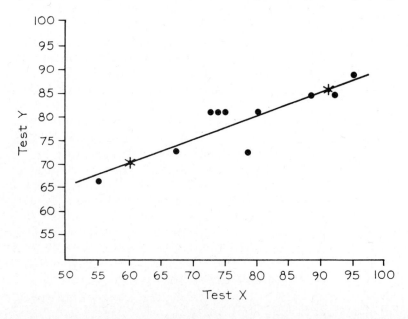

available scores. However, if r is close to zero, even though the regression line is the best fitting line, it may not fit the data well enough. It is important to note that the value of r indicates how well the regression line actually fits the available paired data. If $r = 1$ or $r = -1$, then the regression line fits the data perfectly. If r is near $+1$ or -1, the regression line constitutes a very good approximation of the data. But if r is near zero, the regression line fits poorly. When we must estimate the value of one variable given some value of the other, **we should use the equation of the regression line only if r indicates that there is a significant linear correlation.** However, **in the absence of a significant linear correlation, we should not use the regression equation for projecting or predicting. Instead, our best estimate of the second variable is simply the sample mean of that variable,** regardless of the value assigned to the first variable. To illustrate this concept, let's suppose that we have the two samples of paired data. In both cases we want to estimate y when $x = 5$.

First collection of paired data	Second collection of paired data
$n = 100$ Regression line: $y = 2x + 3$ $\bar{y} = 20$ $r = 0.95$	$n = 100$ Regression line: $y = 2x + 3$ $\bar{y} = 20$ $r = 0.02$
To get the best point estimate of y when $x = 5$, use the equation of the regression line to get $y = 2(5) + 3 = 13$	To get the best point estimate of y for any value of x, simply select $\bar{y} = 20$.

In the first case, $r = 0.95$ indicates that the equation $y = 2x + 3$ will give good results, since the regression line fits the data well. Consequently, the best estimate of y for $x = 5$ can be obtained by substituting 5 for x in the equation of the regression line. The estimated y value of 13 results. However, when we consider the second collection of paired sample data, we see that the linear correlation coefficient of 0.02 reflects a poorly fitting regression line that is useless as a predictor. In this case, the best estimate of y is 20 (the value of $\bar{y}$).

As a practical application of this concept, suppose we wanted to predict the height of an adult male with an I.Q. of 107. Knowing that there is no correlation between height and I.Q. scores, our best prediction of height would be about 5' 10" (the mean). But we do know that there is a positive correlation between the length of a beam as measured in feet and the length of the same beam as measured in inches. The regression equation should reveal, not too surprisingly, that $y = 12x$ where x is the length in feet and y is the length in inches. To predict the number of inches for a beam 8 feet long, we would use the regression equation $y = 12x$ and get $y = 12(8) = 96$ inches. We would not be concerned with the mean length of all beams.

Knowing the qualitative relationship between the value of r and the

Rising to New Heights

The concept of regression was introduced in 1877 by Sir Francis Galton. His research showed that when tall parents have children, those children tend to have heights that "regress," or revert back to the mean height of the population.

In *American Averages*, by Mead and Feinsilber, it is stated that every generation of Americans has been taller than the previous generation by about an inch. An exception is the present generation, which shows no significant change.

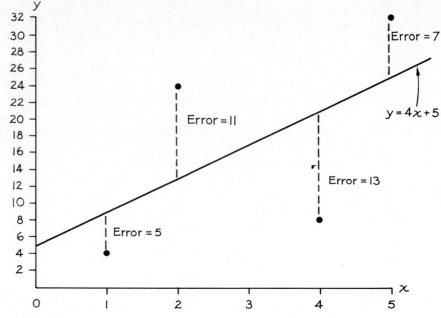

Figure 9-10

goodness of fit of the regression line, we can seek some precise quantitative relationship. For paired data, r is related to m (the slope of the regression line) by

Formula 9-5
$$r = \frac{ms_x}{s_y}$$

where s_x is the standard deviation of the x values and s_y is the standard deviation of the y values.

Formulas 9-2 and 9-3 or 9-4 describe the computations necessary to obtain the regression line equation $y = mx + b$, and we now describe the criterion used to arrive at these particular formulas. To be concise, the regression line obtained through these formulas is unique in that **the sum of the squares of the vertical deviations of the sample points from the regression line is the smallest sum possible.** This property is called the **least-squares property** and can be understood by examining Figure 9-10 and the text that follows. In Figure 9-10 we show the paired data contained in the following table.

x	1	2	4	5
y	4	24	8	32

Application of the formulas for m and b results in the equation $y = 4x + 5$, which is graphed in Figure 9-10. The least-squares property refers to the vertical distances by which the original points miss the regression line. In this figure, these vertical deviations are shown to be 5, 11, 13, and 7 for the four pairs of data. The sum of the squares of these vertical errors is $5^2 + 11^2 + 13^2 + 7^2 = 364$, and the regression line $y = 4x + 5$ is unique in that the value of 364 is the lowest for that particular line. Any other line will yield a sum of squares that exceeds 364, and therfore produce a greater collective error. Thus, using the least-squares property, another line will not fit the data as well as the regression line $y = 4x + 5$.

For example, the line $y = 3x + 8$ will produce vertical errors of 7, 10, 12, and 9, so the sum of the squares of those errors is $49 + 100 + 144 + 81 = 374$. The collective error of 374 exceeds a collective error of 364. Consequently, the line $y = 4x + 5$ provides a better fit to the paired data. Fortunately, we need not deal directly with this least-squares property when we want to obtain the equation of the regression line. Calculus has been used to build the least-squares property into formulas, and our calculations for m and b are a result of that property. Because Formulas 9-2 and 9-3 depend on certain methods of calculus, we have not included their development in this text.

When dealing with larger collections of paired data, the use of a calculator or computer becomes necessary for finding the values of r, m, and b. We have already mentioned that some calculators allow the entry of paired data and provide the values of r, m, and b. Many computer software packages take paired data as input and produce more complete results, such as the STATDISK output shown below. The displayed output corresponds to the speed-mpg data (Table 9-1) introduced in Section 9-1 and used in Sections 9-2 and 9-3. The coefficient of determination and the standard error of estimate will be discussed in the following section.

```
Linear correlation coefficient ........ r =-0.90592
Equation of regression line .. Y = -0.26000 X + 28.62222
Coefficient of determination ............. = 0.82068
Standard error of estimate ............... = 1.77907
Level of significance .................... = .05
CONCLUSION: REJECT the claim of no significant linear correlation
Test statistic is .................... r = -0.90592
Critical value is .................... r =  0.66685
```

9-3 Exercises A
Regression

9-33 Use Formulas 9-2 and 9-4 to verify that the given paired data will produce a regression line whose equation is $y = x + 2$. Plot the scatter diagram and the equation $y = x + 2$.

x	2	3	5
y	4	5	7

9-34 Use Formulas 9-2 and 9-4 to verify that the given paired data will produce a regression line whose equation is $y = 2x + 3$. Plot the scatter diagram and the equation $y = 2x + 3$.

x	2	3	4	6
y	7	9	11	15

In Exercises 9-35 through 9-38, use the data given to obtain the equation of the regression line.

9-35

x	1	1	2	3
y	1	5	4	2

9-36

x	1	2	2	3
y	5	4	3	1

9-37

x	0	1	1	2	5
y	3	3	4	5	6

9-38

x	1	3	3	4	5	5
y	5	3	2	2	0	1

In Exercises 9-39 through 9-50, find the equation of the regression line. (The given data is taken from the preceding set of exercises.)

9-39

Number of drinks	2	2	4	5	8
Blood alcohol concentration	0.05	0.06	0.11	0.13	0.22

9-40

Load	0	1	3	2	6
Length	5	6	8	7	12

9-41

Assessed value	33	34	37	42	40
Appraised value	90	85	95	110	110

9-42

Temperature	80	70	65	50	60
Altitude	0	4	4	10	6

9-43

Age 5	43	41	42	45	46
Age 18	68	66	66	70	71

9-44

Pushups	24	4	40	30	10
Endurance time	4	2	8	12	5

9-45

Exports	16	20	27	39	56	63	66
Incomes on foreign investments	2	3	4	7	11	11	13

9-46

Ages	35	38	40	69	29	42
Blood pressures	89	81	100	90	59	82

9-47

Test X	75	78	88	92	95	67	55	73	74	80
Test Y	81	73	85	85	89	73	66	81	81	81

9-48

Speed	15	23	30	35	42	45	50	54	60	65
MPG	14	17	20	24	26	23	18	15	11	10

9-49

Chirps per minute	66	58	94	120	83	119	121	65	55	50
Temperature	55	53	62	70	59	69	69	55	52	51

9-50

Unemployment	8.6	6.1	6.3	8.4	14.9	12.1	12.7	9.3
Suicide	19.1	12.0	10.2	27.5	29.0	31.4	40.4	23.7

In Exercises 9-51 through 9-54, use the given paired data.

(a) Construct the scatter diagram and sketch the line that appears to fit the data best.

(b) Using the graph from part (a), estimate the coordinates of two different points on the estimated regression line and then use these coordinates to approximate the equation of the regression line.

(c) Determine the exact equation of the regression line by using the formulas for m and b (Formulas 9-2 and 9-4).

9-51

Productivity	63	67	88	44	52	106	99	110	75	58	77	91	101	51	86
Dexterity	2	9	4	5	8	6	9	8	9	7	4	10	7	4	6

9-52

I.Q.	103	113	119	107	78	153	114	101	103	111	105	82	110	90	92
Receptivity to hypnosis	55	55	59	64	45	72	42	63	62	46	41	49	57	52	41

9-53

Cigarette consumption	3522	3597	4171	4258	3993	3971	4042	4053
Percentage of psychiatric admissions (in percentage points)	0.20	0.22	0.23	0.29	0.31	0.33	0.33	0.32

9-54

	Rat	Man	Dolphin	Elephant	Chimp	Lion	Bat	Crow
Body mass	0.2	60	180	8000	60	200	0.02	0.3
Brain mass	0.0025	1.7	1.8	5	0.4	0.25	0.001	0.01

9-55 In each of the following cases, find the best predicted point estimate of y when $x = 5$. The given statistics are summarized from paired sample data.

(a) $n = 40$, $\bar{y} = 6$, $r = 0.01$, and the equation of the regression line is $y = 3x + 2$.

(b) $n = 40$, $\bar{y} = 6$, $r = 0.93$, and the equation of the regression line is $y = 3x + 2$.

(c) $n = 20$, $\bar{y} = 6$, $r = -0.654$, and the equation of the regression line is $y = -3x + 2$.

(d) $n = 20$, $\bar{y} = 6$, $r = 0.432$, and the equation of the regression line is $y = 1.2x + 3.7$.

(e) $n = 100$, $\bar{y} = 6$, $r = -0.175$, and the equation of the regression line is $y = -2.4x + 16.7$.

9-56 In each of the following cases, find the best predicted point estimate of y when $x = 8.0$. The given statistics are summarized from paired sample data.

(a) $n = 10$, $\bar{y} = 8.40$, $r = -0.236$, and the equation of the regression line is $y = -2.0x + 3.5$.

(b) $n = 10, \bar{y} = 8.40, r = -0.654$, and the equation of the regression line is $y = -2.0x + 3.5$.

(c) $n = 10, \bar{y} = 8.40, r = 0.602$, and the equation of the regression line is $y = 2.0x + 3.5$.

(d) $n = 32, \bar{y} = 8.40, r = -.304$, and the equation of the regression line is $y = -2.0x + 3.5$.

(e) $n = 75, \bar{y} = 8.40, r = 0.257$, and the equation of the regression line is $y = 2.0x + 3.5$.

9-3 Exercises B
Regression

9-57 Do Exercise 9-41 after changing 90 to 900. How much of an effect does one exceptional value have on the equation of the regression line?

9-58 Using the paired data in the table below, verify that $r = ms_x/s_y$ where m, s_x, and s_y are as defined in this section and r is computed by using Formula 9-1.

x	1	2	2	3
y	5	4	3	1

9-59 What do you know about s_x and s_y if $r = 0.500$ for paired data having the regression line $y = \frac{1}{2}x + 7.3$?

9-60 Prove that the point $(\bar{x}, \bar{y})$ will always lie on the regression line.

9-61 Prove that r and m have the same sign.

9-62 Show that

$$\frac{ms_x}{s_y} = \frac{n\Sigma xy - (\Sigma x)(\Sigma y)}{\sqrt{n(\Sigma x^2) - (\Sigma x)^2} \; \sqrt{n(\Sigma y^2) - (\Sigma y)^2}}$$

9-63 Using the data in Exercise 9-36 and the equation of the regression line, find the sum of the squares of the vertical deviations for the given points. Show that this sum is less than the corresponding sum obtained by replacing the regression line with $y = -x + 6$.

9-64 If the scatter diagram reveals a nonlinear pattern that we recognize as another type of curve, we may be able to apply the methods of this section. For the data given in the table below, find an equation of the form $y = mx^2 + b$ by using the values of x^2 and y instead of the values of x and y. Use Formulas 9-2 and 9-3 or 9-4, but enter the values of x^2 wherever x occurs.

x	4	5	6	7
y	3	7	11	20

9-4 Variation

In the preceding two sections we introduced the fundamental concepts of linear correlation and regression. We saw that the linear correlation coefficient r could be used to determine whether or not there is a significant statistical relationship between two variables. We interpret r in a very limited way when we make one of the following three conclusions:

1. There is a significant positive linear correlation.
2. There is a significant negative linear correlation.
3. There is not a significant linear correlation.

The actual values of r can provide us with more information. We begin with a sample case, which leads up to an important definition.

Let's assume that we have a large collection of paired data, which yields these results:

1. There is a significant positive linear correlation.
2. The equation of the regression line is $y = 2x + 3$.
3. $\bar{y} = 9$.
4. The scatter diagram contains the point (5, 19), which comes from the original set of observations.

We need a new notation; the convention given below is commonly used on electronic calculators.

NOTATION

We use y' to denote the **predicted value** of y for a given value of x, and y' is obtained by substituting the given x value into the equation of the regression line.

When $x = 5$ we can find the predicted value y' as follows.

$$y' = 2x + 3 = 2(5) + 3 = 13$$

Note that the point (5, 13) is on the regression line, while the point (5, 19) is not. Take the time to examine Figure 9-11 carefully.

DEFINITION

Given a collection of paired data, the **total deviation** (from the mean) of a particular point (x, y) is $y - \bar{y}$.

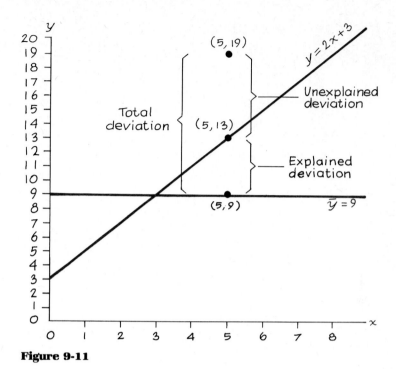

Figure 9-11

For the specific data under consideration, the total deviation of (5, 19) is $y - \bar{y} = 19 - 9 = 10$.

If we were totally ignorant of correlation and regression concepts and we wanted to predict a value of y given a value of x and a collection of paired (x, y) data, our best guess would be $\bar{y}$. But we are not totally ignorant of correlation and regression concepts: We know that in this case the way to predict the value of y when $x = 5$ is to use the regression equation, which yields $y' = 13$. We can explain the discrepancy between $\bar{y} = 9$ and $y' = 13$ by simply noting that there is a significant positive linear correlation best described by the regression line. Consequently, when $x = 5$, y *should* be 13, and not 9. But while y *should* be 13, *it is* 19, and that discrepancy between 13 and 19 cannot be explained by the regression line and is called an unexplained deviation or a residual. This specific case illustrated in Figure 9-11 can be generalized as follows.

(total deviation) = (explained deviation) + (unexplained deviation)

or $(y - \bar{y})$ $=$ $(y' - \bar{y})$ $+$ $(y - y')$

This last expression can be further generalized and modified to include all of the pairs of sample data as follows. (By popular demand, we omit the intermediate algebraic manipulations.)

What SAT Scores Measure

Harvard psychologist David McClellan says that numerous statistical surveys "have shown that no consistent relationship exists between SAT scores in college students and their actual college accomplishments in social leadership, the arts, science, music, writing, speech, and drama." There does appear to be a significant correlation between the SAT scores and the income levels of the tested students' families. Higher scores tend to come from high-income families, while low scores tend to come from low-income families. But a variety of studies have shown that SAT scores have limited value as predictors of college grades and *no* significant relationship to career success. Motivation appears to be a major success factor not measured by the SAT scores.

Formula 9-6

(total variation) = (explained variation) + (unexplained variation)

$$\text{or} \quad \Sigma(y - \bar{y})^2 \quad = \quad \Sigma(y' - \bar{y})^2 \quad + \quad \Sigma(y - y')^2$$

The components of this last expression are used in the next important definition.

DEFINITION

The amount of the variation in y that is explained by the regression line is indicated by the **coefficient of determination,** which is given by

$$r^2 = \frac{\text{explained variation}}{\text{total variation}}$$

We can compute r^2 by using this definition with Formula 9-6 above, or we can simply square the linear correlation coefficient r, which is found by using the methods given in Section 9-2. To make some sense of this last definition, consider the following example.

EXAMPLE

If $r = 0.8$, then the coefficient of determination is $r^2 = 0.8^2 = 0.64$, which means that 64% of the total variation can be explained by the regression line. Thus 36% of the total variation remains unexplained.

Table 9-5 illustrates that the linear correlation coefficient r can be computed with the total variation and explained variation. This procedure for determining r is not recommended since it is longer, less direct, and more likely to produce errors than the method of Section 9-2. The paired xy data are in the first two columns. The third column of y' values is determined by substituting each value of x into the equation of the regression line ($y = 2.09x + 1.56$). Also, $\bar{y} = 7.00$. The computed value of $r = 0.994$ agrees with the value of $r = 0.996$ found by using Formula 9-1 from Section 9-2; the discrepancy is due to rounding errors.

Table 9-5

x	y	y'	$y - \bar{y}$	$(y - \bar{y})^2$	$y' - \bar{y}$	$(y' - \bar{y})^2$
1	3	3.65	−4	16	−3.35	11.22
1	4	3.65	−3	9	−3.35	11.22
2	6	5.74	−1	1	−1.26	1.59
3	8	7.83	1	1	0.83	0.69
6	14	14.10	7	49	7.10	50.41
			Totals:	76		75.13

$\uparrow$ Total variation $\uparrow$ Explained variation

$$r^2 = \frac{\text{explained variation}}{\text{total variation}} = \frac{\Sigma(y' - \bar{y})^2}{\Sigma(y - \bar{y})^2} = \frac{75.13}{76} = 0.989$$

so that

$$r = \sqrt{0.989} = 0.994$$

While the computations shown in the table don't seem too involved, we first had to compute the equation of the regression line, which is why this method for finding r is inferior. Also observe that r is not always the *positive* square root of r^2 since we can have negative values for the linear correlation coefficient. Examination of the scatter diagram should reveal whether r is negative or positive. If the pattern of points is predominantly uphill (from left to right), then r should be positive, but if there is a downward pattern, r should be negative.

Suppose we use the regression equation to predict a value of y from a given value of x. For example, we have seen that with a significant positive linear correlation, the regression equation $y = 2x + 3$ can be used to predict y when $x = 5$; the result is $y' = 13$. Having made that prediction, we should consider how dependable it is. Intuition suggests that if $r = 1$, all sample points lie exactly on the regression line and the regression equation should therefore provide very dependable results, as long as we don't try to project beyond reasonable limits. If r is very close to 1, the sample points are close to the regression line, and again our y' values should be very dependable. What we need is a less subjective measure of the spread of the sample points about the regression line. The next definition provides such a measure.

DEFINITION

The **standard error of estimate** is given by

$$s_e = \sqrt{\frac{\Sigma(y - y')^2}{n - 2}}$$

EXAMPLE

Find the standard error of estimate s_e for the five pairs of data given in the first two columns of Table 9-6.

Solution

We extend the table to include the necessary components.

Table 9-6

x	y	y'	$y - y'$	$(y - y')^2$
1	3	3.65	−0.65	0.42
1	4	3.65	0.35	0.12
2	6	5.74	0.26	0.07
3	8	7.83	0.17	0.03
6	14	14.10	−0.10	0.01
			Total:	0.65

We can now evaluate s_e:

$$s_e = \sqrt{\frac{\Sigma(y - y')^2}{n - 2}} = \sqrt{\frac{0.65}{3}} = 0.465$$

The development of the standard error of estimate closely parallels that of the ordinary standard deviation introduced in Chapter 2, but this standard error involves deviations away from the regression line instead of the mean. The reasoning behind dividing by $n - 2$ is similar to the reasoning that led to division by $n - 1$ for the ordinary standard deviation, and we will not pursue the complex details. Do remember that smaller values of s_e reflect points that stay close to the regression line, while larger values indicate greater dispersion of points away from the regression line.

Formula 9-7 can also be used to compute the standard error of estimate. It is algebraically equivalent to the expression in the definition, but this form is generally easier to work with since it doesn't require that we compute each of the y' values.

Formula 9-7 $$s_e = \sqrt{\frac{\Sigma y^2 - b\Sigma y - m\Sigma xy}{n - 2}}$$

where m and b are the slope and y-intercept of the regression equation.

We can use the standard error of estimate, s_e, to construct interval estimates that will help us to see how dependable our point estimates of y really are. Assume that for each fixed value of x, the corresponding sample values of y are normally distributed about the regression line, and those normal distributions have the same variance. Given a fixed value of x (denoted by x_0), we have the following **interval estimate of the predicted value of y.**

DEFINITION

$$y' - E < y < y' + E$$

where

$$E = s_e \cdot t(\alpha/2) \sqrt{1 + \frac{1}{n} + \frac{n(x_0 - \overline{x})^2}{n(\Sigma x^2) - (\Sigma x)^2}}$$

x_0 represents the given value of x.
$t(\alpha/2)$ has $n - 2$ degrees of freedom.
s_e is found from Formula 9-7.

The regression equation for the data given in Table 9-6 is $y = 2.09x + 1.56$, so that when $x = 4$ we find the best point estimate for the predicted value of y to be $y' = 9.92$. Using that same data, we will construct the 95% confidence interval for the predicted y value corresponding to the given value of $x = 4$. We have already found that $s_e = 0.465$. With $n = 5$ we have 3 degrees of freedom, so that $t(\alpha/2) = 3.182$ for $\alpha = 0.05$ in two tails. With $x_0 = 4$, $\overline{x} = \frac{13}{5} = 2.6$, $\Sigma x^2 = 51$, and $\Sigma x = 13$, we get

$$E = (0.465)(3.182) \sqrt{1 + \frac{1}{5} + \frac{5(4 - 2.6)^2}{5(51) - 169}}$$

$$= (0.465)(3.182)(1.146) = 1.696$$

so that $y' - E < y < y' + E$ becomes $9.92 - 1.696 < y < 9.92 + 1.696$. Given that $x = 4$, we therefore get the following 95% confidence interval for the predicted value of y.

$$8.224 < y < 11.616$$

9-4 Exercises A
Variation

9-65   If $r = 0.333$, find the coefficient of determination and the percentage of the total variation that can be explained by the regression line.

9-66   If $r = -0.444$, find the coefficient of determination and the percentage of the total variation that can be explained by the regression line.

9-67   If $r = 0.800$, find the coefficient of determination and the percentage of the total variation that can be explained by the regression line.

9-68   If $r = -0.700$, find the coefficient of determination and the percentage of the total variation that can be explained by the regression line.

9-69   Use the given data to find the standard error of estimate, s_e, using Formula 9-7. What does the value of s_e indicate about this particular set of data?

x	1	2	3	5	6
y	5	8	11	17	20

(The equation of the regression line is $y = 3x + 2$.)

9-70   Use the given data to find the standard error of estimate, s_e, using Formula 9-7.

x	1	2	3	5	6
y	5	4	11	13	20

(The equation of the regression line is $y = 2.95x + 0.56$.)

9-71   Use the given data to find the standard error of estimate, s_e, using Formula 9-7.

x	4	7	5	2	3
y	13	22	16	7	9

(The equation of the regression line is $y = 3.08x + 0.46$.)

9-72   Use the given data to find the standard error of estimate, s_e, using Formula 9-7.

x	2	5	7	10	15
y	20	15	14	12	10

(The equation of the regression line is $y = -0.717x + 19.789$.)

9-73 Refer to the data given in Exercise 9-69 and assume that the necessary conditions of normality and variance are met.
(a) For $x = 4$, find y', the predicted value of y.
(b) Find s_e. How does this value of s_e affect the construction of the 95% interval estimate of y for $x = 4$?

9-74 Refer to Exercise 9-70 and assume that the necessary conditions of normality and variance are met.
(a) For $x = 4$, find y', the predicted value of y.
(b) Find the 99% interval estimate of y for $x = 4$.

9-75 Refer to the data given in Exercise 9-71, and assume that the necessary conditions of normality and variance are met.
(a) For $x = 6$, find y', the predicted value of y.
(b) Find the 95% interval estimate of y for $x = 6$.

9-76 Refer to the data given in Exercise 9-72, and assume that the necessary conditions of normality and variance are met.
(a) For $x = 1$, find y', the predicted value of y.
(b) Find the 99% interval estimate of y for $x = 1$.

9-4 Exercises B
Variation

9-77 Given the pairs of data below, find each value.
(a) Total variation.
(b) Explained variation.
(c) Use the results of parts (a) and (b) to find the coefficient of determination r^2.

x	2	2	3	4	6
y	5	7	7	9	13

(d) Suppose we have a different collection of paired data for which $r^2 = 0.900$ and the regression equation is $y = -2x + 3$. Find the linear correlation coefficient.

9-78 Assuming that a collection of paired data includes at least three pairs of values, what do you know about the linear correlation coefficient if $s_e = 0$?

9-79 If a collection of paired data is such that the total explained variation is zero, what can be deduced about the slope of the regression line?

9-80 Using the paired data given in Table 9-6, suppose we use each pair of data *twice* instead of once. Is the equation of the regression line affected? Is the value of the standard error of estimate affected? If so, how?

Computer Project
Correlation and
Regression

(a) Develop a computer program that will take a set of paired data as input. Output should consist of the value of the linear correlation coefficient r and the values of the slope (m) and y-intercept (b) for the equation of the regression line.

(b) Use existing software to obtain the linear correlation coefficient and the equation of the regression line for a set of paired data. Use the paired data given in Exercise 9-9.

Review

In this chapter we studied the concepts of **linear correlation** and **regression** so that we could analyze paired sample data. We limited our discussion to linear relationships because consideration of nonlinear relationships requires more advanced mathematics. With correlation, we attempted to decide whether there is a significant linear relationship between the two variables. With regression, we attempted to specify what that relationship is. While a scatter diagram provides a graphic display of the paired data, the linear correlation coefficient r and the equation of the regression line serve as more precise and objective tools for analysis.

Given a list of paired data, we can compute the linear correlation coefficient r by using Formula 9-1. Table A-7 gives critical r values that we compare to the computed r values to decide whether there is a significant linear relationship. The presence of a significant linear correlation does not necessarily mean that there is a direct cause-and-effect relationship between the two variables.

In Section 9-3 we developed procedures for obtaining the equation of the regression line which, by the least-squares criterion, is the straight line that best fits the paired data. When there is a significant linear correlation, the regression line can be used to predict the value of one variable when given some value of the other variable. The regression line has the form $y = mx + b$ where the constants m and b can be found by using the formulas given in this section.

In Section 9-4 we introduced the concept of **total variation,** with components of explained and unexplained variation. We defined the coefficient of determination r^2 to be the quotient of explained variation by total variation. We saw that we could measure the amount of spread of the sample points about the regression line by the standard error of estimate, s_e.

IMPORTANT FORMULAS

$$r = \frac{n\Sigma xy - (\Sigma x)(\Sigma y)}{\sqrt{n(\Sigma x^2) - (\Sigma x)^2}\sqrt{n(\Sigma y^2) - (\Sigma y)^2}}$$

$$m = \frac{n\Sigma xy - (\Sigma x)(\Sigma y)}{n(\Sigma x^2) - (\Sigma x)^2}$$

$$b = \frac{(\Sigma y)(\Sigma x^2) - (\Sigma x)(\Sigma xy)}{n(\Sigma x^2) - (\Sigma x)^2} \quad \text{or} \quad b = \bar{y} - m\bar{x}$$

$$y = mx + b$$

$$r = \frac{ms_x}{s_y}$$

$$r^2 = \frac{\text{explained variation}}{\text{total variation}}$$

$$s_e = \sqrt{\frac{\Sigma(y - y')^2}{n - 2}} \quad \text{or} \quad \sqrt{\frac{\Sigma y^2 - b\Sigma y - m\Sigma xy}{n - 2}}$$

$$y' - E < y < y' + E$$

$$\text{where } E = s_e \cdot t(\alpha/2)\sqrt{1 + \frac{1}{n} + \frac{n(x_0 - \bar{x})^2}{n(\Sigma x^2) - (\Sigma x)^2}}$$

Review Exercises

9-81 In each of the following, determine whether correlation or regression analysis is more appropriate.
(a) Is the value of a car related to its age?
(b) What is the relationship between the age of a car and annual repair costs?
(c) How are Celsius and Fahrenheit temperatures related?
(d) Is the age of a car related to annual repair costs?
(e) Is there a relationship between cigarette smoking and lung cancer?

9-82 (a) What should you conclude if 40 pairs of data produce a linear correlation coefficient of $r = -0.508$?
(b) What should you conclude if ten pairs of data produce a linear correlation coefficient of $r = 0.608$?

In Exercises 9-83 through 9-90, use the given paired data.
(a) Find the value of the linear correlation coefficient r.
(b) Assuming a 5% level of significance, find the critical value of r from Table A-7.
(c) Use the results from parts (a) and (b) to decide whether there is a significant linear correlation.
(d) Find the equation of the regression line.
(e) Plot the regression line on the scatter diagram.

9-83   The following table lists midterm and final exam grades for randomly selected students in a statistics course.

Midterm	82	65	93	70	80
Final	94	77	94	79	91

9-84   The following table lists car rental costs (in dollars) with the corresponding numbers of miles driven.

Cost	40	55	45	80
Mileage	40	100	60	200

9-85   A psychologist suspects that there is a relationship between the scores on a test for motivation and the scores on a personality test. The following sample data is collected.

Motivation test	49	36	62	50	51	42
Personality test	16	14	18	20	17	12

9-86   The table below lists the values (in billions of dollars) of U.S. exports and incomes on foreign investments for various sample years.

Exports	16	20	27	39	56	63
Incomes on foreign investments	2	3	4	7	11	11

9-87   The owner of a bookstore investigates the relationship between outside pedestrian traffic (per hour) and the number of books sold (per hour). The following sample data are obtained.

Traffic	21	37	12	15	21	32	48	25
Books	35	72	25	30	35	63	85	45

9-88   The given paired data lists the heights (in inches) of various males who were measured on their eighth and sixteenth birthdays.

Height (8 yrs)	52	50	48	51	51	48
Height (16 yrs)	69	66	64	67	69	65

9-89   A test designer develops two separate tests, which are intended to measure a person's level of creative thinking. Randomly selected subjects were given both versions and their results follow.

Test A	85	97	100	76	80	116	120	105
Test B	92	109	100	74	85	118	125	90

9-90 A pill designed to lower systolic blood pressure is administered to 10 randomly selected volunteers. The results follow.

Before pill	120	136	160	98	115	110	180	190	138	128
After pill	118	122	143	105	98	98	180	175	105	112

Vocabulary List

Define and give an example of each term.

bivariate data
scatter diagram
correlation
linear correlation
   coefficient
centroid
Pearson's product
   moment
regression line

slope
y-intercept
least-squares
   property
predicted value
total deviation
explained deviation
unexplained
   deviation

explained variation
unexplained
   variation
total variation
coefficient of
   determination
standard error of
   estimate

Chapter 9:
Case Study Activity

Conduct a study to determine whether there is a correlation between two variables. Collect sample paired data consisting of at least 15 different pairs. Identify the population and the method used to obtain the data, and identify any factors suggesting that the sample data are not representative. Calculate the linear correlation coefficient and determine the equation of the regression line. Graph the regression line on the scatter diagram. Assuming a 0.05 level of significance, find the critical value and form a conclusion about the correlation between the two variables.

10-1 **Overview**
We identify chapter **objectives.** This chapter introduces some methods for dealing with data coming from more than two sample groups.

10-2 **Multinomial Experiments**
This section presents a procedure for testing hypotheses made about more than two population proportions. We define and consider **multinomial experiments.**

10-3 **Contingency Tables**
We test hypotheses that the row and column variables of **contingency tables** are independent.

10-4 **Analysis of Variance**
In testing the claim that several population means are equal, we use **analysis of variance** techniques to examine the significance of two estimates of the population variance.

Chi-Square and Analysis of Variance

Because of the different ways of arranging sample data in this chapter, we will find that the techniques used to analyze them may vary dramatically. In Section 10-2 we will consider a typical application involving the distribution of absences on the different days of the week. If a sample of employee work records shows that the Monday through Friday absence totals are 31, 10, 6, 24, and 29, respectively, do those sample results differ significantly from the occurrence of an equal number of absences on the five days? We will present a solution to this problem in Section 10-2.

In Section 10-3 we will consider an application involving 75 randomly selected disabled adults who are classified by sex (male, female) and by the effects of their disabilities on employment (unemployable, limited employment, employable with only minor effects). Based on the given sample results, does it appear that among disabled adults, sex and employability are independent of each other? A solution will be presented in Section 10-3.

In Section 10-4, we will test for equality of means among three or more sample groups. In particular, we will consider the following problem. Three different brands of rechargeable batteries are used to power aircraft radios. For each brand, sample data is obtained as a list of operating times (in hours) before recharging is necessary. Do the sample data suggest that the three brands have significantly different means? We will solve this problem in Section 10-4.

10-1 Overview

Prior to Chapter 8 we considered statistical analyses involving only one sample. In Chapter 8 we introduced methods for comparing the statistics of two different samples, and in Chapter 9 we introduced ways of analyzing the relationship between two variables. In this chapter we develop methods for comparing more than two samples. We also consider ways of dealing with situations in which we draw a single sample from a single population, but partition the sample into several categories.

In Section 10-2 we begin by considering a method for testing a hypothesis made about several population proportions. Instead of working with the sample proportions, we deal directly with the frequencies with which the events occur. Our objective is to test for the significance of the differences between observed frequencies and the frequencies we would expect in a theoretically ideal experiment. We introduce the test statistic that measures the differences between observed frequencies and expected frequencies. In repeated large samplings, the test statistic can be approximated by the chi-square distribution discussed in Section 6-6.

In Section 10-3 we analyze tables of frequencies called **contingency tables.** In these tables the rows represent categories of one variable, while the columns represent categories of another variable. We test the hypothesis that the two classification variables are independent. We again use the chi-square distribution to determine whether there is a significant difference between the observed sample frequencies and the frequencies we would expect in a theoretically ideal experiment involving independent variables.

Section 10-4 introduces a technique called **analysis of variance** to test the claim that several population means are equal. In Section 8-3 we tested the hypothesis $\mu_1 = \mu_2$, but in Section 10-4 we test claims such as $\mu_1 = \mu_2 = \mu_3$. When dealing with more than two samples, we use a fundamentally different approach that incorporates the F distribution instead of the normal or student t distributions.

10-2 Multinomial Experiments

In Chapter 4 we introduced the binomial probability distribution and indicated that each trial must have all outcomes classified into exactly one of two categories. This feature of two categories is reflected in the prefix *bi*, which begins the term *binomial*. In this section we consider **muitinomiai experiments** that require each trial to yield outcomes belonging to one of several categories. Except for this difference, binomial and multinomial experiments are essentially the same.

DEFINITION

A **multinomial experiment** is one in which:

1. There is a fixed number of trials.
2. The trials are independent.
3. Each trial must have all outcomes classified into exactly one of several categories.
4. The probabilities remain constant for each trial.

We have already discussed methods for testing hypotheses made about one population proportion (Section 6-5) and two population proportions (Section 8-4). However, there is often a real need to deal with more than two proportions. Suppose, for example, that an employer randomly selects 100 absence reports and obtains the sample data in Table 10-1.

Table 10-1					
Day of week	Mon	Tues	Wed	Thurs	Fri
Number of absences	31	10	6	24	29

If absences occurred with equal frequencies on different days, Table 10-1 would look like Table 10-2.

Table 10-2					
Day of week	Mon	Tues	Wed	Thurs	Fri
Number of absences	20	20	20	20	20

We know that samples deviate naturally from ideal expectations because of chance fluctuations, so we now pose the key question: Are the differences between the real data of Table 10-1 and the theoretically ideal data of Table 10-2 attributable to chance, or are the differences significant? To answer this question we need some way of measuring the significance of the differences between the observed values and the theoretical values. **The expected frequency of an outcome is the product of the probability of that outcome and the total number of trials.** If there are five possible outcomes that are supposed to be equally likely, the probability of each outcome is $\frac{1}{5}$, or 0.2. For 100 trials and five equally likely outcomes, the expected frequency of each outcome is $0.2 \times 100 = 20$.

In testing for the differences among the five sample proportions

Did Mendel Fudge His Data?

R. A. Fisher statistically analyzed the claimed results of Mendel's experiments in hybridization. Fisher noted that the data were unusually close to theoretically expected outcomes. He says that "the data have evidently been sophisticated systematically, and after examining various possibilities, I have no doubt that Mendel was deceived by a gardening assistant, who knew only too well what his principal expected from each trial made." Fisher used chi-square tests and concluded that there is only about a 0.00004 probability of such close agreement between expected observations and reported observations.

(0.31, 0.10, 0.06, 0.24, 0.29), one approach might be to compare them two at a time by using the methods of Section 8-4. That approach would, however, be very inefficient and would lead to severe problems related to the level of significance. There are 10 different combinations of two samples, and if each individual hypothesis test has a 0.95 probability of not leading to a type I error (rejecting a true null hypothesis) and if the tests were independent, then the probability of no type I errors among the ten tests is only 0.95^{10}, or about 0.599. Instead of pairing off samples and conducting ten separate tests, we develop one comprehensive test, which is based on a statistic that measures the differences between observed values and the corresponding values that we expect in an ideal case. This statistic will incorporate the observed and expected frequencies instead of the proportions. We introduce the following notation.

NOTATION

O represents the **observed frequency** of an outcome.
E represents the theoretical or **expected frequency** of an outcome.

From Table 10-1 we see that the O values are 31, 10, 6, 24, and 29. From Table 10-2 we see that the corresponding values of E are 20, 20, 20, 20, and 20. For the given sample data we have $\Sigma O = \Sigma E = 100$ and, in general, $\Sigma O = \Sigma E = n$. The method generally used in testing for agreement between O and E values is based on the following test statistic.

TEST STATISTIC

$$\chi^2 = \sum \frac{(O - E)^2}{E}$$

Simply summing the differences between observed and expected frequencies would not lead to a good measure, since that sum is always zero.

$$\Sigma(O - E) = \Sigma O - \Sigma E = n - n = 0$$

Squaring the $O - E$ values provides a better statistic, which does reflect the differences between observed and expected frequencies, but $\Sigma(O - E)^2$ grows larger as the sample size increases. We average out that sum of squares through division by the expected frequencies. For the data of our absentee analysis, the Monday values of Table 10-1 and Table 10-2 indicate that 31 is the observed frequency, while 20 is the expected frequency;

$$\frac{(O-E)^2}{E} = \frac{(31-20)^2}{20} = \frac{11^2}{20} = \frac{121}{20} = 6.05$$

Proceeding in a similar manner with the remaining data, we get

$$\chi^2 = \sum \frac{(O-E)^2}{E}$$

$$= \frac{(31-20)^2}{20} + \frac{(10-20)^2}{20} + \frac{(6-20)^2}{20} + \frac{(24-20)^2}{20} + \frac{(29-20)^2}{20}$$

$$= \frac{121}{20} + \frac{100}{20} + \frac{196}{20} + \frac{16}{20} + \frac{81}{20}$$

$$= \frac{514}{20} = 25.7$$

The theoretical distribution of χ^2 is a discrete distribution, since there are a limited number of possible χ^2 values. However, extensive studies have established that, in repeated large samplings, the distribution of χ^2 can be approximated by a chi-square distribution. **This approximation is generally considered to be acceptable, provided that all values of E are at least 5.** In Section 5-5 we saw that the continuous normal probability distribution can reasonably approximate the discrete binomial probability distribution, provided that np and nq are both at least 5. We now see that the continuous chi-square distribution can reasonably approximate the discrete distribution of χ^2 provided that all values of E are at least 5. There are ways of circumventing the problem of an expected frequency that is less than 5, and one procedure requires us to combine categories so that all expected frequencies are at least 5.

When we use the chi-square distribution as an approximation to the distribution of χ^2 values, we obtain the critical test value from Table A-5 after determining the level of significance α and the number of degrees of freedom. In a multinomial experiment with k possible outcomes, the number of degrees of freedom is $k - 1$.

> **degrees of freedom = $k - 1$**

This reflects the fact that, for n trials, the frequencies of $k - 1$ outcomes can be freely varied, but the frequency of the last outcome is determined. Our 100 absences are distributed among five categories or cells, but we can freely vary the frequencies of only four cells, since the last cell would be 100 minus the total of the first four cell frequencies. In this case, we say that the number of degrees of freedom is 4.

Note that close agreement between observed and expected values will lead to a small value of χ^2. A large value of χ^2 will indicate strong

Of Course, Your Mileage May Vary

Two car advertisements were placed only a few pages apart in one issue of a national news magazine. The first ad listed the U.S. Environmental Protection Agency fuel consumption rates for all the major cars tested by a "simulated average trip under city driving conditions." For that year, the Oldsmobile Cutlass was listed as 10.3 miles per gallon. The second ad had the headline: "Oldsmobile Cutlass— 17.6 Miles Per Gallon." In smaller print, the second ad stated that "recent proving-ground tests show 17.6 mpg average at 55 mph." In yet smaller print we are told that "three Cutlass models were driven at a steady 55 mph on a level, straight road at the G.M. Desert Proving Grounds." Broken in and warmed up cars with tuned engines got "an average of 17.6 mpg at a constant speed of 55 mph; 12.8 mpg in Suburban-City tests."

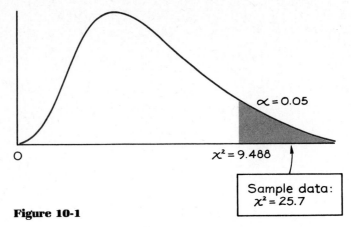

Figure 10-1

disagreement between observed and expected values. A significantly large value of χ^2 will therefore cause rejection of the null hypothesis of no difference between observed and expected frequencies. Our test is therefore right-tailed since the critical value and critical region are located at the extreme right of the distribution.

For the sample absentee data, we have determined the value of the test statistic ($\chi^2 = 25.7$) and the number of degrees of freedom (4). Let's assume a 5% level of significance so that $\alpha = 0.05$. With 4 degrees of freedom and $\alpha = 0.05$, Table A-5 indicates a critical value of 9.488 (see Figure 10-1).

Figure 10-1 indicates that our test statistic of 25.7 falls within the critical region, so we reject the null hypothesis that absences occur on the different weekdays with equal frequencies. From the sample data, it appears that Monday and Friday are the days with lower attendance.

The first example in this section dealt with the implied null hypothesis that the frequencies of absences on the five working days were all equal. However, the theory and methods we present here can also be used in cases where the claimed frequencies are different. The next example illustrates this.

EXAMPLE

A car manufacturer claims that the best selling model has the following color preference rates: 30% of the buyers prefer red, 10% prefer white, 15% prefer green, 25% prefer blue, 5% prefer brown, and 15% prefer yellow. A random survey of 200 buyers produces the following results:

Color	Red	White	Green	Blue	Brown	Yellow
Number prefer	64	14	38	49	6	29

At the $\alpha = 0.01$ significance level, test the claim that the percentages given by the manufacturer are correct.

Solution

The null hypothesis H_0 is the claim that the percentages given are correct. We can state this as follows.

H_0: $p_r = 0.30$ and $p_w = 0.10$ and $p_g = 0.15$ and $p_{bl} = 0.25$ and $p_{br} = 0.05$ and $p_y = 0.15$

H_1: At least one of the above probabilities is incorrect.

If the manufacturer's claim is exactly correct, then the 200 preferences would have occurred with these frequencies:

Color	Red	White	Green	Blue	Brown	Yellow
Number prefer	60	20	30	50	10	30

The first table lists observed values, while this one lists expected values. We now compute χ^2 as a measure of the disagreement between observed and expected values.

$$\chi^2 = \sum \frac{(O - E)^2}{E}$$

$$= \frac{(64 - 60)^2}{60} + \frac{(14 - 20)^2}{20} + \frac{(38 - 30)^2}{30}$$

$$+ \frac{(49 - 50)^2}{50} + \frac{(6 - 10)^2}{10} + \frac{(29 - 30)^2}{30}$$

$$= 5.853$$

This is a right-tailed test with $\alpha = 0.01$ and $6 - 1$, or 5, degrees of freedom, so the critical value from Table A-5 is 15.086. Since the test statistic of $\chi^2 = 5.853$ does not fall within the critical region bounded by 15.086, we fail to reject the null hypothesis and we attribute the observed deviations from the expected values to chance fluctuations. The available sample data do not warrant rejection of the manufacturer's claim.

The techniques of this section can be used to test for how well an observed frequency distribution conforms to some theoretical frequency distribution. For this reason, these methods are sometimes referred to as **goodness-of-fit** tests. For the employee absence data considered in this section, we used a goodness-of-fit test to decide whether the observed absences conformed to a uniform distribution, and we found that the differences were significant. It appears that the observed frequencies do not make a good fit with a uniform distribution. Since many statistical analyses require a normally distributed population, we may use the chi-square test of this section to determine that given samples are drawn from normally distributed populations. (See Exercise 10-18.)

10-2 Exercises A
Multinomial Experiments

10-1   The following table is obtained from a random sample of 100 absences. At the $\alpha = 0.01$ significance level, test the claim that absences occur on the five days with equal frequency.

Day	Mon	Tues	Wed	Thurs	Fri
Number absent	27	19	22	20	12

(a) Find the χ^2 value based on the sample data.
(b) Find the critical value of χ^2.
(c) What conclusion can you draw?

10-2   In an experiment on perception, 50 subjects are asked to select the most pleasant of five different photographs. The results are as follows. At the $\alpha = 0.05$ significance level, test the hypothesis that the photographs are equally pleasant.

Photograph	A	B	C	D	E
Number selected	0	9	19	10	12

10-3   A biochemist conducts an experiment designed to reveal the most effective aspirin. The eight best selling brands of aspirin are tested on 120 random subjects and, for each subject, a best aspirin is selected according to certain technical criteria. The results are as follows. At the $\alpha = 0.05$ significance level, test the claim that the eight brands of aspirin are equally effective.

Brand	A	B	C	D	E	F	G	H
Number of times selected as best	10	14	16	21	17	14	15	13

10-4   A genetics experiment involves 320 mice and is designed to determine whether Mendelian principles hold for a certain list of characteristics. The following table summarizes the actual experimental results and the expected Mendelian results for the five characteristics being considered. At the $\alpha = 0.01$ significance level, test the claim that Mendelian principles hold.

Characteristic	A	B	C	D	E
Observed frequency	30	15	58	83	134
Expected (Mendelian) frequency	20	20	40	120	120

10-5   A psychology course has ten sections with enrollments listed in the following table. Test the claim that students enroll in the various sections with equal frequencies.

Section	1	2	3	4	5	6	7	8	9	10
Enrollment	22	14	12	25	25	20	17	15	21	19

10-6   In an experiment on extrasensory perception, subjects were asked to identify the month showing on a calendar in the next room. If the results are as shown, test the claim that months were selected with equal frequencies. Assume a significance level of 0.05.

Month	Jan.	Feb.	Mar.	Apr.	May	Jun.	Jul.	Aug.	Sept.	Oct.	Nov.	Dec.
Number selected	8	12	9	15	6	12	4	7	11	11	5	20

10-7   A die is loaded and the manufacturer claims that the outcome of 2 occurs in 30% of the rolls, while the other outcomes occur with the same frequency of 14%. Test the claim at the $\alpha = 0.05$ significance level if 600 sample rolls produce the following results.

Outcome	1	2	3	4	5	6
Frequency	102	161	100	79	65	93

10-8   A politician claims that the Republican, Democratic, and Independent mayoral candidates are favored by voters at the rates of 35%, 40%, and 25%, respectively. Test the claim at the 0.05 significance level if a random survey of 30 voters produces the following results.

Candidate	Republican	Democrat	Independent
Number selected	13	11	6

10-9   The manager of an ice cream shop claims that 30% of the customers prefer vanilla, 25% prefer chocolate, 10% prefer strawberry, 20% prefer butter pecan, and 15% prefer maple walnut. A random sample of 100 customers produces the following results. At the 0.05 significance level, test the claim that the manager's percentages are correct (see page 470).

Flavor	Vanilla	Chocolate	Strawberry	Butter pecan	Maple walnut
Number selected	24	24	8	21	23

10-10 A television company is told by a consulting firm that its eight leading shows are favored according to the percentages given in the following table. A separate and independent sample is obtained by another consulting firm. Do the figures agree? Assume a significance level of 0.05.

Show	A	B	C	D	E	F	G	H
First consultant	22%	18%	12%	12%	10%	9%	9%	8%
Second consultant (number of respondents favoring show)	29	30	20	16	9	17	10	19

10-11 In a certain county, 80% of the drivers have no accidents in a given year, 16% have one accident, and 4% have more than one accident. A survey of 200 randomly selected teachers from the county produced 172 with no accidents, 23 with one accident, and 5 with more than one accident. At the 5% level of significance, test the claim that the teachers exhibit the same accident rate as the countywide population.

10-12 A college dean expects 40% of the students to register on Monday, 45% on Tuesday, and 15% on Wednesday. Of 1850 students, 1073 register on Monday, 555 register on Tuesday, and 222 register on Wednesday. At the 1% level of significance, test the claim that the observed results are compatible with the dean's expectation.

10-13 A teacher divides a course into four units, expecting to spend an equal amount of time on each unit. Units I, II, III, and IV required 13, 15, 8, and 9 class hours, respectively. At the 5% level of significance, test the claim that the actual number of class hours agreed with the teacher's expectation.

10-14 For the past several years, the percentage of A's, B's, C's, D's, and F's for a certain statistics course have been 8%, 17%, 45%, 22%, and 8%, respectively. A new testing method was used in the last semester and there were 10 A's, 20 B's, 10 C's, 5 D's, and 5 F's. At the 5% level of significance, test the claim that the grades obtained through the new testing method are in the same proportion as before.

10-15 A pair of dice has 36 possible outcomes that are supposed to be equally likely. If we plan to test a certain pair of dice for fairness using the chi-square distribution, what is the minimum number of rolls necessary if each outcome is to have an expected frequency of at least 5? Suppose we conduct the minimum number of rolls and obtain a χ^2 value of 67.2. What do we conclude at the 0.01 significance level?

10-16 A roulette wheel has 38 possible outcomes that are supposed to be equally likely. If we plan to test a roulette wheel for fairness using the chi-square distribution, what is the minimum number of trials we must make if we are to satisfy the requirement that each expected frequency must be at least 5? Suppose we conduct the minimum number of trials and compute χ^2 to be 49.6. What do we conclude at the 0.01 significance level?

10-2 Exercises B
Multinomial Experiments

10-17 In testing for agreement between observed and expected frequencies in the following table, we cannot use the chi-square distribution since all expected values are not at least 5. However, we can combine some columns so that all expected values do equal or exceed 5. Use this suggestion to test the claim that the observed and expected frequencies are compatible.

Characteristic	A	B	C	D	E	F	G	H	I	J
Observed frequency	2	8	8	9	3	5	3	0	12	3
Expected frequency	4	5	8	7	4	6	5	2	9	3

10-18 An observed frequency distribution of I.Q. scores is given below.
(a) Assuming a normal distribution with $\mu = 100$ and $\sigma = 15$, use the methods of Chapter 5 to find the probability of a randomly selected subject belonging to each class.
(b) Using the probabilities found in part (a), find the expected frequency for each category.
(c) Use a 0.01 level of significance to test the claim that the I.Q. scores were randomly selected from a normally distributed population with $\mu = 100$ and $\sigma = 15$.

I.Q. score	Under 80	80–95	96–110	111–120	above 120
Frequency	20	20	80	40	40

10-19 An observed frequency distribution is given on the top of page 472.
(a) Assuming a binomial distribution with $n = 3$ and $p = 1/3$, use the binomial probability formula to find the probability corresponding to each category of the table.
(b) Using the probabilities found in part (a), find the expected frequency for each category.
(c) Use a 0.05 level of significance to test the claim that the observed frequencies fit a binomial distribution for which $n = 3$ and $p = 1/3$.

Number of successes	0	1	2	3
Frequency	89	133	52	26

10-20   In executing a chi-square test in this section, suppose we multiply each observed frequency by the same positive integer that is greater than 1. How is the critical value affected? How is the test statistic affected?

10-3 Contingency Tables

Suppose we collect the sample data relating to the employability of disabled adults and summarize them in Table 10-3. Note that the entries in the table are *frequencies*.

Table 10-3

Effect of a disability on employment.

	Unemployable	Limited employment	Employable with only minor effects
Males	8	11	16
Females	18	10	12

This table indicates that 75 disabled subjects were classified according to their ability to work. Tables similar to this one are generally called **contingency tables,** or **two-way tables.** In this context, the word *contingency* refers to dependence, and the contingency table serves as a useful medium for analyzing the dependence of one variable on another. This is only a statistical dependence that cannot be used to establish an inherent cause-and-effect relationship.

We need to test the *null hypothesis that the two variables in question are independent.* That is, we will test the claim that, among disabled adults, sex and employability are independent. Let's select a significance level of $\alpha = 0.05$. We can now compute a test statistic based on the data and then compare that test statistic to the appropriate critical test value. As in the previous section, we use the chi-square distribution where the test statistic is given by χ^2.

TEST STATISTIC

$$\chi^2 = \sum \frac{(O - E)^2}{E}$$

This test statistic allows us to measure the degree of disagreement between the frequencies actually observed and those that we would theoretically expect when the two variables are independent. The reasons underlying the development of the χ^2 statistic in the previous section also apply here. In repeated large samplings, **the distribution of the test statistic χ^2 can be approximated by the chi-square distribution provided that all expected frequencies are at least 5.**

In the preceding section we knew the corresponding probabilities and could easily determine the expected values, but the typical contingency table does not come with the relevant probabilities. Consequently, we need to devise a method for obtaining the corresponding expected values. We will first describe the procedure for finding the values of the expected frequencies, and we will then proceed to justify that procedure. For each cell in the frequency table, the expected frequency E can be calculated by using the following.

EXPECTED FREQUENCY

$$\text{expected frequency } E = \frac{(\text{row total}) \cdot (\text{column total})}{(\text{grand total})}$$

where *grand total* refers to the total number of observations in the table. For example, in the lower right cell of Table 10-3, we see the observed frequency of 12. The total of the frequencies for that row is 40, the total of the column frequencies is 28, and the total of all frequencies in the table is 75, so we get an expected frequency of

$$E = \frac{40 \cdot 28}{75} = 14.93$$

In the lower right cell, the *observed* frequency is 12 while the *expected* frequency is 14.93. Table 10-4 reproduces Table 10-3, with the expected frequencies inserted in parentheses. As in Section 10-2, we require that all expected frequencies be at least 5 before we can conclude that the chi-square distribution serves as a suitable approximation to the distribution of χ^2 values.

Table 10-4	Unemployable	Limited employment	Employable with only minor effects	
Males	8 (12.13)	11 (9.8)	16 (13.07)	35
Females	18 (13.87)	10 (11.2)	12 (14.93)	40
	26	21	28	(Grand total: 75)

Survey Solicits Contributions

The American Institute for Cancer Research distributed a survey on diet and breast cancer. The stated purpose of the survey was to develop a "statistical profile of the eating habits of American women—and to help find the link between diet and breast cancer." However, the survey ended with an option for enclosing a gift ranging from $5 to $500. Two fundamental questions arise. What was the real purpose of the survey? Does the contribution request affect the validity of the results they receive?

To better understand the rationale for this procedure, let's pretend that we know only the row and column totals and that we must fill in the cell frequencies by assuming that there is no relationship between the two variables involved (see Table 10-5).

Table 10-5

	Unemployable	Limited employment	Employable with only minor effects	
Males				35
Females				40
	26	21	28	(Grand total: 75)

We begin with the empty cell in the upper left-hand corner that corresponds to unemployable males. Since 35 of the 75 subjects are males, if one of the subjects is randomly selected, $P(\text{male}) = \frac{35}{75}$. Similarly, 26 of the 75 subjects are unemployable, so if one subject is randomly selected, $P(\text{unemployable}) = \frac{26}{75}$. Since we assume that sex and employability are independent, we conclude that $P(\text{male and unemployable}) = P(\text{male}) \times P(\text{unemployable}) = \frac{35}{75} \times \frac{26}{75}$. This follows from a basic principle of probability whereby $P(A \text{ and } B) = P(A) \times P(B)$ if A and B are independent events. To obtain the expected value of the upper left cell, we simply multiply the probability for that cell by the total number of subjects available to get

$$\frac{35}{75} \times \frac{26}{75} \times 75 = 12.13$$

The form of this product suggests a general way to obtain the expected frequency of a cell:

$$\text{expected frequency } E = \frac{(\text{row total})}{(\text{grand total})} \cdot \frac{(\text{column total})}{(\text{grand total})} \cdot (\text{grand total})$$

This expression can be simplified to

$$E = \frac{(\text{row total}) \cdot (\text{column total})}{(\text{grand total})}$$

as given above.

Using the observed and expected frequencies shown in Table 10-5, we can now compute the χ^2 test statistic based on the sample data.

$$\chi^2 = \sum \frac{(O - E)^2}{E}$$

$$\chi^2 = \frac{(8 - 12.13)^2}{12.13} + \frac{(11 - 9.8)^2}{9.8} + \frac{(16 - 13.07)^2}{13.07}$$
$$+ \frac{(18 - 13.87)^2}{13.87} + \frac{(10 - 11.2)^2}{11.2} + \frac{(12 - 14.93)^2}{14.93}$$
$$= 1.406 + 0.147 + 0.657 + 1.230 + 0.129 + 0.575$$
$$= 4.144$$

With $\alpha = 0.05$ we proceed to Table A-5 to obtain the critical value of χ^2, but we must first know where the critical region lies and the number of degrees of freedom. **Tests of independence with contingency tables involve only right-tailed critical** regions since small values of χ^2 support the claimed independence of the two variables. That is, χ^2 is small if observed and expected frequencies are close. Large values of χ^2 are to the right of the chi-square distribution, and they reflect significant differences between observed and expected frequencies.

In a contingency table with r rows and c columns, the number of degrees of freedom is given by

$$\text{degrees of freedom} = (r - 1)(c - 1)$$

Thus Table 10-3 has $(2 - 1) \cdot (3 - 1)$, or 2, degrees of freedom. In a right-tailed test with $\alpha = 0.05$ and with 2 degrees of freedom we refer to Table A-5 to get a critical χ^2 value of 5.991. Since the calculated χ^2 value of 4.144 does not fall in the critical region bounded by $\chi^2 = 5.991$, we fail to reject the null hypothesis of independence between the two variables. See Figure 10.2. Sex and employability appear to be independent. That is, sex seems to have no effect on the impact of a disability on employment.

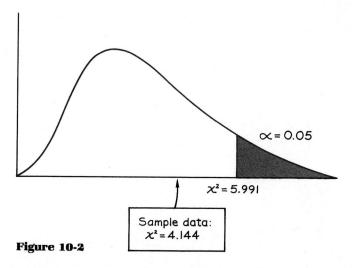

Figure 10-2

High Hopes

In his book *The Height of Your Life,* author Ralph Keyes observes that taller people tend to have higher incomes. He notes, for example, that a study of several thousand people showed that those taller than 6 feet earned about 8% more than those shorter than $5\frac{1}{2}$ feet. Keyes also refers to studies of presidential races in which voters show a distinct preference for taller candidates.

In the next example, we test for independence between the categories of scientists and their views on nuclear energy.

EXAMPLE

After being polled about their views on how to proceed with the development of nuclear energy, the responses of various randomly selected scientists were summarized in Table 10-6 below. At the 0.05 level of significance, test the claim that the categories of scientists are independent of their views.

Table 10-6			
Scientist	Proceed rapidly	Proceed slowly	Stop development
General	160	94	41
Energy expert	108	37	10
Nuclear expert	47	9	0

Solution

H_0: A scientist's view on nuclear energy is independent of his or her scientific field.

H_1: A scientist's view and field are dependent.

H_0 will be rejected only if the sample data exhibit a significant difference between the observed frequencies of Table 10-6 and the frequencies we would expect with two independent variables. In order to evaluate the chi-square test statistic of $\Sigma(O - E)^2/E$, we must first determine the expected frequency E for each of the nine cells in the table. In Table 10-7 we show the row and column totals used to calculate the expected values shown in parentheses. Each expected frequency is calculated using

$$E = \frac{(\text{row total}) \cdot (\text{column total})}{(\text{grand total})}$$

Note that each cell does have an expected frequency of at least 5 so that the chi-square distribution can be used. The zero in Table 10-6 does not violate this "rule of 5" because it is an *observed* frequency, not an *expected* frequency.

Having found the expected frequencies shown in Table 10-7, we can now compute the test statistic.

Table 10-7			
Scientist	Proceed rapidly	Proceed slowly	Stop development
General	160 (183.6)	94 (81.6)	41 (29.7)	295
Energy expert	108 (96.5)	37 (42.9)	10 (15.6)	155
Nuclear expert	47 (34.9)	9 (15.5)	0 (5.6)	56

315   140   51   (Grand total: 506)

$$\chi^2 = \sum \frac{(O - E)^2}{E}$$

$$= \frac{(160 - 183.6)^2}{183.6} + \frac{(94 - 81.6)^2}{81.6} + \ldots + \frac{(0 - 5.6)^2}{5.6}$$

$$= 25.930$$

We obtain the critical value of χ^2 by noting that $\alpha = 0.05$, the test is right-tailed, and

$$\text{degrees of freedom} = (r - 1)(c - 1) = (3 - 1)(3 - 1) = 4$$

Referring to Table A-5, we get a critical value of 9.488. The test statistic of $\chi^2 = 25.930$ does fall within the critical region bounded by the critical value of 9.488, so we reject the null hypothesis. It appears that the categories of scientists and their views are *not* independent. Further examination of the data would reveal that 84% (47 ÷ 56) of the nuclear experts favor proceeding rapidly while only 54% (160 ÷ 295) of the general scientists have that view. Such percentages appear to be significantly different.

Because of the nature of the calculations, it is often helpful to use a statistics software package for tests such as those included in this section. The two screens below show the output when STATDISK is used with the data of the last example. The first screen shows the observed frequencies originally entered, along with the corresponding expected frequencies, which were determined by the computer. The second screen shows the test results. The discrepancy in the test statistic is due to the rounding error included in the above calculations. The discrepancy in the critical values can be attributed to the approximate nature of the calculations used in the program.

```
      A    [ 160 ]   [  94 ]   [  41 ]
           (183.6)   (81.6 )   (29.7 )

      B    [ 108 ]   [  37 ]   [  10 ]
           (96.5 )   (42.9 )   (15.6 )

      C    [  47 ]   [   9 ]   [   0 ]
           (34.9 )   (15.5 )   ( 5.6 )
```

```
   Test statistic .. Chi Square = 25.98784
   Critical value .. Chi Square = 9.45785
   P-value ......................... = 3.66864E-05
   Significance level .......... = .05
   Degrees of freedom .......... = 4
   CONCLUSION: REJECT the null hypothesis
 that row and column variables are independent.
```

10-3 Exercises A
Contingency Tables

10-21   A survey on a local gun-control proposal produced the data summarized in the following table. At the $\alpha = 0.05$ significance level, test the claim that voting on the bill is independent of the political party of the voter.
(a) Assume independence of party and voter opinion and determine the expected frequency for each of the four cells.
(b) Compute the χ^2 test statistic based on the sample data.
(c) Determine the critical value of χ^2.
(d) What conclusion can you draw?

	In favor	Opposed
Democrat	20	30
Republican	30	10

10-22   A presidential candidate employs a polling organization to find out whether his popularity differs between rural and urban residents. The results follow. At the $\alpha = 0.05$ significance level, test the claim of no significant difference between the responses of rural and urban residents.

	In favor	Opposed
Rural	5	45
Urban	15	55

10-23 A particular job in an assembly process involves considerable stress, and a study is done to determine whether there is any difference between the adjustment of men and women to the job. The results of a random survey follow. At the $\alpha = 0.01$ significance level, test the claim of no difference.

	Well adjusted	Not well adjusted
Males	5	20
Females	30	10

10-24 Two teaching methods are used to train bank employees. We want to determine whether there is any relationship between the method used and the quality of learning. The following table summarizes the sample data. At the $\alpha = 0.01$ level of significance, test the claim that the quality of learning and the teaching method are independent.

	Average	Superior
Method A	12	18
Method B	20	24

10-25 The claim has been made that males do better in mathematics courses than females. At the $\alpha = 0.01$ level of significance, use the sample data in the following table to test that claim.

	Math grades		
	Low	Average	Superior
Males	15	60	25
Females	10	35	5

10-26 Researchers studied 90 jurors to find the effect of sex on punitive attitudes. They administered psychological tests and classified each subject as having liberal, fair, or vindictive tendencies. The results follow. At the $\alpha = 0.05$ level of significance, test the claim that the presence of these tendencies is independent of the sex of the juror.

	Liberal	Fair	Vindictive
Males	5	18	7
Females	21	24	15

10-27   The manager of an assembly operation wants to determine whether the number of defective parts is dependent on the day of the week. She develops the following sample data. Test the claim of the union representative that the day of the week makes no difference in the number of defects.

	Mon	Tues	Wed	Thurs	Fri
Acceptable products	80	100	95	93	82
Defective products	15	5	5	7	12

10-28   A candidate for national office wants to know whether there are regional differences in his popularity and conducts a survey to find out. At the $\alpha = 0.05$ level of significance, test the claim that the candidate's popularity is independent of geographic region. The sample results are given in the following table.

	In favor	Against	No opinion
Northeast	10	20	5
South	12	36	8

10-29   The following table summarizes sample grades for two subjects. At the $\alpha = 0.05$ level of significance, test the claim that grade distribution is independent of the subject.

	Grades				
	A	B	C	D	F
Math	13	16	13	8	9
English	6	15	15	5	6

10-30   Use the data in the following table to test the claim that the number of cigarettes smoked per day is independent of the age of the smoker. Assume a significance level of $\alpha = 0.05$.

Age	Number of cigarettes smoked per day		
	1–20	21–40	More than 40
17–30	10	7	5
31–45	7	9	5
46–60	8	12	6

10-31 Data drawn from randomly selected students is summarized in the table below. At the 0.10 level of significance, test the claim that the subject and grade are independent.

Course	Grade				
	A	B	C	D	F
Math	22	24	28	10	16
Art	40	64	44	9	22
English	65	150	160	38	62

10-32 A sociologist conducts a study of attitudes and obtains results for the first question in a survey. Those results are listed in the table below. At the 0.01 level of significance, test the claim that age and response are independent.

Age	Response				
	Strongly agree	Agree	Undecided	Disagree	Strongly disagree
Under 21	20	27	10	15	15
21–30	40	52	30	20	10
31–40	8	8	12	16	15
Over 40	10	12	15	20	8

10-3 Exercises B
Contingency Tables

10-33   The only sample data available on attitudes about capital punishment are summarized as follows. We wish to test the independence of geographic region and opinion, but each cell does not have an expected value of at least 5. Combine rows in a reasonable way so that the expected value of each cell is at least 5, and then complete the test using a significance level of $\alpha = 0.05$.

	In favor	Opposed	No opinion
Northeast	4	8	3
Southeast	7	8	6
North Central	2	4	4
South Central	9	3	5
Northwest	12	20	4
Southwest	2	14	3

10-34   The chi-square distribution is continuous while the test statistic used in this section is actually discrete. Some statisticians use **Yates' correction for continuity** in cells with an expected frequency less than 10 or in all cells of a contingency table with two rows and two columns. With Yates' correction, we replace

$$\Sigma \frac{(O - E)^2}{E} \quad \text{with} \quad \Sigma \frac{(|O - E| - 0.5)^2}{E}$$

Given the contingency table below, find the value of the χ^2 test statistic with and without Yates' correction. In general, what effect does Yates' correction have on the value of the test statistic?

	X	Y
A	5	25
B	65	5

10-35   If each observed frequency in a contingency table is multiplied by a positive integer K (where $K \geq 2$), how is the value of the test statistic affected?

10-36 For the contingency table given below, verify that the test statistic becomes

$$\chi^2 = \frac{(a + b + c + d)\ (ad - bc)^2}{(a + b)(c + d)(b + d)(a + c)}$$

	Column 1	Column 2
Row 1	a	b
Row 2	c	d

10-4 Analysis of Variance

In Section 10-2 we developed a procedure for testing the hypothesis that the differences among several sample proportions are due to chance. In this section we develop a procedure for testing the hypothesis that differences among several sample **means** are due to chance. We now require that the populations under consideration have distributions that are essentially normal. Also, the variances of the populations must be equal, and the samples must be independent. The requirements of normality and equal variances are somewhat loose since the methods in this section work reasonably well unless there is a very nonnormal distribution or unless the population variances differ by very large amounts. (If the samples are independent but the distributions are not normal, we can use the Kruskal-Wallis test presented in Section 11-5.) The method we will describe in this section is called **analysis of variance** (often referred to as ANOVA) because it is based on a comparison of two different estimates of the variance that is common to the different populations.

Let's assume that we have k different populations and that a sample of size n is drawn from each population. From the sample data we compute the sample means $\bar{x}_1, \bar{x}_2, \ldots, \bar{x}_k$ and the sample standard deviations $s_1, s_2, \ldots, s_k$. Our method of testing the null hypothesis $\mu_1 = \mu_2 = \ldots = \mu_k$ requires that we obtain two different estimates of the common population variance σ^2 and then compare those estimates by using the F distribution. Let's consider a specific example.

EXAMPLE

A pilot does extensive bad weather flying and decides to buy a battery-powered radio as an independent backup for her regular radios, which depend on the airplane's electrical system. She has a choice of three brands of rechargeable batteries that vary in cost. She obtains the sample data in the following table. She randomly

Statistics and Baseball Strategy

We are all well aware of trite sports statistics such as baseball batting averages. But beyond such descriptive statistics lies another dimension. Statisticians are now using computers to develop very sophisticated measures of performance and strategy. They have found, for example, that sacrifice bunts, sacrifice fly balls, and stolen bases rarely help win games and that it is seldom wise to have a pitcher intentionally walk the batter. They can identify the ballparks that favor pitchers, as well as those that favor batters. Instead of simply comparing the batting averages of two different players, they can develop better measures of offensive strength by taking into account such relevant factors as pitchers faced, position in the lineup, ballparks played, and weather conditions.

selects five batteries for each brand, and tests them for the operating time (in hours) before recharging is necessary. Do the three brands have the same mean usable time before recharging is required?

Brand X	Brand Y	Brand Z
26.0	29.0	30.0
28.5	28.8	26.3
27.3	27.6	29.2
25.9	28.1	27.1
28.2	27.0	29.8
$\bar{x}_1 = 27.18$	$\bar{x}_2 = 28.10$	$\bar{x}_3 = 28.48$
$s_1^2 = 1.46$	$s_2^2 = 0.69$	$s_3^2 = 2.81$

Solution

$$H_0: \mu_x = \mu_y = \mu_z$$
$$H_1: \text{The preceding means are not all equal.}$$

Using these sample results we develop two separate estimates of σ^2, where σ^2 is assumed to be the population variance common to all three brands. We employ the F distribution in the comparison of the two separate estimates. In Section 8-2, we discussed the use of the F distribution in testing for equality of two variances. We saw that when two independent samples were drawn from populations that were approximately normally distributed, the sampling distribution of $F = s_1^2/s_2^2$ was the F distribution. Here we again use the F distribution, but we replace the two sample variances by two different estimates of the population variance common to all of the populations under consideration. Specifically, we use

TEST STATISTIC

For cases with equal sample sizes:

$$F = \frac{\text{variance between samples}}{\text{variance within samples}} = \frac{ns_{\bar{x}}^2}{\mu_{s_i^2}}$$

The variance *between* samples is obtained by finding the product of n and $s_{\bar{x}}^2$; n is the number of scores in each sample and $s_{\bar{x}}^2$ is the variance of the sample means. For the data in the table, $n = 5$ since each sample consists of 5 scores, and $s_{\bar{x}}^2 = 0.45$ is found by computing the variance of the sample means (27.18, 28.10, 28.48). Thus $ns_{\bar{x}}^2 = (5)(0.45) = 2.25$. The expression $ns_{\bar{x}}^2$ is justified as an estimate of σ^2 since $\sigma_{\bar{x}} = \sigma/\sqrt{n}$. Squaring both sides of this last expression and solving for σ^2, we get $\sigma^2 = n\sigma_{\bar{x}}^2$, which indicates that σ^2 can be estimated by $ns_{\bar{x}}^2$.

The variance *within* samples is obtained by computing $\mu_{s_i^2}$, which denotes the mean of the sample variances. Since we have three sets of samples, $\mu_{s_i^2}$ becomes

$$\frac{s_1^2 + s_2^2 + s_3^2}{3} = \frac{1.46 + 0.69 + 2.81}{3} = 1.65$$

This approach seems reasonable since we assume that equal population variances imply that representative samples yield sample variances which, when pooled, provide a good estimate of the value of the common population variance.

We can now proceed to include both estimates of σ^2 in the determination of the value of the F statistic based on the data.

$$F = \frac{\text{variance between samples}}{\text{variance within samples}} = \frac{ns_{\bar{x}}^2}{\mu_{s_i^2}} = \frac{2.25}{1.65} = 1.36$$

If the two estimates of variance are close, the calculated value of F will be close to 1 and we would conclude that there are no significant differences among the sample means. But if the value of F is excessively *large*, then we would reject the claim of equal means. The estimate of variance in the denominator ($\mu_{s_i^2}$) depends only on the sample variances and is not affected by differences among the sample means. However, if there are extreme differences among the sample means, the numerator will be larger so the value of F will be larger. In general, as the sample means move farther apart, the value of $ns_{\bar{x}}^2$ grows larger and the value of F itself grows larger. Since excessively large values of F reflect unequal means, the test is right tailed.

The critical value of F that separates excessive values from acceptable values is found in Table A-6, where α is again the level of significance and the numbers of degrees of freedom are as follows. Assuming that there are k sets of separate samples with n scores in each set,

For cases with equal sample sizes:

numerator degrees of freedom $= k - 1$
denominator degrees of freedom $= k(n - 1)$

Our battery example involves $k = 3$ separate sets of samples and there are $n = 5$ scores in each set, so

$$\text{numerator degrees of freedom} = k - 1 = 3 - 1 = 2$$
$$\text{denominator degrees of freedom} = k(n - 1) = 3(5 - 1) = 12$$

With $\alpha = 0.05$, the critical F value corresponding to these degrees of freedom is 3.8853. The computed test statistic of $F = 1.36$ does not fall within the right-tailed critical region bounded by $F = 3.8853$, so we fail to reject the null hypothesis of equal population means. That is, there are no significant differences among the means of the three given sets of samples. The pilot should purchase the least expensive battery.

94022

ZIP Codes Reveal Much

The Claritas Corporation has developed a way of obtaining much information about people from their ZIP codes. ZIP code data was extracted from a variety of mailing lists, purchase orders, warranty cards, census data, and market research surveys. With people of the same social and economic levels tending to live in the same areas, it became possible to match ZIP codes with such factors as purchase patterns, types of cars driven, foods preferred, leisure time activities, and television viewing choices. The company helps clients target their advertising efforts to regions that are most likely to accept particular products. They can identify, for example, a region where 2% of the national population buys 10% of the outdoor power tools sold nationwide. Such information is obviously valuable to advertisers and campaigning politicians.

The preceding analysis led to a right-tailed critical region, and every other similar situation will also involve a right-tailed critical region. Since the individual components of n, $s_{\bar{x}}^2$, and $\mu_{s_i^2}$ are all positive, F is always positive, and we know that values of F near 1 correspond to relatively close sample means. The only values of the F test statistic that are indicative of significant differences among the sample means are those F values that exceed 1 beyond the critical value obtained from Table A-6.

It may seem strange that we are testing equality of several means by analyzing only variances, but the sample means directly affect the value of the test statistic F. To illustrate this important point, we add 10 to each score listed under brand X, but we leave the brand Y and brand Z values unchanged. The revised sample statistics follow.

Brand X	Brand Y	Brand Z
$\bar{x}_1 = 37.18$	$\bar{x}_2 = 28.10$	$\bar{x}_3 = 28.48$
$s_1^2 = 1.46$	$s_2^2 = 0.69$	$s_3^2 = 2.81$

We again assume a significance level of $\alpha = 0.05$ and test the claim that the three brands have the same population mean. In this way, we can see the effect of increasing the brand X scores by 10. The F statistic based on the revised sample data follows.

$$F = \frac{\text{variance between samples}}{\text{variance within samples}} = \frac{ns_{\bar{x}}^2}{\mu_{s_i^2}} = \frac{(5)(26.38)}{(1.46 + 0.69 + 2.81)/3}$$
$$= \frac{131.90}{1.65} = 79.94$$

We get $n = 5$ since each brand yields 5 sample values. The value of $s_{\bar{x}}^2 = 26.38$ is obtained by calculating the variance of the sample means 37.18, 28.10, and 28.48. We find $\mu_{s_i^2}$ to be 1.65 by computing the mean of the three sample variances. With three samples of five scores each, we have $3 - 1$, or 2, degrees of freedom for the numerator and $3(5 - 1)$, or

12, degrees of freedom for the denominator. With $\alpha = 0.05$ we therefore obtain a critical F value of 3.8853. Since the computed F statistic of 79.94 far exceeds the critical F value of 3.8853, we reject the null hypothesis of equal population means. Further examination of the data reveals that Brand X appears to last significantly longer than either Brands Y or Z, which don't appear to be significantly different.

Before adding 10 to each score for brand X, we obtained a test statistic of $F = 1.36$, but the increased brand X values cause F to become 79.94. Note that, in both cases, the brand X variance is 1.46, so the difference in the F test statistic is attributable only to the change in $\bar{x}_1$. By changing $\bar{x}_1$ from 27.18 to 37.18 and retaining the same values of $\bar{x}_2$, $\bar{x}_3$, s_1^2, s_2^2, and s_3^2, we find that the F test statistic changes from 1.36 to 79.94. This illustrates that the F statistic is very sensitive to sample means even though it is obtained through two different estimates of the common population variance.

Up to this point, our discussion of analysis of variance has involved only examples with the same number of values in each sample. We will now proceed to consider the treatment of samples with unequal sizes, and we will introduce some of the common terminology and notation used in analysis of variance. We will again proceed by analyzing the significance of the difference between variation due to **treatment** (variance between samples) and variation due to **error** (variation within samples). When we refer to "treatments," we are actually referring to the different samples as somehow being the results of different factors that we are analyzing. An agriculture study might involve treatments consisting of different fertilizer mixtures. An athlete might analyze treatments consisting of different training techniques. In each case, we obtain sample data from the different treatments, and we proceed to analyze the results in order to decide whether the effects of the treatments are different.

We begin by presenting the notation and procedure, which will be illustrated in a solved example, and we will then present the underlying rationale. In the table that follows on the next page we identify the components used in our calculations.

EXAMPLE

Three different computer programming languages (A, B, and C) are used by different students to solve a problem; the times (in hours) required for the solution are listed below each language designation. At the 0.05 level of significance, test the claim that the mean times for all three languages are the same. (Note that the three sample means are 4.0, 5.7, and 3.8, respectively.)

k = number of different samples (or the number of columns of data)

n_i = number of values in the ith sample

c_i = total of the values in the ith sample

$N = \Sigma n_i$ = total number of values in all samples combined. (N has been used previously to represent the size of a population.)

$\Sigma x = \Sigma c_i$ = sum of the scores of all samples combined

Σx^2 = sum of the squares of all scores from all samples combined

$SS(\text{total}) = \Sigma x^2 - \dfrac{(\Sigma x)^2}{N}$

Total sum of squares. This expression is algebraically the same as $\Sigma(x - \bar{x})^2$

$SS(\text{treatment}) = \left(\Sigma \dfrac{c_i^2}{n_i}\right) - \dfrac{(\Sigma c_i)^2}{N}$

Sum of squares that represents variation among the different samples (sometimes called *explained variation*).

$SS(\text{error}) = SS(\text{total}) - SS(\text{treatment})$

Sum of squares that represents the variation within samples that is due to chance (sometimes called *unexplained variation*).

$df(\text{treatment}) = k - 1$

Degrees of freedom associated with the different treatments.

$df(\text{error}) = N - k$

Degrees of freedom associated with the errors within samples.

$MS(\text{treatment}) = \dfrac{SS(\text{treatment})}{df(\text{treatment})}$

Mean square or variance estimate explained by the different treatments.

$MS(\text{error}) = \dfrac{SS(\text{error})}{df(\text{error})}$

Mean square or variance estimate that is unexplained (due to chance).

$F = \dfrac{MS(\text{treatment})}{MS(\text{error})}$

Test statistic representing the ratio of two estimates of variance.

ANOVA table

Source of variation	Sum of squares SS	Degrees of freedom	Mean square MS	Test statistic
Treatments	$SS(\text{treatment})$	$k - 1$	$MS(\text{treatment}) = \dfrac{SS(\text{treatment})}{df(\text{treatment})}$	$F = \dfrac{MS(\text{treatment})}{MS(\text{error})}$
Error	$SS(\text{error})$	$N - k$	$MS(\text{error}) = \dfrac{SS(\text{error})}{df(\text{error})}$	
Total	$SS(\text{total})$	$N - 1$		

A	B	C
5	8	2
4	3	3
4	6	5
3		3
		6

Total: $c_1 = 16$ $c_2 = 17$ $c_3 = 19$
Size: $n_1 = 4$ $n_2 = 3$ $n_3 = 5$

Solution

The null hypothesis is the claim that the samples come from populations with the same mean.

We have

H_0: $\mu_A = \mu_B = \mu_C$

H_1: The preceding means are not all equal.

With $\alpha = 0.05$, we proceed to calculate the test statistic as follows.

Number of samples: $k = 3$

Total number of values: $N = n_1 + n_2 + n_3 = 4 + 3 + 5 = 12$

$\Sigma x = \Sigma c_i = c_1 + c_2 + c_3 = 16 + 17 + 19 = 52$

$(\Sigma x)^2 = 52^2 = 2704$

$\Sigma x^2 = 5^2 + 4^2 + 4^2 + 3^2 + 8^2 + 3^2 + 6^2 + 2^2 + 3^2 + 5^2 + 3^2$
$$+ 6^2 = 258$$

$$SS(\text{total}) = \Sigma x^2 - \frac{(\Sigma x)^2}{N} = 258 - \frac{2704}{12} = 32.667$$

$$SS(\text{treatment}) = \left(\Sigma \frac{c_i^2}{n_i}\right) - \frac{(\Sigma c_i)^2}{N} = \left(\frac{16^2}{4} + \frac{17^2}{3} + \frac{19^2}{5}\right)$$
$$- \frac{52^2}{12} = 7.200$$

$$SS(\text{error}) = SS(\text{total}) - SS(\text{treatment}) = 32.667 - 7.200$$
$$= 25.467$$

$$df(\text{treatment}) = k - 1 = 3 - 1 = 2$$
$$df(\text{error}) = N - k = 12 - 3 = 9$$

$$MS(\text{treatment}) = \frac{SS(\text{treatment})}{df(\text{treatment})} = \frac{7.200}{2} = 3.600$$

$$MS(\text{error}) = \frac{SS(\text{error})}{df(\text{error})} = \frac{25.467}{9} = 2.830$$

$$F = \frac{MS(\text{treatment})}{MS(\text{error})} = \frac{3.600}{2.830} = 1.272$$

The analysis of variance table summarizes the important components used in determining the value of the test statistic F.

Source of variation	Sum of squares SS	Degrees of freedom df	Mean square MS	Test statistic F
Treatments	7.200	2	$\frac{7.200}{2} = 3.600$	$F = \frac{3.600}{2.830}$
Error	25.467	9	$\frac{25.467}{9} = 2.830$	$= 1.272$
Total	32.667			

With a test statistic of $F = 1.272$, we now proceed to obtain the critical value. The significance level indicates that $\alpha = 0.05$, and we have 2 degrees of freedom for the numerator and 9 degrees of freedom for the denominator so that the critical value from Table A-5 is found to be $F = 4.2565$. Since the test statistic $F = 1.272$ does not exceed the critical value of $F = 4.2565$, we fail to reject the null hypothesis that the means are equal. It appears that differences among the three sample means are not significant.

After first conducting one of these analysis of variance tests, one is not generally inclined to be overwhelmed with the simplicity of the calculations. Fortunately, many computer software packages do the necessary arithmetic. The STATDISK display for the preceding example is shown below.

```
D.F.     (TREATMENT) = 2          D.F.     (ERROR) = 9
Sum sq.  (TREATMENT) = 7.2000     Sum sq.  (ERROR) = 25.4667
            Total sum of squares = 32.6667
Mean sq. (TREATMENT) = 3.6000     Mean sq. (ERROR) = 2.8296
            Significance level. = .05
            F................. = 1.2723
            P-value........... = 0.3261
CONCLUSION: FAIL TO REJECT the null hypothesis of equal means
```

As efficient and reliable as such computer programs may be, they are totally worthless if we don't understand the relevant concepts. We should recognize that the methods of this section are used to test the claim that several samples come from populations with the same mean.

Pollster Lou Harris

Lou Harris has been in the polling business since 1947, and he has been involved with about 250 political campaigns. While George Gallup believes that pollsters should remain detached and objective, Lou Harris prefers a more personal involvement. He advised John F. Kennedy to openly attack the prejudice against a Catholic becoming president. Kennedy followed that advice and won the nomination.

These methods require normally distributed populations with the same variance, and the samples must be independent. We reject or fail to reject the null hypothesis of equal means by analyzing the two estimates of variance. The MS(treatment) is an estimate of the variation between samples, while the MS(error) is an estimate of the variation within samples. If MS(treatment) is significantly greater than MS(error), then we reject the claim of equal means, otherwise we fail to reject that claim.

The procedure of analysis of variance, as presented in this section, can be used to decide whether differences among sample means are significant or attributable to chance. This procedure is referred to as **one-way analysis of variance** or **single-factor analysis of variance** to indicate that the data are classified into groups according to a single criterion. In the last example, the three samples were categorized according to the single factor of programming language used. A more complex analysis might involve samples with the times classified according to more than one variable, such as experience of the programmer, education of the programmer, and other relevant factors. Methods for dealing with any number of classification variables have been developed, but this text considers only cases involving a single factor.

When we use the single-factor analysis of variance technique and conclude that the differences among the means are significant, we cannot necessarily conclude that the given factor is responsible for those differences. In the last example, if the differences were significant, we would reject the null hypothesis of equal means. However, that rejection would not necessarily imply that the differences were due to the programming language used. Perhaps all the newer programmers used the same language, but their inexperience made them slower, so that significant differences are the results of experience rather than the programming language used. One way to reduce the effect of extraneous factors (such as experience of the programmer) is to design the experiment so that it has a **completely randomized design.** That is, each element is given the same chance of belonging to the different categories or treatments. Another way to reduce the effect of extraneous factors is to use a **rigorously controlled design** in which all other factors are forced to be constant. To eliminate the effect of experience and education, for example, we might use only programmers with the same experience and education. Obviously, this approach is often difficult or impossible to implement. In any event, we should carefully scrutinize the way in which the sample data were collected so that we avoid serious errors in attributing significant differences among means to factors that might not be responsible for those differences. The design of the experiment is critically important, and no statistical calisthenics can salvage a poor design.

10-4 Exercises A
Analysis of Variance

10-37   The dean of a college wants to compare grade-point averages of resident, commuting, and part-time students. A random sample of each group is selected, and the results are as follows. At the $\alpha = 0.05$ level of significance, test the claim that the three populations have equal means.

Residents	Commuters	Part-time
$n = 15$	$n = 15$	$n = 15$
$\bar{x} = 2.60$	$\bar{x} = 2.55$	$\bar{x} = 2.30$
$s^2 = 0.30$	$s^2 = 0.25$	$s^2 = 0.16$

10-38   Do Exercise 10-37 after changing the sample mean grade-point average for resident students from 2.60 to 3.00.

10-39   Five socioeconomic classes are being studied by a sociologist, and members from each class are rated for their adjustments to society. The sample data are summarized as follows. At the $\alpha = 0.05$ level of significance, test the claim that the five populations have equal means.

A	B	C	D	E
$n_1 = 10$	$n_2 = 10$	$n_3 = 10$	$n_4 = 10$	$n_5 = 10$
$\bar{x}_1 = 103$	$\bar{x}_2 = 97$	$\bar{x}_3 = 102$	$\bar{x}_4 = 100$	$\bar{x}_5 = 110$
$s_1^2 = 230$	$s_2^2 = 75$	$s_3^2 = 200$	$s_4^2 = 150$	$s_5^2 = 100$

10-40   A unit on elementary algebra is taught to five different classes of randomly selected students with the same academic backgrounds. A different method of teaching is used in each class, and the final averages of the 20 students in each class are compiled. The results yield the following data. At the $\alpha = 0.05$ level of significance, test the claim that the five population means are equal.

Traditional	Programmed	Audio	Audiovisual	Visual
$n = 20$	$n = 20$	$n = 20$	$n = 20$	$n = 20$
$\bar{x} = 76$	$\bar{x} = 74$	$\bar{x} = 70$	$\bar{x} = 75$	$\bar{x} = 74$
$s^2 = 60$	$s^2 = 50$	$s^2 = 100$	$s^2 = 36$	$s^2 = 40$

10-41   Five car models are studied in a test that involves four of each model. For each of the four cars in each of the five samples, exactly 1 gallon of gas is placed in the tank and the car is driven until the gas is used up. The results follow. At the $\alpha = 0.05$ significance level, test the claim that the five population means are all equal.

Distance traveled in miles				
A	B	C	D	E
16	18	18	19	15
22	23	18	21	16
17	15	20	22	20
17	20	20	22	17

10-42 A sociologist randomly selects subjects from three types of family structure: stable families, divorced families, and families in transition. The selected subjects are interviewed and rated for their social adjustment, and the sample results are as follows. At the $\alpha = 0.05$ level of significance, test the claim that the three population means are equal.

Stable	Divorced	Transition
110	115	90
105	105	120
100	110	125
95	130	100
120	105	105

10-43 Do Exercise 10-42 after adding 30 to each score in the stable group.

10-44 Readability studies are conducted to determine the clarity of four different texts, and the sample scores follow. At the $\alpha = 0.05$ level of significance, test the claim that the four texts produce the same mean readability score.

Text A	Text B	Text C	Text D
50	59	48	60
51	60	51	65
53	58	47	62
58	57	49	68
53	61	50	70

10-45 An introductory calculus course is taken by students with varying high-school records. The sample results from each of three groups follow. The values given are the final numerical averages in the calculus course. At the $\alpha = 0.05$ level of significance, test the claim that the mean scores are equal in the three groups.

Good high-school record	Fair high-school record	Poor high-school record
90	80	60
86	70	60
88	61	55
93	52	62
80	73	50
	65	70
	83	

10-46 A preliminary study is conducted to determine whether there is any relationship between education and income. The sample results are as follows. The figures represent, in thousands of dollars, the lifetime incomes of randomly selected workers from each category. At the $\alpha = 0.05$ level of significance, test the claim that the samples come from populations with equal means (see the top of page 494).

Years of education				
8 years or less	9–11 years	12 years	13–15 years	16 or more years
300	270	400	420	570
210	330	430	480	640
260	380	370	510	590
330	310	390	390	700
		420	470	620
				660

10-47   Three car models are studied in tests that involve several cars of each model. In each case, the car is run on exactly 1 gallon of gas until the fuel supply is exhausted, and the distances traveled are as follows. Test the claim that the three population means are equal.

A	B	C
16	14	20
20	16	21
18	16	19
18	17	22
	15	18
	18	24
		18
		20

10-48   Three groups of adult men were selected for an experiment designed to measure their blood alcohol levels after consuming five drinks. Members of group A were tested after one hour, members of group B were tested after two hours, and members of group C were tested after four hours, and the results are given below. At the 0.05 level of significance, test the claim that the three groups have the same mean level.

A	B	C
0.11	0.08	0.04
0.10	0.09	0.04
0.09	0.07	0.05
0.09	0.07	0.05
0.10	0.06	0.06
		0.04
		0.05

10-49

Medical researchers use three different treatments in an experiment involving diseased rabbits, and the recovery times (in days) are given at right. Using a 0.01 significance level, test the claim that the different treatments result in the same means.

A	B	C
6	9	11
8	8	9
12	7	10
9	6	8
7	9	11
2		9
		12
		14

10-50 A dental research team investigating a new tooth filling material experiments with four different hardening methods, and the sample results are given below. At the 0.01 significance level, test the claim that the four methods yield the same mean index of hardness.

A	B	C	D
8.2	6.9	8.2	8.0
7.9	7.3	8.5	7.2
8.4	7.5	8.9	7.3
8.0	8.2	8.7	7.1
8.0	6.3	8.6	7.9
	6.8	8.4	7.3
	6.7	8.8	7.1
			7.4

10-51 The numbers of program errors are recorded for four different programmers on randomly selected days. Test the claim that they produce the same mean number of errors. Use a 0.01 level of significance.

1	2	3	4
14	3	17	16
16	5	20	18
18	12	22	20
14	8	24	17
22	7	26	21
9	6	18	
	6	9	
	4	11	
	7		
	16		

10-52 Five different machines are used to produce floppy disks and the numbers of defects are recorded for batches randomly selected on different days. Use a 0.01 significance level to test the claim that the machines produce the same mean number of defects.

1	2	3	4	5
8	11	14	32	10
9	11	13	33	8
6	8	9	26	11
10	10	10	15	14
12	13	12	18	22
	12		25	
			31	
			40	

10-4 Exercises B
Analysis of Variance

10-53   A study is made of three police precincts to determine the time required for a police car to be dispatched after a crime is reported. Sample results are as follows.

Precinct 1	Precinct 2	Precinct 3
$n_1 = 50$	$n_2 = 50$	$n_3 = 50$
$\bar{x}_1 = 170$ seconds	$\bar{x}_2 = 202$ seconds	$\bar{x}_3 = 165$ seconds
$s_1 = 18$ seconds	$s_2 = 20$ seconds	$s_3 = 23$ seconds

(a) At the 5% level of significance, test the claim that $\mu_1 = \mu_2$. Use the methods discussed in Chapter 8.

(b) At the 5% level of significance, test the claim that $\mu_2 = \mu_3$. Use the methods discussed in Chapter 8.

(c) At the 5% level of significance, test the claim that $\mu_1 = \mu_3$. Use the methods discussed in Chapter 8.

(d) At the 5% level of significance, test the claim that $\mu_1 = \mu_2 = \mu_3$. Use analysis of variance.

(e) Compare the methods and results of parts (a), (b), and (c) to part (d).

10-54   Five independent samples of 50 scores are randomly drawn from populations that are normally distributed with equal variances. We wish to test the claim that $\mu_1 = \mu_2 = \mu_3 = \mu_4 = \mu_5$.

(a) If we use only the methods of Chapter 8, we would test the individual claims $\mu_1 = \mu_2$, $\mu_1 = \mu_3$, . . . , $\mu_4 = \mu_5$. What is the number of claims? That is, how many ways can we pair off five means?

(b) Assume that for each test of equality between two means, there is a 0.95 probability of not making a type I error. If all possible pairs of means are tested for equality, what is the probability of making no type I errors? (Although the tests are not actually independent, assume that they are.)

(c) If we use analysis of variance to test the claim $\mu_1 = \mu_2 = \mu_3 = \mu_4 = \mu_5$ at the 5% level of significance, what is the probability of not making a type I error?

(d) Compare the results of parts (b) and (c).

10-55   Five independent samples of 50 scores are randomly drawn from populations that are normally distributed with equal variances, and the values of n, $\bar{x}$, and s are obtained in each case. Analysis of variance is then used to test the claim that $\mu_1 = \mu_2 = \mu_3 = \mu_4 = \mu_5$.

(a) If a constant is added to each of the five sample means, how is the value of the test statistic affected?

(b) If each of the five means is multiplied by a constant, how is the value of the test statistic affected?

10-56   Each of six different people was given four different tests. In testing the hypothesis that the tests produce the same mean scores, the test statistic is found to be $F = 4.000$ and the estimate of variation within samples is calculated as 60. Construct the corresponding ANOVA table.

Computer Project
Chi-Square and Analysis of Variance

(a)   Use existing software to solve Exercise 10-21. Repeat that exercise by adding 10 to each frequency in the table, and note the effect of that change. Finally, repeat Exercise 10-21 a third time after multiplying each frequency by 10, and note the effect of that change.

(b)   Use existing software to employ analysis of variance in solving Exercise 10-47. Then repeat that exercise after adding 10 to each score and note the effect of that change. Finally, repeat Exercise 10-47 a third time after multiplying each score by 10, and note the effect of that change.

Review

We began this chapter by developing methods for testing hypotheses made about more than two population proportions. For **multinomial experiments** we tested for agreement between observed and expected frequencies by using the chi-square test statistic given in the table below. In repeated large samplings, the distribution of the χ^2 test statistic can be approximated by the chi-square distribution. This approximation is generally considered acceptable as long as all expected frequencies are at least 5. In a multinomial experiment with k cells or categories, the number of degrees of freedom is $k - 1$.

In Section 10-3 we used the sample χ^2 test statistic to measure disagreement between observed and expected frequencies in **contingency tables.** A contingency table contains frequencies; the rows correspond to categories of one variable while the columns correspond to categories of another variable. With contingency tables, we test the hypothesis that the two variables of classification are independent. We can again approximate the sampling distribution of that statistic by the chi-square distribution as long as all expected frequencies are at least 5. In a contingency table with r rows and c columns, the number of degrees of freedom is $(r - 1)(c - 1)$.

In Section 10-4 we used **analysis of variance** to determine whether differences among three or more sample means are due to chance fluctuations or whether the differences are significant. This method requires normally distributed populations with equal variances. Our comparison of sample means is based on two different estimates of the common population variance. In repeated samplings, the distribution of the F test statistic can be approximated by the F distribution, which has critical values given in Table A-6.

IMPORTANT FORMULAS

Application	Applicable distribution	Test statistic	Degrees of freedom	Table of critical values
Multinomial	chi-square	$\chi^2 = \sum \dfrac{(O - E)^2}{E}$	$k - 1$	Table A-5
Contingency table	chi-square	$\chi^2 = \sum \dfrac{(O - E)^2}{E}$ where $E = \dfrac{(\text{row total})(\text{column total})}{(\text{grand total})}$	$(r - 1)(c - 1)$	Table A-5
Analysis of variance (equal sample sizes only)	F	$F = \dfrac{ns_{\bar{x}}^2}{\mu_{s_i^2}}$	num: $k - 1$ den: $k(n - 1)$	Table A-6
(all cases)	F	$F = \dfrac{MS(\text{treatment})}{MS(\text{error})}$ (see below)	num: $k - 1$ den: $N - k$	Table A-6

For analysis of variance:

k = number of samples

n_i = number of values in the ith sample

Σx = sum of all sample values

$SS(\text{total}) = \Sigma x^2 - \dfrac{(\Sigma x)^2}{N}$

$SS(\text{error}) = SS(\text{total}) - SS(\text{treatment})$

$df(\text{error}) = N - k$

$MS(\text{error}) = \dfrac{SS(\text{error})}{df(\text{error})}$

N = total number of values

c_i = total of values in the ith sample

Σx^2 = sum of the squares of all sample values

$SS(\text{treatment}) = \left(\sum \dfrac{c_i^2}{n_i} \right) - \dfrac{(\sum c_i)^2}{N}$

$df(\text{treatment}) = k - 1$

$MS(\text{treatment}) = \dfrac{SS(\text{treatment})}{df(\text{treatment})}$

$F = \dfrac{MS(\text{treatment})}{MS(\text{error})}$

Review Exercises

10-57 The owner of a new grocery store records the number of customers arriving on the different days for one week. The results are as follows. At the $\alpha = 0.05$ significance level, test the claim that customers arrive on the different days with equal frequencies.

Sun	Mon	Tues	Wed	Thurs	Fri	Sat
97	72	55	68	70	88	110

10-58 A random number generator produces the outcomes listed in the following table. At the 0.05 significance level, test the claim that the values of 1, 2, 3, 4, 5, 6 are equally likely.

Outcome	1	2	3	4	5	6
Frequency	16	13	8	9	6	8

10-59 At the 0.01 level of significance, test the claim that four separate and distinct genetic characteristics are equally likely. Sample data consist of 100 randomly selected subjects, and the four characteristics occurred with the frequencies of 20, 30, 15, and 35, respectively.

10-60 In a certain region, a survey is made of companies that officially declared bankruptcy during the past year. Of the 120 small bankrupt businesses, 72 advertised in weekly newspapers. Of the 65 medium-sized bankrupt businesses, 25 advertised in weekly newspapers, while 8 of the 15 large bankrupt businesses did so. At the 5% level of significance, test the claim that the three proportions of bankrupt businesses that used weekly newspaper ads are equal.

10-61 Five years ago, a survey was made of a group of students. The same survey was made of a group of students this year. The first question dealt with sexual permissiveness, and there were five possible responses. At the $\alpha = 0.01$ level of significance, test the claim that the responses are independent of the student group.

	A	B	C	D	E
Student group five years ago	25	32	18	10	5
Student group today	40	73	44	28	20

10-62 A psychologist conducted studies on the relationship between forgetfulness and I.Q. scores. The results are in the following table. At the 0.05 level of significance, test the claim that I.Q. scores and levels of forgetfulness are independent.

	Forgets infrequently	Forgets occasionally	Forgets often
Low I.Q.	15	10	5
Medium I.Q.	20	30	10
High I.Q.	15	25	15

10-63   A marketing study is conducted in order to determine if a product's appeal is affected by geographic region. Given the sample data in the following table, use a 0.01 significance level to test the claim that the consumer's opinion is independent of region.

	Like	Dislike	Uncertain
Northeast	30	15	15
Southeast	10	30	20
West	40	60	15

10-64   Given the data in the following table, test the claim that a high-school student's program is independent of his or her family's social class. Assume a significance level of 0.05.

High school program	Social class of family			
	Lower	Lower middle	Upper middle	Upper
Academic	5	10	40	30
General	20	40	50	15
Vocational	15	50	20	5

10-65   A lawyer is studying punishments for a certain crime and wants to compare the sentences imposed by three different judges. Randomly selected results follow. At the $\alpha = 0.05$ level of significance, test the claim that the three judges impose sentences that have the same mean.

Judge A	Judge B	Judge C
$n = 36$	$n = 36$	$n = 36$
$\bar{x} = 5.2$ years	$\bar{x} = 4.1$ years	$\bar{x} = 5.5$ years
$s = 1.4$ years	$s = 1.1$ years	$s = 1.5$ years

10-66   Three teaching methods are used with three groups of randomly selected students and the results follow. At the 0.05 level of significance, test the claim that the samples came from populations with equal means.

Method A	Method B	Method C
$n = 20$	$n = 20$	$n = 20$
$\bar{x} = 72.0$	$\bar{x} = 76.0$	$\bar{x} = 71.0$
$s = 9.0$	$s = 10.0$	$s = 12.0$

10-67 In testing the effectiveness of four different diets, subjects with the same overweight characteristics are randomly selected for each diet and the weight losses are listed below. At the 0.01 level of significance, test the claim that the diets produce the same mean weight loss.

Diet 1	Diet 2	Diet 3	Diet 4	
8	10	6	21	35
12	14	24	23	12
14	14	12	16	19
16	21	10	19	15
3	5	10	27	40

Vocabulary List

Define and give an example of each term.

multinomial experiment
observed frequency
expected frequency
contingency table
two-way table
analysis of variance
ANOVA

variance between samples
variance within samples
variation due to treatment
variation due to error
one-way analysis of variance

single-factor analysis of variance
completely randomized design
rigorously controlled design

Chapter 10: Case Study Activity

Conduct a survey by asking the question "Do you favor or oppose the death penalty for people convicted of murder?" Record each response (yes, no, undecided) along with the sex (male, female) of the respondent. Include in your data set only responses of "yes" or "no." At the 0.05 level of significance, test the claim that the opinion is independent of the sex of the respondent. Be sure to survey enough people so that the expected frequency of each cell in the resulting contingency table is at least 5. Identify any factors suggesting that your sample is not representative of the people in your region.

11-1 Overview

We present the general nature and the advantages and disadvantages of **nonparametric** methods, introduce the concept of **ranked data,** and identify chapter **objectives.**

11-2 Sign Test

The sign test is a nonparametric method that can be used to test the claim that two sets of dependent data have the same median.

11-3 Wilcoxon Signed-Ranks Test for Two Dependent Samples

The Wilcoxon signed-ranks test can be used to test the claim that two sets of dependent data come from identical populations. This test takes into account the magnitudes of the numbers.

11-4 Wilcoxon Rank-Sum Test for Two Independent Samples

The Wilcoxon rank-sum test can be used to test the claim that two independent samples come from identical populations.

11-5 Kruskal-Wallis Test

The Kruskal-Wallis test can be used to test the claim that several independent samples come from identical populations.

11-6 Rank Correlation

The rank correlation coefficient can be used to test for an association between two sets of paired data.

11-7 Runs Test for Randomness

The runs test can be used to test for randomness in the way data are selected.

Nonparametric Statistics

Suppose seven teachers are being ranked for promotion by a faculty committee and by a separate committee of administrators.

Teacher	Faculty rank	Administrator's rank
Arnold	1	2
Bennet	7	7
Cohen	2	6
Davis	5	3
Ellis	4	5
Farrell	3	1
Gallo	6	4

In analyzing this particular promotion procedure, we might investigate whether there is a correlation between the rankings of the two committees. However, the linear correlation coefficient (see Section 9-2) cannot be used because of the nature of the data. The data in this problem consist of ranks, and such data sets do not satisfy the requirement of the linear correlation coefficient introduced in Chapter 9. The linear correlation coefficient from Chapter 9 requires that the two variables be normally distributed, and ranks do not satisfy that requirement. We cannot use the methods of Chapter 9, but we will introduce an alternative way to solve this problem.

11-1   Overview

Most of the methods of inferential statistics covered before Chapter 10 can be called **parametric methods,** because their validity is based on sampling from a particular parametric family of means, standard deviations, variances, proportions, and so on. Parametric methods can usually be applied only to circumstances in which some fairly strict requirements are met. One typical requirement is that the sample data must come from a normally distributed population. What do we do when the necessary requirements are not satisfied? It is very possible that there may be an alternative approach among the many methods that are classified as **nonparametric.** In addition to being an alternative to parametric methods, nonparametric techniques are frequently valuable in their own right.

In this chapter, we introduce six of the more popular nonparametric methods that are in current use. These methods have advantages and disadvantages.

Advantages of Nonparametric Methods

1. Nonparametric methods can be applied to a wider variety of situations, because they do not have the more rigid requirements of their parametric counterparts. In particular, nonparametric methods do not require normally distributed populations. For this reason, nonparametric tests of hypotheses are often called **distribution-free** tests.
2. Unlike the parametric methods, nonparametric methods can often be applied to nominal data that lack exact numerical values.
3. Nonparametric methods usually involve computations that are simpler than the corresponding parametric methods.
4. Since nonparametric methods tend to require simpler computations, they tend to be easier to understand.

If all of these terrific advantages could be accrued without any significant disadvantages, we could ignore the parametric methods and enjoy much less-complicated procedures. Unfortunately, there are some disadvantages.

Disadvantages of Nonparametric Methods

1. Nonparametric methods tend to waste information, since exact numerical data are often reduced to a qualitative form.
2. Nonparametric methods are generally less sensitive than the corresponding parametric methods. This means that we need stronger evidence before we reject a null hypothesis.

As an example of the way that information is wasted, there is one nonparametric method in which weight losses by dieters are recorded simply as negative signs. With this particular method, a weight loss of only 1 pound receives the same representation as a weight loss of 50 pounds. This would not thrill dieters.

Although nonparametric tests are less sensitive than their parametric counterparts, this can be compensated by an increased sample size. The **efficiency** of a nonparametric method is one concrete measure of its sensitivity. Section 11-6 deals with a concept called the *rank correlation coefficient,* which has an efficiency rating of 0.91 when compared to the linear correlation coefficient of Chapter 9. This means that with all things being equal, this nonparametric approach would require 100 sample observations to achieve the same results as 91 sample observations analyzed through the parametric approach, assuming the stricter requirements for using the parametric method are met. Not bad! The point, though, is that an increased sample size can overcome lower sensitivity. Table 11-1 lists nonparametric methods covered in this chapter, along with the corresponding parametric approach and efficiency rating. You can see from this table that the lower efficiency might not be a critical factor.

In choosing between a parametric method and a nonparametric method, the key factors that should govern our decision are cost, time, efficiency, amount of data available, type of data available, method of sampling, nature of the population, and probabilities (α and β) of making errors. In one experiment we might have abundant data with strong assurances that all of the requirements of a parametric test are satisfied, and we would probably be wise to choose that parametric test. But given another experiment with relatively few cases drawn from some mys-

Table 11-1

Application	Parametric test	Nonparametric test	Efficiency of nonparametric test with normal population
Two dependent samples	t test or z test	Sign test Wilcoxon signed-ranks	0.63 0.95
Two independent samples	t test or z test	Wilcoxon rank-sum	0.95
Several independent samples	Analysis of variance (F test)	Kruskal-Wallis test	0.95
Correlation	Linear correlation	Rank correlation	0.91
Randomness	No parametric test	Runs test	No basis for comparison

terious population, we would probably fare better with a nonparametric test. Sometimes we don't really have a choice. If we want to test data to see if they are randomly selected, the only available test happens to be nonparametric. As another example, only nonparametric methods can be used on data consisting of observations that can only be ranked. Some of the methods included in this chapter are based on ranks. Instead of describing ranks in each of those sections or making some sections dependent on others, we will now discuss ranks so that we will be prepared to use them wherever they are required.

Data are **ranked** when they are arranged according to some criterion, such as smallest to largest or best to worst. The first item in the arrangement is given a rank of 1, the second item is given a rank of 2, and so on. For example, the numbers 5, 3, 40, 10, and 12 can be arranged from lowest to highest as 3, 5, 10, 12, and 40, so that 3 has rank 1, 5 has rank 2, 10 has rank 3, 12 has rank 4, and 40 has rank 5. If a tie in ranks should occur, the usual procedure is to find the mean of the ranks involved and then assign that mean rank to each of the tied items. This might sound complicated, but it's really quite simple. For example, the numbers 3, 5, 5, 10, and 12 would be given ranks of 1, 2.5, 2.5, 4, and 5, respectively. In this case, there is a tie for ranks 2 and 3, so we find the mean of 2 and 3 (which is 2.5) and assign it to the scores that created the tie. As another example, the scores 3, 5, 5, 7, 10, 10, 10, and 15 would be ranked 1, 2.5, 2.5, 4, 6, 6, 6, and 8, respectively. From these examples we can see how to convert numbers to ranks, but there are many situations in which the original data consist of ranks. If a judge ranks five piano contestants, we get ranks of 1, 2, 3, 4, 5 corresponding to five names; it's this type of data that precludes the use of parametric methods and enhances the importance of nonparametric methods.

11-2 Sign Test

The **sign test** is one of the easiest nonparametric tests to use, and it is applicable to a few different types of situations. One application is to the paired data that form the basis for hypothesis tests involving two dependent samples. We considered such cases by using parametric tests in Section 8-3. One example from Section 8-3 involves an educational service that offered a course designed to improve scores on I.Q. tests. The sample results for ten randomly selected subjects appear in Table 11-2. We will test the claim that the course is effective.

In Section 8-3 we used the parametric student t test, but in this section we apply the nonparametric sign test, which can be used to test for equality between two medians. The key concept underlying the sign test is this: **If the two sets of data have equal medians, the number of positive signs should be approximately equal to the number of negative**

Table 11-2

Subject	I.Q. before course	I.Q. after course	Sign of change from before to after
A	96	99	+
B	110	112	+
C	98	107	+
D	113	110	−
E	88	88	0
F	92	101	+
G	106	107	+
H	119	123	+
I	100	91	−
J	97	99	+

signs. For the data in Table 11-2, we can conclude that the course is effective if there is an excess of positive signs and a deficiency of negative signs.

The sign test requires that we exclude ties (represented by zeros), so we are left with this specific question: Do the seven positive signs in Table 11-2 "significantly" outnumber the two negative signs? Or, to put it another way, is the number of negative signs small enough? The answer to this question depends on the level of significance we require, so let's use $\alpha = 0.05$, as we did in Section 8-3. When we assume the null hypothesis of no increase in I.Q. scores, we assume that positive signs and negative signs occur with equal frequency, so $P(\text{positive sign}) = P(\text{negative sign}) = \frac{1}{2}$. (The null hypothesis of no increase also includes the possibility of a decrease, but we continue to assume that positive signs and negative signs are equally likely.)

H_0:   There is no increase in I.Q. scores.

H_1:   There is an increase in I.Q. scores.

Since our results fall into two categories (positive or negative) and we have a fixed number of independent cases, we could use the binomial probability distribution to determine the likelihood of getting two or fewer negative signs among the nine subjects. Instead, we have constructed a separate table (Table A-8), which lists critical values for the sign test. For consistency, we will stipulate that **the test statistic x is the number of times the less-frequent sign occurs.** With seven positive signs and two negative signs, the test statistic x is then the lesser of 7 and 2, so $x = 2$. After discarding the zero, we are left with nine sample cases so that $n = 9$. Our test is one-tailed with $\alpha = 0.05$, and Table A-8 indicates that the critical value is 1. Thus we should reject the null hypothesis only if the test statistic is less than or equal to 1. With a test statistic of $x = 2$, we fail to reject the null hypothesis of no increase in I.Q. scores.

Air Is Healthier Than Tobacco

A consumer testing group studied the tar and nicotine levels of various brands of cigarettes. Now cigarettes were advertised as being lowest in tar and nicotine, but they seemed to burn more quickly than other brands. Sam-ples of Now and Winston were randomly selected and weighed. Both brands were products of the R. J. Reynolds Tobacco Company, and Winston was their best seller at the time of the comparison. Results showed that the average weight of the tobacco in a Now was about two-thirds the average weight of the tobacco in a Winston. With one-third less tobacco, it isn't difficult to get lower tar and nicotine levels. The study also noted that smokers tend to consume more cigarettes when they are of the low tar and nicotine variety. The net effect for Now was that more cigarettes could be sold at lower production costs.

Because of the way that we are determining the value of the test statistic, we should check to ensure that our conclusion is consistent with the circumstances. It is only when the sense of the sample data is *against* the null hypothesis that we should even consider rejecting it. If the sense of the data supports the null hypothesis, we should fail to reject it regardless of the test statistic and critical value. Figure 11-1 summarizes the procedure for the sign test and includes this check for consistency of results.

An examination of Figure 11-1 shows that when $n \leq 25$, we should use Table A-8 to find the critical value, but for $n > 25$, we use a normal approximation to obtain the critical values.

In the preceding example, we arrived at the conclusion obtained in Section 8-3. Table 11-3 illustrates how the sign test can waste information. The sign test again involves seven positive signs and two negative signs, so once again we fail to reject the null hypothesis of no increase in I.Q. scores. But if we employ the student t test, as in Section 8-3, we get

$$t = \frac{\bar{d} - 0}{s_d/\sqrt{n}} = \frac{22.8}{18.8/\sqrt{10}} = 3.835$$

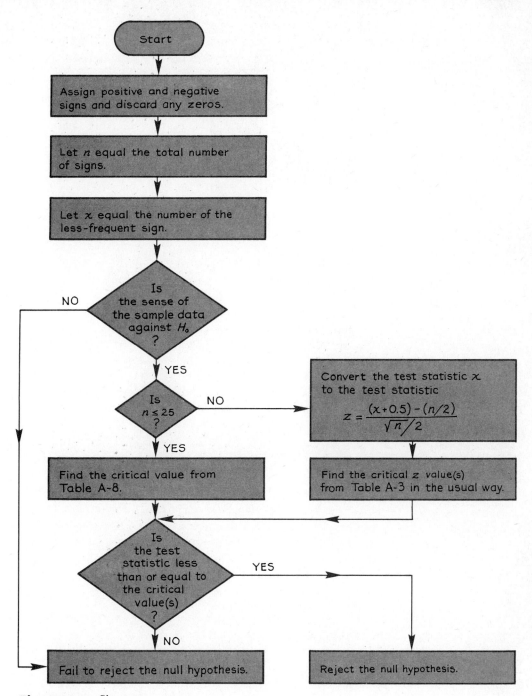

Figure 11-1   Sign test

Table 11-3

Subject	I.Q. before course	I.Q. after course	Difference	Sign of change
A	96	129	+33	+
B	110	142	+32	+
C	98	137	+39	+
D	113	110	−3	−
E	88	88	0	0
F	92	131	+39	+
G	106	137	+31	+
H	119	153	+34	+
I	100	91	−9	−
J	97	129	+32	+

which causes *rejection* of the null hypothesis because the critical t value is found to be 1.833. Using the data of Table 11-3, we fail to reject H_0 if we use the sign test, but we reject H_0 if we use the student t test.

An intuitive analysis of Table 11-3 suggests that the course seems to be effective, but the sign test is blind to the magnitude of the changes, so we fail to reject the null hypothesis—which is probably false. This illustrates the previous assertion that nonparametric tests lack the sensitivity of parametric tests, with the resulting tendency that stronger evidence is required before a null hypothesis is rejected.

The next example illustrates the fact that nonparametric methods can be used to work with nominal data.

EXAMPLE

Each year, two security firms bid for service contracts with a power generating plant. In the past 40 years, State Company won 30 contracts, while the rival Gunn Company won 10. Test the null hypothesis that both companies are equal in their ability to win the contracts. Use a significance level of $\alpha = 0.05$.

Solution

H_0: $p_1 = p_2$ (both proportions are the same)
H_1: $p_1 \neq p_2$
If we denote State Company wins by $+$ and Gunn Company wins by $-$, we have 30 positive signs and 10 negative signs. Refer now to the flowchart of Figure 11-1. The test statistic x is the smaller of 30 and 10, so $x = 10$. This test involves two tails, since a disproportionately low number of contracts won by either company would cause us to reject the claim of equal contract-winning abilities. The sense of the sample data is against the null hypothesis as

Figure 11-2

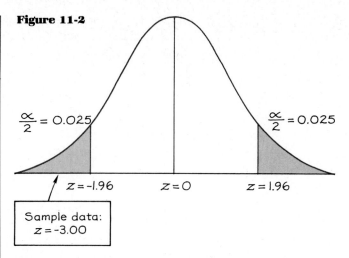

$\frac{\alpha}{2} = 0.025$ $\frac{\alpha}{2} = 0.025$

z = -1.96   z = 0   z = 1.96

Sample data:
z = -3.00

30 and 10 aren't exactly equal. Continuing with the procedure of Figure 11-1, we note that the value of $n = 40$ is above 25, so the test statistic x is converted to the test statistic z as follows:

$$z = \frac{(x + 0.5) - (n/2)}{\sqrt{n}/2}$$

$$= \frac{(10 + 0.5) - (40/2)}{\sqrt{40}/2}$$

$$= \frac{10.5 - 20}{\sqrt{40}/2} = \frac{-9.5}{\sqrt{40}/2} = -3.00$$

With $\alpha = 0.05$ in a two-tailed test, the critical values are $z = -1.96$ and 1.96. The test statistic $z = -3.00$ is less than these critical values (see Figure 11-2), so we reject the null hypothesis of equal abilities to win contracts. The State Company wins significantly more contracts than the Gunn Company.

When $n > 25$, the test statistic z is based on a normal approximation to the binomial probability distribution with $p = q = \frac{1}{2}$. In Section 5-5 we saw that the normal approximation to the binomial distribution is acceptable when both $np \geq 5$ and $nq \geq 5$. Also, in Section 4-5 we saw that $\mu = np$ and $\sigma = \sqrt{n \cdot p \cdot q}$ for binomial experiments. Since this sign test assumes that $p = q = \frac{1}{2}$, we meet the $np \geq 5$ and $nq \geq 5$ prerequisites whenever $n \geq 10$; we have a table of critical values (Table A-8) for n up to 25, so that we need the normal approximation only for values of n above 25. Also, with the assumption that $p = q = \frac{1}{2}$, we get $\mu = np = n/2$ and $\sigma = \sqrt{n \cdot p \cdot q} = \sqrt{n/4} = \sqrt{n}/2$, so that $z = (x - \mu)/\sigma$ becomes

$$z = \frac{x - (n/2)}{\sqrt{n}/2}$$

Finally, we replace x by $x + 0.5$ as a correction for continuity. That is, the values of x are discrete, but since we are using a continuous probability distribution, a discrete value such as 10 is actually represented by the interval from 9.5 to 10.5. Because x represents the less-frequent sign, we need to concern ourselves only with $x + 0.5$; we thus get the test statistic z as given above and in Figure 11-1.

The previous examples involved application of the sign test to a comparison of *two* sets of data, but we can sometimes use the sign test to investigate a claim made about one set of data, as the next example shows.

E X A M P L E

Use the sign test to test the claim that the median I.Q. of pilots is at least 100 if a sample of 50 pilots contained exactly 22 members with I.Q.'s of 100 or higher.

Solution

The null hypothesis is the claim that the median is equal to or greater than 100; the alternative hypothesis is the claim that the median is less than 100.

H_0:   Median is at least 100. (Median ≥ 100)

H_1:   Median is less than 100. (Median < 100)

We select a significance level of 0.05, and we use $+$ to denote each I.Q. score that is at least 100. We therefore have 22 positive signs and 28 negative signs. We can now determine the significance of getting 22 positive signs out of a possible 50. Referring to Figure 11-1, we note that $n = 50$ and $x = 22$ (the smaller of 22 and 28). The sense of the data is against the null hypothesis, since a median of at least 100 would require at least 25 (half of 50) scores of 100 or higher. The value of n exceeds 25, so we convert the test statistic x to the test statistic z.

$$z = \frac{(x + 0.5) - (n/2)}{\sqrt{n}/2}$$

$$= \frac{(22 + 0.5) - (50/2)}{\sqrt{50}/2}$$

$$= \frac{22.5 - 25}{\sqrt{50}/2} = -0.71$$

In this one-tailed test with $\alpha = 0.05$, we use Table A-3 to get the critical z value of -1.645. From Figure 11-3, we can see that the computed value of -0.71 does not fall within the critical region. We

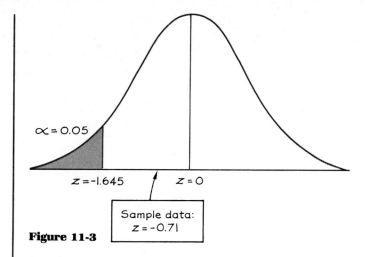

Figure 11-3

therefore fail to reject the null hypothesis. Based on the available sample evidence, we cannot reject the claim that the median I.Q. is at least 100. A corresponding parametric test may or may not lead to the same conclusion, depending on the specific values of the 50 sample scores.

We have shown that the sign test wastes information because it uses only information about the direction of the differences between pairs of data, while the magnitudes of those differences are ignored. The next section introduces the Wilcoxon signed-ranks test, which largely overcomes that disadvantage.

11-2 Exercises A
Sign Test

In the following exercises, use the sign test.

11-1 At the 0.05 level of significance, test the claim that the sample x and y values come from the same population. Assume that the x and y values are paired as shown.

x	1	2	2	3	5	6	8	4	6	7	2	5	3	2
y	2	0	1	4	4	4	9	3	5	7	2	6	2	1

11-2 Two different firms design their own I.Q. tests, and a psychologist administers both tests to randomly selected subjects with the results given below. At the 0.05 level of significance, test the claim that there is no significant difference between the two tests.

Subject	A	B	C	D	E	F	G	H	I	J
Test I	98	94	111	102	108	105	92	88	100	99
Test II	105	103	113	98	112	109	97	95	107	103

11-3   A pill designed to lower systolic blood pressure is administered to ten randomly selected volunteers. The results follow. At the $\alpha = 0.05$ significance level, test the claim that systolic blood pressure is not affected by the pill.

Before pill	120	136	160	98	115	110	180	190	138	128
After pill	118	122	143	105	98	98	180	175	105	112

11-4   A test of driving ability is given to a random sample of ten student drivers before and after they completed a formal driver education course. The results follow. At the $\alpha = 0.05$ significance level, test the claim that the course does not affect scores.

Before course	100	121	93	146	101	109	149	130	127	120
After course	136	129	125	150	110	138	136	130	125	129

11-5   A student hears that fish is a "brain food" that helps to make people more intelligent. She participates in an experiment involving 12 randomly selected volunteers who take an I.Q. test and then begin an intensive fish diet. A second I.Q. test is given at the end of the experiment, with the results given in the table below. At the 0.05 significance level, test the claim that the fish diet has no effect on I.Q. scores.

I.Q. score before diet	98	110	105	121	100	88	112	92	99	109	103	104
I.Q. score after diet	98	112	106	118	102	97	115	90	99	110	105	109

11-6   A course is designed to increase readers' speed and comprehension. To evaluate the effectiveness of this course, a test is given both before and after the course, and sample results follow. At the 0.05 significance level, test the claim that the scores are higher after the course.

Before	100	110	135	167	200	118	127	95	112	116
After	136	160	120	169	200	140	163	101	138	129

11-7   The following chart lists a random sampling of the ages of married couples. The age of each husband is listed above the age of his wife. At the 0.01 significance level, test the claim that there is no difference between the ages of husbands and wives.

Husband	28.1	33.0	29.8	53.1	56.7	41.6	50.6	21.4	62.0	19.7
Wife	28.4	27.6	32.7	52.0	58.1	41.2	50.7	20.6	61.1	18.1

11-8 Ten randomly selected volunteers test a new diet, with the following results. At the 0.01 level of significance, test the claim that the diet is effective. All the weights are given in kilograms.

Subject	A	B	C	D	E	F	G	H	I	J
Weight before diet	68	54	59	60	57	62	62	65	88	76
Weight after diet	65	52	52	60	58	59	60	63	78	75

11-9 When 20 students are asked if they understand the purpose of their student senate, 15 respond affirmatively and 5 respond negatively. At the 0.05 level of significance, test the claim that most (more than half) students feel that they understand the purpose of their student senate.

11-10 A standardized aptitude test yields a mathematics score M and a verbal score V for each person. Among 15 male subjects, $M - V$ is positive in 12 cases, negative in 2 cases, and zero in 1 case. At the 0.05 level of significance, use the sign test to test the claim that males do better on the mathematics portion of the test.

11-11 A political party preference poll is taken among 20 randomly selected voters. If 7 prefer the Republican Party while 13 prefer the Democratic Party, apply the sign test to test the claim that both parties are preferred equally. Use a 0.05 level of significance.

11-12 A television commercial advertises that seven out of ten dentists surveyed prefer Covariant toothpaste over the leading competitor. Assume that ten dentists are surveyed and seven do prefer Covariant, while three favor the other brand. Is this a reasonable basis for making the claim that most (more than half) dentists favor Covariant toothpaste? Use the sign test with a significance level of 0.05.

11-13 Use the sign test to test the claim that the median life of a battery is at least 40 hours if a random sample of 75 includes exactly 32 that last 40 hours or more. Assume a significance level of 0.05.

11-14 A college aptitude test is given to 100 randomly selected high school seniors. After a period of intensive training, another similar test is given to the same students, and 59 receive higher grades, 36 receive lower grades, and 5 students receive the same grades. At the 0.05 level of significance, use the sign test to test the claim that the training is effective.

11-15 A company is experimenting with a new fertilizer at 50 different locations. In 32 of the locations there is an increase in production, while in 18 locations there is a decrease. At the 0.05 level of significance, use the sign test to test the claim that production is increased by the new fertilizer.

11-16 A new diet is designed to lower cholesterol levels. In six months, 36 of the 60 subjects on the diet have lower cholesterol levels, 22 have slightly higher levels, and 2 register no change. At the 0.01 level of significance, use the sign test to test the claim that the diet produces no change in cholesterol levels.

11-17 After 30 drivers are tested for reaction times, they are given two drinks and tested again, with the result that 22 have slower reaction times, 6 have faster reaction times, and 2 receive the same scores as before the drinks. At the 0.01 significance level, use the sign test to test the claim that the drinks had no effect on the reaction times.

11-18 In target practice, 40 police academy students use two different pistols. Analysis of the scores shows that 24 students get higher scores with the more expensive pistol, while 16 students get better scores with the less expensive pistol. At the 0.05 level of significance, use the sign test to test the claim that both pistols are equally effective.

11-19 Of 50 voters surveyed, 28 favor a tax revision bill before Congress, while all the others are opposed. At the 0.10 level of significance, use the sign test to test the claim that the majority (more than half) of voters favor the bill.

11-20 Use the sign test to test the claim that the median I.Q. score of Philadelphians is at least 100 if a sample of 200 Philadelphians contains 86 with I.Q.'s of 100 or more.

11-2 Exercises B
Sign Test

11-21 Given n sample scores sorted in ascending order $(x_1, x_2, \ldots, x_n)$, if we wish to find the approximate $1 - \alpha$ confidence interval for the population median M, we get

$$x_{k+1} < M < x_{n-k}$$

where k is the critical value (Table A-8) for the number of signs in a two-tailed hypothesis test conducted at the significance level of α. Find the approximate 95% confidence interval for the sample scores listed below.

3 8 6 2 1 7 9 11 17 23 25 10 14 8 30

11-22 Of the voters surveyed, 57 favor passage of a certain bill, and they constitute a majority. At the 0.05 significance level, we apply the sign test and reject the claim that voters are equally split on the bill. Given the preceding information, what is the largest sample size possible?

11-23 Of n subjects tested for high blood pressure, a majority of exactly 50 provided negative results. (That is, their blood pressure is not high.) This is sufficient for us to apply the sign test and reject (at the 0.01 level of significance) the claim that the median blood pressure level is high. Find the largest value n can assume.

11-24 Table A-8 lists critical values for limited choices of α. Use Table A-2 to add a new column in Table A-8 that would represent a significance level of 0.03 in one tail or 0.06 in two tails. For any particular n we use $p = 0.5$, since the sign test requires the assumption that

$$P(\text{positive sign}) = P(\text{negative sign}) = 0.5$$

The probability of x or fewer like signs is the sum of the probabilities up to and including x.

11-3 Wilcoxon Signed-Ranks Test for Two Dependent Samples

In the preceding section, we used the sign test to analyze the differences between paired data. The sign test used only the signs of the differences, while ignoring their actual magnitudes. In this section we introduce the Wilcoxon signed-ranks test, which takes the magnitudes into account. Because this test incorporates and uses more information than the ordinary sign test, it tends to yield better results than the sign test. However, the Wilcoxon signed-ranks test requires that the two sets of data come from populations with a common distribution. Unlike the t test for paired data (see Section 8-3), the Wilcoxon signed-ranks test does *not* require normal distributions.

Consider the data given in Table 11-4. The 13 subjects are given a test for logical thinking. They are then given a tranquilizer and retested. We will use the Wilcoxon signed-ranks test to test the claim that the tranquilizer has no effect, so that there is no significant difference between before and after scores. We will assume a 0.05 level of significance.

H_0: The tranquilizer has no effect on logical thinking.

H_1: The tranquilizer has an effect on logical thinking.

Table 11-4

Subject	Before	After	Difference	Ranks of differences	Signed-ranks
A	67	68	−1	1	−1
B	78	81	−3	2	−2
C	81	85	−4	3	−3
D	72	60	+12	10	+10
E	75	75	0	—	—
F	92	81	+11	8.5	+8.5
G	84	73	+11	8.5	+8.5
H	83	78	+5	4	+4
I	77	84	−7	5	−5
J	65	56	+9	6	+6
K	71	61	+10	7	+7
L	79	64	+15	11	+11
M	80	63	+17	12	+12

We summarize here the procedure for using the Wilcoxon signed-ranks test with paired data.

PROCEDURE

1. For each pair of data, find the difference d by subtracting the second score from the first. Retain signs, but discard any pairs for which $d = 0$.
2. Ignoring the signs of those differences, rank them from lowest to highest. When differences have the same numerical value, assign to them the mean of the ranks involved in the tie.
3. Assign to each rank the sign of the difference from which it came.
4. Find the sum of the absolute values of the negative ranks. Also find the sum of the positive ranks.
5. Let T be the smaller of the two sums found in Step 4.
6. Let n be the number of pairs of data for which the difference d is not zero.
7. If $n \le 30$, use Table A-9 to find the critical value of T. Reject the null hypothesis if the sample data yield a value of T less than or equal to the value in Table A-9. Otherwise, fail to reject the null hypothesis. If $n > 30$, compute the test statistic z by using Formula 11-1.

Formula 11-1
$$z = \frac{T - n(n + 1)/4}{\sqrt{\dfrac{n(n + 1)(2n + 1)}{24}}}$$

When Formula 11-1 is used, the critical z values are found from Table A-3 in the usual way. Again reject the null hypothesis if the test statistic z is less than or equal to the critical value(s) of z. Otherwise, fail to reject the null hypothesis.

We will now use this procedure to test the null hypothesis that the tranquilizer has no effect on logical thinking.

Step 1. In Table 11-4, the column of differences is obtained by subtracting each *after* score from the corresponding *before* score. Differences of zero are discarded.

Step 2. Ignoring their signs, the differences are then ranked from lowest to highest, with ties being treated in the manner described at the end of Section 11-1. See the fifth column of Table 11-4.

Step 3. The signed-ranks column is then created by applying to each rank the sign of the corresponding difference. The results are listed in the last column of Table 11-4. If the tranquilizer really has no effect, we would expect the number of positive ranks to be approximately equal to the number of negative ranks. If the tranquilizer tends to lower scores, then ranks of positive sign would tend to outnumber ranks of negative sign. If the tranquilizer tends to raise scores, then ranks of negative sign would tend to outnumber ranks of positive sign. We can detect a domination by either sign through analysis of the rank sums.

Step 4. Now find the sum of the absolute values of the negative ranks and the sum of the positive ranks. For the data of Table 11-4, we get

$$\text{sum of absolute values of negative ranks} = 1 + 2 + 3 + 5 = 11$$
$$\text{sum of positive ranks} = 10 + 8.5 + 8.5 + 4 + 6$$
$$+ 7 + 11 + 12 = 67$$

Step 5. We will base our test on the smaller of those two sums, denoted by T. For the given data we have $T = 11$. We use the following notation.

NOTATION

T Smaller of these two sums:
1. The sum of the absolute values of the negative ranks.
2. The sum of the positive ranks.

n Number of *pairs* of data after excluding any pairs in which both values are the same.

How Valid Are Unemployment Figures?

The actions of presidents, economists, budget directors, corporation heads, and many other key decision-makers are often based on the monthly unemployment figures. Yet these figures have been criticized as being exaggerated or understated.

Julius Shickin, a commissioner of the Bureau of Labor Statistics, says that "if you're thinking of unemployment in terms of economic potential, then the answer is no, the figure doesn't overstate the problem. But if you're thinking in terms of hardships then the answer is yes, it does. It all depends upon your value judgments."

Critics point out that, in developing unemployment figures, families are asked whether anyone out of work and aged 16 years or older has actually looked for work sometime within the last four weeks. However, some people have been out of work so long that they have abandoned efforts to find a job, while others do ask about work, but they do it in very casual ways. At the other end of the scale is a chemical engineer who is counted as employed even though she is merely driving a cab 35 hours a week while she looks for meaningful employment. Such people distort the unemployment figure, which is arrived at through a survey involving about 50,000 families each month. The survey data are sent to Washington, processed by the Census Bureau, and then given to the Bureau of Labor Statistics. This procedure, in effect since 1940, costs about $5 million per year.

Step 6. $n = 12$ since there are 12 pairs of data with nonzero differences.

Step 7. Whenever $n \leq 30$, we use Table A-9 to find the critical values. If $n > 30$ we can use a normal approximation with the test statistic given in Formula 11-1 and critical values given in Table A-3. As in Section 11-2, we would be justified in using the normal approximation whenever $n \geq 10$, but we have a table of critical values for values of n up to 30 so that we really need the normal approximation only for $n > 30$. Because the data of Table 11-4 yield $n = 12$, we use Table A-9 to get the critical value of 14. ($\alpha = 0.05$ and the test is two-tailed since our null hypothesis is the claim that the scores have not changed significantly.) Since $T = 11$ is less than or equal to the critical value of 14, we reject the null hypothesis. It appears that the drug does affect scores.

In this last example, the unsigned ranks go from 1 through 12 and the sum of those 12 integers is 78. If the two sets of data have no significant differences, each of the two signed-rank totals should be in the neighborhood of $78 \div 2$, or 39. However, for the given sample data, the negative ranks were the smaller values, while the positive ranks tended to the larger values. We got 11 for one total and 67 for the other; this 11-67 split was a significant departure from the 39-39 split expected with a true null hypothesis. The table of critical values shows that at the 0.05 level of significance with 12 pairs of data, a 14-64 split represents a significant departure from the null hypothesis, and any split farther apart (such as 13-65 or 12-66) will also represent a significant departure from the null hypothesis. Conversely, splits like 15-63, 16-62, or 38-40 do not represent significant departures away from a 39-39 split, and they would not be a basis for rejecting the null hypothesis. The Wilcoxon signed-ranks test is based on the lower rank total, so that instead of analyzing both numbers that constitute the split, it is necessary to analyze only the lower number.

In general, the sum $1 + 2 + 3 + \cdots + n$ is equal to $n(n + 1)/2$; if this is a rank sum to be divided equally between two categories (positive and negative), each of the two totals should be near $n(n + 1)/4$, which is $n(n + 1)/2$ after it is halved. Recognition of this principle forms a basis for understanding the rationale behind Formula 11-1. The denominator in that formula represents a standard deviation of T and is based on the principle that $1^2 + 2^2 + 3^2 + \cdots + n^2 = n(n + 1)(2n + 1)/6$.

If we were to apply the ordinary sign test (Section 11-2) to the example given in this section, we would fail to reject the null hypothesis of no change in before and after scores. This is not the conclusion reached through the Wilcoxon signed-ranks test, which is more sensitive to the magnitudes of the differences and is therefore more likely to be correct.

This section can be used for paired data only, but the next section involves a rank-sum test that can be applied to two sets of data that are not paired.

11-3 Exercises A
Wilcoxon Signed-Ranks Test
for Two Dependent Samples

In Exercises 11-25 through 11-32, first arrange the given data in order of lowest to highest and then find the rank of each entry.

11-25   5, 8, 12, 15, 10

11-26   1, 3, 6, 8, 99

11-27   150, 600, 200, 100, 50, 400

11-28   47, 53, 46, 57, 82, 63, 90, 55

11-29   6, 8, 8, 9, 12, 20

11-30   6, 8, 8, 8, 9, 12, 20

11-31   16, 13, 16, 13, 13, 14, 15, 18, 12

11-32   36, 27, 27, 27, 41, 39, 58, 63, 63

In Exercises 11-33 through 11-36, use the given before and after test scores in the Wilcoxon signed-ranks test procedure to:

(a) Find the differences *d*.
(b) Rank the differences while ignoring their signs.
(c) Find the signed ranks.
(d) Find *T*.

11-33

Before	103	98	112	94	118	99	90	101
After	100	105	114	98	119	99	100	116

11-34

Before	66	58	59	58	63	52	54	60
After	58	51	56	53	53	51	45	64

11-35

Before	83	76	91	59	62	75	80	66	73
After	82	77	89	62	68	70	90	86	73

11-36

Before	52	49	37	45	50	48	39	49	55	42	40
After	44	46	40	35	41	43	41	34	35	35	40

In Exercises 11-37 and 11-38, assume a 0.05 level of significance in a two-tailed hypothesis test. Use the given statistics to find the critical score from Table A-9 and then form a conclusion about the null hypothesis H_0.

11-37
(a) $T = 24$, $n = 15$ (b) $T = 25$, $n = 15$
(c) $T = 26$, $n = 15$ (d) $T = 81$, $n = 24$
(e) $T = 40$, $n = 17$

11-38
(a) $T = 25$, $n = 17$ (b) $T = 8$, $n = 10$
(c) $T = 5$, $n = 10$ (d) $T = 15$, $n = 12$
(e) $T = 24$, $n = 12$

In Exercises 11-39 through 11-48, use the Wilcoxon signed-ranks test.

11-39   A psychologist wants to test the claim that two different I.Q. tests produce the same results. Both tests are given to a sample of nine randomly selected students, with the results given below. At the 0.05 level of significance, test the claim that both tests produce the same results.

Test A	100	111	93	92	99	85	117	110	98
Test B	106	112	95	90	107	100	126	105	110

11-40   A biomedical researcher wants to test the effectiveness of a synthetic antitoxin. The 12 randomly selected subjects are tested for resistance to a particular poison. They are retested after receiving the antitoxin, with the results given below. At the 0.05 level of significance, test the claim that the antitoxin is not effective and produces no change.

Before	18.2	21.6	23.5	22.9	16.3	19.2	21.6	21.8	20.3	19.5	18.9	20.3
After	18.4	20.3	21.5	20.2	17.6	18.5	21.7	22.3	19.4	18.6	20.1	19.7

11-41   A researcher devises a test of depth perception while the subject has one eye covered. The test is repeated with the other eye covered. Results are given below for eight randomly selected subjects. At the 0.05 level of significance, test the claim that depth perception is the same for both eyes.

Right eye	14.7	16.3	12.4	8.1	21.6	13.9	14.2	15.8
Left eye	15.2	16.7	12.6	10.4	24.1	17.2	11.9	18.4

11-42   To test the effect of smoking on pulse rate, a researcher compiled data consisting of pulse rate before and after smoking. The results are given below. At the 0.05 level of significance, test the claim that smoking does not affect pulse rate.

Before smoking	68	72	69	70	70	74	66	71
After smoking	69	76	68	73	72	76	66	71

11-43   An anxiety-level index is invented for third grade students, and results for each of 14 randomly selected students are obtained in a classroom setting and a recess situation. The results follow. At the 0.05 level of significance, test the claim that anxiety levels are the same for both situations.

Classroom	7.2	7.8	6.0	5.1	3.9	4.7	8.2	9.1	8.7	4.1	3.8	6.7	5.8	4.9
Recess	8.7	7.1	5.8	4.2	4.0	3.6	7.0	8.7	7.9	2.5	2.4	5.0	4.0	3.4

11-44   Two types of cooling systems are being tested in preparation for construction of a nuclear power plant. Eight different standard experimental situations yield the temperature in degrees Fahrenheit of water expelled by each of the cooling systems, and the results follow. At the 0.05 level of significance, test the claim that both cooling systems produce the same results.

Type A	72	78	81	77	84	76	79	74
Type B	75	71	71	72	73	74	70	74

11-45 Randomly selected voters are given two different tests designed to measure their attitudes about conservatism. Both tests supposedly use the same rating scale and the same criteria. At the 0.05 level of significance, test the claim that both tests produce the same results. The sample data follow.

Test A	237	215	312	190	217	250	341	380	270	245
Test B	217	190	307	192	220	233	314	367	249	238

11-46 Two different ways of computing a consumer price index are developed. These indices are used in randomly selected trial cities, with the following sample results. At the 0.05 level of significance, test the claim that the indices yield equal results.

Index A	110.6	109.1	97.6	117.4	121.0	103.7	105.2	98.6
Index B	108.3	107.2	97.7	115.3	124.8	105.9	108.4	100.0

11-47 Randomly selected executives are surveyed in an attempt to measure their attitudes toward two different minority groups, and the sample results follow. At the 0.05 level of significance, test the claim that there is no difference in their attitudes towards the two groups.

Group A	420	490	380	570	630	710	425	576	550	610	580	575
Group B	520	510	450	530	600	705	415	600	625	730	500	530

11-48 Randomly selected married couples were given questionnaires about family attitudes, and their scores are summarized below. Test the claim that husbands and wives produce the same results.

Husband	62	68	54	73	72	83	76	74	73	74
Wife	54	65	69	55	61	71	79	79	73	81

11-3 Exercises B
Wilcoxon Signed-Ranks Test
for Two Dependent Samples

11-49 Two checkout systems are being tested at a department store. One system uses an optical scanner to record prices while the other system has prices manually entered by the clerk. Randomly selected customers are paid to use both checkout systems, and their processing times are recorded. Listed below are the differences (in seconds) obtained when the times for the scanner system are subtracted from the corresponding times for the manual system. At the 0.01 significance level, use the Wilcoxon signed-ranks test to test the claim that both systems require the same times.

$$
\begin{array}{rrrrrrrrrrrrr}
30 & 33 & 27 & 0 & -5 & -3 & 18 & 10 & 16 & 12 & 3 & 52 & 14 & -8 \\
-27 & 0 & 42 & 26 & 19 & 35 & 72 & 14 & 5 & 1 & 12 & -6 & 23 & 52 \\
47 & 33 & 19 & 16 & 0 & -12 & 44 & 40 & 29 & 59 & 38 & & &
\end{array}
$$

11-50 Do Exercise 11-49 after subtracting 20 from each value listed. How does that change affect the results?

11-51 (a) With $n = 8$ pairs of data, find the lowest and highest possible values of T.

   (b) With $n = 10$ pairs of data, find the lowest and highest possible values of T.

   (c) With $n = 50$ pairs of data, find the lowest and highest possible values of T.

11-52 Use Formula 11-1 to find the critical value of T for a two-tailed hypothesis test with a significance level of 0.05. Assume that there are $n = 100$ pairs of data with no differences of zero.

11-4 Wilcoxon Rank-Sum Test for Two Independent Samples

While Section 11-3 uses ranks to analyze dependent or paired data, this section introduces **Wilcoxon's rank-sum test,** which can be applied to situations involving two samples that are independent and not paired. (This test is equivalent to the **Mann-Whitney U test** found in some other books. See Exercise 11-66.)

We will test the null hypothesis that two independent samples come from populations with the same distribution. The alternative hypothesis is the claim that the two distributions are different in some way. This procedure requires that each sample size be greater than 10. For cases involving samples with 10 or fewer values, special tables are available in other reference books. Being typical of nonparametric tests, this procedure does not require that any distribution must be normal. The type of problem we consider will closely resemble many of the problems given in Section 8-3, where we tested hypotheses made about the means of two populations. However, the methods of Section 8-3 required that the samples come from normally distributed populations. The Wilcoxon rank-sum test of this section is a very reasonable alternative to those situations where the prerequisite of a normally distributed population is in question.

In Section 11-1, we noted that this test has a 0.95 efficiency rating when compared with the parametric t test or z test. Because this test has such a high efficiency rating and involves easier calculations, it is often preferred over the Section 8-3 parametric tests, even when the condition of normality is satisfied.

We will illustrate the procedure of the Wilcoxon rank-sum test with the following example. The basis of the procedure is the principle that if two samples are drawn from identical populations and the individual scores are all ranked as one combined collection of values, then the high and low ranks should be dispersed evenly between the two samples. If we find that the low (or high) ranks are found predominantly in one of the samples, we suspect that the two populations are not identical. While this is called a rank-sum test, it is actually the means of the ranks that will affect the results.

E X A M P L E

Random samples of teachers' salaries from New York State and Florida are as follows. At the 0.05 level of significance, test the claim that the salaries of teachers are the same in both states.

New York		Florida	
$25,300	(28)	$15,500	(8)
21,600	(24)	16,200	(10)
20,100	(21)	14,800	(5.5)
20,200	(22)	16,000	(9)
22,800	(25)	17,300	(13)
17,500	(15)	13,800	(1)
18,200	(16)	14,300	(4)
17,100	(12)	17,400	(14)
17,000	(11)	14,800	(5.5)
23,000	(26)	15,300	(7)
29,000	(29)	19,000	(17)
		21,300	(23)
		14,200	(3)
		14,100	(2)
		19,700	(18)
		23,200	(27)
		19,900	(19)
		20,000	(20)

$$n_1 = 11 \qquad n_2 = 18$$
$$R_1 = 229 \qquad R_2 = 206$$
$$\frac{R_1}{n_1} = 20.8 \qquad \frac{R_2}{n_2} = 11.4$$

Solution

H_0: The populations of salaries are identical.

H_1: The populations are not identical.

We may be tempted to use the student t test to compare the means of two independent samples (as in Section 8-3), but we cannot meet the prerequisite of having normally distributed populations, since salaries are not normally distributed. We therefore require a non-parametric method; Wilcoxon's rank-sum test is appropriate here.

We rank all 29 salaries, beginning with a rank of 1 (assigned to the lowest salary of \$13,800), a rank of 2 (assigned to the second lowest salary of \$14,100), and so forth. The ranks corresponding to the various salaries are shown in parentheses in the preceding table. Note that there is a tie between the fifth and sixth scores. We treat ties in the manner described at the end of Section 11-1, by computing the mean of the ranks involved in the tie and assigning that mean rank to each of the tying values. In the case of our tie between the fifth and sixth scores, we assign the rank of 5.5 to each of those two salaries since 5.5 is the mean of 5 and 6. We denote by R the sum of the ranks for one of the two samples. If we choose the New York salaries we get

$$R = 28 + 24 + 21 + 22 + 25 + 15 + 16 + 12 + 11 + 26 + 29$$
$$= 229$$

With a null hypothesis of identical populations and with both sample sizes greater than 10, the sampling distribution of R is approximately normal. Denoting the mean of the sample R values by μ_R and the standard deviation of the sample R values by σ_R, we are able to use the test statistic

Formula 11-2
$$z = \frac{R - \mu_R}{\sigma_R}$$

where

$$\mu_R = \frac{n_1(n_1 + n_2 + 1)}{2}$$

$$\sigma_R = \sqrt{\frac{n_1 n_2(n_1 + n_2 + 1)}{12}}$$

n_1 = size of the sample from which the rank sum R is found

n_2 = size of the other sample

R = sum of ranks of the sample with size n_1

The expression for μ_R is a variation of a result of mathematical induction, which states that the sum of the first n positive integers is given by $1 + 2 + 3 + \cdots + n = n(n + 1)/2$, and the expression for σ_R is a variation of a result that states that the integers 1, 2, 3, . . . , n have standard deviation $\sqrt{n(n + 1)/12}$.

For the salary data given in the table we have already found that the rank sum R for the New York salaries is 229. Since there are 11 New York salaries, we have $n_1 = 11$. Also, $n_2 = 18$ since there are 18 Florida salaries. We can now determine the values of μ_R, σ_R, and z.

$$\mu_R = \frac{n_1(n_1 + n_2 + 1)}{2} = \frac{11(11 + 18 + 1)}{2} = 165.0$$

$$\sigma_R = \sqrt{\frac{n_1 n_2(n_1 + n_2 + 1)}{12}} = \sqrt{\frac{(11)(18)(11 + 18 + 1)}{12}} = 22.2$$

$$z = \frac{R - \mu_R}{\sigma_R} = \frac{229 - 165.0}{22.2} = 2.88$$

The significance of the test statistic z can now be treated in the same manner as in previous chapters. We are now testing (with $\alpha = 0.05$) the hypothesis that the two populations are the same, so we have a two-tailed test with critical z values of 1.96 and −1.96. The test statistic of $z = 2.88$ falls within the critical region and we therefore reject the null hypothesis that the salaries are the same in both states. New York appears to have significantly higher teacher salaries than Florida.

We can verify that if we interchange the two sets of salaries, we will find that $R = 206$, $\mu_R = 270.0$, $\sigma_R = 22.2$, and $z = -2.88$, so that the same conclusion will be reached.

The Case of Coke versus Pepsi

In an advertising war between Coca-Cola and Pepsi-Cola, a television commercial showed a taste test in which the majority of regular Coke drinkers preferred Pepsi (labeled M) over Coke (labeled Q). The Coke camp responded with research results indicating that people prefer the letter M over the letter Q. One Coke commercial had an announcer saying that they ran a taste test with Coca-Cola in two glasses labeled M and Q. He went on to say that since most people chose M as the better-tasting drink, the test proves that people prefer the letter M over the letter Q. After Pepsi switched letters to L (for Pepsi) and S (for Coke), another Coke commercial had the taster saying that "when it comes to letters, I will pick L every time over S. L stands for liberty and lunch . . . all the things that really made our country great. Could we do numbers now? My favorite number is six."

Like the Wilcoxon signed-ranks test, this test also considers the relative magnitudes of the sample data, whereas the sign test does not. In the sign test, a weight loss of 1 pound or 50 pounds receives the same sign, so the actual magnitude of the loss is ignored. While rank-sum tests do not directly involve quantitative differences between data from two samples, changes in magnitude do cause changes in rank, and these in turn affect the value of the test statistic.

For example, if we change the Florida salary of $15,500 to $22,000, then the value of the rank-sum R will change, and the value of the z test statistic will also change.

11-4 Exercises A
Wilcoxon Rank-Sum Test for Two Independent Samples

11-53   The following scores were randomly selected from last year's college entrance examination scores. Use the Wilcoxon rank-sum test to test the claim that the performance of New Yorkers equals that of Californians. Assume a significance level of 0.05.

New York	520	490	571	398	602	475	557	621	737	403	511	598
California	508	563	385	617	704	401	409	527	393	478	521	536

11-54   The following scores represent the reaction times (in seconds) of randomly selected subjects from two age groups. Use the Wilcoxon rank-sum approach at the 0.05 level of significance to test the claim that both groups have the same reaction times.

18 years of age	1.96	0.94	0.96	1.51	1.36	1.41	1.03	1.12
	2.12	0.86	0.79	1.17	1.13	1.00	1.01	
50 years of age	1.03	1.42	1.75	2.01	0.93	1.92	2.00	1.87
	2.09	1.73	1.49	1.82				

11-55   A class of statistics students rated the president's performance using several different criteria, and the composite scores follow. Use the Wilcoxon rank-sum test at the 0.05 level of significance to test the claim that the ratings of both sexes are the same.

Males	27	36	42	57	88	92	60	60	43	29	76	79
Females	21	43	38	40	40	60	87	72	73	10	12	

11-56   Job applicants from two cultural backgrounds are tested to determine the number of trials they need to learn a certain task. The results are as follows. Use the Wilcoxon rank-sum test at the 0.05 level of significance to test the claim that the number of trials is the same for both groups.

Group A	7	12	18	15	13	14	22	9	11	10	10	10
Group B	21	19	17	8	16	16	20	24	6	19	23	

11-57   An auto parts supplier must send many shipments from the central warehouse to the city in which the assembly takes place, and she wants to determine the faster of two railroad routes. A search of past records provides the following data. (The shipment times are in hours.) At the 0.05 level of significance, use the Wilcoxon rank-sum test to determine whether or not there is a significant difference between the routes.

Route A	98	102	83	117	128	92	112	108	108	100	93	72	95	91
Route B	96	132	121	87	106	102	116	95	99	76	97	104	115	114

11-58   A large city police department offers a refresher course on arrest procedures. The effectiveness of this course is examined by testing 15 randomly selected officers who have recently completed the course. The same test is given to 15 randomly selected officers who have not had the refresher course. The results are as follows. At the 0.05 level of significance, test the claim that the course has no effect on the test grades by using the Wilcoxon rank-sum approach.

Group completing the course	173	141	219	157	163	165	178	200
	154	189	192	201	157	168	181	
Group without the course	159	124	170	148	135	133	137	189
	181	111	144	127	138	151	162	

11-59   A study is conducted to determine whether a drug affects eye movements. A standardized scale is developed and the drug is administered to one group, while a control group is given a placebo that produces no effects. The eye movement ratings of subjects are as follows. At the 0.01 level of significance, test the claim that the drug has no effect on eye movements. Use the Wilcoxon rank-sum test.

Drugged group	652	512	711	621	508	603	787	747	516	624	627	777	729
Control group	674	676	821	830	565	821	837	652	549	668	772	563	703
	789	800	711	598									

11-60   Two coffee-vending machines are studied to determine whether they distribute the same amounts. Samples are obtained and the contents (in liters) are as follows. At the 0.05 level of significance, use the Wilcoxon rank-sum test to test the claim that the machines distribute the same amount.

Machine A	0.210	0.213	0.206	0.195	0.180	0.250	0.212	0.217
	0.213	0.222	0.201	0.205	0.209			
Machine B	0.229	0.224	0.221	0.247	0.270	0.233	0.237	0.235
	0.238	0.200	0.198	0.216	0.241	0.273	0.205	

11-61 Groups of randomly selected men and women are given questionnaires designed to measure their attitudes toward capital punishment, and the results are as follows. At the 0.05 level of significance, test the claim that there is no difference between the attitudes of men and women concerning capital punishment. Use the Wilcoxon rank-sum test.

Men	67	72	48	30	92	15	5	87	91	54	66	72	98	97	75	74
Women	20	40	37	42	51	15	68	35	12	31	85					

11-62 In a study of longevity, two groups of adult males are randomly selected; their longevity data (in years) are summarized below. Use the Wilcoxon rank-sum test to test the claim that there is no difference between the two groups.

Group A	65	66	73	78	54	39	47	59	67	67
	69	71	74	77	62	73	75	76	68	67
Group B	64	67	73	70	58	69	72	71	63	64
	63	63	55	43	35	50	74	61	62	69

11-63 In a study of crop yields, two different fertilizer treatments are tested on parcels with the same area and soil conditions. Listed below are the yields (in bushels of corn) for sample plots. Use a 0.05 significance level and apply the Wilcoxon rank-sum test to determine whether there is a difference between the two treatments.

Treatment A	132	137	129	142	160	139	143	147	145	140	131	136
Treatment B	162	180	149	157	159	159	152	167	163	165	180	156
	158	151										

11-64 A consumer investigator obtains prices from mail order companies and computer stores. Listed below are the prices (in dollars) quoted for boxes of ten floppy disks from various manufacturers. Use a 0.05 level of significance to test the claim that there is no difference between mail order and store prices. Use the Wilcoxon rank-sum test (see page 532).

Mail order	23.00	26.00	27.99	31.50	32.75	27.00
	27.98	24.50	24.75	28.15	29.99	29.99
Computer store	30.99	33.98	37.75	38.99	35.79	33.99
	34.79	32.99	29.99	33.00	32.00	

11-4 Exercises B
Wilcoxon Rank-Sum Test
for Two Independent Samples

11-65 (a) The *ranks* for Group A are 1, 2, . . . , 15 and the *ranks* for Group B are 16, 17, . . . , 30. At the 0.05 level of significance, use the Wilcoxon rank-sum test to test the claim that both groups come from the same population.

(b) The *ranks* for Group A are 1, 3, 5, 7, . . . , 29 and the *ranks* for Group B are 2, 4, 6, . . . , 30. At the 0.05 level of significance, use the Wilcoxon rank-sum test to test the claim that both groups come from the same population.

(c) Compare parts (a) and (b).

(d) What changes occur when the rankings of the two groups in part (a) are interchanged?

(e) Use the two groups in part (a) and interchange the ranks of 1 and 30 and then note the changes that occur.

11-66 The Mann-Whitney U test is equivalent to the Wilcoxon rank-sum test for independent samples in the sense that they both apply to the same situations and they always lead to the same conclusions. In the Mann-Whitney U test we calculate

$$z = \frac{U - \dfrac{n_1 n_2}{2}}{\sqrt{\dfrac{n_1 n_2 (n_1 + n_2 + 1)}{12}}}$$

where

$$U = n_1 n_2 + \frac{n_1(n_1 + 1)}{2} + R$$

Show that if the expression for U is substituted into the above expression for z, we get the same test statistic (with opposite sign) used in the Wilcoxon rank-sum test for two independent samples.

11-67 Assume that we have two treatments (A and B) that produce measurable results, and we have only two observations for treatment A and two observations for treatment B. We cannot use Formula 11-2 because both sample sizes do not exceed 10.

(a) Complete the table below by listing the other five rows corresponding to the other five cases, and enter the corresponding rank sums for treatment A.

	Rank			(Rank sum for treatment A)
1	2	3	4	R
A	A	B	B	3
	.			.
	.			.
	.			.

(b) List the possible values of R along with their corresponding probabilities. (Assume that the rows of the table from part (a) are equally likely.)

(c) Is it possible, at the 0.10 significance level, to reject the null hypothesis that there is no difference between treatments A and B? Explain.

11-68 Do Exercise 11-67 for the case involving a sample of size three for treatment A and a sample of size three for treatment B.

11-5 Kruskal-Wallis Test

In Section 10-4 we used analysis of variance to test hypotheses that differences among several sample means are due to chance. That parametric F test required that all the involved populations possess normal distributions with variances that were approximately equal. In this section we introduce the **Kruskal-Wallis test** as a nonparametric alternative that does not have these more rigid requirements. In using the Kruskal-Wallis test (also called the H **test**), we test the null hypothesis that samples come from the same or identical populations. We compute the test statistic H, **which has a distribution that can be approximated by the chi-square distribution as long as each sample has at least five observations.** (For cases involving samples with fewer than five observations, we must refer to special tables of critical values.) When we use the chi-square distribution in this context, the number of degrees of freedom is $k - 1$, where k is the number of samples.

degrees of freedom = $k - 1$

Formula 11-3 $\quad H = \dfrac{12}{N(N + 1)}\left(\dfrac{R_1^2}{n_1} + \dfrac{R_2^2}{n_2} + \ldots + \dfrac{R_k^2}{n_k}\right) - 3(N + 1)$

where

N = total number of observations in all samples combined

R_1 = sum of ranks for the first sample

R_2 = sum of ranks for the second sample

R_k = sum of ranks for the kth sample

k = the number of samples

In using the Kruskal-Wallis test, we replace the original scores by their corresponding ranks. We then proceed to calculate the test statistic H, which is basically a measure of the variance of the rank sums R_1, $R_2, \ldots, R_k$. If the ranks are distributed evenly among the sample groups, then H should be a relatively small number. If the samples are very different, then the ranks will be excessively low in some groups and high in others, with the net effect that H will be large. Consequently, only large values of H lead to rejection of the null hypothesis that the samples come from identical populations. The Kruskal-Wallis test is therefore a right-tailed test.

Begin by considering all observations together, and then assign a rank to each one. We rank from lowest to highest, and we also treat ties as we did in the previous sections of this chapter—the mean value of the ranks is assigned to each of the tied observations. Then take each individual sample and find the sum of the ranks and the corresponding sample size. We will illustrate the Kruskal-Wallis test by considering the following example, which is similar to the one given in Section 10-4. The data here have been modified to include additional information about four more observations. We will see similarities between the Kruskal-Wallis test and Wilcoxon's rank-sum test since both are based on rank sums.

EXAMPLE

A pilot wants to buy a battery-powered radio as an alternative to his regular radios, which are powered by the airplane's electrical system. He has a choice of three brands of rechargeable batteries, which vary in cost. Sample data for each brand are obtained and are given below. The numbers represent the running time (in hours) before recharging becomes necessary. At the 0.05 level of significance, test the claim that the three samples come from identical populations.

Brand X	Brand Y	Brand Z
26.0	29.0	30.0
28.5	28.8	26.3
27.3	27.6	29.2
25.9	28.1	27.1
28.2	27.0	29.8
	26.9	29.7
		30.3
		30.3

Solution

H_0: The populations are identical.
H_1: The populations are not identical.

Step 1. Rank the combined samples from lowest to highest. Begin with the lowest observation of 25.9, which is assigned a rank of 1. The original data are repeated below with the appropriate ranks given in parentheses.

Brand X		Brand Y		Brand Z	
26.0	(2)	29.0	(13)	30.0	(17)
28.5	(11)	28.8	(12)	26.3	(3)
27.3	(7)	27.6	(8)	29.2	(14)
25.9	(1)	28.1	(9)	27.1	(6)
28.2	(10)	27.0	(5)	29.8	(16)
		26.9	(4)	29.7	(15)
				30.3	(18.5)
				30.3	(18.5)
↓		↓		↓	
$n_1 = 5$		$n_2 = 6$		$n_3 = 8$	
$R_1 = 31$		$R_2 = 51$		$R_3 = 108$	

(Note that Brands X, Y, and Z have mean ranks of 6.2, 8.5, and 13.5, respectively.)

Step 2. For each individual sample, find the number of observations and the sum of the ranks. The first sample (brand X) has five observations, so $n_1 = 5$ and $R_1 = 2 + 11 + 7 + 1 + 10 = 31$. The values of n_2, n_3, R_2, and R_3 are shown above. Since the total number of observations is 19, $N = 19$.

Step 3. Compute the value of the test statistic H. Using Formula 11-3 for the given data, we get

$$H = \frac{12}{N(N + 1)}\left(\frac{R_1^2}{n_1} + \frac{R_2^2}{n_2} + \frac{R_3^2}{n_3}\right) - 3(N + 1)$$

$$= \frac{12}{19(20)}\left(\frac{31^2}{5} + \frac{51^2}{6} + \frac{108^2}{8}\right) - 3(20)$$

$$= \frac{12}{380}(192.2 + 433.5 + 1458) - 60$$

$$= 65.801 - 60 = 5.801$$

Step 4. Since each sample has at least five observations, the distribution of H is approximately a chi-square distribution with $k - 1$ degrees of freedom. The number of samples is $k = 3$, so we get $3 - 1$, or 2, degrees of freedom. Refer to Table A-5 to find the critical value of 5.991, which corresponds to 2 degrees of freedom and a significance level of $\alpha = 0.05$. (This use of the chi-square distribution is always right-tailed, since only large values of H reflect disparity in the distribution of ranks among the samples.) If any sample has fewer than five observations, use the special tables found in texts devoted exclusively to nonparametric statistics. Such cases are not included in this text.

Step 5. The test statistic $H = 5.801$ is less than the critical value of 5.991, so we fail to reject the null hypothesis of identical populations. We reject the null hypothesis of identical populations only when H exceeds the critical value. The three brands of batteries are not significantly different.

In comparing the procedures of the parametric F test for analysis of variance and the nonparametric Kruskal-Wallis test, we see that the Kruskal-Wallis test is much simpler to apply. We need not compute the sample variances and sample means. We do not require normal population distributions. Life becomes so much easier. However, the Kruskal-Wallis test is not as efficient as the F test, and it may require more dramatic differences for the null hypothesis to be rejected.

Seat Belts Save Lives

Statistical analysis of data sometimes evolves into changes in public policy. The car seat belt issue is one such example. The Highway Users Federation recently estimated that if everyone in the United States used seat belts, the number of

highway deaths would drop by 12,000 each year, and there would be 330,000 fewer disabling injuries each year. Similar statistics have convinced some state legislatures to pass laws mandating the use of seat belts.

One study of 1126 accidents showed that riders wearing seat belts had 86% fewer life-threatening injuries. Another study of 28,780 accident victims showed that riders not wearing seat belts died in crashes involving speeds as low as 12 miles per hour, but no one wearing a seat belt and shoulder harness was killed at a speed under 60 miles per hour. It is

estimated that accident victims wearing seat belts are half as likely to be killed when compared to beltless accident victims.

Some people question the use of seat belts because they know of cases where serious injury was avoided when an unbelted rider was thrown clear of a wreck. There are cases where the seat belt had a negative effect, but such cases are far outnumbered by incidents where the seat belt was helpful. Clearly, the wisest strategy is to buckle up.

11-5 Exercises A
Kruskal-Wallis Test

In Exercises 11-69 through 11-80, use the Kruskal-Wallis test.

11-69 An experiment involves raising samples of corn under identical conditions except for the type of fertilizer used. The yields are obtained for three different fertilizers, and those values are ranked with the results shown below. Find the value of the test statistic H, where H is given in Formula 11-3.

Treatment		
A	B	C
1	2	3
6	4	5
7	8	9
12	11	10
14	15	13

11-70 Do Exercise 11-69 after replacing the given ranks with those listed below.

Treatment		
A	B	C
1	6	11
2	7	12
3	8	13
4	9	14
5	10	15

11-71 A sociologist randomly selects subjects from three different types of family structure: stable families, divorced families, and families in transition. The selected subjects are interviewed and rated for their social adjustment, and the sample results are given below. (The numbers in parentheses are the ranks.) At the 0.05 level of significance, test the claim that the samples come from identical populations.

Stable		Divorced		Transition	
108	(8.5)	113	(10)	92	(1)
104	(4)	106	(7)	123	(13)
103	(3)	108	(8.5)	126	(14)
97	(2)	127	(15)	105	(5.5)
118	(12)	114	(11)	105	(5.5)

11-72 Readability studies are conducted to determine the clarity of four different texts, and the sample scores follow. (The numbers in parentheses are the ranks.) At the 0.05 level of significance, test the claim that the four texts have the same readability level.

Text A	Text B	Text C	Text D
50 (3.5)	59 (10.5)	45 (1)	62 (13)
50 (3.5)	60 (12)	48 (2)	64 (15)
53 (6)	63 (14)	51 (5)	68 (18)
58 (9)	65 (16)	54 (7)	70 (19)
59 (10.5)	67 (17)	55 (8)	72 (20)

11-73 An introductory calculus course is taken by students with varying high-school records. The sample results of six students from each of the three groups follow. The values given are the final numerical averages in the calculus course. At the 0.05 level of significance, test the claim that the three groups come from identical populations.

Good high-school record	Fair high-school record	Poor high-school record
90	80	60
86	70	60
88	61	55
93	52	62
80	73	50
96	65	70

11-74 A store owner records the gross receipts for days randomly selected from times during which she used only newspaper advertising, only radio advertising, or no advertising. The results are listed below. At the 0.05 level of significance, test the claim that the receipts are the same, regardless of advertising.

Newspaper	Radio	None
845	811	612
907	782	574
639	749	539
883	863	641
806	872	666

11-75 Three methods of instruction are used to train air traffic controllers. With method A, an experienced controller is assigned to a trainee for practical field experience. With method B, trainees are given extensive classroom instruction and are then placed without supervision. Method C requires a moderate amount of classroom training and then several trainees are supervised on the job by an experienced instructor. A standardized test is used to measure levels of competency, and sample results are given below. At the 0.01 level of significance, test the claim that the three methods are equally effective.

Method A	Method B	Method C
195	187	193
198	210	212
223	222	215
240	238	231
251	256	252
		260
		267

11-76 Do Exercise 11-75 after adding 100 to each value of method A.

11-77 A unit on basic consumer economics is taught to five different classes of randomly selected students. A different teaching method is used for each group, and sample final test data are given below. The scores represent the final averages of the individual students. Test the claim that the five methods are equally effective.

Traditional	Programmed	Audio	Audiovisual	Visual
76.2	85.2	67.3	75.8	50.5
78.3	74.3	60.1	81.6	70.2
85.1	76.5	55.4	90.3	88.8
63.7	80.3	72.3	78.0	67.1
91.6	67.4	40.0	67.8	77.7
87.2	67.9		57.6	73.9
	72.1			
	60.4			

11-78 A study is made of the response time of police cars after dispatching occurs. Sample results for three precincts are given below in seconds. At the 0.05 level of significance, test the claim that the three precincts have the same response time.

Precinct 1	Precinct 2	Precinct 3
160	165	162
172	174	175
176	180	177
176	181	179
176	184	187
178	186	195
	190	210
	200	215
		216
		220

11-79 Given the measurements randomly obtained for the five samples below, can we conclude that the samples come from identical populations? Use a 0.05 level of significance. The given values are levels of algae in different ponds located in the same county.

Sample A	Sample B	Sample C	Sample D	Sample E
67.2	67.7	70.2	72.6	72.2
67.4	68.3	70.3	73.0	72.3
67.9	68.5	70.5	73.8	72.5
69.3	68.7	71.0	74.0	72.8
69.5	68.8	71.6	75.0	73.4
69.8	68.9	71.7	75.6	73.7
		71.9	75.7	74.6
		72.0	75.9	74.9

11-80   (a) If 20 is added to each observed sample value of Exercise 11-79, how is the value of the test statistic affected?

   (b) If each observed sample value in Exercise 11-79 is multiplied by 5, how is the value of the test statistic affected?

11-5 Exercises B
Kruskal-Wallis Test

11-81   Simplify Formula 11-3 for the special case of eight samples, all consisting of exactly six observations each.

11-82   For three samples, each of size five, what are the largest and smallest possible values of H?

11-83   In using the Kruskal-Wallis test, there is a correction factor that should be applied whenever there are many ties: Divide H by

$$1 - \frac{\Sigma T}{N^3 - N}$$

where $T = t^3 - t$. For each group of tied scores, find the number of observations that are tied and represent this number by t. Then compute $t^3 - t$ to find the value of T. Repeat this procedure for all cases of ties and find the total of the T values, which is ΣT. Find the corrected value of H for Exercise 11-73 by using this procedure.

11-84   Construct three sets of sample data with five scores in each sample. Assume that we want to test the null hypothesis (at the 0.05 significance level) that the samples come from the same population. Arrange the data so that the Kruskal-Wallis test conclusion is failure to reject the null hypothesis, while analysis of variance (see Section 10-4) leads to rejection of the null hypothesis.

11-6 Rank Correlation

In Chapter 9 we considered the concept of correlation, and we introduced the *linear correlation coefficient* as a measure of the strength of the association between two variables. In this section we will study **rank**

Does Television Watching Cause Violence?

A study of 1650 British youngsters yielded a strong correlation between the commission of violent acts and the number of hours of television viewed. A separate California study showed a very strong negative correlation between academic skills and television viewing; more television watching tended to correspond to lower skill levels. More than 6000 research studies on the effects of television on children have not led to much agreement, but the following conclusions seem to emerge from many of those studies.

- Prolonged television watching by children tends to correspond to aggressive behavior.

- Prolonged television watching by children tends to correspond to lower vocabulary and reading abilities.

- Children who view television extensively tend to view the world as being more violent than it really is.

Studies may show a correlation between television watching and aggressive behavior, but that doesn't imply that television is the cause of the aggressive behavior. Perhaps aggressive children watch television because it complements their existing behavior. Perhaps some other factor encourages aggressive behavior and television viewing.

correlation, the nonparametric counterpart of that parametric measure. In Chapter 9 we computed values for the linear correlation coefficient r, but in this section we will be computing values for the rank correlation coefficient. One major advantage of this nonparametric approach is that it allows us to analyze some types of data that can be ranked but not measured; yet such data could not be considered with the parametric linear correlation coefficient r of Chapter 9.

A second major advantage of rank correlation is that it can be used to detect some relationships that are not linear. An example illustrating this will be given later in this section.

Another possible advantage of this nonparametric approach is that the computations are much simpler than those for the linear correlation coefficient r. This can be readily seen by comparing Formula 11-4 (given later in this section) to Formula 9-1. With certain calculators, you can get the value of r easily, but if you do not have a calculator or computer, you would probably find that the rank correlation coefficient is easier to compute.

A fourth advantage of the rank correlation approach is that it can be used when some of the more restrictive requirements of the linear correlation approach are not met. That is, the nonparametric approach can be used in a wider variety of circumstances than can the parametric method. For example, the parametric approach requires that the involved populations have normal distributions; the nonparametric approach does not require normality.

As an example, observe the data in Table 11-5. Seven teachers are being ranked for promotion by two separate committees.

Table 11-5		
Teacher	Faculty rank	Administrator's rank
Arnold	1	2
Bennet	7	7
Cohen	2	6
Davis	5	3
Ellis	4	5
Farrell	3	1
Gallo	6	4

Even though we lack any of the real numerical data that led to the above table, we can still use the given ranks to obtain a rank correlation coefficient by using Formula 11-4.

Formula 11-4 $$r_s = 1 - \frac{6\Sigma d^2}{n(n^2 - 1)}$$

where

r_s = rank correlation coefficient

n = number of *pairs* of data

d = difference between ranks for the two observations within a pair

We use the notation r_s for the **rank correlation coefficient** so that we don't confuse it with the linear correlation coefficient r. The subscript s is commonly used in honor of Charles Spearman (1863-1945), who originated the rank correlation approach. In fact, r_s is often called **Spearman's rank correlation coefficient.** The subscript s has nothing to do with standard deviation, and for that we should be thankful. Just as r is a sample statistic that can be considered an estimate of the population parameter ρ, we can also consider r_s to be an estimate of ρ_s.

We now calculate r_s for the data given in Table 11-5. Since there are seven pairs of data, $n = 7$. We obtain the differences d for each pair by subtracting the lower rank from the higher rank.

Faculty rank	Administrator's rank	d (difference)	d^2
1	2	1	1
7	7	0	0
2	6	4	16
5	3	2	4
4	5	1	1
3	1	2	4
6	4	2	4
			Total 30

With $n = 7$ and $\Sigma d^2 = 30$, we get

$$r_s = 1 - \frac{6\Sigma d^2}{n(n^2 - 1)}$$

$$= 1 - \frac{6(30)}{7(7^2 - 1)}$$

$$= 1 - 0.536$$

$$= 0.464$$

For practical reasons, we are omitting the theoretical derivation of Formula 11-4, but we can gain some insight by considering the following three cases. If we intuitively examine Formula 11-4, we can see that strong agreement between the two sets of ranks will lead to values of d near zero, so that r_s will be close to 1 (see Case I). Conversely, when the ranks of one set tend to be at opposite extremes when compared to the ranks of the second set, then the values of d tend to be high, which will cause r_s to be near -1 (see Case II). If there is no relationship between the two sets of ranks, then the values of d will be neither high nor low and r_s will tend to be near zero (see Case III).

Case I: Perfect Positive Correlation

	Subject				
	A	B	C	D	E
Rank X	1	3	5	4	2
Rank Y	1	3	5	4	2
d	0	0	0	0	0

$\Sigma d^2 = 0$ and

$$r_s = 1 - \frac{6(0)}{5(5^2 - 1)} = 1$$

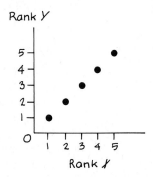

Study Criticized as Misleading

A study sponsored by the National Association of Elementary School Principals and the Kettering Foundation suggested that there was a *correlation* between children with one parent and children who were low achievers. The media coverage of the final report implied that the *cause* of lower achievement was the absence of one of the parents. Critics charged that the media confused correlation with cause and effect, and the study itself was misleading since it emphasized negative factors. Also, some conclusions were based on samples too small to be statistically significant. Critics cited the conclusion that children with single parents were three times more likely to be suspended, and charged that this conclusion was based on a study of only 11 children.

Case II: Perfect Negative Correlation

	Subject				
	A	B	C	D	E
Rank X	1	2	3	4	5
Rank Y	5	4	3	2	1
d	4	2	0	2	4

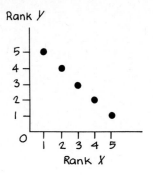

$\Sigma d^2 = 40$ and

$$r_s = 1 - \frac{6(40)}{5(5^2 - 1)} = -1$$

Case III: No Correlation

	Subject				
	A	B	C	D	E
Rank X	1	2	3	4	5
Rank Y	2	5	3	1	4
d	1	3	0	3	1

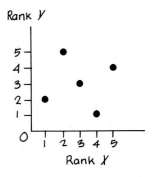

$\Sigma d^2 = 20$ and

$$r_s = 1 - \frac{6(20)}{5(5^2 - 1)} = 0$$

Cases I and II illustrate the most extreme cases, so that the following property applies.

$$-1 \le r_s \le 1$$

Recall that this same property applies to the linear correlation coefficient r described in Chapter 9. We can formulate conclusions about correlation by comparing the computed rank correlation coefficient r_s to a critical value. Suppose we have somehow established that at the 0.05 level of significance, the critical value of r_s for twelve pairs of ranks is 0.591. Figure 11-4 shows us how to interpret the computed rank correlation coefficient r_s for the critical value $r_s = 0.591$. Note that this interpretation follows the precedent set by the interpretation of the linear correlation coefficient r in Chapter 9. We have a table of critical values of r_s in Table A-10, which stops at $n = 30$ pairs of ranks. (Be careful to avoid confusion between Table A-7 for the linear correlation coefficient r and Table A-10 for Spearman's rank correlation coefficient r_s.) When the number of pairs of ranks, n, exceeds 30, the sampling distribution of r_s is approxi-

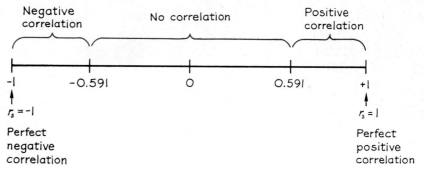

Figure 11-4

mately a normal distribution with mean zero and standard deviation $1/\sqrt{n-1}$. We therefore get

$$z = \frac{r_s - 0}{1/\sqrt{n-1}} = r_s\sqrt{n-1}$$

In a two-tailed case we would use the positive and negative z values. Solving for r_s, we then get

Formula 11-5
$$r_s = \frac{\pm z}{\sqrt{n-1}}$$

which we can use to get critical values of r_s when n exceeds 30. The value of z would correspond to the significance level.

E X A M P L E

Find the critical values of Spearman's rank correlation coefficient r_s when the data consist of 40 pairs of ranks. Assume a two-tailed case with a 0.05 significance level.

Solution

Since there are 40 pairs of data, $n = 40$. Because n exceeds 30, we use Formula 11-5 instead of Table A-10. With $\alpha = 0.05$ in two tails, we let $z = 1.96$ to get

$$r_s = \frac{\pm 1.96}{\sqrt{40-1}} = \pm 0.314$$

In Figure 11-5 we summarize the procedure to be followed when using Spearman's nonparametric approach in testing for correlation. This next example illustrates that procedure.

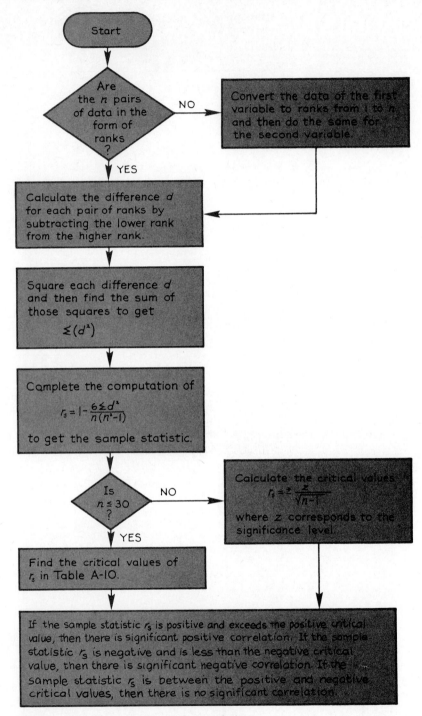

Figure 11-5 Spearman's rank correlation

EXAMPLE

Ten different students study for a test; the table below lists the number of hours studied (x) and the corresponding number of correct answers (y). At the 0.05 level of significance, use Spearman's rank correlation approach to determine if there is a correlation between hours studied and the number of correct answers.

x	5	9	17	1	2	21	3	29	7	100
y	6	16	18	1	3	21	7	20	15	22

Solution

$$H_0: \quad \rho_s = 0$$
$$H_1: \quad \rho_s \neq 0$$

Refer to Figure 11-5, which we will follow in this solution. The given data are not ranks, so we convert the table into ranks, as follows. (For x, 5 is the fourth score, 9 is the sixth score, 17 is the seventh score, and so on.)

x	4	6	7	1	2	8	3	9	5	10
y	3	6	7	1	2	9	4	8	5	10
d	1	0	0	0	0	1	1	1	0	0
d^2	1	0	0	0	0	1	1	1	0	0

After expressing all data as ranks, we next calculate the differences, d, and then we square them. The sum of the d^2 values is 4. We now calculate

$$r_s = 1 - \frac{6\Sigma d^2}{n(n^2 - 1)}$$

$$= 1 - \frac{6(4)}{10(10^2 - 1)}$$

$$= 1 - 0.024$$

$$= 0.976$$

Proceeding with Figure 11-5, $n = 10$, so we answer yes when asked if $n \leq 30$. We use Table A-10 to get the critical values of -0.648 and 0.648. Finally, the sample statistic of 0.976 exceeds 0.648, so we conclude that there is significant positive correlation. More hours of study appear to be associated with higher grades. (You didn't really think we would suggest otherwise, did you?)

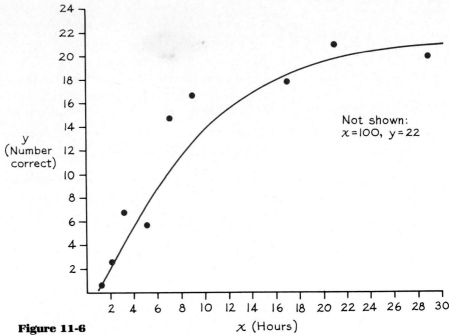

Not shown:
$x = 100$, $y = 22$

Figure 11-6

y (Number correct)

x (Hours)

Dangerous to Your Health

In 1965, the first warning labels were put on cigarette packs and, in 1969, cigarette commercials were banned from television and radio. Surgeon General C. Everett Koop calls cigarette smoking "the chief preventable cause of death in our so-ciety and the most important public health issue of our time." Reports of the surgeon general relate a plethora of convincing statistics showing that smoking greatly increases the danger of cancer. The Tobacco Institute contests these reports. Horace Kornegay, the Tobacco Institute's chairman, says that "while many people believe a causal link between smoking and cancer is a given, scientific research has not been able to establish that link." Critics note that there is not a single case in which smoking was determined to be the cause of cancer, but the available evidence clearly indicates a significant correlation.

If we compute the linear correlation coefficient r (using Formula 9-1) for the original data in this last example, we get $r = 0.629$, which leads to the conclusion that there is no significant linear correlation at the 0.05 level of significance. If we examine the scatter diagram of Figure 11-6, we can see that there does seem to be a relationship, but it's not linear. This last example was intended to illustrate two more advantages of the nonparametric approach over the parametric approach:

1. Spearman's rank correlation coefficient r_s is less sensitive to a value that is very far out of line, such as the 100 hours in the preceding data.
2. Spearman's rank correlation coefficient r_s sometimes indicates significant correlation even when the relationship between the variables is not linear. For example, r_s detected the curved relationship shown in Figure 11-6.

11-6 Exercises A
Rank Correlation

In Exercises 11-85 and 11-86, find the critical value for r_s by using Table A-10 or Formula 11-5, as appropriate. Assume two-tailed cases where α represents the level of significance and n represents the number of pairs of data.

11-85   (a) $n = 20$, $\alpha = 0.05$ (b) $n = 50$, $\alpha = 0.05$
            (c) $n = 40$, $\alpha = 0.02$ (d) $n = 25$, $\alpha = 0.02$
            (e) $n = 37$, $\alpha = 0.04$

11-86   (a) $n = 82$, $\alpha = 0.04$ (b) $n = 15$, $\alpha = 0.01$
            (c) $n = 50$, $\alpha = 0.01$ (d) $n = 37$, $\alpha = 0.01$
            (e) $n = 43$, $\alpha = 0.05$

In Exercises 11-87 and 11-88, rewrite the table of paired data so that each entry is replaced by its corresponding rank in accordance with the procedure described in this section.

11-87

x	2	5	8	6	10
y	30	25	20	15	5

11-88

x	15	14	10	9	3
y	3	7	10	11	20

In Exercises 11-89 through 11-92, compute the sample statistic r_s by using Formula 11-4.

11-89

x	1	2	3	4
y	4	2	3	1

11-90

x	10	20	30	40
y	40	20	30	10

11-91

x	63	68	71	55	70	75
y	43	44	39	30	28	20

11-92

x	28	28	35	37	40
y	16	17	12	19	20

In Exercises 11-93 through 11-104:

(a) Compute the rank correlation coefficient r_s for the given sample data.

(b) Assume that $\alpha = 0.05$ and find the critical value of r_s from Table A-10 or Formula 11-5.

(c) Based on the results of parts (a) and (b), decide whether there is significant positive correlation, significant negative correlation, or no significant correlation. In each case, assume a significance level of 0.05

11-93 Two of the judges in a gymnastics competition rank ten athletes, with the results given below.

	A	B	C	D	E	F	G	H	I	J
Judge Allen	8	6	1	3	5	10	9	7	2	4
Judge Zeller	6	7	1	4	3	10	9	8	5	2

11-94 A company ranks its sales personnel after giving them a test designed to measure aggressiveness. They are also ranked in order of sales for the last year. All these results are summarized below for 12 randomly selected employees.

	A	B	C	D	E	F	G	H	I	J	K	L
Aggressiveness	6	4	7	3	12	5	10	2	11	1	8	9
Sales	2	1	8	9	11	10	3	4	12	5	6	7

11-95 Several cities were ranked according to the number of hotel rooms and the amount of office space. The results are listed below.

City	NY	Ch	SF	Ph	LA	At	Mi	KC	NO	Da	Ba	Bo	Se	Ho	SL
Office rank	1	3	2	7	6	10	14	15	11	9	12	4	8	5	13
Hotel rank	1	2	3	8	6	7	14	15	4	10	13	5	9	11	12

11-96 The following table lists absences and corresponding final exam grades for randomly selected students.

Absences	5	12	1	7	7	40	9	10	4	2	14	3	6
Final exam grades	65	60	89	72	55	0	25	78	80	64	50	60	84

11-97 The following table lists the grade point averages and corresponding weights in pounds of 12 randomly selected students.

Grade-point average	3.60	2.96	3.20	2.63	2.80	2.01	3.25	3.00	3.20	3.00	2.53	2.94
Weight	136	127	148	167	153	148	121	145	155	160	110	130

11-98 A farmer notices that crickets seem to chirp faster on warm days. The accompanying table lists the number of chirps made by a cricket in 1 minute, along with the corresponding temperature in degrees Fahrenheit.

Chirps per minute	66	58	94	120	83	119	121	65	55	50
Temperature	55	53	62	70	59	69	69	55	52	51

11-99 Two different tests are designed to measure a person's understanding of a certain topic. Two tests are given to ten different subjects, with the results listed in the following table.

Text X	73	77	86	91	96	68	54	76	72	80
Test Y	45	48	50	51	54	38	31	41	51	57

11-100 A manager in a factory randomly selects 15 assembly line workers and develops scales to measure their dexterity and productivity levels. The results are listed in the following table.

Productivity	63	67	88	44	52	106	99	110	75	58	77	91	101	51	86
Dexterity	12	19	14	15	18	16	20	17	21	27	24	30	26	13	28

11-101 Randomly selected subjects are given a standard I.Q. test and then tested for their receptivity to hypnosis. The results are listed in the following table.

I.Q.	103	113	119	107	78	153	114	101	103	111	105	82	110	90	92
Receptivity to hypnosis	55	55	59	64	45	72	42	63	62	46	41	49	57	52	41

11-102   The following table lists the ages and diastolic blood pressures (in millimeters of mercury) of six randomly selected subjects.

Age	35	38	40	69	29	42
Blood pressure	89	81	100	90	59	82

11-103   The table below lists the weights of some popular cars along with their highway gas mileage consumption levels.

Weight (pounds)	1924	3185	4160	2250	3417	3751	5080	2225	2040
Highway (MPG)	33	22	15	42	20	21	14	36	36

11-104   A medical researcher tests the effects of a drug on the time it takes a patient to perform a standard manual task. The drug amounts in milligrams and corresponding times in seconds for randomly selected patients are given in the table.

Drug amount	15	20	25	30	35	40	45	50	55	60	65	70	75
Time	48	46	55	54	60	58	73	74	82	90	105	130	200

11-6   Exercises B
Rank Correlation

11-105   Two judges each rank three contestants, and the ranks for the first judge are 1, 2, and 3, respectively.
  (a)   List all possible ways that the second judge can rank the same three contestants. (No ties allowed.)
  (b)   Compute r_s for each of the cases found in part (a).
  (c)   Assuming that all of the cases from part (a) are equally likely, find the probability that the sample statistic, r_s, is greater than 0.9.

11-106   One alternative to using Table A-10 involves an approximation of critical values for r_s given as

$$r_s = \pm \sqrt{\frac{t^2}{t^2 + n - 2}}$$

where t is the t score from Table A-4 corresponding to the significance level and $n - 2$ degrees of freedom. Apply this approximation to find critical values of r_s for the following cases.

(a) $n = 8$, $\alpha = 0.05$ (b) $n = 15$, $\alpha = 0.05$
(c) $n = 30$, $\alpha = 0.05$ (d) $n = 30$, $\alpha = 0.01$
(e) $n = 8$, $\alpha = 0.01$

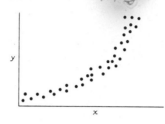

11-107 Given the bivariate data depicted in the scatter diagram, which would be more likely to detect the relationship between x and y: the linear correlation coefficient r or the rank correlation coefficient r_s? Explain.

11-108 (a) Find the sample statistic r_s for the sample data given below and find the critical value of r_s from Table A-10. What do you conclude? (Assume a 0.05 significance level.)

(b) For the same data, find the linear correlation coefficient r and the critical value of r from Table A-7. What do you conclude? (Assume a 0.05 significance level.)

(c) Change the y value of 36 to 36,000 and repeat parts (a) and (b).

(d) What do you conclude about the effect of one extreme score on r and r_s?

x	1	2	3	4	5	6
y	1	4	9	16	25	36

11-7   Runs Test for Randomness

A classic example of the use of statistics involves a manufacturer of pantyhose who obtained sample data describing the production output and later became convinced of employee stealing when fewer pairs of pantyhose were produced. As it turned out, the problem was not employee theft. Instead, the initial sampling was done with newly serviced machinery, which produced a finer mesh and more items than the same machinery used for many hours. The initial sampling was not random, and it led to misleading results and embarassment for the poor president who proclaimed that pantyhose were being pilfered.

In many of the examples and exercises already discussed in this book, it was explicitly assumed that data were randomly selected. But how can we really check for randomness? If a pollster reports back with a sequence of 20 interviews with females followed by a string of 20 interviews with males, we might strongly suspect a lack of randomness.

Magazine Survey Results Reflect Readership

Magazines often boost sales through surveys of their readers, but such surveys are typically biased and reflect only the views of the respondents. A *Time* article on magazine surveys noted that when wives were asked if they have ever had an extramarital affair, the results were 21% yes for *Ladies Home Journal,* 34% yes for *Playboy,* and 54% yes for *Cosmopolitan.* One pollster suggested that a *Reader's Digest* survey "would probably find that *nobody* had any extramarital affairs." Readers should realize that the results of such surveys are typically very biased because the surveys do not follow good techniques for sampling. While the results may be interesting, they are generally unreliable.

But where do we draw the line? We need a systematic and standard procedure for testing for the randomness of data. We can use the **runs test** as a test for randomness. We will work with sequences of data that fall into two distinct categories.

DEFINITION

A **run** is a sequence of data that exhibit the same characteristic; the sequence is preceded and followed by different data or no data at all.

As an example, consider the political party of the sequence of ten voters interviewed by a pollster. We let D denote a Democrat, while R indicates a Republican. The following sequence contains exactly four runs.

EXAMPLE 1

$$\underbrace{D,\ D,\ D,\ D,}_{\text{1st run}}\ \underbrace{R,\ R,}_{\text{2nd run}}\ \underbrace{D,\ D,\ D,}_{\text{3rd run}}\ \underbrace{R}_{\text{4th run}}$$

We would use the runs test in this situation to test for the randomness with which Democrats and Republicans occur. Let's use common sense to see how runs relate to randomness. Examine the sequence in Example 2 and then stop to consider how randomly Democrats and Republicans occur. Also count the number of runs.

EXAMPLE 2

$$D, D, D, D, D, D, D, D, D, D, R, R, R, R, R, R, R, R, R, R$$

In Example 2, it is reasonable to conclude that Democrats and Republicans occur in a sequence that is *NOT* random. Note that in the sequence of 20 data, there are only two runs. This example might suggest that if the number of runs is very low, randomness may be lacking. Now consider the sequence of 20 data given in Example 3. Try again to form your own conclusion about randomness before you continue reading.

EXAMPLE 3

$$D, R, D, R, D, R, D, R, D, R, D, R, D, R, D, R, D, R, D, R$$

In Example 3, it should be apparent that the sequence of Democrats and Republicans is again *NOT* random, since there is a distinct, predictable pattern. In this case, the number of runs is 20; this example suggests that randomness is lacking when the number of runs is too high.

It is important to note that this test for randomness is based on the *order* in which the data occur. This test is *NOT* based on the *frequency* of the data. For example, a particular sequence containing 3 Democrats and 20 Republicans might lead to the conclusion that the sequence is random. The issue of whether or not 3 Democrats and 20 Republicans is a *biased* sample is another issue not addressed by the runs test.

The sequences in Examples 2 and 3 are obvious in their lack of randomness, but most sequences are not so obvious; we therefore need more sophisticated techniques for analysis. We begin by introducing some notation.

NOTATION

n_1 Number of elements in the sequence that have the same characteristic.

n_2 Number of elements in the sequence that have the other characteristic.

G Number of runs.

We use G to represent the number of runs because n and r have already been used for other statistics, and G is a relatively innocuous letter that deserves more attention.

EXAMPLE

In the sequence $D, D, D, D, R, R, D, D, D, R, R, R, R, D$, we obtain the following values for n_1, n_2, and G.

$$n_1 = 8 \quad \text{since there are eight Democrats}$$
$$n_2 = 6 \quad \text{since there are six Republicans}$$
$$G = 5 \quad \text{since there are five runs}$$

We can now revert to our standard procedure for hypothesis testing. **The null hypothesis H_0 will be the claim that the sequence is random;** the alternative hypothesis H_1 will be the claim that the sequence is *NOT* random. The flowchart in Figure 11-7 summarizes the mechanics of the procedure. That flowchart directs us to a table (Table A-11) of critical G values when the following three conditions are all met:

Figure 11-7 The runs
test of randomness

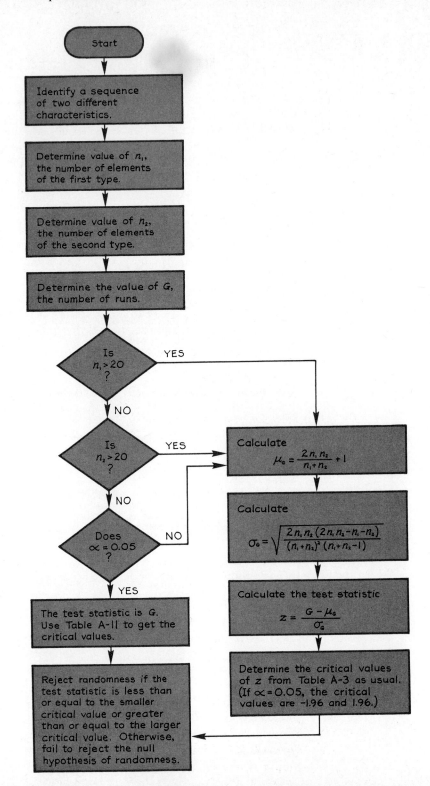

Uncle Sam Wants You, if You're Randomly Selected

A considerable amount of controversy was created when a lottery was used to determine who would be drafted into the U.S. Army. In 1970, the lottery approach was instituted in an attempt to make the selection process random, but many claimed that the outcome was unfair because men born later in the year had a better chance of being drafted.

The 1970 lottery involved 366 capsules cor-responding to the dates in a leap year. (The first dates selected would be the birthdays of the first men drafted.) First, the 31 January capsules were placed in a box. The 29 February capsules were added and the two months were mixed. Then the 31 March capsules were added and the three months were mixed. This process continued; one result was that January capsules were mixed 11 times, while December was mixed only once. Later arguments claimed that this process tended to place early dates closer to the bottom while dates later in the year tended to be near the top. The first ten dates selected, in order of priority, were September 14, April 24, December 30, February 14, October 18, September 6, October 26, September 7, November 22, and December 6. Statisticians used the runs test for randomness and concluded that the lottery was random. Yet there was enough criticism to cause a revised lottery the next year.

In 1971, statisticians used two drums of capsules. One drum contained capsules with dates, which were deposited in a random order according to a table of random numbers. A second drum contained the draft priority numbers, which were also deposited randomly according to a table of random numbers. Both drums were rotated for an hour before the selection process was begun. September 16 was drawn from one drum and 139 was drawn from the other. This meant that men born on September 16 had draft priority number 139. This process continued until all dates had a priority number. The randomness of this procedure was significantly less controversial.

1. $\alpha = 0.05$, and
2. $n_1 \leq 20$, and
3. $n_2 \leq 20$.

If all these conditions are not satisfied, we use the fact that G has a distribution that is approximately normal with mean and standard deviation as follows.

Formula 11-6
$$\mu_G = \frac{2n_1 n_2}{n_1 + n_2} + 1$$

Formula 11-7
$$\sigma_G = \sqrt{\frac{(2n_1 n_2)(2n_1 n_2 - n_1 - n_2)}{(n_1 + n_2)^2 (n_1 + n_2 - 1)}}$$

When this normal approximation is used, the test statistic is

Formula 11-8
$$z = \frac{G - \mu_G}{\sigma_G}$$

and the critical values are found by using the procedures introduced in Chapter 6. This normal approximation is quite good. If the entire table of critical values (Table A-11) had been computed using this normal approximation, no critical value would be off by more than one unit!

We now illustrate the use of the runs test for randomness by presenting examples of complete tests of hypotheses.

E X A M P L E

The president of an investment firm has observed that men and women have been hired in the following sequence: *M, M, M, W, M, M, M, M, W, W, W, M*. At the 0.05 level of significance, test the personnel officer's claim that the sequence of men and women is random. (Note that we are not testing for a *bias* in favor of one sex over the other. There are eight men and four women, but we are testing only for the *randomness* in the way they appear in the given sequence.)

Solution

The null hypothesis is the claim of randomness so we get

H_0: The eight men and four women have been hired in a random sequence.

H_1: The sequence is not random.

The significance level is $\alpha = 0.05$. Figure 11-7 summarizes the procedure for the runs test, so we refer to that flowchart. We now determine the values of n_1, n_2, and G for the given sequence.

$$n_1 = \text{number of men} = 8$$
$$n_2 = \text{number of women} = 4$$
$$G = \text{number of runs} = 5$$

Continuing with Figure 11-7, we answer no when asked if $n_1 > 20$ (since $n_1 = 8$), no when asked if $n_2 > 20$ (since $n_2 = 4$), and yes when asked if $\alpha = 0.05$. The test statistic is $G = 5$, and we refer to Table A-11 to find the critical values of 3 and 10. Figure 11-8 shows that the test statistic $G = 5$ does not fall in the critical region. We therefore fail to reject the null hypothesis that the given sequence of men and women is random.

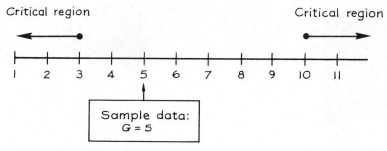

Figure 11-8

The next example will illustrate the procedure to be followed when Table A-11 cannot be used and the normal approximation must be used instead.

EXAMPLE

A machine produces defective (D)items and acceptable (A)items in the following sequence.

$$D, A, A, A, A, D, A, A, A, A, D, A, A, A, A,$$
$$D, A, A, A, A, D, A, A, A, A, D, A, A, A, A$$

At the 0.05 level of significance, test the claim that the sequence is random.

H_0: The sequence is random.

H_1: The sequence is not random.

Here $\alpha = 0.05$, and the statistic relevant to this test is G. Referring to the flowchart in Figure 11-7, we determine the values of n_1, n_2, and G as follows:

n_1 = number of defective items = 6
n_2 = number of acceptable items = 24
G = number of runs = 12

Continuing with the flowchart, we answer no when asked if $n_1 > 20$ (since $n_1 = 6$) and yes when asked if $n_2 > 20$ (since $n_2 = 24$). The flowchart now directs us to calculate μ_G. Because of the sample sizes involved, we are on our way to using a normal approximation.

$$\mu_G = \frac{2n_1 n_2}{n_1 + n_2} + 1$$

$$= \frac{2(6)(24)}{6 + 24} + 1$$

$$= \frac{288}{30} + 1 = 9.6 + 1 = 10.6$$

The next step requires the computation of σ_G.

$$\sigma_G = \sqrt{\frac{(2n_1 n_2)(2n_1 n_2 - n_1 - n_2)}{(n_1 + n_2)^2(n_1 + n_2 - 1)}}$$

$$= \sqrt{\frac{2(6)(24)[2(6)(24) - 6 - 24]}{(6 + 24)^2(6 + 24 - 1)}}$$

$$= \sqrt{\frac{288(258)}{900(29)}} = \sqrt{\frac{74{,}304}{26{,}100}}$$

$$= \sqrt{2.847} = 1.69$$

We now calculate the test statistic z.

$$z = \frac{G - \mu_G}{\sigma_G} = \frac{12 - 10.6}{1.69} = 0.83$$

At the 0.05 level of significance, the critical values of z are -1.96 and 1.96. The test statistic $z = 0.83$ does not fall in the critical region, so we fail to reject the null hypothesis that the sequence is random. This is a good example of why we do not say "accept the null hypothesis" because examination of the data reveals that there is a consistent pattern that is clearly not random.

In each of the preceding examples, the data clearly fit into two categories, but we can also test for randomness in the way numerical data fluctuate above or below a mean or median. That is, in addition to analyzing sequences of nominal data of the type already discussed, we can also analyze sequences of data at the interval or ratio levels of measurement. Consider the following sequence of times.

$$3,\ 2,\ 5,\ 2,\ 4,\ 6,\ 3,\ 7,\ 8,\ 5,\ 6,\ 9,\ 10,\ 4,\ 7$$

Let B denote a value *below* the sample mean of 5.4 and let A represent a value *above* 5.4. We can rewrite the given sequence as follows. (When a value is equal to the mean, we simply delete it from the sequence.)

$$B,\ B,\ B,\ B,\ B,\ A,\ B,\ A,\ A,\ B,\ A,\ A,\ \ A,\ B,\ A$$
$$3,\ 2,\ 5,\ 2,\ 4,\ 6,\ 3,\ 7,\ 8,\ 5,\ 6,\ 9,\ 10,\ 4,\ 7$$

Hint: it is helpful to write the A's and B's directly above the numbers they represent. This makes checking easier and it also reduces the chance of having the wrong number of letters. After finding the sequence of letters, we can apply the runs test in the usual way. We get

$$n_1 = \text{number of } B\text{'s} = 8$$
$$n_2 = \text{number of } A\text{'s} = 7$$
$$G = \text{number of runs} = 8$$

Consumer Price Index as a Measure of Inflation

The government uses the consumer price index (CPI) as a measure of inflation. Each month, the cost of certain goods and services is determined for a typical family. The CPI was constructed so that the cost of mortgages and home purchase prices accounted for about 25% of the total monthly cost. The effect was that radical changes in the cost of houses and mortgages caused radical changes in the monthly CPI, even though a small percentage of families actually buy or mortgage a house in any given month. The Bureau of Labor Statistics revised the formula for determining CPI so that home purchase costs were replaced by housing rental costs. In one particular month, that change lowered the CPI from 15.4% to 10%.

Since almost 50 million Americans receive wages or benefits affected by the CPI, its value has dramatic effects. For example, an increase in the CPI of only 1 percentage point will cost the government about $3 billion in increased spending.

From Table A-11, the critical values of G are found to be 4 and 13. At the 0.05 level of significance, we would reject a null hypothesis of randomness above and below the mean if the number of runs G is 4 or lower, or 13 or more. Since $G = 8$, we fail to reject the null hypothesis of randomness above and below the mean.

We could also test for randomness above and below the median by following the same procedure (after we have deleted from the sequence any values that are equal to the median).

Economists use the runs test for randomness above and below the median in an attempt to identify trends or cycles. An upward economic trend would contain a predominance of B's in the beginning and A's at the end, so that the number of runs would be small. A downward trend would have A's dominating the beginning and B's at the end, with a low number of runs. A cyclical pattern would yield a sequence that systematically changes, so that the number of runs would tend to be large.

11-7 Exercises A
Runs Test for Randomness

In Exercises 11-109 through 11-116, use the given sequence to determine the values of n_1, n_2, the number of runs, G, and the appropriate critical values from Table A-11. (Assume a 0.05 significance level.)

11-109 A, B, B, B, A, A, A, A

11-110 A, A, A, B, A, B, A, A, B, B

11-111 A, A, A, A, B, B, B, B, B, B

11-112 A, A, B, B, A, B, A, B, A, B, A

11-113 A, A, A, A, A, A, A, B, B

11-114 A, A, A, B, A, A, B, B, A, B, B, B, B

11-115 A, B, B, B, A, B, B, B, B, B, A, B, A, A, A, A, A, B, A, A

11-116 A, A, A, B, B, B, B, A, A, A, A, A, B, B, B, B, B, A, A, A, B, B, B, B, B, B, A, A, A, A, A

In Exercises 11-117 through 11-120, use the given sequence to answer the following.
(a) Find the mean.
(b) Let B represent a value below the mean and let A represent a value above the mean. Rewrite the given numerical sequence as a sequence of A's and B's.
(c) Find the values of n_1, n_2, and G.
(d) Assuming a 0.05 level of significance, use Table A-11 to find the appropriate critical values of G.
(e) What do you conclude about the randomness of the values above and below the mean?

11-117 3, 8, 7, 7, 9, 12, 10, 16, 20, 18

11-118 2, 2, 3, 2, 4, 4, 5, 8, 4, 6, 7, 9, 9, 12

11-119 15, 12, 12, 10, 17, 11, 8, 7, 7, 5, 6, 5, 5, 9

11-120 3, 3, 4, 4, 4, 8, 8, 8, 9, 7, 10, 4, 4, 3, 2, 4, 4, 10, 10, 12, 9, 10, 10, 11, 5, 4, 3, 3, 4, 5

In Exercises 11-121 through 11-124, use the given sequence to answer the following.
(a) Find the median.
(b) Let B represent a value below the median and let A represent a value above the median. Rewrite the given numerical sequence as a sequence of A's and B's.
(c) Find the values of n_1, n_2, and G.

(d) Assuming a 0.05 level of significance, use Table A-11 to find the appropriate critical values of G.

(e) What do you conclude about the randomness of the values above and below the median?

11-121 3, 8, 7, 7, 9, 12, 10, 16, 20, 18

11-122 2, 2, 3, 2, 4, 4, 5, 8, 4, 6, 7, 9, 9, 12

11-123 3, 3, 4, 5, 8, 8, 8, 9, 9, 9, 10, 15, 14, 13, 16, 18, 20

11-124 19, 17, 3, 2, 2, 15, 4, 12, 14, 13, 9, 7, 7, 7, 8, 8, 8

11-125 A pollster is hired to determine the sex and income of the heads of households. The sexes, in order of occurrence, are given below. At the 0.05 level of significance, test the claim that males and females occur randomly.

F, F, F, F, F, F, M, M, M, F, F, F,

M, M, M, M, M, M, M, M, M, M, M, M, M, M

11-126 At the 0.05 level of significance, test for randomness of odd (O) and even (E) numbers on a roulette wheel that yielded the results given in the sequence below.

O, E, O, O, O, O, E, E, O, O, E, O, O, O, O, O, O,

E, E, E, O, O, O, E, O, O, O

11-127 A machine produces defective (D) items and acceptable (A) items in the sequence given below. At the 0.05 level of significance, test the claim that the sequence is random.

A, A, A, D, A, A, A, A, D, A, A, A, A, A, D, A, A, A, A, A, A, D,

A, A, A, A, A, A, A, D, D, D, A, D, D, D, D, A, A, A, A, D, D, D, D

11-128 The Apgar rating scale is used to rate the health of newborn babies. A sequence of births results in the following values.

9, 6, 9, 8, 7, 9, 5, 8, 9, 10, 8, 9, 7, 5, 9, 8, 8

At the 0.05 level of significance, test the claim that the ratings above and below the sample mean are random.

11-129 The number of housing starts for the most recent years in one particular county are listed below. At the 0.05 level of significance, test the claim of randomness of those values above and below the *median.*

973, 1067, 856, 971, 1456, 903, 899, 905,

812, 630, 720, 676, 731, 655, 598, 617

11-130 Recent daily lottery drawings in a certain state are listed below, in their order of occurrence. At the 0.05 level of significance, test the claim that the occurrence of *even* and *odd* numbers in that sequence is random.

429, 103, 911, 093, 802, 635, 488, 697, 612, 754, 597, 587, 791,
845, 209, 228, 861, 531, 483, 953, 871, 646, 947, 987, 436, 308,
494, 695, 363, 386, 922, 244, 308, 527, 814, 322, 191, 687, 615,
367, 116, 667, 750, 619, 162, 645, 612, 874, 323, 574, 042, 634

11-131 An elementary-school nurse records the grade level of students that develop chicken pox; the sequential list is given below. Consider children in grades 1, 2, and 3 to be younger, while those in grades 4, 5, and 6 are older. At the 0.05 level of significance, test the claim that the sequence is random with respect to the categories of younger and older.

3, 3, 3, 2, 3, 1, 4, 3, 6, 5, 3, 3, 2, 1, 4, 4, 4, 4, 4, 4, 4, 4, 4,
5, 5, 5, 5, 4, 3, 2, 1, 6, 6, 6, 6, 1, 3, 4, 2, 5, 6, 6, 6, 6, 6

11-132 A computer program is used to generate random numbers from 1 through 6, and the sequence of results is listed below. Test the claim of randomness of the given values above and below the median. Assume a 0.02 level of significance.

1, 6, 5, 4, 2, 3, 1, 6, 5, 4, 2, 3, 3, 1, 6, 5, 4, 2, 2, 3, 1, 6, 5, 4,
4, 2, 3, 1, 6, 5, 5, 4, 2, 3, 1, 6, 5, 5, 4, 2, 3, 1, 5, 4, 2, 3, 1, 6

11-133 A manufacturer of ball bearings analyzes the sequence of output on a particular machine. Observation of the sequence of the first 60 bearings results in 30 below the median and 30 above. The number of runs is found to be eight. At the 0.05 level of significance, test for randomness above and below the median.

11-134 A sequence of 50 altimeters produced by a manufacturer yields 30 altimeters that read too low and 20 that read too high. The number of runs is determined to be 18. At the 0.05 level of significance, test for randomness of high and low readings in the sequence of 50 altimeters.

11-135 An economist analyzes a sequence of 80 consumer price indices. In testing for randomness above and below the median, six runs were found. What do you conclude? Assume a 0.05 level of significance.

11-136 A sequence of dates is selected through a process that has the outward appearance of being random. At the 0.05 level of significance, test the resulting sequence for randomness before and after the middle of the year.

Nov. 27, July 7, Aug. 3, Oct. 19, Dec. 19, Sept. 21, Apr. 1,
Mar. 5, June 10, May 21, June 27, Jan. 5

11-7 Exercises B
Runs Test for Randomness

11-137 Using A, A, B, B, what is the minimum number of possible runs that can be arranged? What is the maximum number of runs? Now refer to Table A-11 to find the critical G values for $n_1 = n_2 = 2$. What do you conclude about this case?

11-138 Let $z = 1.96$ and $n_1 = n_2 = 20$, and then compute μ_G and σ_G. Use those values in Formula 11-8 to solve for G. What is the importance of this result? How does this result compare to the corresponding value found in Table A-11? How do you explain any discrepancy?

11-139 (a) Using all of the elements $A, A, A, B, B, B, B, B, B$, list the 84 different possible sequences.
(b) Find the number of runs for each of the 84 sequences.
(c) Assuming that each sequence has a probability of $\frac{1}{84}$, find $P(2 \text{ runs})$, $P(3 \text{ runs})$, $P(4 \text{ runs})$, and so on.
(d) Use the results of part (c) to establish your own critical G values in a two-tailed test with 0.025 in each tail.
(e) Compare your results to those given in Table A-11.

11-140 Using all of the elements $A, A, A, B, B, B, B, B, B, B, B, B$, it is possible to arrange 220 different sequences.
(a) List all of those sequences having exactly three runs.
(b) Using your result from part (a), find $P(3 \text{ runs})$.
(c) Using your answer to part (b), should $G = 3$ be in the critical region?
(d) Find the lower critical value from Table A-11.
(e) Find the lower critical value by using the normal approximation.

Computer Project
Nonparametric Statistics

Most computer systems have a random-number generator. Use such a system to generate 100 numbers and then use the runs test to test for randomness above and below the mean.

Review

In this chapter we examined six different **nonparametric** methods for analyzing statistics. Besides excluding involvement with population parameters like μ and σ, nonparametric methods are not encumbered by many of the restrictions placed on parametric methods, such as the requirement that data must come from normally distributed populations. While nonparametric methods generally lack the sensitivity of their parametric counterparts, they can be used in a wider variety of circumstances and can accommodate more types of data. Also, the computations required in nonparametric tests are generally much simpler than the computations required in the corresponding parametric tests.

In Table 11-6, we list the nonparametric tests of this chapter, along with their functions. The table also lists the corresponding parametric tests. Following Table 11-6 is a listing of the important formulas from this chapter.

Table 11-6

Nonparametric test	Function	Parametric test
Sign test (Section 11-2)	Test for claimed value of average with one sample.	t test or z test (Sections 6-2, 6-3, 6-4)
	Test for a difference between two dependent samples.	t test or z test (Section 8-3)
Wilcoxon signed-ranks test (Section 11-3)	Test for difference between two dependent samples.	t test or z test (Section 8-3)
Wilcoxon rank-sum test (Section 11-4)	Test for difference between two independent samples.	t test or z test (Section 8-3)
Kruskal-Wallis test (Section 11-5)	Test for more than two independent samples coming from identical populations.	Analysis of variance (Section 10-4)
Rank correlation (Section 11-6)	Test for a relationship between two variables.	Linear correlation (Section 9-2)
Runs test (Section 11-7)	Test for randomness of sample data.	(No parametric test)

Review Exercises

11-141 Final exam grades are listed below according to the order in which they were completed. At the 0.05 level of significance, test for randomness above and below the mean.

45 50 92 87 79 89 93 75 76 74 76 73 65 68 69

70 78 60 60 60 55 100

11-142 An annual art award was won by a woman in 30 out of 40 presentations. At the 0.05 significance level, use the sign test to test the claim that men and women are equal in their abilities to win this award.

IMPORTANT FORMULAS

Test	Test statistic	Distribution
Sign test	Test statistic when $n > 25$: $$z = \frac{(x + 0.5) - (n/2)}{\sqrt{n}/2}$$	If $n \leq 25$, see Table A-8. If $n > 25$, use the normal distribution (Table A-3).
Wilcoxon signed-ranks test for two dependent samples	Test statistic when $n > 30$: $$z = \frac{T - n(n + 1)/4}{\sqrt{\dfrac{n(n + 1)(2n + 1)}{24}}}$$ (T = smaller of the rank sums)	If $n \leq 30$, see Table A-9. If $n > 30$, use the normal distribution (Table A-3).
Wilcoxon rank-sum test for two independent samples	$$z = \frac{R - \mu_R}{\sigma_R}$$ where R = sum of ranks of the sample with size n_1 $$\mu_R = \frac{n_1(n_1 + n_2 + 1)}{2}$$ $$\sigma_R = \sqrt{\frac{n_1 n_2(n_1 + n_2 + 1)}{12}}$$	Normal distribution (Table A-3) (Requires that $n_1 > 10$ and $n_2 > 10$.)
Kruskal-Wallis test	$$H = \frac{12}{n(n + 1)}\left(\frac{R_1^2}{n_1} + \frac{R_2^2}{n_2} + \cdots + \frac{R_k^2}{n_k}\right) - 3(n + 1)$$ where N = total number of sample values R_k = sum of ranks for the kth sample k = the number of samples	Chi-square with $k - 1$ degrees of freedom (Requires that each sample has at least five values.)
Rank correlation	$$r_s = 1 - \frac{6\Sigma d^2}{n(n^2 - 1)}$$	If $n \leq 30$, use Table A-10. If $n > 30$, critical values are $$r_s = \frac{\pm z}{\sqrt{n - 1}}$$ where z is from the normal distribution.
Runs test for randomness	Test statistic when $n > 20$: $$z = \frac{G - \mu_G}{\sigma_G}$$ where $\mu_G = \dfrac{2n_1 n_2}{n_1 + n_2} + 1$ $$\sigma_G = \sqrt{\frac{(2n_1 n_2)(2n_1 n_2 - n_1 - n_2)}{(n_1 + n_2)^2(n_1 + n_2 - 1)}}$$	If $n \leq 20$, use Table A-11. If $n > 20$, use the normal distribution (Table A-3).

11-143 A dentist records the ages (in years) and total annual bills (in dollars) for several randomly selected patients, and the results are listed below. Is there a correlation between age and the amount billed?

Age	7	9	12	36	42	15	16	17	27	53	55	8
Dental bill	60	72	95	312	108	220	90	40	65	120	30	430

11-144 Three machines are programmed to produce keychains, and the daily outputs are listed below for randomly selected days. Test the claim that the machines produce the same amount. Use a 0.05 level of significance.

Machine A: 660   690   690   672   683
Machine B: 590   588   560   570   592
Machine C: 520   572   578   553   564

11-145 A study was conducted of drivers on similar sections of an interstate highway in Iowa and Nevada, and randomly selected speeds are listed below. At the 0.05 level of significance, use the Wilcoxon rank-sum test to test the claim that there is no difference between the speeds in the two states.

Iowa	56	58	62	54	47	68	70	55	56	52	53	51
Nevada	63	60	59	57	67	64	65	72	50	55	55	49
	48	46										

11-146 Randomly selected cars are tested for fuel consumption and then re-tested after a tune-up. The measures of fuel consumption follow. At the 0.05 significance level, use the sign test to evaluate the claim that the tune-up has no effect on fuel consumption.

Before tune-up	16	23	12	13	7	31	27	18	19	19	19	11	9	15
After tune-up	18	23	16	17	8	29	31	21	19	20	24	13	14	18

11-147 Do Exercise 11-146 using the Wilcoxon signed-ranks test instead of the sign test.

11-148 A pollster is hired to collect data from 30 randomly selected adults. As the data are turned in, the sex of the interviewed subjects is noted and the sequence below is obtained. At the 0.05 level of significance, test the claim that the sequence is random.

M, M, M, M, M, M, M, M, F, M, M, M, M,

F, F, F, F, F, F, F, F, F, M, M, M, M, M, M, M, M

11-149 Sample entrance exam scores are given below for applicants randomly selected from three counties. Test the claim that the three county populations are identical. Use a 0.05 level of significance.

Kings County	76	52	39	27	88	73	75	92	99	83	79	
County	67	48	52	40	53	91	30	20	18	23	61	63	62
Bronx County	78	76	52	63	85	77	68					

11-150 A college dean randomly selects several faculty who earn money for extra duties and lists those earnings (in dollars) along with their corresponding ages (in years). Is there a correlation between age and extra service income? Use the rank correlation coefficient at the 0.05 level of significance.

Age	27	42	29	30	32	55	57	33
Income	3,600	120	2,800	1,200	1,200	50	140	900

11-151 A police academy gives an entrance exam; sample results for applicants from two different counties follow. Use the Wilcoxon rank-sum test to test the claim that there is no difference between the scores from the two counties. Assume a significance level of 0.05.

Kings County	76	52	39	27	88	73	75	92	99	83	79	
Queens County	67	48	52	40	53	91	30	20	18	23	61	63	62

11-152 An index of civic awareness is obtained for ten randomly selected subjects before and after they take a short course. The results follow. Use the Wilcoxon signed-ranks test to test the claim that the course does not affect the index. Use a 0.05 level of significance.

Before course	63	65	67	70	71	77	78	80	85	91
After course	73	64	72	74	77	80	79	82	96	91

11-153 Test the claim that the median age of a teacher in Montana is 38 years. The ages of a sample group of teachers from Montana are 35, 31, 27, 42, 38, 39, 56, 64, 61, 33, 35, 24, 25, 28, 37, 36, 40, 43, 54, and 50. Use a 0.05 significance level.

11-154 A telephone solicitor writes a report listing N for unanswered calls and Y for answered calls. At the 0.05 level of significance, test the claim that the sample listed below is random.

$N N N N N N N N Y Y N N N N N N Y Y Y Y Y Y Y N N N N N N N N N N N N$

11-155 Samples of equal amounts of two brands of a food substance were randomly selected and the amounts of carbohydrates were measured (in grams). The results are listed below. Use the Wilcoxon rank-sum test to test the claim that both brands are the same when compared on the basis of carbohydrate content. Use a 0.05 level of significance (see page 570).

Brand x	20.3	21.2	19.3	19.2	19.1	19.0	22.6	23.6	22.9	20.7	20.7
Brand y	18.9	18.8	19.1	21.0	20.0	18.6	20.4	23.3	20.1	17.9	17.7

11-156 Three judges grade entries in a science fair. The grades for eight randomly selected contestants are as follows. Use the rank correlation coefficient to determine if there is significant positive correlation, significant negative correlation, or no significant correlation between the scores of Judge A and Judge B. Use a significance level of 0.05.

Judge A	6.3	7.2	6.6	8.5	9.7	7.0	7.3	8.8
Judge B	7.1	6.5	8.2	8.6	9.0	6.1	6.3	8.8
Judge C	7.0	6.4	8.3	8.5	8.9	6.0	6.2	8.7

11-157 Use the data for Judge A and Judge B in Exercise 11-156 to test the claim that there is no difference between their scores. Use the Wilcoxon signed-ranks test at the 0.05 level of significance.

11-158 Use the data from all three judges listed in Exercise 11-156 and test the claim that they give the same scores. Use a 0.05 level of significance.

Vocabulary List

Define and give an example of each term.

parametric methods
nonparametric
   methods
efficiency
rank
sign test

Wilcoxon
   signed-ranks test
Wilcoxon rank-sum
   test
Kruskal-Wallis test
H test
rank correlation
   coefficient

Spearman's rank
   correlation
   coefficient
runs test
run

Chapter 11
Case Study Activity

Obtain a collection of sample data listed in the order of selection, and test for randomness. You might refer to summaries of state lottery numbers, or stock market changes, or priority numbers used in the draft (see *U.S. News and World Report,* page 34 of the December 15, 1969, issue, or page 27 of the July 13, 1970, issue).

12-1 **Designing the Experiment**
The complete plan for data collection must be defined before the collection process is started.

12-2 **Sampling and Collecting Data**
We present different methods of sampling: random sampling, stratified sampling, systematic sampling, and cluster sampling.

12-3 **Analyzing Data and Drawing Conclusions**
The conclusions must be developed according to the previously developed design of the experiment.

12-4 **Writing the Report**
We present a suggested outline for a written report.

12-5 **A Do-It-Yourself Project**
We present suggestions for a student project.

Design, Sampling, and Report Writing

IDENTIFYING OBJECTIVES

In setting up a statistical experiment, we must begin by determining exactly what question we want answered. Beginning researchers are often overcome by enthusiasm as they set out to collect facts without considering *precisely* what they are investigating and which facts are truly relevant. The original statement of the problem is often too vague or too broad. When this happens, too many different directions are sometimes pursued. For example, a social scientist may want to know how attitudes toward racial integration have changed in the past 30 years. This broad problem must be condensed to some specific areas that can be objectively measured and analyzed. One method is to repeat earlier surveys used to determine the percentages of people with prointegration responses to such questions as:

Do you think white students and black students should attend separate schools?

Do you think marriages between blacks and whites should be prohibited by law?

It is important at this stage of the experiment to identify the population that will be considered. If we are measuring changing attitudes toward racial integration, we must identify exactly the group whose attitudes we want to know. The population may be adult Americans over the age of 18, Southern whites, middle-class adult white males, and so on.

12-1 Designing the Experiment

In order to obtain meaningful data in an efficient way, we must develop a complete plan for collecting data *before* the collection is actually begun. Researchers are often frustrated and discouraged if they learn that the method of collection or the data themselves cannot be used to answer their questions. Will the experiment be conducted on the entire population (in which case we will use descriptive statistics) or will a sample be drawn from the population (in which case we will use inferential statistics)? The population size usually requires us to make inferences on the basis of sample data. We need to determine the size of the sample and the method of sampling in the very beginning.

Some aspects of sample size determination were discussed in Chapter 7, and we discuss some types of sampling later in this chapter. In addition to determining sample size and sampling plan, we need to obtain a source from which to draw our samples. This source may be a telephone directory, voter registration files, class roster, car registration list, etc.

Finally, we should describe how we intend to analyze the data after sampling is completed. We should note which statistics are to be computed and which formulas will be used in making estimates and conducting tests of hypotheses.

12-2 Sampling and Collecting Data

Sampling and data collection usually require the most time, effort, and money. Careful planning will minimize the expenditure of those precious resources. Take care that the sampling is done according to plan and the data are recorded in a complete and accurate manner. Some different methods of sampling follow.

Random Sampling

In random sampling, each member of the population has an equal chance of being selected. Random sampling is also called representative or proportionate sampling since all groups of the population should be proportionately represented in the sample.

Random sampling is not the same as haphazard or unsystematic sampling. Much effort and planning must be invested in order to avoid any bias. For example, if a list of all elements from the population is available, the names of those elements can be placed in capsules and put in a bowl; those capsules can then be mixed and samples selected.

Another approach involves numbering the list and using a table of random numbers to determine which specific elements are to be selected.

A major problem with these approaches is the difficulty of finding a complete list of *all* elements in the population.

Survey Medium Can Affect Results

In a survey of Catholics in Boston, the subjects were asked if contraceptives should be made available to unmarried women. In personal interviews, 44% of the respondents said yes. But for a similar group contacted by mail or telephone, 75% of the respondents answered yes to the same question.

Stratified Sampling

After classifying the population into at least two different strata (or classes) that share the same characteristics, we draw a sample from each stratum. In surveying views on an equal rights amendment to the Constitution, we might use sex as a basis for creating two strata. After obtaining a list of men and a list of women, we use some suitable method (such as random sampling) to select a certain number of people from each list. If it should happen that some strata are not represented in the proper proportion, then the results can be adjusted or weighted accordingly. Stratified sampling is often the most efficient of the various types.

Systematic Sampling

In systematic sampling we select some starting point and then select every nth element. For example, we use a telephone directory of 10,000 names as our population, and we must choose 200 of those names. We can randomly select one of the first 50 names and then choose every 50th name after that. This method is simple and is used frequently.

Phone Surveys

Each year, about 185 million Americans get telephone calls from organizations that appear to be research oriented. About 45% of these calls are really attempts to sell products or services, but the remaining 55% are from legitimate research firms. Ideally, telephone numbers are randomly selected so that the 16% of American homes with unlisted numbers will be included. Here is a typical case in which the Gallup organization did research for Zenith: The United States was first partitioned into four categories, according to community populations. These four strata were further broken down by location. Using this process, sample locations were drawn, and in each location about eight telephone starts (the first five digits) were drawn from a working bank of telephone numbers. These starts determined the list of people to be called. For example, the start 987-65?? would suggest 100 telephone numbers, beginning with 987-6500 and ending with 987-6599. Each number was called until an interview was completed, except that no number was called more than four times.

Invisible Ink Deceives Subscribers

Newspapers and magazines routinely survey subscribers so that they can better serve advertisers and readers. The *National Observer* once hired a firm to conduct a confidential mail survey. The *National Observer* editor, Henry Gemmill, wrote in a letter accompanying the survey: "Each individual reply will be kept confidential, of course, but when your reply is combined with others from all over this land, we'll have a composite picture of our subscribers." The questionnaire did not ask for name or address, but one very clever subscriber used an ultraviolet light to detect a code written with invisible ink. This code could be used to identify the person who answered the "confidential survey." Gemmill was unaware that this procedure was used, and he publicly apologized to all subscribers. In this situation, confidentiality was observed as promised, but anonymity was not directly promised and it was certainly not observed.

Cluster Sampling

In cluster sampling we first divide the population area into sections and then randomly select a few of those sections. We can then choose all the members from those sections. For example, in conducting a pre-election poll, we could randomly select 30 election precincts and then survey 50 people from each of those chosen precincts. This would be much more efficient than selecting one person from each of the 1500 precincts. We can again adjust or weight the results to correct for any disproportionate representations of groups. Cluster sampling is used extensively by the government and by private research organizations.

Importance of Sampling

Even experienced and reputable research organizations sometimes get erroneous results due to biased sampling or poor methodology. A classic example is the 1936 telephone survey that indicated that Landon would defeat Roosevelt in the presidential election. The samples were drawn from a population of telephone owners and, in 1936, that population consisted of a disproportionately large number of people with high incomes. It was not representative of the population of all voters.

In 1948 the Gallup poll was wrong when it predicted Truman would lose to Dewey. That mistake led to a revision of Gallup's polling methods. A quota system had been used to obtain the opinions of a proportionate number of men, women, rich, poor, Catholics, Protestants, Jews, and so on. After the 1948 fiasco, Gallup abandoned the quota system and instituted random sampling based on clusters of interviews in several hundred areas throughout the nation.

These examples highlight the importance of the sample.

12-3 Analyzing Data and Drawing Conclusions

The data must be analyzed and the conclusions drawn according to the methods specified when the experiment was designed. After the data have been analyzed and the conclusions have been formed, it is important to note the level of confidence used in any hypothesis test or estimation.

12-4 Writing the Report

In writing the final report, the author should consider the people who will read the report. Statisticians will expect specific and detailed results accompanied by fairly complete descriptions of methodology. However, a lay audience would not benefit from this type of report so a different approach is necessary. A suggested outline for the written report follows.

1. Front matter
   (a) Title page
   (b) Table of contents
   (c) List of tables and illustrations
   (d) Preface
   (e) Summary of results and conclusions
2. Body of report
   (a) Statement of objectives
   (b) Description of procedure and methods
   (c) Analysis of data
   (d) Conclusions drawn from data
3. Supplementary material
   (a) Appendices
   (b) Bibliography
   (c) Glossary of terms requiring definition
   (d) Index

Code of Ethics for Survey Research

The American Association for Public Opinion Research has developed a voluntary code of ethics to be used in news reports of survey results. This code of ethics requires that the following items be included in such reports.

1. Identification of the person, group, or organization that sponsored the survey.
2. The date the survey was conducted.
3. The size of the sample.
4. The nature of the population sampled.
5. The type of survey used.
6. The exact wording of survey questions.

Surveys funded by the U.S. government are subject to a prescreening that assesses the risk to those surveyed, the scientific merit of the survey, and the guarantee of the subject's consent to participate.

12-5   A Do-It-Yourself Project

An ideal way to gain insight into statistical methods is to conduct a statistical experiment from beginning to end. It might be advantageous to relate the subject of such an experiment to another course or discipline. The project should involve the actual collection of raw data and should conclude with a typewritten report that generally follows the outline given in Section 12-4. A title page should list your name, the course, the instructor's name, and the title of the report. An example of a typical title is "Statistical Analysis of the Hours Worked by Full-Time Students." A suggested outline and format for the table of contents is listed below.

Table of Contents

Summary 2

Objectives.. 3

Procedure 3

Analysis of Data 4

Conclusion.. 4

Appendix of Raw Data 5–6

Ethics in Experiments

Sample data can often be obtained by simply observing or surveying members selected from the population. Many other situations require that we somehow manipulate circumstances to obtain sample data. In both cases, ethical questions may arise. Researchers in Tuskegee, Alabama, withheld the effective penicillin treatment to syphilis victims so that the disease could be studied. This continued for a period of 27 years!

Be sure to identify the type of sampling (such as random or systematic). Identify the features of the sampling process that tend to make your sample representative of the population. Also be sure to identify the population from which the sample was drawn.

Caution: It may be necessary to gain approval and permission before collecting certain types of data. Reference to published statistical almanacs and abstracts can normally be made at your own discretion, but there may be regulations governing the process of polling people in person or by telephone. We now know that some past studies have adversely affected the people being studied. In one experiment, a psychologist had subjects play a game in which it was impossible to win without cheating. Some of the subjects underwent some behavior modification when they decided to cheat and, in the process, altered their attitudes about cheating in general. As another example, if we were to give people an in-depth survey on suicide, we might run the risk of pushing a potential suicide victim over the edge. Any survey should be seriously and carefully constructed with consideration for the sensitivities and emotions of those surveyed. If our data source is people, we should take every necessary precaution and obtain any necessary approval.

In the future, you may be required to conduct statistical experiments as part of job or course requirements. It is natural and common to have feelings of apprehension (or even outright fear or dread) because you might sense that the subject matter of this course was not mastered to the extent necessary. To keep things in a proper perspective, you should recognize that nobody expects an introductory statistics course to make you an expert statistician. It is also very important to recognize that you can get help from those who have much more extensive training and experience with statistical methods. Even when not mastered fully, this course will help you open a dialogue with a professional statistician who can be of help. At the very least, this course gives you a literacy so necessary in this era of data proliferation.

Exercises A
Design, Sampling, and Report Writing

12-1 What is random sampling?

12-2 What is stratified sampling?

12-3 What is systematic sampling?

12-4 What is cluster sampling?

12-5   Identify the type of sampling used in each of the following:
(a) A teacher selects every fifth student in the class for a test.
(b) A teacher writes the name of each student on a card, shuffles the cards, and then draws five names.
(c) A teacher selects five students from each of 12 classes.
(d) A teacher selects five men and five women from each of four classes.

12-6   You plan to conduct a survey to determine the percentage of students who favor a pub on campus. Describe the population and a method of sampling that should give reasonable results.

12-7   You plan to estimate the mean weight of all passenger cars used in the United States. Is there universal agreement about what a "passenger car" is? Are there any factors that might lead to regional differences among the weights of passenger cars? How can you obtain a sample?

12-8   Two categories of survey questions are *open* and *closed*. An open question allows a free response, while a closed question allows only fixed responses. Here are examples based on Gallup surveys.

Open question: What do you think can be done to reduce crime?
Closed question: Which of the following approaches would be most effective in reducing crime?
• Hire more police officers.
• Get parents to discipline children more.
• Correct social and economic conditions in slums.
• Improve rehabilitation efforts in jails.
• Give convicted criminals tougher sentences.
• Reform courts.

What are the advantages and disadvantages of open questions? What are the advantages and disadvantages of closed questions? Which type is easier to analyze with formal statistical procedures?

Exercises B
Design, Sampling, and Report Writing

12-9   Find an article that deals with some hypothesis test in a professional journal. Use the given information to write a report that follows, as closely as possible, the outline for a written report. Include the name of the journal, the author, and the title of the article.

12-10   Conduct a complete statistical experiment following the outline given in Section 12-5. Collect sample data consisting of single numbers. Include a frequency table, histogram, point estimate of the population mean, point estimate of the population standard deviation, and interval estimate of the population mean; then state and test some appropriate null hypothesis.

12-11 Conduct a complete statistical experiment following the outline given in Section 12-5. Collect sample data consisting of pairs of numbers. Include a scatter diagram, the correlation coefficient, the regression equation, and the coefficient of determination. Form a conclusion about the relationship between the two variables that produced the pairs of data. (See Chapter 9.)

12-12 Conduct a complete statistical experiment following the outline given in Section 12-5. Collect sample data that can be summarized as a contingency table. Form a conclusion about the independence of the two variables used in classifying the table. (See Section 10-3.)

12-13 Conduct a complete statistical experiment following the outline given in Section 12-5. Collect sample data that can be used to establish some population proportion or percentage p. Include the point estimate of p and an interval estimate of p; then state and test some appropriate hypothesis.

12-14 Conduct a complete statistical experiment following the outline given in Section 12-5. Select a topic of your own, or use one of the following general topics for suggestions.
(a) The use of videocassette recorders.
(b) Television viewing habits.
(c) Voting habits.
(d) Spending habits.
(e) Career goals of college students.
(f) Health status of college students.
(g) Smoking and/or drinking habits of college students.
(h) Analysis of the college registration process.

Tables

Table A-1

Factorials of Numbers 1 to 20	
n	$n!$
0	1
1	1
2	2
3	6
4	24
5	120
6	720
7	5040
8	40320
9	362880
10	3628800
11	39916800
12	479001600
13	6227020800
14	87178291200
15	1307674368000
16	20922789888000
17	355687428096000
18	6402373705728000
19	121645100408832000
20	2432902008176640000

Table A-2

Binomial Probabilities

n	x	.01	.05	.10	.20	.30	.40	p .50	.60	.70	.80	.90	.95	.99	x
2	0	980	902	810	640	490	360	250	160	090	040	010	002	0+	0
	1	020	095	180	320	420	480	500	480	420	320	180	095	020	1
	2	0+	002	010	040	090	160	250	360	490	640	810	902	980	2

(continued)

Table A-2

Binomial Probabilities

n	x	.01	.05	.10	.20	.30	.40	p .50	.60	.70	.80	.90	.95	.99	x
3	0	970	857	729	512	343	216	125	064	027	008	001	0+	0+	0
	1	029	135	243	384	441	432	375	288	189	096	027	007	0+	1
	2	0+	007	027	096	189	288	375	432	441	384	243	135	029	2
	3	0+	0+	001	008	027	064	125	216	343	512	729	857	970	3
4	0	961	815	656	410	240	130	062	026	008	002	0+	0+	0+	0
	1	039	171	292	410	412	346	250	154	076	026	004	0+	0+	1
	2	001	014	049	154	265	346	375	346	265	154	049	014	001	2
	3	0+	0+	004	026	076	154	250	346	412	410	292	171	039	3
	4	0+	0+	0+	002	008	026	062	130	240	410	656	815	961	4
5	0	951	774	590	328	168	078	031	010	002	0+	0+	0+	0+	0
	1	048	204	328	410	360	259	156	077	028	006	0+	0+	0+	1
	2	001	021	073	205	309	346	312	230	132	051	008	001	0+	2
	3	0+	001	008	051	132	230	312	346	309	205	073	021	001	3
	4	0+	0+	0+	006	028	077	156	259	360	410	328	204	048	4
	5	0+	0+	0+	0+	002	010	031	078	168	328	590	774	951	5
6	0	941	735	531	262	118	047	016	004	001	0+	0+	0+	0+	0
	1	057	232	354	393	303	187	094	037	010	002	0+	0+	0+	1
	2	001	031	098	246	324	311	234	138	060	015	001	0+	0+	2
	3	0+	002	015	082	185	276	312	276	185	082	015	002	0+	3
	4	0+	0+	001	015	060	138	234	311	324	246	098	031	001	4
	5	0+	0+	0+	002	010	037	094	187	303	393	354	232	057	5
	6	0+	0+	0+	0+	001	004	016	047	118	262	531	735	941	6
7	0	932	698	478	210	082	028	008	002	0+	0+	0+	0+	0+	0
	1	066	257	372	367	247	131	055	017	004	0+	0+	0+	0+	1
	2	002	041	124	275	318	261	164	077	025	004	0+	0+	0+	2
	3	0+	004	023	115	227	290	273	194	097	029	003	0+	0+	3
	4	0+	0+	003	029	097	194	273	290	227	115	023	004	0+	4
	5	0+	0+	0+	004	025	077	164	261	318	275	124	041	002	5
	6	0+	0+	0+	0+	004	017	055	131	247	367	372	257	066	6
	7	0+	0+	0+	0+	0+	002	008	028	082	210	478	698	932	7
8	0	923	663	430	168	058	017	004	001	0+	0+	0+	0+	0+	0
	1	075	279	383	336	198	090	031	008	001	0+	0+	0+	0+	1
	2	003	051	149	294	296	209	109	041	010	001	0+	0+	0+	2
	3	0+	005	033	147	254	279	219	124	047	009	0+	0+	0+	3
	4	0+	0+	005	046	136	232	273	232	136	046	005	0+	0+	4
	5	0+	0+	0+	009	047	124	219	279	254	147	033	005	0+	5
	6	0+	0+	0+	001	010	041	109	209	296	294	149	051	003	6
	7	0+	0+	0+	0+	001	008	031	090	198	336	383	279	075	7
	8	0+	0+	0+	0+	0+	001	004	017	058	168	430	663	923	8
9	0	914	630	387	134	040	010	002	0+	0+	0+	0+	0+	0+	0
	1	083	299	387	302	156	060	018	004	0+	0+	0+	0+	0+	1
	2	003	063	172	302	267	161	070	021	004	0+	0+	0+	0+	2
	3	0+	008	045	176	267	251	164	074	021	003	0+	0+	0+	3
	4	0+	001	007	066	172	251	246	167	074	017	001	0+	0+	4

NOTE: 0+ represents a positive probability less than 0.0005.

(continued)

Table A-2 (continued)

Binomial Probabilities

n	x	.01	.05	.10	.20	.30	.40	p .50	.60	.70	.80	.90	.95	.99	x
	5	0+	0+	001	017	074	167	246	251	172	066	007	001	0+	5
	6	0+	0+	0+	003	021	074	164	251	267	176	045	008	0+	6
	7	0+	0+	0+	0+	004	021	070	161	267	302	172	063	003	7
	8	0+	0+	0+	0+	0+	004	018	060	156	302	387	299	083	8
	9	0+	0+	0+	0+	0+	0+	002	010	040	134	387	630	914	9
10	0	904	599	349	107	028	006	001	0+	0+	0+	0+	0+	0+	0
	1	091	315	387	268	121	040	010	002	0+	0+	0+	0+	0+	1
	2	004	075	194	302	233	121	044	011	001	0+	0+	0+	0+	2
	3	0+	010	057	201	267	215	117	042	009	001	0+	0+	0+	3
	4	0+	001	011	088	200	251	205	111	037	006	0+	0+	0+	4
	5	0+	0+	001	026	103	201	246	201	103	026	001	0+	0+	5
	6	0+	0+	0+	006	037	111	205	251	200	088	011	001	0+	6
	7	0+	0+	0+	001	009	042	117	215	267	201	057	010	0+	7
	8	0+	0+	0+	0+	001	011	044	121	233	302	194	075	004	8
	9	0+	0+	0+	0+	0+	002	010	040	121	268	387	315	091	9
	10	0+	0+	0+	0+	0+	0+	001	006	028	107	349	599	904	10
11	0	895	569	314	086	020	004	0+	0+	0+	0+	0+	0+	0+	0
	1	099	329	384	236	093	027	005	001	0+	0+	0+	0+	0+	1
	2	005	087	213	295	200	089	027	005	001	0+	0+	0+	0+	2
	3	0+	014	071	221	257	177	081	023	004	0+	0+	0+	0+	3
	4	0+	001	016	111	220	236	161	070	017	002	0+	0+	0+	4
	5	0+	0+	002	039	132	221	226	147	057	010	0+	0+	0+	5
	6	0+	0+	0+	010	057	147	226	221	132	039	002	0+	0+	6
	7	0+	0+	0+	002	017	070	161	236	220	111	016	001	0+	7
	8	0+	0+	0+	0+	004	023	081	177	257	221	071	014	0+	8
	9	0+	0+	0+	0+	001	005	027	089	200	295	213	087	005	9
	10	0+	0+	0+	0+	0+	001	005	027	093	236	384	329	099	10
	11	0+	0+	0+	0+	0+	0+	0+	004	020	086	314	569	895	11
12	0	886	540	282	069	014	002	0+	0+	0+	0+	0+	0+	0+	0
	1	107	341	377	206	071	017	003	0+	0+	0+	0+	0+	0+	1
	2	006	099	230	283	168	064	016	002	0+	0+	0+	0+	0+	2
	3	0+	017	085	236	240	142	054	012	001	0+	0+	0+	0+	3
	4	0+	002	021	133	231	213	121	042	008	001	0+	0+	0+	4
	5	0+	0+	004	053	158	227	193	101	029	003	0+	0+	0+	5
	6	0+	0+	0+	016	079	177	226	177	079	016	0+	0+	0+	6
	7	0+	0+	0+	003	029	101	193	227	158	053	004	0+	0+	7
	8	0+	0+	0+	001	008	042	121	213	231	133	021	002	0+	8
	9	0+	0+	0+	0+	001	012	054	142	240	236	085	017	0+	9
	10	0+	0+	0+	0+	0+	002	016	064	168	283	230	099	006	10
	11	0+	0+	0+	0+	0+	0+	003	017	071	206	377	341	107	11
	12	0+	0+	0+	0+	0+	0+	0+	002	014	069	282	540	886	12

NOTE: 0+ represents a positive probability less than 0.0005.

(continued)

Table A-2 (continued)

Binomial Probabilities

n	x	.01	.05	.10	.20	.30	.40	.50	.60	.70	.80	.90	.95	.99	x
13	0	878	513	254	055	010	001	0+	0+	0+	0+	0+	0+	0+	0
	1	115	351	367	179	054	011	002	0+	0+	0+	0+	0+	0+	1
	2	007	111	245	268	139	045	010	001	0+	0+	0+	0+	0+	2
	3	0+	021	100	246	218	111	035	006	001	0+	0+	0+	0+	3
	4	0+	003	028	154	234	184	087	024	003	0+	0+	0+	0+	4
	5	0+	0+	006	069	180	221	157	066	014	001	0+	0+	0+	5
	6	0+	0+	001	023	103	197	209	131	044	006	0+	0+	0+	6
	7	0+	0+	0+	006	044	131	209	197	103	023	001	0+	0+	7
	8	0+	0+	0+	001	014	066	157	221	180	069	006	0+	0+	8
	9	0+	0+	0+	0+	003	024	087	184	234	154	028	003	0+	9
	10	0+	0+	0+	0+	001	006	035	111	218	246	100	021	0+	10
	11	0+	0+	0+	0+	0+	001	010	045	139	268	245	111	007	11
	12	0+	0+	0+	0+	0+	0+	002	011	054	179	367	351	115	12
	13	0+	0+	0+	0+	0+	0+	0+	001	010	055	254	513	878	13
14	0	869	488	229	044	007	001	0+	0+	0+	0+	0+	0+	0+	0
	1	123	359	356	154	041	007	001	0+	0+	0+	0+	0+	0+	1
	2	008	123	257	250	113	032	006	001	0+	0+	0+	0+	0+	2
	3	0+	026	114	250	194	085	022	003	0+	0+	0+	0+	0+	3
	4	0+	004	035	172	229	155	061	014	001	0+	0+	0+	0+	4
	5	0+	0+	008	086	196	207	122	041	007	0+	0+	0+	0+	5
	6	0+	0+	001	032	126	207	183	092	023	002	0+	0+	0+	6
	7	0+	0+	0+	009	062	157	209	157	062	009	0+	0+	0+	7
	8	0+	0+	0+	002	023	092	183	207	126	032	001	0+	0+	8
	9	0+	0+	0+	0+	007	041	122	207	196	086	008	0+	0+	9
	10	0+	0+	0+	0+	001	014	061	155	229	172	035	004	0+	10
	11	0+	0+	0+	0+	0+	003	022	085	194	250	114	026	0+	11
	12	0+	0+	0+	0+	0+	001	006	032	113	250	257	123	008	12
	13	0+	0+	0+	0+	0+	0+	001	007	041	154	356	359	123	13
	14	0+	0+	0+	0+	0+	0+	0+	001	007	044	229	488	869	14
15	0	860	463	206	035	005	0+	0+	0+	0+	0+	0+	0+	0+	0
	1	130	366	343	132	031	005	0+	0+	0+	0+	0+	0+	0+	1
	2	009	135	267	231	092	022	003	0+	0+	0+	0+	0+	0+	2
	3	0+	031	129	250	170	063	014	002	0+	0+	0+	0+	0+	3
	4	0+	005	043	188	219	127	042	007	001	0+	0+	0+	0+	4
	5	0+	001	010	103	206	186	092	024	003	0+	0+	0+	0+	5
	6	0+	0+	002	043	147	207	153	061	012	001	0+	0+	0+	6
	7	0+	0+	0+	014	081	177	196	118	035	003	0+	0+	0+	7
	8	0+	0+	0+	003	035	118	196	177	081	014	0+	0+	0+	8
	9	0+	0+	0+	001	012	061	153	207	147	043	002	0+	0+	9
	10	0+	0+	0+	0+	003	024	092	186	206	103	010	001	0+	10
	11	0+	0+	0+	0+	001	007	042	127	219	188	043	005	0+	11
	12	0+	0+	0+	0+	0+	002	014	063	170	250	129	031	0+	12
	13	0+	0+	0+	0+	0+	0+	003	022	092	231	267	135	009	13
	14	0+	0+	0+	0+	0+	0+	0+	005	031	132	343	366	130	14
	15	0+	0+	0+	0+	0+	0+	0+	0+	005	035	206	463	860	15

NOTE: 0+ represents a positive probability less than 0.0005.

(continued)

Table A-2 (continued)

Binomial Probabilities

n	x	.01	.05	.10	.20	.30	.40	p .50	.60	.70	.80	.90	.95	.99	x
16	0	851	440	185	028	003	0+	0+	0+	0+	0+	0+	0+	0+	0
	1	138	371	329	113	023	003	0+	0+	0+	0+	0+	0+	0+	1
	2	010	146	275	211	073	015	002	0+	0+	0+	0+	0+	0+	2
	3	0+	036	142	246	146	047	009	001	0+	0+	0+	0+	0+	3
	4	0+	006	051	200	204	101	028	004	0+	0+	0+	0+	0+	4
	5	0+	001	014	120	210	162	067	014	001	0+	0+	0+	0+	5
	6	0+	0+	003	055	165	198	122	039	006	0+	0+	0+	0+	6
	7	0+	0+	0+	020	101	189	175	084	019	001	0+	0+	0+	7
	8	0+	0+	0+	006	049	142	196	142	049	006	0+	0+	0+	8
	9	0+	0+	0+	001	019	084	175	189	101	020	0+	0+	0+	9
	10	0+	0+	0+	0+	006	039	122	198	165	055	003	0+	0+	10
	11	0+	0+	0+	0+	001	014	067	162	210	120	014	001	0+	11
	12	0+	0+	0+	0+	0+	004	028	101	204	200	051	006	0+	12
	13	0+	0+	0+	0+	0+	001	009	047	146	246	142	036	0+	13
	14	0+	0+	0+	0+	0+	0+	002	015	073	211	275	146	010	14
	15	0+	0+	0+	0+	0+	0+	0+	003	023	113	329	371	138	15
	16	0+	0+	0+	0+	0+	0+	0+	0+	003	028	185	440	851	16
17	0	843	418	167	023	002	0+	0+	0+	0+	0+	0+	0+	0+	0
	1	145	374	315	096	017	002	0+	0+	0+	0+	0+	0+	0+	1
	2	012	158	280	191	058	010	001	0+	0+	0+	0+	0+	0+	2
	3	001	041	156	239	125	034	005	0+	0+	0+	0+	0+	0+	3
	4	0+	008	060	209	187	080	018	002	0+	0+	0+	0+	0+	4
	5	0+	001	017	136	208	138	047	008	001	0+	0+	0+	0+	5
	6	0+	0+	004	068	178	184	094	024	003	0+	0+	0+	0+	6
	7	0+	0+	001	027	120	193	148	057	009	0+	0+	0+	0+	7
	8	0+	0+	0+	008	064	161	185	107	028	002	0+	0+	0+	8
	9	0+	0+	0+	002	028	107	185	161	064	008	0+	0+	0+	9
	10	0+	0+	0+	0+	009	057	148	193	120	027	001	0+	0+	10
	11	0+	0+	0+	0+	003	024	094	184	178	068	004	0+	0+	11
	12	0+	0+	0+	0+	001	008	047	138	208	136	017	001	0+	12
	13	0+	0+	0+	0+	0+	002	018	080	187	209	060	008	0+	13
	14	0+	0+	0+	0+	0+	0+	005	034	125	239	156	041	001	14
	15	0+	0+	0+	0+	0+	0+	001	010	058	191	280	158	012	15
	16	0+	0+	0+	0+	0+	0+	0+	002	017	096	315	374	145	16
	17	0+	0+	0+	0+	0+	0+	0+	0+	002	023	167	418	843	17
18	0	835	397	150	018	002	0+	0+	0+	0+	0+	0+	0+	0+	0
	1	152	376	300	081	013	001	0+	0+	0+	0+	0+	0+	0+	1
	2	013	168	284	172	046	007	001	0+	0+	0+	0+	0+	0+	2
	3	001	047	168	230	105	025	003	0+	0+	0+	0+	0+	0+	3
	4	0+	009	070	215	168	061	012	001	0+	0+	0+	0+	0+	4
	5	0+	001	022	151	202	115	033	004	0+	0+	0+	0+	0+	5
	6	0+	0+	005	082	187	166	071	015	001	0+	0+	0+	0+	6
	7	0+	0+	001	035	138	189	121	037	005	0+	0+	0+	0+	7
	8	0+	0+	0+	012	081	173	167	077	015	001	0+	0+	0+	8
	9	0+	0+	0+	003	039	128	185	128	039	003	0+	0+	0+	9

NOTE: 0+ represents a positive probability less than 0.0005.

(continued)

Table A-2 (continued)

Binomial Probabilities

n	x	.01	.05	.10	.20	.30	.40	p .50	.60	.70	.80	.90	.95	.99	x
	10	0+	0+	0+	001	015	077	167	173	081	012	0+	0+	0+	10
	11	0+	0+	0+	0+	005	037	121	189	138	035	001	0+	0+	11
	12	0+	0+	0+	0+	001	015	071	166	187	082	005	0+	0+	12
	13	0+	0+	0+	0+	0+	004	033	115	202	151	022	001	0+	13
	14	0+	0+	0+	0+	0+	001	012	061	168	215	070	009	0+	14
	15	0+	0+	0+	0+	0+	0+	003	025	105	230	168	047	001	15
	16	0+	0+	0+	0+	0+	0+	001	007	046	172	284	168	013	16
	17	0+	0+	0+	0+	0+	0+	0+	001	013	081	300	376	152	17
	18	0+	0+	0+	0+	0+	0+	0+	0+	002	018	150	397	835	18
19	0	826	377	135	014	001	0+	0+	0+	0+	0+	0+	0+	0+	0
	1	159	377	285	068	009	001	0+	0+	0+	0+	0+	0+	0+	1
	2	014	179	285	154	036	005	0+	0+	0+	0+	0+	0+	0+	2
	3	001	053	180	218	087	017	002	0+	0+	0+	0+	0+	0+	3
	4	0+	011	080	218	149	047	007	001	0+	0+	0+	0+	0+	4
	5	0+	002	027	164	192	093	022	002	0+	0+	0+	0+	0+	5
	6	0+	0+	007	095	192	145	052	008	001	0+	0+	0+	0+	6
	7	0+	0+	001	044	153	180	096	024	002	0+	0+	0+	0+	7
	8	0+	0+	0+	017	098	180	144	053	008	0+	0+	0+	0+	8
	9	0+	0+	0+	005	051	146	176	098	022	001	0+	0+	0+	9
	10	0+	0+	0+	001	022	098	176	146	051	005	0+	0+	0+	10
	11	0+	0+	0+	0+	008	053	144	180	098	017	0+	0+	0+	11
	12	0+	0+	0+	0+	002	024	096	180	153	044	001	0+	0+	12
	13	0+	0+	0+	0+	001	008	052	145	192	095	007	0+	0+	13
	14	0+	0+	0+	0+	0+	002	022	093	192	164	027	002	0+	14
	15	0+	0+	0+	0+	0+	001	007	047	149	218	080	011	0+	15
	16	0+	0+	0+	0+	0+	0+	002	017	087	218	180	053	001	16
	17	0+	0+	0+	0+	0+	0+	0+	005	036	154	285	179	014	17
	18	0+	0+	0+	0+	0+	0+	0+	001	009	068	285	377	159	18
	19	0+	0+	0+	0+	0+	0+	0+	0+	001	014	135	377	826	19
20	0	818	358	122	012	001	0+	0+	0+	0+	0+	0+	0+	0+	0
	1	165	377	270	058	007	0+	0+	0+	0+	0+	0+	0+	0+	1
	2	016	189	285	137	028	003	0+	0+	0+	0+	0+	0+	0+	2
	3	001	060	190	205	072	012	001	0+	0+	0+	0+	0+	0+	3
	4	0+	013	090	218	130	035	005	0+	0+	0+	0+	0+	0+	4
	5	0+	002	032	175	179	075	015	001	0+	0+	0+	0+	0+	5
	6	0+	0+	009	109	192	124	037	005	0+	0+	0+	0+	0+	6
	7	0+	0+	002	055	164	166	074	015	001	0+	0+	0+	0+	7
	8	0+	0+	0+	022	114	180	120	035	004	0+	0+	0+	0+	8
	9	0+	0+	0+	007	065	160	160	071	012	0+	0+	0+	0+	9
	10	0+	0+	0+	002	031	117	176	117	031	002	0+	0+	0+	10
	11	0+	0+	0+	0+	012	071	160	160	065	007	0+	0+	0+	11
	12	0+	0+	0+	0+	004	035	120	180	114	022	0+	0+	0+	12
	13	0+	0+	0+	0+	001	015	074	166	164	055	002	0+	0+	13
	14	0+	0+	0+	0+	0+	005	037	124	192	109	009	0+	0+	14

NOTE: 0+ represents a positive probability less than 0.0005.

(continued)

Table A-2 (continued)

Binomial Probabilities

n	x	.01	.05	.10	.20	.30	.40	p .50	.60	.70	.80	.90	.95	.99	x
	15	0+	0+	0+	0+	0+	001	015	075	179	175	032	002	0+	15
	16	0+	0+	0+	0+	0+	0+	005	035	130	218	090	013	0+	16
	17	0+	0+	0+	0+	0+	0+	001	012	072	205	190	060	001	17
	18	0+	0+	0+	0+	0+	0+	0+	003	028	137	285	189	016	18
	19	0+	0+	0+	0+	0+	0+	0+	0+	007	058	270	377	165	19
	20	0+	0+	0+	0+	0+	0+	0+	0+	001	012	122	358	818	20
21	0	810	341	109	009	001	0+	0+	0+	0+	0+	0+	0+	0+	0
	1	172	376	255	048	005	0+	0+	0+	0+	0+	0+	0+	0+	1
	2	017	198	284	121	022	002	0+	0+	0+	0+	0+	0+	0+	2
	3	001	066	200	192	058	009	001	0+	0+	0+	0+	0+	0+	3
	4	0+	016	100	216	113	026	003	0+	0+	0+	0+	0+	0+	4
	5	0+	003	038	183	164	059	010	001	0+	0+	0+	0+	0+	5
	6	0+	0+	011	122	188	105	026	003	0+	0+	0+	0+	0+	6
	7	0+	0+	003	065	172	149	055	009	0+	0+	0+	0+	0+	7
	8	0+	0+	001	029	129	174	097	023	002	0+	0+	0+	0+	8
	9	0+	0+	0+	010	080	168	140	050	006	0+	0+	0+	0+	9
	10	0+	0+	0+	003	041	134	168	089	018	001	0+	0+	0+	10
	11	0+	0+	0+	001	018	089	168	134	041	003	0+	0+	0+	11
	12	0+	0+	0+	0+	006	050	140	168	080	010	0+	0+	0+	12
	13	0+	0+	0+	0+	002	023	097	174	129	029	001	0+	0+	13
	14	0+	0+	0+	0+	0+	009	055	149	172	065	003	0+	0+	14
	15	0+	0+	0+	0+	0+	003	026	105	188	122	011	0+	0+	15
	16	0+	0+	0+	0+	0+	001	010	059	164	183	038	003	0+	16
	17	0+	0+	0+	0+	0+	0+	003	026	113	216	100	016	0+	17
	18	0+	0+	0+	0+	0+	0+	001	009	058	192	200	066	001	18
	19	0+	0+	0+	0+	0+	0+	0+	002	022	121	284	198	017	19
	20	0+	0+	0+	0+	0+	0+	0+	0+	005	048	255	376	172	20
	21	0+	0+	0+	0+	0+	0+	0+	0+	001	009	109	341	810	21
22	0	802	324	098	007	0+	0+	0+	0+	0+	0+	0+	0+	0+	0
	1	178	375	241	041	004	0+	0+	0+	0+	0+	0+	0+	0+	1
	2	019	207	281	107	017	001	0+	0+	0+	0+	0+	0+	0+	2
	3	001	073	208	178	047	006	0+	0+	0+	0+	0+	0+	0+	3
	4	0+	018	110	211	096	019	002	0+	0+	0+	0+	0+	0+	4
	5	0+	003	044	190	149	046	006	0+	0+	0+	0+	0+	0+	5
	6	0+	001	014	134	181	086	018	001	0+	0+	0+	0+	0+	6
	7	0+	0+	004	077	177	131	041	005	0+	0+	0+	0+	0+	7
	8	0+	0+	001	036	142	164	076	014	001	0+	0+	0+	0+	8
	9	0+	0+	0+	014	095	170	119	034	003	0+	0+	0+	0+	9
	10	0+	0+	0+	005	053	148	154	066	010	0+	0+	0+	0+	10
	11	0+	0+	0+	001	025	107	168	107	025	001	0+	0+	0+	11
	12	0+	0+	0+	0+	010	066	154	148	053	005	0+	0+	0+	12
	13	0+	0+	0+	0+	003	034	119	170	095	014	0+	0+	0+	13
	14	0+	0+	0+	0+	001	014	076	164	142	036	001	0+	0+	14

NOTE: 0+ represents a positive probability less than 0.0005.

(continued)

Table A-2 (continued)

Binomial Probabilities

n	x	.01	.05	.10	.20	.30	.40	.50	.60	.70	.80	.90	.95	.99	x
	15	0+	0+	0+	0+	0+	005	041	131	177	077	004	0+	0+	15
	16	0+	0+	0+	0+	0+	001	018	086	181	134	014	001	0+	16
	17	0+	0+	0+	0+	0+	0+	006	046	149	190	044	003	0+	17
	18	0+	0+	0+	0+	0+	0+	002	019	096	211	110	018	0+	18
	19	0+	0+	0+	0+	0+	0+	0+	006	047	178	208	073	001	19
	20	0+	0+	0+	0+	0+	0+	0+	001	017	107	281	207	019	20
	21	0+	0+	0+	0+	0+	0+	0+	0+	004	041	241	375	178	21
	22	0+	0+	0+	0+	0+	0+	0+	0+	0+	007	098	324	802	22
23	0	794	307	089	006	0+	0+	0+	0+	0+	0+	0+	0+	0+	0
	1	184	372	226	034	003	0+	0+	0+	0+	0+	0+	0+	0+	1
	2	020	215	277	093	013	001	0+	0+	0+	0+	0+	0+	0+	2
	3	001	079	215	163	038	004	0+	0+	0+	0+	0+	0+	0+	3
	4	0+	021	120	204	082	014	001	0+	0+	0+	0+	0+	0+	4
	5	0+	004	051	194	133	035	004	0+	0+	0+	0+	0+	0+	5
	6	0+	001	017	145	171	070	012	001	0+	0+	0+	0+	0+	6
	7	0+	0+	005	088	178	113	029	003	0+	0+	0+	0+	0+	7
	8	0+	0+	001	044	153	151	058	009	0+	0+	0+	0+	0+	8
	9	0+	0+	0+	018	109	168	097	022	002	0+	0+	0+	0+	9
	10	0+	0+	0+	006	065	157	136	046	005	0+	0+	0+	0+	10
	11	0+	0+	0+	002	033	123	161	082	014	0+	0+	0+	0+	11
	12	0+	0+	0+	0+	014	082	161	123	033	002	0+	0+	0+	12
	13	0+	0+	0+	0+	005	046	136	157	065	006	0+	0+	0+	13
	14	0+	0+	0+	0+	002	022	097	168	109	018	0+	0+	0+	14
	15	0+	0+	0+	0+	0+	009	058	151	153	044	001	0+	0+	15
	16	0+	0+	0+	0+	0+	003	029	113	178	088	005	0+	0+	16
	17	0+	0+	0+	0+	0+	001	012	070	171	145	017	001	0+	17
	18	0+	0+	0+	0+	0+	0+	004	035	133	194	051	004	0+	18
	19	0+	0+	0+	0+	0+	0+	001	014	082	204	120	021	0+	19
	20	0+	0+	0+	0+	0+	0+	0+	004	038	163	215	079	001	20
	21	0+	0+	0+	0+	0+	0+	0+	001	013	093	277	215	020	21
	22	0+	0+	0+	0+	0+	0+	0+	0+	003	034	226	372	184	22
	23	0+	0+	0+	0+	0+	0+	0+	0+	0+	006	089	307	794	23
24	0	786	292	080	005	0+	0+	0+	0+	0+	0+	0+	0+	0+	0
	1	190	369	213	028	002	0+	0+	0+	0+	0+	0+	0+	0+	1
	2	022	223	272	081	010	001	0+	0+	0+	0+	0+	0+	0+	2
	3	002	086	221	149	031	003	0+	0+	0+	0+	0+	0+	0+	3
	4	0+	024	129	196	069	010	001	0+	0+	0+	0+	0+	0+	4
	5	0+	005	057	196	118	027	003	0+	0+	0+	0+	0+	0+	5
	6	0+	001	020	155	160	056	008	0+	0+	0+	0+	0+	0+	6
	7	0+	0+	006	100	176	096	021	002	0+	0+	0+	0+	0+	7
	8	0+	0+	001	053	160	136	044	005	0+	0+	0+	0+	0+	8
	9	0+	0+	0+	024	122	161	078	014	001	0+	0+	0+	0+	9

NOTE: 0+ represents a positive probability less than 0.0005.

(continued)

Table A-2 (continued)

Binomial Probabilities

n	x	.01	.05	.10	.20	.30	.40	p .50	.60	.70	.80	.90	.95	.99	x
	10	0+	0+	0+	009	079	161	117	032	003	0+	0+	0+	0+	10
	11	0+	0+	0+	003	043	137	149	061	008	0+	0+	0+	0+	11
	12	0+	0+	0+	001	020	099	161	099	020	001	0+	0+	0+	12
	13	0+	0+	0+	0+	008	061	149	137	043	003	0+	0+	0+	13
	14	0+	0+	0+	0+	003	032	117	161	079	009	0+	0+	0+	14
	15	0+	0+	0+	0+	001	014	078	161	122	024	0+	0+	0+	15
	16	0+	0+	0+	0+	0+	005	044	136	160	053	001	0+	0+	16
	17	0+	0+	0+	0+	0+	002	021	096	176	100	006	0+	0+	17
	18	0+	0+	0+	0+	0+	0+	008	056	160	155	020	001	0+	18
	19	0+	0+	0+	0+	0+	0+	003	027	118	196	057	005	0+	19
	20	0+	0+	0+	0+	0+	0+	001	010	069	196	129	024	0+	20
	21	0+	0+	0+	0+	0+	0+	0+	003	031	149	221	086	002	21
	22	0+	0+	0+	0+	0+	0+	0+	001	010	081	272	223	022	22
	23	0+	0+	0+	0+	0+	0+	0+	0+	002	028	213	369	190	23
	24	0+	0+	0+	0+	0+	0+	0+	0+	0+	005	080	292	786	24
25	0	778	277	072	004	0+	0+	0+	0+	0+	0+	0+	0+	0+	0
	1	196	365	199	024	001	0+	0+	0+	0+	0+	0+	0+	0+	1
	2	024	231	266	071	007	0+	0+	0+	0+	0+	0+	0+	0+	2
	3	002	093	226	136	024	002	0+	0+	0+	0+	0+	0+	0+	3
	4	0+	027	138	187	057	007	0+	0+	0+	0+	0+	0+	0+	4
	5	0+	006	065	196	103	020	002	0+	0+	0+	0+	0+	0+	5
	6	0+	001	024	163	147	044	005	0+	0+	0+	0+	0+	0+	6
	7	0+	0+	007	111	171	080	014	001	0+	0+	0+	0+	0+	7
	8	0+	0+	002	062	165	120	032	003	0+	0+	0+	0+	0+	8
	9	0+	0+	0+	029	134	151	061	009	0+	0+	0+	0+	0+	9
	10	0+	0+	0+	012	092	161	097	021	001	0+	0+	0+	0+	10
	11	0+	0+	0+	004	054	147	133	043	004	0+	0+	0+	0+	11
	12	0+	0+	0+	001	027	114	155	076	011	0+	0+	0+	0+	12
	13	0+	0+	0+	0+	011	076	155	114	027	001	0+	0+	0+	13
	14	0+	0+	0+	0+	004	043	133	147	054	004	0+	0+	0+	14
	15	0+	0+	0+	0+	001	021	097	161	092	012	0+	0+	0+	15
	16	0+	0+	0+	0+	0+	009	061	151	134	029	0+	0+	0+	16
	17	0+	0+	0+	0+	0+	003	032	120	165	062	002	0+	0+	17
	18	0+	0+	0+	0+	0+	001	014	080	171	111	007	0+	0+	18
	19	0+	0+	0+	0+	0+	0+	005	044	147	163	024	001	0+	19
	20	0+	0+	0+	0+	0+	0+	002	020	103	196	065	006	0+	20
	21	0+	0+	0+	0+	0+	0+	0+	007	057	187	138	027	0+	21
	22	0+	0+	0+	0+	0+	0+	0+	002	024	136	226	093	002	22
	23	0+	0+	0+	0+	0+	0+	0+	0+	007	071	266	231	024	23
	24	0+	0+	0+	0+	0+	0+	0+	0+	001	024	199	365	196	24
	25	0+	0+	0+	0+	0+	0+	0+	0+	0+	004	072	277	778	25

NOTE: 0+ represents a positive probability less than 0.0005.

Frederick Mosteller, Robert E. K. Rourke, and George B. Thomas, Jr., *Probability with Statistical Applications,* 2nd ed. (Reading, Mass.: Addison-Wesley, 1961 and 1970). Reprinted with permission of the publisher.

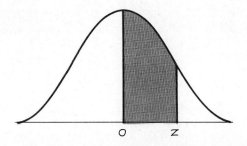

Table A-3

The Standard Normal (z) Distribution

z	.00	.01	.02	.03	.04	.05	.06	.07	.08	.09
0.0	.0000	.0040	.0080	.0120	.0160	.0199	.0239	.0279	.0319	.0359
0.1	.0398	.0438	.0478	.0517	.0557	.0596	.0636	.0675	.0714	.0753
0.2	.0793	.0832	.0871	.0910	.0948	.0987	.1026	.1064	.1103	.1141
0.3	.1179	.1217	.1255	.1293	.1331	.1368	.1406	.1443	.1480	.1517
0.4	.1554	.1591	.1628	.1664	.1700	.1736	.1772	.1808	.1844	.1879
0.5	.1915	.1950	.1985	.2019	.2054	.2088	.2123	.2157	.2190	.2224
0.6	.2257	.2291	.2324	.2357	.2389	.2422	.2454	.2486	.2517	.2549
0.7	.2580	.2611	.2642	.2673	.2704	.2734	.2764	.2794	.2823	.2852
0.8	.2881	.2910	.2939	.2967	.2995	.3023	.3051	.3078	.3106	.3133
0.9	.3159	.3186	.3212	.3238	.3264	.3289	.3315	.3340	.3365	.3389
1.0	.3413	.3438	.3461	.3485	.3508	.3531	.3554	.3577	.3599	.3621
1.1	.3643	.3665	.3686	.3708	.3729	.3749	.3770	.3790	.3810	.3830
1.2	.3849	.3869	.3888	.3907	.3925	.3944	.3962	.3980	.3997	.4015
1.3	.4032	.4049	.4066	.4082	.4099	.4115	.4131	.4147	.4162	.4177
1.4	.4192	.4207	.4222	.4236	.4251	.4265	.4279	.4292	.4306	.4319
1.5	.4332	.4345	.4357	.4370	.4382	.4394	.4406	.4418	.4429	.4441
1.6	.4452	.4463	.4474	.4484	.4495	.4505	.4515	.4525	.4535	.4545
1.7	.4554	.4564	.4573	.4582	.4591	.4599	.4608	.4616	.4625	.4633
1.8	.4641	.4649	.4656	.4664	.4671	.4678	.4686	.4693	.4699	.4706
1.9	.4713	.4719	.4726	.4732	.4738	.4744	.4750	.4756	.4761	.4767
2.0	.4772	.4778	.4783	.4788	.4793	.4798	.4803	.4808	.4812	.4817
2.1	.4821	.4826	.4830	.4834	.4838	.4842	.4846	.4850	.4854	.4857
2.2	.4861	.4864	.4868	.4871	.4875	.4878	.4881	.4884	.4887	.4890
2.3	.4893	.4896	.4898	.4901	.4904	.4906	.4909	.4911	.4913	.4916
2.4	.4918	.4920	.4922	.4925	.4927	.4929	.4931	.4932	.4934	.4936
2.5	.4938	.4940	.4941	.4943	.4945	.4946	.4948	.4949	.4951	.4952
2.6	.4953	.4955	.4956	.4957	.4959	.4960	.4961	.4962	.4963	.4964
2.7	.4965	.4966	.4967	.4968	.4969	.4970	.4971	.4972	.4973	.4974
2.8	.4974	.4975	.4976	.4977	.4977	.4978	.4979	.4979	.4980	.4981
2.9	.4981	.4982	.4982	.4983	.4984	.4984	.4985	.4985	.4986	.4986
3.0	.4987	.4987	.4987	.4988	.4988	.4989	.4989	.4989	.4990	.4990

Frederick Mosteller and Robert E. K. Rourke, *Sturdy Statistics* Table A-1. (Reading, Mass.: Addison-Wesley, 1973.) Reprinted with permission.

NOTE: For values of z above 3.09, use 0.4999.

t DISTRIBUTION

One-tailed values

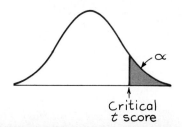

Two-tailed values

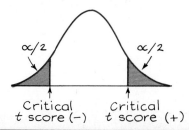

Table A-4

t Distribution

Degrees of freedom	α					
	.005 (one tail) .01 (two tails)	.01 (one tail) .02 (two tails)	.025 (one tail) .05 (two tails)	.05 (one tail) .10 (two tails)	.10 (one tail) .20 (two tails)	.25 (one tail) .50 (two tails)
1	63.657	31.821	12.706	6.314	3.078	1.000
2	9.925	6.965	4.303	2.920	1.886	.816
3	5.841	4.541	3.182	2.353	1.638	.765
4	4.604	3.747	2.776	2.132	1.533	.741
5	4.032	3.365	2.571	2.015	1.476	.727
6	3.707	3.143	2.447	1.943	1.440	.718
7	3.500	2.998	2.365	1.895	1.415	.711
8	3.355	2.896	2.306	1.860	1.397	.706
9	3.250	2.821	2.262	1.833	1.383	.703
10	3.169	2.764	2.228	1.812	1.372	.700
11	3.106	2.718	2.201	1.796	1.363	.697
12	3.054	2.681	2.179	1.782	1.356	.696
13	3.012	2.650	2.160	1.771	1.350	.694
14	2.977	2.625	2.145	1.761	1.345	.692
15	2.947	2.602	2.132	1.753	1.341	.691
16	2.921	2.584	2.120	1.746	1.337	.690
17	2.898	2.567	2.110	1.740	1.333	.689
18	2.878	2.552	2.101	1.734	1.330	.688
19	2.861	2.540	2.093	1.729	1.328	.688
20	2.845	2.528	2.086	1.725	1.325	.687
21	2.831	2.518	2.080	1.721	1.323	.686
22	2.819	2.508	2.074	1.717	1.321	.686
23	2.807	2.500	2.069	1.714	1.320	.685
24	2.797	2.492	2.064	1.711	1.318	.685
25	2.787	2.485	2.060	1.708	1.316	.684
26	2.779	2.479	2.056	1.706	1.315	.684
27	2.771	2.473	2.052	1.703	1.314	.684
28	2.763	2.467	2.048	1.701	1.313	.683
29	2.756	2.462	2.045	1.699	1.311	.683
Large (z)	2.575	2.327	1.960	1.645	1.282	.675

Table A-5

The Chi-Square (χ^2) Distribution

Degrees of freedom	α (Area to the Right of the Critical Value)									
	0.995	0.99	0.975	0.95	0.90	0.10	0.05	0.025	0.01	0.005
1	–	–	0.001	0.004	0.016	2.706	3.841	5.024	6.635	7.879
2	0.010	0.020	0.051	0.103	0.211	4.605	5.991	7.378	9.210	10.597
3	0.072	0.115	0.216	0.352	0.584	6.251	7.815	9.348	11.345	12.838
4	0.207	0.297	0.484	0.711	1.064	7.779	9.488	11.143	13.277	14.860
5	0.412	0.554	0.831	1.145	1.610	9.236	11.071	12.833	15.086	16.750
6	0.676	0.872	1.237	1.635	2.204	10.645	12.592	14.449	16.812	18.548
7	0.989	1.239	1.690	2.167	2.833	12.017	14.067	16.013	18.475	20.278
8	1.344	1.646	2.180	2.733	3.490	13.362	15.507	17.535	20.090	21.955
9	1.735	2.088	2.700	3.325	4.168	14.684	16.919	19.023	21.666	23.589
10	2.156	2.558	3.247	3.940	4.865	15.987	18.307	20.483	23.209	25.188
11	2.603	3.053	3.816	4.575	5.578	17.275	19.675	21.920	24.725	26.757
12	3.074	3.571	4.404	5.226	6.304	18.549	21.026	23.337	26.217	28.299
13	3.565	4.107	5.009	5.892	7.042	19.812	22.362	24.736	27.688	29.819
14	4.075	4.660	5.629	6.571	7.790	21.064	23.685	26.119	29.141	31.319
15	4.601	5.229	6.262	7.261	8.547	22.307	24.996	27.488	30.578	32.801
16	5.142	5.812	6.908	7.962	9.312	23.542	26.296	28.845	32.000	34.267
17	5.697	6.408	7.564	8.672	10.085	24.769	27.587	30.191	33.409	35.718
18	6.265	7.015	8.231	9.390	10.865	25.989	28.869	31.526	34.805	37.156
19	6.844	7.633	8.907	10.117	11.651	27.204	30.144	32.852	36.191	38.582
20	7.434	8.260	9.591	10.851	12.443	28.412	31.410	34.170	37.566	39.997
21	8.034	8.897	10.283	11.591	13.240	29.615	32.671	35.479	38.932	41.401
22	8.643	9.542	10.982	12.338	14.042	30.813	33.924	36.781	40.289	42.796
23	9.260	10.196	11.689	13.091	14.848	32.007	35.172	38.076	41.638	44.181
24	9.886	10.856	12.401	13.848	15.659	33.196	36.415	39.364	42.980	45.559
25	10.520	11.524	13.120	14.611	16.473	34.382	37.652	40.646	44.314	46.928
26	11.160	12.198	13.844	15.379	17.292	35.563	38.885	41.923	45.642	48.290
27	11.808	12.879	14.573	16.151	18.114	36.741	40.113	43.194	46.963	49.645
28	12.461	13.565	15.308	16.928	18.939	37.916	41.337	44.461	48.278	50.993
29	13.121	14.257	16.047	17.708	19.768	39.087	42.557	45.722	49.588	52.336
30	13.787	14.954	16.791	18.493	20.599	40.256	43.773	46.979	50.892	53.672
40	20.707	22.164	24.433	26.509	29.051	51.805	55.758	59.342	63.691	66.766
50	27.991	29.707	32.357	34.764	37.689	63.167	67.505	71.420	76.154	79.490
60	35.534	37.485	40.482	43.188	46.459	74.397	79.082	83.298	88.379	91.952
70	43.275	45.442	48.758	51.739	55.329	85.527	90.531	95.023	100.425	104.215
80	51.172	53.540	57.153	60.391	64.278	96.578	101.879	106.629	112.329	116.321
90	59.196	61.754	65.647	69.126	73.291	107.565	113.145	118.136	124.116	128.299
100	67.328	70.065	74.222	77.929	82.358	118.498	124.342	129.561	135.807	140.169

Donald B. Owen, *Handbook of Statistical Tables*, U.S. Department of Energy (Reading, Mass.: Addison-Wesley, 1962). Reprinted with permission of the publisher.

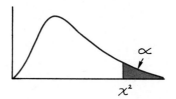

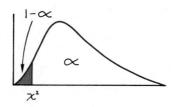

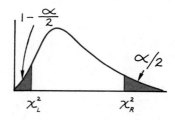

Table A-6

F Distribution ($\alpha = 0.01$ in the right tail)

Numerator degrees of freedom

df_2 \ df_1	1	2	3	4	5	6	7	8	9
1	4052.2	4999.5	5403.4	5624.6	5763.6	5859.0	5928.4	5981.1	6022.5
2	98.503	99.000	99.166	99.249	99.299	99.333	99.356	99.374	99.388
3	34.116	30.817	29.457	28.710	28.237	27.911	27.672	27.489	27.345
4	21.198	18.000	16.694	15.977	15.522	15.207	14.976	14.799	14.659
5	16.258	13.274	12.060	11.392	10.967	10.672	10.456	10.289	10.158
6	13.745	10.925	9.7795	9.1483	8.7459	8.4661	8.2600	8.1017	7.9761
7	12.246	9.5466	8.4513	7.8466	7.4604	7.1914	6.9928	6.8400	6.7188
8	11.259	8.6491	7.5910	7.0061	6.6318	6.3707	6.1776	6.0289	5.9106
9	10.561	8.0215	6.9919	6.4221	6.0569	5.8018	5.6129	5.4671	5.3511
10	10.044	7.5594	6.5523	5.9943	5.6363	5.3858	5.2001	5.0567	4.9424
11	9.6460	7.2057	6.2167	5.6683	5.3160	5.0692	4.8861	4.7445	4.6315
12	9.3302	6.9266	5.9525	5.4120	5.0643	4.8206	4.6395	4.4994	4.3875
13	9.0738	6.7010	5.7394	5.2053	4.8616	4.6204	4.4410	4.3021	4.1911
14	8.8616	6.5149	5.5639	5.0354	4.6950	4.4558	4.2779	4.1399	4.0297
15	8.6831	6.3589	5.4170	4.8932	4.5556	4.3183	4.1415	4.0045	3.8948
16	8.5310	6.2262	5.2922	4.7726	4.4374	4.2016	4.0259	3.8896	3.7804
17	8.3997	6.1121	5.1850	4.6690	4.3359	4.1015	3.9267	3.7910	3.6822
18	8.2854	6.0129	5.0919	4.5790	4.2479	4.0146	3.8406	3.7054	3.5971
19	8.1849	5.9259	5.0103	4.5003	4.1708	3.9386	3.7653	3.6305	3.5225
20	8.0960	5.8489	4.9382	4.4307	4.1027	3.8714	3.6987	3.5644	3.4567
21	8.0166	5.7804	4.8740	4.3688	4.0421	3.8117	3.6396	3.5056	3.3981
22	7.9454	5.7190	4.8166	4.3134	3.9880	3.7583	3.5867	3.4530	3.3458
23	7.8811	5.6637	4.7649	4.2636	3.9392	3.7102	3.5390	3.4057	3.2986
24	7.8229	5.6136	4.7181	4.2184	3.8951	3.6667	3.4959	3.3629	3.2560
25	7.7698	5.5680	4.6755	4.1774	3.8550	3.6272	3.4568	3.3239	3.2172
26	7.7213	5.5263	4.6366	4.1400	3.8183	3.5911	3.4210	3.2884	3.1818
27	7.6767	5.4881	4.6009	4.1056	3.7848	3.5580	3.3882	3.2558	3.1494
28	7.6356	5.4529	4.5681	4.0740	3.7539	3.5276	3.3581	3.2259	3.1195
29	7.5977	5.4204	4.5378	4.0449	3.7254	3.4995	3.3303	3.1982	3.0920
30	7.5625	5.3903	4.5097	4.0179	3.6990	3.4735	3.3045	3.1726	3.0665
40	7.3141	5.1785	4.3126	3.8283	3.5138	3.2910	3.1238	2.9930	2.8876
60	7.0771	4.9774	4.1259	3.6490	3.3389	3.1187	2.9530	2.8233	2.7185
120	6.8509	4.7865	3.9491	3.4795	3.1735	2.9559	2.7918	2.6629	2.5586
∞	6.6349	4.6052	3.7816	3.3192	3.0173	2.8020	2.6393	2.5113	2.4073

Denominator degrees of freedom

From Maxine Merrington and Catherine M. Thompson, "Tables of Percentage Points of the Inverted Beta (F) Distribution," *Biometrika 33* (1943): 80–84. Reproduced by permission of Professor E. S. Pearson.

Numerator degrees of freedom

df_2 \ df_1	10	12	15	20	24	30	40	60	120	∞
1	6055.8	6106.3	6157.3	6208.7	6234.6	6260.6	6286.8	6313.0	6339.4	6365.9
2	99.399	99.416	99.433	99.449	99.458	99.466	99.474	99.482	99.491	99.499
3	27.229	27.052	26.872	26.690	26.598	26.505	26.411	26.316	26.221	26.125
4	14.546	14.374	14.198	14.020	13.929	13.838	13.745	13.652	13.558	13.463
5	10.051	9.8883	9.7222	9.5526	9.4665	9.3793	9.2912	9.2020	9.1118	9.0204
6	7.8741	7.7183	7.5590	7.3958	7.3127	7.2285	7.1432	7.0567	6.9690	6.8800
7	6.6201	6.4691	6.3143	6.1554	6.0743	5.9920	5.9084	5.8236	5.7373	5.6495
8	5.8143	5.6667	5.5151	5.3591	5.2793	5.1981	5.1156	5.0316	4.9461	4.8588
9	5.2565	5.1114	4.9621	4.8080	4.7290	4.6486	4.5666	4.4831	4.3978	4.3105
10	4.8491	4.7059	4.5581	4.4054	4.3269	4.2469	4.1653	4.0819	3.9965	3.9090
11	4.5393	4.3974	4.2509	4.0990	4.0209	3.9411	3.8596	3.7761	3.6904	3.6024
12	4.2961	4.1553	4.0096	3.8584	3.7805	3.7008	3.6192	3.5355	3.4494	3.3608
13	4.1003	3.9603	3.8154	3.6646	3.5868	3.5070	3.4253	3.3413	3.2548	3.1654
14	3.9394	3.8001	3.6557	3.5052	3.4274	3.3476	3.2656	3.1813	3.0942	3.0040
15	3.8049	3.6662	3.5222	3.3719	3.2940	3.2141	3.1319	3.0471	2.9595	2.8684
16	3.6909	3.5527	3.4089	3.2587	3.1808	3.1007	3.0182	2.9330	2.8447	2.7528
17	3.5931	3.4552	3.3117	3.1615	3.0835	3.0032	2.9205	2.8348	2.7459	2.6530
18	3.5082	3.3706	3.2273	3.0771	2.9990	2.9185	2.8354	2.7493	2.6597	2.5660
19	3.4338	3.2965	3.1533	3.0031	2.9249	2.8442	2.7608	2.6742	2.5839	2.4893
20	3.3682	3.2311	3.0880	2.9377	2.8594	2.7785	2.6947	2.6077	2.5168	2.4212
21	3.3098	3.1730	3.0300	2.8796	2.8010	2.7200	2.6359	2.5484	2.4568	2.3603
22	3.2576	3.1209	2.9779	2.8274	2.7488	2.6675	2.5831	2.4951	2.4029	2.3055
23	3.2106	3.0740	2.9311	2.7805	2.7017	2.6202	2.5355	2.4471	2.3542	2.2558
24	3.1681	3.0316	2.8887	2.7380	2.6591	2.5773	2.4923	2.4035	2.3100	2.2107
25	3.1294	2.9931	2.8502	2.6993	2.6203	2.5383	2.4530	2.3637	2.2696	2.1694
26	3.0941	2.9578	2.8150	2.6640	2.5848	2.5026	2.4170	2.3273	2.2325	2.1315
27	3.0618	2.9256	2.7827	2.6316	2.5522	2.4699	2.3840	2.2938	2.1985	2.0965
28	3.0320	2.8959	2.7530	2.6017	2.5223	2.4397	2.3535	2.2629	2.1670	2.0642
29	3.0045	2.8685	2.7256	2.5742	2.4946	2.4118	2.3253	2.2344	2.1379	2.0342
30	2.9791	2.8431	2.7002	2.5487	2.4689	2.3860	2.2992	2.2079	2.1108	2.0062
40	2.8005	2.6648	2.5216	2.3689	2.2880	2.2034	2.1142	2.0194	1.9172	1.8047
60	2.6318	2.4961	2.3523	2.1978	2.1154	2.0285	1.9360	1.8363	1.7263	1.6006
120	2.4721	2.3363	2.1915	2.0346	1.9500	1.8600	1.7628	1.6557	1.5330	1.3805
∞	2.3209	2.1847	2.0385	1.8783	1.7908	1.6964	1.5923	1.4730	1.3246	1.0000

Denominator degrees of freedom

(continued)

Table A-6 (continued)

F Distribution ($\alpha = 0.025$ in the right tail)

0.025

F

					Numerator degrees of freedom				
df_2 \ df_1	1	2	3	4	5	6	7	8	9
1	647.79	799.50	864.16	899.58	921.85	937.11	948.22	956.66	963.28
2	38.506	39.000	39.165	39.248	39.298	39.331	39.335	39.373	39.387
3	17.443	16.044	15.439	15.101	14.885	14.735	14.624	14.540	14.473
4	12.218	10.649	9.9792	9.6045	9.3645	9.1973	9.0741	8.9796	8.9047
5	10.007	8.4336	7.7636	7.3879	7.1464	6.9777	6.8531	6.7572	6.6811
6	8.8131	7.2599	6.5988	6.2272	5.9876	5.8198	5.6955	5.5996	5.5234
7	8.0727	6.5415	5.8898	5.5226	5.2852	5.1186	4.9949	4.8993	4.8232
8	7.5709	6.0595	5.4160	5.0526	4.8173	4.6517	4.5286	4.4333	4.3572
9	7.2093	5.7147	5.0781	4.7181	4.4844	4.3197	4.1970	4.1020	4.0260
10	6.9367	5.4564	4.8256	4.4683	4.2361	4.0721	3.9498	3.8549	3.7790
11	6.7241	5.2559	4.6300	4.2751	4.0440	3.8807	3.7586	3.6638	3.5879
12	6.5538	5.0959	4.4742	4.1212	3.8911	3.7283	3.6065	3.5118	3.4358
13	6.4143	4.9653	4.3472	3.9959	3.7667	3.6043	3.4827	3.3880	3.3120
14	6.2979	4.8567	4.2417	3.8919	3.6634	3.5014	3.3799	3.2853	3.2093
15	6.1995	4.7650	4.1528	3.8043	3.5764	3.4147	3.2934	3.1987	3.1227
16	6.1151	4.6867	4.0768	3.7294	3.5021	3.3406	3.2194	3.1248	3.0488
17	6.0420	4.6189	4.0112	3.6648	3.4379	3.2767	3.1556	3.0610	2.9849
18	5.9781	4.5597	3.9539	3.6083	3.3820	3.2209	3.0999	3.0053	2.9291
19	5.9216	4.5075	3.9034	3.5587	3.3327	3.1718	3.0509	2.9563	2.8801
20	5.8715	4.4613	3.8587	3.5147	3.2891	3.1283	3.0074	2.9128	2.8365
21	5.8266	4.4199	3.8188	3.4754	3.2501	3.0895	2.9686	2.8740	2.7977
22	5.7863	4.3828	3.7829	3.4401	3.2151	3.0546	2.9338	2.8392	2.7628
23	5.7498	4.3492	3.7505	3.4083	3.1835	3.0232	2.9023	2.8077	2.7313
24	5.7166	4.3187	3.7211	3.3794	3.1548	2.9946	2.8738	2.7791	2.7027
25	5.6864	4.2909	3.6943	3.3530	3.1287	2.9685	2.8478	2.7531	2.6766
26	5.6586	4.2655	3.6697	3.3289	3.1048	2.9447	2.8240	2.7293	2.6528
27	5.6331	4.2421	3.6472	3.3067	3.0828	2.9228	2.8021	2.7074	2.6309
28	5.6096	4.2205	3.6264	3.2863	3.0626	2.9027	2.7820	2.6872	2.6106
29	5.5878	4.2006	3.6072	3.2674	3.0438	2.8840	2.7633	2.6686	2.5919
30	5.5675	4.1821	3.5894	3.2499	3.0265	2.8667	2.7460	2.6513	2.5746
40	5.4239	4.0510	3.4633	3.1261	2.9037	2.7444	2.6238	2.5289	2.4519
60	5.2856	3.9253	3.3425	3.0077	2.7863	2.6274	2.5068	2.4117	2.3344
120	5.1523	3.8046	3.2269	2.8943	2.6740	2.5154	2.3948	2.2994	2.2217
∞	5.0239	3.6889	3.1161	2.7858	2.5665	2.4082	2.2875	2.1918	2.1136

Denominator degrees of freedom

(continued)

Numerator degrees of freedom

df_2 \ df_1	10	12	15	20	24	30	40	60	120	∞
1	968.63	976.71	984.87	993.10	997.25	1001.4	1005.6	1009.8	1014.0	1018.3
2	39.398	39.415	39.431	39.448	39.456	39.465	39.473	39.481	39.490	39.498
3	14.419	14.337	14.253	14.167	14.124	14.081	14.037	13.992	13.947	13.902
4	8.8439	8.7512	8.6565	8.5599	8.5109	8.4613	8.4111	8.3604	8.3092	8.2573
5	6.6192	6.5245	6.4277	6.3286	6.2780	6.2269	6.1750	6.1225	6.0693	6.0153
6	5.4613	5.3662	5.2687	5.1684	5.1172	5.0652	5.0125	4.9589	4.9044	4.8491
7	4.7611	4.6658	4.5678	4.4667	4.4150	4.3624	4.3089	4.2544	4.1989	4.1423
8	4.2951	4.1997	4.1012	3.9995	3.9472	3.8940	3.8398	3.7844	3.7279	3.6702
9	3.9639	3.8682	3.7694	3.6669	3.6142	3.5604	3.5055	3.4493	3.3918	3.3329
10	3.7168	3.6209	3.5217	3.4185	3.3654	3.3110	3.2554	3.1984	3.1399	3.0798
11	3.5257	3.4296	3.3299	3.2261	3.1725	3.1176	3.0613	3.0035	2.9441	2.8828
12	3.3736	3.2773	3.1772	3.0728	3.0187	2.9633	2.9063	2.8478	2.7874	2.7249
13	3.2497	3.1532	3.0527	2.9477	2.8932	2.8372	2.7797	2.7204	2.6590	2.5955
14	3.1469	3.0502	2.9493	2.8437	2.7888	2.7324	2.6742	2.6142	2.5519	2.4872
15	3.0602	2.9633	2.8621	2.7559	2.7006	2.6437	2.5850	2.5242	2.4611	2.3953
16	2.9862	2.8890	2.7875	2.6808	2.6252	2.5678	2.5085	2.4471	2.3831	2.3163
17	2.9222	2.8249	2.7230	2.6158	2.5598	2.5020	2.4422	2.3801	2.3153	2.2474
18	2.8664	2.7689	2.6667	2.5590	2.5027	2.4445	2.3842	2.3214	2.2558	2.1869
19	2.8172	2.7196	2.6171	2.5089	2.4523	2.3937	2.3329	2.2696	2.2032	2.1333
20	2.7737	2.6758	2.5731	2.4645	2.4076	2.3486	2.2873	2.2234	2.1562	2.0853
21	2.7348	2.6368	2.5338	2.4247	2.3675	2.3082	2.2465	2.1819	2.1141	2.0422
22	2.6998	2.6017	2.4984	2.3890	2.3315	2.2718	2.2097	2.1446	2.0760	2.0032
23	2.6682	2.5699	2.4665	2.3567	2.2989	2.2389	2.1763	2.1107	2.0415	1.9677
24	2.6396	2.5411	2.4374	2.3273	2.2693	2.2090	2.1460	2.0799	2.0099	1.9353
25	2.6135	2.5149	2.4110	2.3005	2.2422	2.1816	2.1183	2.0516	1.9811	1.9055
26	2.5896	2.4908	2.3867	2.2759	2.2174	2.1565	2.0928	2.0257	1.9545	1.8781
27	2.5676	2.4688	2.3644	2.2533	2.1946	2.1334	2.0693	2.0018	1.9299	1.8527
28	2.5473	2.4484	2.3438	2.2324	2.1735	2.1121	2.0477	1.9797	1.9072	1.8291
29	2.5286	2.4295	2.3248	2.2131	2.1540	2.0923	2.0276	1.9591	1.8861	1.8072
30	2.5112	2.4120	2.3072	2.1952	2.1359	2.0739	2.0089	1.9400	1.8664	1.7867
40	2.3882	2.2882	2.1819	2.0677	2.0069	1.9429	1.8752	1.8028	1.7242	1.6371
60	2.2702	2.1692	2.0613	1.9445	1.8817	1.8152	1.7440	1.6668	1.5810	1.4821
120	2.1570	2.0548	1.9450	1.8249	1.7597	1.6899	1.6141	1.5299	1.4327	1.3104
∞	2.0483	1.9447	1.8326	1.7085	1.6402	1.5660	1.4835	1.3883	1.2684	1.0000

Denominator degrees of freedom

(continued)

Table A-6 (continued)

F Distribution ($\alpha = 0.05$ in the right tail)

Numerator degrees of freedom

df_2 \ df_1	1	2	3	4	5	6	7	8	9
1	161.45	199.50	215.71	224.58	230.16	233.99	236.77	238.88	240.54
2	18.513	19.000	19.164	19.247	19.296	19.330	19.353	19.371	19.385
3	10.128	9.5521	9.2766	9.1172	9.0135	8.9406	8.8867	8.8452	8.8123
4	7.7086	6.9443	6.5914	6.3882	6.2561	6.1631	6.0942	6.0410	5.9988
5	6.6079	5.7861	5.4095	5.1922	5.0503	4.9503	4.8759	4.8183	4.7725
6	5.9874	5.1433	4.7571	4.5337	4.3874	4.2839	4.2067	4.1468	4.0990
7	5.5914	4.7374	4.3468	4.1203	3.9715	3.8660	3.7870	3.7257	3.6767
8	5.3177	4.4590	4.0662	3.8379	3.6875	3.5806	3.5005	3.4381	3.3881
9	5.1174	4.2565	3.8625	3.6331	3.4817	3.3738	3.2927	3.2296	3.1789
10	4.9646	4.1028	3.7083	3.4780	3.3258	3.2172	3.1355	3.0717	3.0204
11	4.8443	3.9823	3.5874	3.3567	3.2039	3.0946	3.0123	2.9480	2.8962
12	4.7472	3.8853	3.4903	3.2592	3.1059	2.9961	2.9134	2.8486	2.7964
13	4.6672	3.8056	3.4105	3.1791	3.0254	2.9153	2.8321	2.7669	2.7144
14	4.6001	3.7389	3.3439	3.1122	2.9582	2.8477	2.7642	2.6987	2.6458
15	4.5431	3.6823	3.2874	3.0556	2.9013	2.7905	2.7066	2.6408	2.5876
16	4.4940	3.6337	3.2389	3.0069	2.8524	2.7413	2.6572	2.5911	2.5377
17	4.4513	3.5915	3.1968	2.9647	2.8100	2.6987	2.6143	2.5480	2.4943
18	4.4139	3.5546	3.1599	2.9277	2.7729	2.6613	2.5767	2.5102	2.4563
19	4.3807	3.5219	3.1274	2.8951	2.7401	2.6283	2.5435	2.4768	2.4227
20	4.3512	3.4928	3.0984	2.8661	2.7109	2.5990	2.5140	2.4471	2.3928
21	4.3248	3.4668	3.0725	2.8401	2.6848	2.5727	2.4876	2.4205	2.3660
22	4.3009	3.4434	3.0491	2.8167	2.6613	2.5491	2.4638	2.3965	2.3419
23	4.2793	3.4221	3.0280	2.7955	2.6400	2.5277	2.4422	2.3748	2.3201
24	4.2597	3.4028	3.0088	2.7763	2.6207	2.5082	2.4226	2.3551	2.3002
25	4.2417	3.3852	2.9912	2.7587	2.6030	2.4904	2.4047	2.3371	2.2821
26	4.2252	3.3690	2.9752	2.7426	2.5868	2.4741	2.3883	2.3205	2.2655
27	4.2100	3.3541	2.9604	2.7278	2.5719	2.4591	2.3732	2.3053	2.2501
28	4.1960	3.3404	2.9467	2.7141	2.5581	2.4453	2.3593	2.2913	2.2360
29	4.1830	3.3277	2.9340	2.7014	2.5454	2.4324	2.3463	2.2783	2.2229
30	4.1709	3.3158	2.9223	2.6896	2.5336	2.4205	2.3343	2.2662	2.2107
40	4.0847	3.2317	2.8387	2.6060	2.4495	2.3359	2.2490	2.1802	2.1240
60	4.0012	3.1504	2.7581	2.5252	2.3683	2.2541	2.1665	2.0970	2.0401
120	3.9201	3.0718	2.6802	2.4472	2.2899	2.1750	2.0868	2.0164	1.9588
∞	3.8415	2.9957	2.6049	2.3719	2.2141	2.0986	2.0096	1.9384	1.8799

Denominator degrees of freedom

0.05

F

(continued)

Numerator degrees of freedom

df_2	10	12	15	20	24	30	40	60	120	∞
1	241.88	243.91	245.95	248.01	249.05	250.10	251.14	252.20	253.25	254.31
2	19.396	19.413	19.429	19.446	19.454	19.462	19.471	19.479	19.487	19.496
3	8.7855	8.7446	8.7029	8.6602	8.6385	8.6166	8.5944	8.5720	8.5494	8.5264
4	5.9644	5.9117	5.8578	5.8025	5.7744	5.7459	5.7170	5.6877	5.6581	5.6281
5	4.7351	4.6777	4.6188	4.5581	4.5272	4.4957	4.4638	4.4314	4.3985	4.3650
6	4.0600	3.9999	3.9381	3.8742	3.8415	3.8082	3.7743	3.7398	3.7047	3.6689
7	3.6365	3.5747	3.5107	3.4445	3.4105	3.3758	3.3404	3.3043	3.2674	3.2298
8	3.3472	3.2839	3.2184	3.1503	3.1152	3.0794	3.0428	3.0053	2.9669	2.9276
9	3.1373	3.0729	3.0061	2.9365	2.9005	2.8637	2.8259	2.7872	2.7475	2.7067
10	2.9782	2.9130	2.8450	2.7740	2.7372	2.6996	2.6609	2.6211	2.5801	2.5379
11	2.8536	2.7876	2.7186	2.6464	2.6090	2.5705	2.5309	2.4901	2.4480	2.4045
12	2.7534	2.6866	2.6169	2.5436	2.5055	2.4663	2.4259	2.3842	2.3410	2.2962
13	2.6710	2.6037	2.5331	2.4589	2.4202	2.3803	2.3392	2.2966	2.2524	2.2064
14	2.6022	2.5342	2.4630	2.3879	2.3487	2.3082	2.2664	2.2229	2.1778	2.1307
15	2.5437	2.4753	2.4034	2.3275	2.2878	2.2468	2.2043	2.1601	2.1141	2.0658
16	2.4935	2.4247	2.3522	2.2756	2.2354	2.1938	2.1507	2.1058	2.0589	2.0096
17	2.4499	2.3807	2.3077	2.2304	2.1898	2.1477	2.1040	2.0584	2.0107	1.9604
18	2.4117	2.3421	2.2686	2.1906	2.1497	2.1071	2.0629	2.0166	1.9681	1.9168
19	2.3779	2.3080	2.2341	2.1555	2.1141	2.0712	2.0264	1.9795	1.9302	1.8780
20	2.3479	2.2776	2.2033	2.1242	2.0825	2.0391	1.9938	1.9464	1.8963	1.8432
21	2.3210	2.2504	2.1757	2.0960	2.0540	2.0102	1.9645	1.9165	1.8657	1.8117
22	2.2967	2.2258	2.1508	2.0707	2.0283	1.9842	1.9380	1.8894	1.8380	1.7831
23	2.2747	2.2036	2.1282	2.0476	2.0050	1.9605	1.9139	1.8648	1.8128	1.7570
24	2.2547	2.1834	2.1077	2.0267	1.9838	1.9390	1.8920	1.8424	1.7896	1.7330
25	2.2365	2.1649	2.0889	2.0075	1.9643	1.9192	1.8718	1.8217	1.7684	1.7110
26	2.2197	2.1479	2.0716	1.9898	1.9464	1.9010	1.8533	1.8027	1.7488	1.6906
27	2.2043	2.1323	2.0558	1.9736	1.9299	1.8842	1.8361	1.7851	1.7306	1.6717
28	2.1900	2.1179	2.0411	1.9586	1.9147	1.8687	1.8203	1.7689	1.7138	1.6541
29	2.1768	2.1045	2.0275	1.9446	1.9005	1.8543	1.8055	1.7537	1.6981	1.6376
30	2.1646	2.0921	2.0148	1.9317	1.8874	1.8409	1.7918	1.7396	1.6835	1.6223
40	2.0772	2.0035	1.9245	1.8389	1.7929	1.7444	1.6928	1.6373	1.5766	1.5089
60	1.9926	1.9174	1.8364	1.7480	1.7001	1.6491	1.5943	1.5343	1.4673	1.3893
120	1.9105	1.8337	1.7505	1.6587	1.6084	1.5543	1.4952	1.4290	1.3519	1.2539
∞	1.8307	1.7522	1.6664	1.5705	1.5173	1.4591	1.3940	1.3180	1.2214	1.0000

Denominator degrees of freedom

Table A-7

Critical Values of the Pearson Correlation Coefficient r

n	$\alpha = .05$	$\alpha = .01$
4	.950	.999
5	.878	.959
6	.811	.917
7	.754	.875
8	.707	.834
9	.666	.798
10	.632	.765
11	.602	.735
12	.576	.708
13	.553	.684
14	.532	.661
15	.514	.641
16	.497	.623
17	.482	.606
18	.468	.590
19	.456	.575
20	.444	.561
25	.396	.505
30	.361	.463
35	.335	.430
40	.312	.402
45	.294	.378
50	.279	.361
60	.254	.330
70	.236	.305
80	.220	.286
90	.207	.269
100	.196	.256

To test H_0: $\rho = 0$ against H_1: $\rho \neq 0$, reject H_0 if the absolute value of r is greater than the critical value in the table.

Table A-8

Critical Values for the Sign Test

	α			
n	.005 (one tail) .01 (two tails)	.01 (one tail) .02 (two tails)	.025 (one tail) .05 (two tails)	.05 (one tail) .10 (two tails)
1	*	*	*	*
2	*	*	*	*
3	*	*	*	*
4	*	*	*	*
5	*	*	*	0
6	*	*	0	0
7	*	0	0	0
8	0	0	0	1
9	0	0	1	1
10	0	0	1	1
11	0	1	1	2
12	1	1	2	2
13	1	1	2	3
14	1	2	2	3
15	2	2	3	3
16	2	2	3	4
17	2	3	4	4
18	3	3	4	5
19	3	4	4	5
20	3	4	5	5
21	4	4	5	6
22	4	5	5	6
23	4	5	6	7
24	5	5	6	7
25	5	6	7	7

NOTES:

1. * indicates that it is not possible to get a value in the critical region.
2. The null hypothesis is rejected if the number of the less-frequent sign (x) is less than or equal to the value in the table.
3. For values of n greater than 25, a normal approximation is used with

$$z = \frac{(x + 0.5) - (n/2)}{\sqrt{n}/2}$$

Table A-9

Critical Values of T for the Wilcoxon Signed-Rank Test

n	.005 (one tail) .01 (two tails)	.01 (one tail) .02 (two tails)	.025 (one tail) .05 (two tails)	.05 (one tail) .10 (two tails)
5				1
6			1	2
7		0	2	4
8	0	2	4	6
9	2	3	6	8
10	3	5	8	11
11	5	7	11	14
12	7	10	14	17
13	10	13	17	21
14	13	16	21	26
15	16	20	25	30
16	19	24	30	36
17	23	28	35	41
18	28	33	40	47
19	32	38	46	54
20	37	43	52	60
21	43	49	59	68
22	49	56	66	75
23	55	62	73	83
24	61	69	81	92
25	68	77	90	101
26	76	85	98	110
27	84	93	107	120
28	92	102	117	130
29	100	111	127	141
30	109	120	137	152

Reject the null hypothesis if the test statistic T is less than or equal to the critical value found in this table. Fail to reject the null hypothesis if the test statistic T is greater than the critical value found in this table.

"Table of Critical Values of T for the Wilcoxon Signed Rank Test" from *Some Rapid Approximate Statistical Procedures,* Copyright © 1949, 1964, Lederle Laboratories Division of American Cyanamid Company, All Rights Reserved, and Reprinted With Permission.

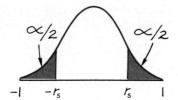

Table A-10			

Critical Values of Spearman's Rank Correlation Coefficient r_s			
n	$\alpha = 0.10$	$\alpha = 0.05$	$\alpha = 0.02$	$\alpha = 0.01$
---	---	---	---	---
5	.900	–	–	–
6	.829	.886	.943	–
7	.714	.786	.893	–
8	.643	.738	.833	.881
9	.600	.683	.783	.833
10	564	.648	.745	.794
11	.523	.623	.736	.818
12	.497	.591	.703	.780
13	.475	.566	.673	.745
14	.457	.545	.646	.716
15	.441	.525	.623	.689
16	.425	.507	.601	.666
17	.412	.490	.582	.645
18	.399	.476	.564	.625
19	.388	.462	.549	.608
20	.377	.450	.534	.591
21	.368	.438	.521	.576
22	.359	.428	.508	.562
23	.351	.418	.496	.549
24	.343	.409	.485	.537
25	.336	.400	.475	.526
26	.329	.392	.465	.515
27	.323	.385	.456	.505
28	.317	.377	.448	.496
29	.311	.370	.440	.487
30	.305	.364	.432	.478

For $n > 30$ use $r_s = \pm z/\sqrt{n-1}$, where z corresponds to the level of significance. For example, if $\alpha = 0.05$, then $z = 1.96$.

To test H_0: $\rho_s = 0$

against H_1: $\rho_s \neq 0$

E. G. Olds, "Distribution of sums of squares of rank differences to small numbers of individuals," *Annals of Statistics*, 9 (1938): 133–148, and Olds, E. G., with amendment in 20, 1949, 117–118 of *Annals of Statistics*. Reprinted with permission.

Table A-11

Critical Values for Number of Runs G

Value of n_2

n_1	2	3	4	5	6	7	8	9	10	11	12	13	14	15	16	17	18	19	20
2	1	1	1	1	1	1	1	1	1	1	2	2	2	2	2	2	2	2	2
	6	6	6	6	6	6	6	6	6	6	6	6	6	6	6	6	6	6	6
3	1	1	1	1	2	2	2	2	2	2	2	2	2	3	3	3	3	3	3
	6	8	8	8	8	8	8	8	8	8	8	8	8	8	8	8	8	8	8
4	1	1	1	2	2	2	3	3	3	3	3	3	3	3	4	4	4	4	4
	6	8	9	9	9	10	10	10	10	10	10	10	10	10	10	10	10	10	10
5	1	1	2	2	3	3	3	3	3	4	4	4	4	4	4	4	5	5	5
	6	8	9	10	10	11	11	12	12	12	12	12	12	12	12	12	12	12	12
6	1	2	2	3	3	3	3	4	4	4	4	5	5	5	5	5	5	6	6
	6	8	9	10	11	12	12	13	13	13	13	14	14	14	14	14	14	14	14
7	1	2	2	3	3	3	4	4	5	5	5	5	5	6	6	6	6	6	6
	6	8	10	11	12	13	13	14	14	14	14	15	15	15	16	16	16	16	16
8	1	2	3	3	3	4	4	5	5	5	6	6	6	6	6	7	7	7	7
	6	8	10	11	12	13	14	14	15	15	16	16	16	16	17	17	17	17	17
9	1	2	3	3	4	4	5	5	5	6	6	6	7	7	7	7	8	8	8
	6	8	10	12	13	14	14	15	16	16	16	17	17	18	18	18	18	18	18
10	1	2	3	3	4	5	5	5	6	6	7	7	7	7	8	8	8	8	9
	6	8	10	12	13	14	15	16	16	17	17	18	18	18	19	19	19	20	20
11	1	2	3	4	4	5	5	6	6	7	7	7	8	8	8	9	9	9	9
	6	8	10	12	13	14	15	16	17	17	18	19	19	19	20	20	20	21	21
12	2	2	3	4	4	5	6	6	7	7	7	8	8	8	9	9	9	10	10
	6	8	10	12	13	14	16	16	17	18	19	19	20	20	21	21	21	22	22
13	2	2	3	4	5	5	6	6	7	7	8	8	9	9	9	10	10	10	10
	6	8	10	12	14	15	16	17	18	19	19	20	20	21	21	22	22	23	23
14	2	2	3	4	5	5	6	7	7	8	8	9	9	9	10	10	10	11	11
	6	8	10	12	14	15	16	17	18	19	20	20	21	22	22	23	23	23	24
15	2	3	3	4	5	6	6	7	7	8	8	9	9	10	10	11	11	11	12
	6	8	10	12	14	15	16	18	18	19	20	21	22	22	23	23	24	24	25
16	2	3	4	4	5	6	6	7	8	8	9	9	10	10	11	11	11	12	12
	6	8	10	12	14	16	17	18	19	20	21	21	22	23	23	24	25	25	25
17	2	3	4	4	5	6	7	7	8	9	9	10	10	11	11	11	12	12	13
	6	8	10	12	14	16	17	18	19	20	21	22	23	23	24	25	25	26	26
18	2	3	4	5	5	6	7	8	8	9	9	10	10	11	11	12	12	13	13
	6	8	10	12	14	16	17	18	19	20	21	22	23	24	25	25	26	26	27
19	2	3	4	5	6	6	7	8	8	9	10	10	11	11	12	12	13	13	13
	6	8	10	12	14	16	17	18	20	21	22	23	23	24	25	26	26	27	27
20	2	3	4	5	6	6	7	8	9	9	10	10	11	12	12	13	13	13	14
	6	8	10	12	14	16	17	18	20	21	22	23	24	25	25	26	27	27	28

Value of n_1

The entries in this table are the critical G values assuming a two-tailed test with a significance level of $\alpha = 0.05$. The null hypothesis of randomness is rejected if the total number of runs G is less than or equal to the smaller entry or greater than or equal to the larger entry.

Adapted from C. Eisenhardt and F. Swed, "Tables for testing randomness of grouping in a sequence of alternatives," *The Annals of Statistics*, 14 (1943): 83–86. Reprinted with permission.

B

Glossary

Addition rule. Rule for determining the probability that, on a single trial, either event A occurs, or event B occurs, or they both occur.

Alternative hypothesis. Denoted by H_1, the statement that is equivalent to the negation of the null hypothesis.

Analysis of variance. A method analyzing population variance in order to make inferences about the population.

ANOVA. See analysis of variance.

Average. Any one of several measures designed to reveal the central tendency of a collection of data.

Bimodal. Having two modes.

Binomial experiment. An experiment with a fixed number of independent trials. Each outcome falls into exactly one of two categories.

Binomial probability formula. See Formula 4-4 in Section 4-4.

Bivariate data. Data arranged in pairs.

Box-and-whisker diagram. Graphic method of showing the spread of a set of data.

Central limit theorem. Theorem stating that sample means tend to be normally distributed.

Centroid. The point $(\bar{x}, \bar{y})$ determined from a collection of bivariate data.

Chi-square distribution. A continuous probability distribution with selected values given in Table A-5.

Class boundaries. Values obtained from a frequency table by increasing the upper class limits and decreasing the lower class limits by the same amount so that there are no gaps between consecutive classes.

Classical approach to probability. Determining the probability of an event by dividing the number of ways the event can occur by the total number of possible outcomes.

Class marks. The midpoints of the classes in a frequency table.

Class width. The difference between two consecutive lower class limits in a frequency table.

Coefficient of determination. The amount of the variation in y that is explained by the regression line.

Combinations rule. Rule for determining the number of different combinations of selected items.

Complement of an event. All outcomes in which the original event does not occur.

Completely randomized design. In analysis of variance, each element is given the same chance of belonging to the different categories or treatments.

Compound event. A combination of simple events.

Confidence interval. A range of values used to estimate some population parameter with a specific level of confidence.

Confidence interval limits. The two numbers that are used as the high and low boundaries of a confidence interval.

Contingency table. A table of observed frequencies where the rows correspond to one variable of classification and the columns correspond to another variable of classification.

Continuity correction. An adjustment made when a discrete random variable is being approximated by a continuous random variable (see Section 5-5).

Continuous random variable. A random variable with infinitely many values that can be associated with points on a continuous line interval.

Correlation. Statistical association between two variables.

Correlation coefficient. A measurement of the strength of the relationship between two variables.

Countable set. A set with either a finite number of values or values that can be made to correspond to the positive integers.

Critical region.   The area under a curve containing the values that lead to rejection of the null hypothesis.

Critical value.   Value separating the critical region from the values of the test statistic that would not lead to rejection of the null hypothesis.

Cumulative frequency table.   Frequency table in which each class and frequency represents cumulative data up to and including that class.

Data.   The numbers or information collected in an experiment.

Decile.   The nine deciles divide the ranked data into ten groups with 10% of the scores in each group.

Degree of confidence.   Probability that a population parameter is contained within a particular confidence interval.

Degrees of freedom.   The number of values that are free to vary after certain restrictions have been imposed on all values.

Denominator degrees of freedom.   The degrees of freedom corresponding to the denominator of the F test statistic.

Dependent events.   Events that are not independent. (See independent events.)

Dependent samples.   The values in one sample are related to the values in another sample.

Descriptive statistics.   The methods used to summarize the key characteristics of known population data.

Discrete random variable.   A random variable with either a finite number of values or a countable number of values.

Efficiency.   Measure of the sensitivity of a nonparametric test in comparison to a corresponding parametric test.

Empirical approximation of probability.   Estimated value of probability based on actual observations.

Event.   A result or outcome of an experiment.

Expected frequency.   The theoretical frequency for a cell of a contingency table or multinomial table.

Expected value.   For a discrete random variable, the sum of the products obtained by multiplying each value of the random variable by the corresponding probability.

Experiment.   Process that allows observations to be made.

Explained deviation.   For one pair of values in a collection of bivariate data, the difference between the predicted y value and the mean of the y values.

Explained variation.   The sum of the squares of the explained deviations for all pairs of bivariate data in a sample.

Exploratory data analysis (EDA).   Branch of statistics emphasizing the investigation of data.

Factorial rule. n different items can be arranged $n!$ different ways.

F distribution.   A continuous probability distribution with selected values given in Table A-6.

Finite population correction factor.   Factor for correcting the standard error of the mean when a sample size exceeds 5% of the size of a finite population.

Frequency polygon.   Graphical method for representing the distribution of data using connected straight-line segments.

Frequency table.   A list of categories of scores along with their corresponding frequencies.

Fundamental counting rule.   For a sequence of two events in which the first event can occur m ways and the second can occur n ways, the events together can occur a total of $m \cdot n$ ways.

Hinge.   The median value of the bottom (or top) half of a set of ranked data.

Histogram.   A graph of vertical bars representing the frequency distribution of a set of data.

H test.   See the Kruskal-Wallis test.

Hypothesis.   A statement or claim that some population characteristic is true.

Hypothesis test.   A method for testing claims made about populations. Also called test of significance.

Independent events.   The case when the occurrence of any one of the events does not affect the probabilities of the occurrences of the other events.

Independent samples.   The values in one sample are not related to the values in another sample.

Inferential statistics.   The methods of using sample data to make generalizations or inferences about a population.

Interquartile range.   The difference between the first and third quartiles.

Interval.   Level of measurement of data: data can be arranged in order, and differences between data values are meaningful.

Interval estimate.   (See confidence interval.)

Kruskal-Wallis test.   A nonparametric hypothesis test used to compare three or more samples.

Least squares property.   For a regression line, the sum of the squares of the vertical deviations of the sample points from the regression line is the smallest sum possible.

Left-tailed test.   Hypothesis test in which the critical region is located in the extreme left area of the probability distribution.

Linear correlation coefficient.   Measure of the strength of the relationship between two variables.

Lower class limits.   The smallest numbers that can actually belong to the different classes in a frequency table.

Maximum error of estimate.   The largest difference between a point estimate and the true value of a population parameter.

Mean.   The sum of a set of scores divided by the number of scores.

Mean deviation.   The measure of dispersion equal to the sum of the deviations of each score from the mean, divided by the number of scores.

Measure of central tendency.   Value intended to indicate the center of the scores in a collection of data.

Measure of dispersion.   Any of several measures designed to reflect the amount of variability among a set of scores.

Median.   The middle value of a set of scores arranged in order of magnitude.

Midquartile.   One-half of the sum of the first and third quartiles.

Midrange.   One-half the sum of the highest and lowest scores.

Mode.   The score that occurs most frequently.

Multimodal.   Having more than two modes.

Multinomial experiment.   An experiment with a fixed number of independent trials and each outcome falls into exactly one of several categories.

Multiplication rule.   Rule for determining the probability that event A will occur on one trial while event B occurs on a second trial.

Mutually exclusive events.   Events that cannot occur simultaneously.

Nominal.   Level of measurement of data: data consist of names, labels, or categories only.

Nonparametric methods.   Statistical procedures for testing hypotheses or estimating parameters; they are not based on population parameters and do not require many of the restrictions of parametric tests.

Normal distribution.   A bell-shaped probability distribution described algebraically by Equation 5-1 in Section 5-1.

Null hypothesis.   Denoted by H_0, it is the claim made about some population characteristic. It usually involves the case of no difference.

Numerator degrees of freedom.   The degrees of freedom corresponding to the numerator of the F test statistic.

Observed frequency.   The actual frequency count recorded in one cell of a contingency table or multinomial table.

Ogive.   Graphical method of representing a cumulative frequency table.

One-way analysis of variance.   Analysis of variance involving data classified into groups according to a single criterion only.

Ordinal.   Level of measurement of data: data may be arranged in order, but differences between data values either cannot be determined or they are meaningless.

Parameter.   A measured characteristic of a population.

Parametric methods.   Statistical procedures for testing hypotheses or estimating parameters; based on population parameters.

Pearson's product moment.   (See linear correlation coefficient.)

Percentile.   The 99 percentiles divide the ranked data into 100 groups with 1% of the scores in each group.

Permutation rule.   Rule for determining the number of different arrangements of selected items.

Pie chart.   Graphical method for representing data in the form of a circle containing wedges.

Point estimate.   A single value that serves as an estimate of a population parameter.

Pooled estimate of p_1 and p_2.   The probability obtained by combining the data from two sample proportions and dividing the total number of successes by the total number of observations.

Population.   The complete and entire collection of elements to be studied.

Predicted value.   Using a regression equation, the value of one variable given a value for the other variable.

Probability distribution.   Collection of values of a random variable along with their corresponding probabilities.

P-value.   The probability that a test statistic in a hypothesis test is at least as extreme as the one actually obtained.

Quartile. The three quartiles divide the ranked data into four groups with 25% of the scores in each group.

Random sample. A sample selected in a way that allows every member of the population to have the same chance of being chosen.

Random variable. The values that correspond to the numbers associated with events in a sample space.

Range. The measure of dispersion that is the difference between the highest and lowest scores.

Rank. The numerical position of an item in a sample set arranged in order.

Rank correlation coefficient. Measure of the strength of the relationship between two variables; based on the ranks of the values.

Ratio. Level of measurement of data: data can be arranged in order, differences between data values are meaningful, and there is an inherent zero starting point.

Regression line. A straight line that summarizes the relationship between two variables.

Right-tailed test. Hypothesis test in which the critical region is located in the extreme right area of the probability distribution.

Rigorously controlled design. In analysis of variance, all factors are forced to be constant so that effects of extraneous factors are eliminated.

Run. Used in the runs test for randomness, a sequence of data exhibiting the same characteristic.

Runs test. Nonparametric method used to test for randomness.

Sample. A subset of a population.

Sample space. In an experiment, the set of all possible outcomes or events that cannot be further broken down.

Scatter diagram. Graphical method for displaying bivariate data.

Semi-interquartile range. One-half of the difference between the first and third quartiles.

Significance level. The probability that serves as a cutoff between results attributed to chance and results attributed to significant differences.

Sign test. A nonparametric hypothesis test used to compare samples from two populations.

Simple event. An experimental outcome that cannot be further broken down.

Single factor analysis of variance. (See one-way analysis of variance.)

Slope. Measure of steepness of a straight line.

Spearman's rank correlation coefficient. (See rank correlation coefficient.)

Standard deviation. The measure of dispersion equal to the square root of the variance.

Standard error of estimate. Measure of the spread of the sample points about the regression line.

Standard error of the mean. The standard deviation of all possible sample means $\bar{x}$.

Standard normal distribution. A normal distribution with a mean of zero and a standard deviation equal to one.

Standard score. Also called z score, it is the number of standard deviations that a given value is above or below the mean.

Statistic. A measured characteristic of a sample.

Statistics. The collection, organization, description, and analysis of data.

Stem-and-leaf plot. Method of sorting and arranging data in a way that reveals the distribution.

Student t distribution. (See t distribution.)

t distribution. A bell-shaped distribution usually associated with small sample experiments. Also called the student t distribution.

10–90 percentile range. The difference between the 10th and 90th percentiles.

Test of significance. (See hypothesis test.)

Test statistic. Used in hypothesis testing, it is the sample statistic based on the sample data.

Total deviation. The sum of the explained deviation and unexplained deviation for a given pair of values in a collection of bivariate data.

Total variation. The sum of the squares of the total deviation for all pairs of bivariate data in a sample.

Tree diagram. Graphical depiction of the different possible outcomes in a compound event.

Two-tailed test. Hypothesis test in which the critical region is divided between the left and right extreme areas of the probability distribution.

Two-way table. (See contingency table.)

Type I error. The mistake of rejecting the null hypothesis when it is true.

Type II error. The mistake of failing to reject the null hypothesis when it is false.

Unexplained deviation. For one pair of values in a collection of bivariate data, the difference between the y coordinate and the predicted value.

Unexplained variation. The sum of the squares of the unexplained deviations for all pairs of bivariate data in a sample.

Uniform distribution. A distribution of values evenly distributed over the range of possibilities.

Upper class limits. The largest numbers that can actually belong to the different classes in a frequency table.

Variance. The measure of dispersion found by using Formula 2-4 in Section 2-5.

Variance between samples. In analysis of variance, the variation among the different samples.

Variation due to error. In analysis of variance, the variation within samples that is due to chance.

Variation due to treatment. (See variance between samples.)

Weighted mean. Mean of a collection of scores that have been assigned different degrees of importance.

Wilcoxon rank-sum test. A nonparametric hypothesis test used to compare two independent samples.

Wilcoxon signed-ranks test. A nonparametric hypothesis used to compare two dependent samples.

y-intercept. Point at which a straight line crosses the y-axis.

z score. (See standard score.)

Bibliography

Adler, I. 1966. *Probability and Statistics for Everyman.* New York: New American Library.

Anderson, R., and T. Bancroft. 1952. *Statistical Theory in Research.* New York: McGraw-Hill.

Armore, S. 1975. *Statistics: A Conceptual Approach.* Columbus, Ohio: Charles E. Merrill.

Bacheller, M. 1978. *The Hammond Almanac.* Maplewood, New Jersey: Hammond Almanac, Inc.

Balsley, H. Editor. 1978. *Basic Statistics for Business and Economics.* Columbus, Ohio: Grid.

Barnett, V. 1973. *Comparative Statistical Inference.* New York: John Wiley.

Bashaw, W. 1969. *Mathematics for Statistics.* New York: John Wiley.

Braverman, J. 1978. *Fundamentals of Business Statistics.* New York: Academic Press.

Chou, Y. 1975. *Statistical Analysis.* 2nd ed. New York: Holt, Rinehart and Winston.

Christensen, H. 1977. *Statistics Step by Step.* Boston: Houghton Mifflin.

Cochran, W. 1982. *Contributions to Statistics.* New York: John Wiley.

Congelosi, V., P. Taylor, and P. Rice. 1979. *Basic Statistics.* 2nd ed. St. Paul, Minnesota: West.

Conover, W. 1971. *Practical Nonparametric Statistics.* New York: John Wiley.

Dixon, W., and F. Massey. 1969. *Introduction to Statistical Analysis.* 2nd ed. New York: McGraw-Hill.

Draper, N., and H. Smith. 1966. *Applied Regression Analysis.* New York: John Wiley.

Dyckman, T., and L. Thomas, 1977. *Fundamental Statistics for Business and Economics.* Englewood Cliffs, New Jersey: Prentice-Hall.

Elzey, F. 1966. *A Programmed Introduction to Statistics.* Belmont, California: Brooks/Cole.

Fairley, W., and F. Mosteller. 1977. *Statistics and Public Policy.* Reading, Massachusetts: Addison-Wesley.

Fisher, R. 1966. *The Design of Experiments.* 8th ed. New York: Hafner.

Freedman, D., R. Pisani, and R. Purves. 1978. *Statistics.* New York: W. W. Norton.

Freund, J. 1979. *Modern Elementary Statistics.* 5th ed. Englewood Cliffs, New Jersey: Prentice-Hall.

Freund, J. 1976. *Statistics, A First Course.* 2nd ed. Englewood Cliffs, New Jersey: Prentice-Hall.

Grant, E. 1964. *Statistical Quality Control.* 3rd ed. New York: McGraw-Hill.

Guenther, W. 1973. *Concepts of Statistical Inference.* 2nd ed. New York: McGraw-Hill.

Haber, A., and R. Runyon. 1973. *General Statistics.* 2nd ed. Reading, Massachusetts: Addison-Wesley.

Hamburg, M. 1977. *Statistical Analysis for Decision Making.* 2nd ed. New York: Harcourt Brace Jovanovich.

Hauser, P. 1975. *Social Statistics in Use.* New York: Russell Sage Foundation.

Heerman, E., and L. Braskam. 1970. *Readings in Statistics for the Behavioral Sciences.* Englewood Cliffs, New Jersey: Prentice-Hall.

Hoaglin, D., F. Mosteller, and J. Tukey. Editors. 1983. *Understanding Robust and Exploratory Data Analysis.* New York: John Wiley.

Hoel, P. 1976. *Elementary Statistics.* 4th ed. New York: John Wiley.

Hollander, M., and D. Wolfe. 1973. *Nonparametric Statistical Methods.* New York: John Wiley.

Hooke, R. 1983. *How to Tell the Liars from the Statisticians.* New York: Marcel Dekker.

Huff, D. 1954. *How to Lie with Statistics.* New York: W. W. Norton.

Johnson, R. 1984. *Elementary Statistics.* North Scituate, Massachusetts: Duxbury.

Kimble, G. 1978. *How to Use (and Misuse) Statistics.* Englewood Cliffs, New Jersey: Prentice-Hall.

King, R., and B. Julstrom. 1982. *Applied Statistics Using the Computer.* Sherman Oaks, California: Alfred Publishing.

Kirk, R. Editor. 1972. *Statistical Issues: A Reader for the Behavioral Sciences*. Belmont, California: Brooks/Cole.

Langley, R. 1970. *Practical Statistics Simply Explained*. New York: Dover.

Lapin, L. 1975. *Statistics: Meaning and Method*. New York: Harcourt Brace Jovanovich.

Lindley, D. 1971. *Making Decisions*. New York: John Wiley.

McClave, J., and P. Benson. 1982. *Statistics for Business and Economics*. San Francisco: Dellen.

Mendenhall, W. 1975. *Introduction to Probability and Statistics*. 4th ed. North Scituate, Massachusetts: Duxbury.

Mood, A., et al. 1974. *Introduction to the Theory of Statistics*. 3rd. ed. New York: McGraw-Hill.

Moore, D. 1979. *Statistics: Concepts and Controversies*. San Francisco: W. H. Freeman.

Mosteller, F., R. Rourke, and G. Thomas. 1970. *Probability with Statistical Applications*. 2nd ed. Reading, Massachusetts: Addison-Wesley.

Neter, J., W. Wasserman, and G. Whitmore. 1973. *Fundamental Statistics for Business and Economics*. 4th ed. Boston: Allyn and Bacon.

Nobile, P., and J. Deedy. Editors. 1972. *The Complete Ecology Fact Book*. Garden City, New York: Doubleday.

Noether, G. 1967. *Elements of Nonparametric Statistics*. New York: John Wiley.

Owen, D. 1962. *Handbook of Statistical Tables*. Reading, Massachusetts: Addison-Wesley.

Raiffa, H. 1968. *Decision Analysis: Introductory Lectures on Choices Under Uncertainty*. Reading, Massachusetts: Addison-Wesley.

Reichard, R. 1974. *The Figure Finaglers*. New York: McGraw-Hill.

Reichmann, W. 1962. *Use and Abuse of Statistics*. New York: Oxford University Press.

Roscoe, J. 1975. *Fundamental Research Statistics for the Behavioral Sciences*. 2nd ed. New York: Holt, Rinehart and Winston.

Schmid, C. 1983. *Statistical Graphics*. New York: John Wiley.

Siegal, S. 1956. *Nonparametric Statistics for the Behavioral Sciences*. New York: McGraw-Hill.

Snedecor, G., and W. Cochran. 1967. *Statistical Methods*. 6th ed. Ames, Iowa: Iowa State University Press.

Spear, M. 1969. *Practical Charting Techniques*. New York: McGraw-Hill.

Steger, J. Editor. 1971. *Readings in Statistics for the Behavioral Sciences*. New York: Holt, Rinehart and Winston.

Tanur, J. Editor. 1972. *Statistics: A Guide to the Unknown*. San Francisco: Holden-Day.

Tukey, J. 1977. *Exploratory Data Analysis*. Reading, Massachusetts: Addison-Wesley.

Ukena, A. 1978. *Statistics Today*. New York: Harper & Row.

Walker, H., and J. Lev. 1969. *Elementary Statistical Methods*. 3rd ed. New York: Holt, Rinehart and Winston.

Wayne, D. 1978. *Applied Nonparametric Statistics*. Boston: Houghton-Mifflin.

Weinberg, G., and J. Schumaker. 1969. *Statistics, An Intuitive Approach*. 2nd ed. Monterey, California: Brooks/Cole.

Winkler, R., and W. Hays. 1975. *Statistics: Probability, Inference and Decision*. 2nd ed. New York: Holt, Rinehart and Winston.

Yamane, T. 1973. *Statistics: An Introductory Analysis*. 3rd ed. New York: Harper & Row.

Zeisel, H. 1968. *Say It with Figures*. 5th ed. New York: Harper & Row.

Zuwaylif, F. 1979. *General Applied Statistics*. 3rd ed. Reading, Massachusetts: Addison-Wesley.

D

Answers to Selected Exercises

CHAPTER 1

1-1 The number of male births exceeded the number of female births by about a nineteenth part.

1-3 Most other branches establish conclusions through deductive proofs, while statistics uses an inductive approach to make inferences about populations.

1-5 Deductive reasoning is analogous to using known population characteristics to form conclusions about a sample, whereas inductive reasoning is analogous to using sample characteristics to form conclusions about a population.

1-7 The figure is very precise, but it is probably not that accurate. The use of the precise number may incorrectly suggest that it is also accurate.

1-9 It is not likely that the amount of money paid in bribes can ever be known with any degree of accuracy. Also, the given number is too precise since it probably represents a very crude estimate or guess.

1-11 Respondents sometimes tend to round off to a nice even number like 50.

1-13 They probably based their figure on the retail selling price, but they could have used cost, wholesale price, and so on. They might want to exaggerate in order to appear more effective.

1-15 Alumni with lower salaries would be less inclined to respond, so that the computed median will tend to be higher than it should be. Also, those who cheat on taxes might not want to reveal to anyone their true income, and others may resent the invasion of privacy.

1-17 50%. Not necessarily. No matter how good the operating levels are, about 50% will be below average.

1-19 Part (b) is probably more accurate. It relates factual data without jumping to a broader conclusion which might not be justified on the basis of the available data.

1-21 a, c, d

1-23 A family is a group of two or more people related by birth, marriage, or adoption and living together in a household. A household consists of one or more people sharing a housing unit (house, apartment, etc.). A person living alone would count as a household but not as a family.

1-25   Nominal: c,d; ordinal: a,g; interval: h,i,j; ratio: b,e,f.

1-27   (a) ratio level of measurement   (b) interval level of measurement.

CHAPTER 2

2-1   5, 8, 11, 14, 17

2-3   2.4, 4.9, 7.4, 9.9, 12.4

2-5   30, 50, 70, 90

2-7   10.0, 12.5, 15.0, 17.5, 20.0, 22.5, 25.0

2-9   5

2-11   0.25

2-13   8

2-15   5.0

2-17   83.5, 91.5, 99.5, 107.5, 115.5

2-19   18.65, 23.65, 28.65, 33.65, 38.65

2-21   Lower: 79.5, 87.5, 95.5, 103.5, 111.5

   Upper: 87.5, 95.5, 103.5, 111.5, 119.5

2-23   Lower: 16.15, 21.15, 26.15, 31.15, 36.15

   Upper: 21.15, 26.15, 31.15, 36.15, 41.15

2-25

I.Q.	Cum. freq.
Less than 88	16
Less than 96	53
Less than 104	103
Less than 112	132
Less than 120	146

2-27

Weight	Cum. freq.
Less than 21.2	16
Less than 26.2	31
Less than 31.2	43
Less than 36.2	51
Less than 41.2	54

2-29   80–84, 85–89, 90–94, 95–99, 100–104, 105–109, 110–114, 115–119

2-31 16.0–19.9, 20.0–23.9, 24.0–27.9, 28.0–31.9, 32.0–35.9, 36.0–39.9, 40.0–43.9

2-33 110–129

2-35 17.3–19.4

2-37

x	f
34–54	5
55–75	4
76–96	6
97–117	5
118–138	17
139–159	6
160–180	4
181–201	1
202–222	1
223–243	1

2-39

x	f
4.92	17
4.93	8
4.94	12
4.95	11
4.96	8
4.97	8
4.98	15
4.99	9
5.00	10
5.01	13
5.02	8
5.03	14
5.04	17

2-41 The distributions appear to be very different. The distributions are very difficult to observe from the raw data.

2-43 Class width is 4; possible class limits are 5.5–8.5, 9.5–12.5, 13.5–16.5, 17.5–20.5, 21.5–24.5, 25.5–28.5.

2-45 (a) 49.5, 59.5, 69.5, 79.5, 89.5, 99.5
(b) 54.5, 64.5, 74.5, 84.5, 94.5
(c) 49.5, 59.5, 69.5, 79.5, 89.5, 99.5

2-47

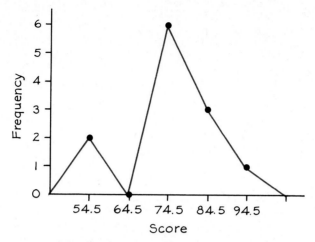

2-49   (a) 0.5, 10.5, 20.5, 30.5, 40.5, 50.5, 60.5
(b) 5.5, 15.5, 25.5, 35.5, 45.5, 55.5
(c) 0.5, 10.5, 20.5, 30.5, 40.5, 50.5, 60.5

2-51

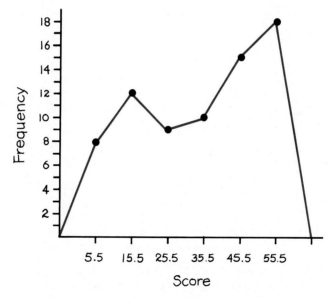

2-53   (a) 0.5, 30.5, 60.5, 90.5, 120.5, 150.5, 180.5, 210.5, 240.5
(b) 15.5, 45.5, 75.5, 105.5, 135.5, 165.5, 195.5, 225.5
(c) 0.5, 30.5, 60.5, 90.5, 120.5, 150.5, 180.5, 210.5, 240.5

2-55

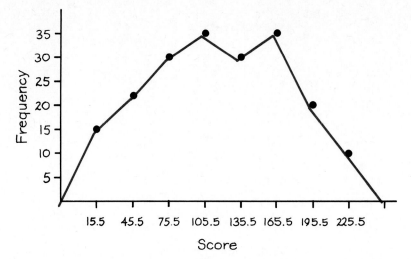

2-57 (a)−0.5, 99.5, 199.5, 299.5, 399.5, 499.5, 599.5, 699.5, 799.5, 899.5, 999.5

(b) 49.5, 149.5, 249.5, 349.5, 449.5, 549.5, 649.5, 749.5, 849.5, 949.5

(c) −0.5, 99.5, 199.5, 299.5, 399.5, 499.5, 599.5, 699.5, 799.5, 899.5, 999.5

2-59

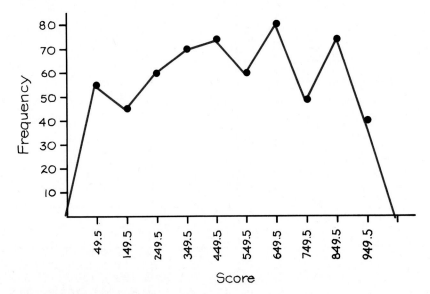

2-61 20 20 23 25 28

2-63 406 406 407 408 410 419 419 419 419 421 423 424
426 426 430 438 438

2-65   6. | 889
          7. | 011346
          8. | 017
          9. | 247

2-67   52. | 3568
          53. | 36
          54. | 0116
          55. | 23468
          56. | 4
          57. | 09

2-69

x	f
0–9	28
10–19	22
20–29	20
30–39	22
40–49	21
50–59	11
60–69	8
70–79	0
80–89	4
90–99	1
100–109	2

2-71

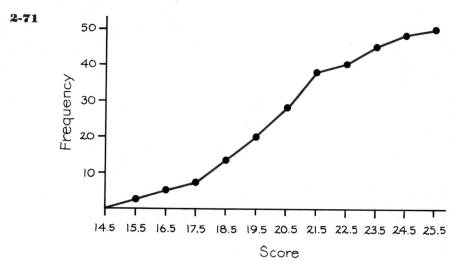

(continued)

(a)

x	f
15	1
16	3
17	2
18	7
19	6
20	9
21	8
22	3
23	5
24	2
25	1

(c) 28

(d) 8.5%

2-73 (a) Start the vertical scale at 7 instead of 0, and stretch out the vertical scale from 7 to 18.

(b) Let the vertical scale range from 0 to 30.

2-75

0–1	0000444446889*234555
2–3	1127*256689
4–5	13456779*1557778
6–7	456677*0112333456678899
8–9	00001455777889*01222466889
10–11	022345568*01145666889
12–13	0227789*0134468
14–15	0123579*014479
16–17	34577*1
18–19	29*2
20–21	
22–23	11*4

	Mean	Median	Mode	Midrange
2-77	12.6	10.0	23	13.5
2-79	104.9	106.0	107	103.5
2-81	25.1	25.0	23	25.5
2-83	142.5	139.5	138	145.0
2-85	71.5	72.0	77	71.0
2-87	0.709	0.725	0.57, 0.71, 0.74, 0.85	0.700
2-89	73.1	72.5	78	77.0
2-91	249.3	250.0	248	246.0

2-93   Class marks: 0.105, 0.125, 0.145, 0.165, 0.185. The mean is 0.131.

2-95   Class marks: 2.45, 7.45, 12.45, 17.45, 22.45, 27.45. The mean is 14.62.

2-97   Class marks: 29.5, 89.5, 149.5, 209.5, 269.5, 329.5, 389.5. The mean is 264.7.

2-99   Class marks: 4.5, 14.5, 24.5, 34.5, 44.5. The mean is 22.7.

2-101   (a) Mean is 1029.66; median is 1026.70; there is no mode; midrange is 1031.95.
(b) Mean is 29.66; median is 26.70; there is no mode; midrange is 31.95.
(c) They change by 1000.
(d) and (e) Some unmanageable data sets can be made manageable by adding or subtracting a constant k. After computing these averages, the answers can then be corrected by the same amount k.

2-103   The averages are the same for both sets and do not reveal any differences, but the lists are different in the degree of variation among the scores.

2-105   72.6; the result of 71.5 from Exercise 2–100 is likely to be better.

2-107   1.092

	Range	Variance	Standard deviation
2-109	19.0	66.3	8.1
2-111	9.0	9.9	3.1
2-113	5.0	3.4	1.9
2-115	40.0	142.6	11.9
2-117	12.0	22.7	4.8
2-119	0.580	0.026	0.161
2-121	52.0	210.6	14.5
2-123	40.0	143.2	12.0
2-125	499.0	16193.6	127.3 (same)
2-127	4990.0	1619355.2	1272.5

(The range and standard deviation are multiplied by 10 while the variance is multiplied by 100.)

2-129		0.000385	0.0196
2-131		67.80	8.23
2-133		8498.9	92.2
2-135		78.0	8.8

2-137   The statistics students are a more homogeneous group and should therefore have a smaller variance.

2-139   If all scores are the same, the variance is zero, but it cannot be negative.

2-141   Group C: range is 19, variance is 36.1, standard deviation is 6.0.
Group D: range is 16, variance is 42.2, standard deviation is 6.5.

2-143 $\bar{x} = 4.56$, $s = 1.67$, $s^2 = 2.78$

2-145   Mean is 100.0 and standard deviation is 47.6.

2-147   a. At least 3/4 of the scores are between 70 and 130.
b. At least 8/9 of the scores are between 55 and 145.
c. At least 3/4 of the scores are between 300 and 700.
d. At least 8/9 of the scores are between 200 and 800.

2-149   (a) 1   (b) 1.89   (c) -1

2-151   (a) 1   (b) -1 (c) 2   (d) -1.5 (e) 3

2-153   (a) 1.08   (b) -0.63 (c) 2.22

2-155   (a) 2   (b)1.25   (c) -1.25 (d) -3.75 (e) 0

2-157   Test b since $z = 0.6$ is greater than $z = 0.3$.

2-159   Test b since $z = 3$ is greater than $z = 2$ and $z = 2.67$.

2-161   18th percentile

2-163   95th percentile

2-165   59.5

2-167   70

2-169   39.5

2-171   41.5

2-173   (a) 36.5   (b) 57.8   (c) 63.5

2-175   63,   75,   84.5

2-177   (a) 31.5   (b) 30.0   (c) 26   (d) 33.0   (e) 14.0
(f) 35.1   (g) 5.9

2-179   (a) 179.8   (b) 182.0   (c) 183   (d) 179.5   (e) 23.0
(f) 51.4   (g) 7.2

2-181   Mean is 8.6 and standard deviation is 4.9.

2-183

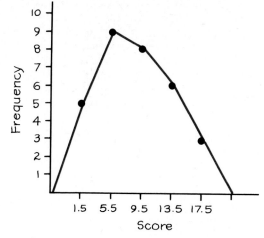

2-185   2 | 668
             3 | 27
             4 | 0

2-187   1.43

2-189

x	f
0–9	2
10–19	4
20–29	10
30–39	5
40–49	7
50–59	2
60–69	6
70–79	2
80–89	7
90–99	5

2-191   (a) 27   (b) 44   (c) 32

2-193   (a) 50.6   (b) 3.0   (c) Error   (d) False   (e) True

CHAPTER 3

3-1   2, −1/2, 5, 1.11, 1.0001

3-3   1/20

3-5   2899/3756

3-7   44/123

3-9   3/8

3-11   7/25

3-13   11/19

3-15   13/100

3-17   11/20

3-19   163/1045

3-21   18/35

3-23   26/55

3-25   (a) bb, bg, gb, gg
(b) 1/4   (c) 1/2

3-27   (a) ttt, ttf, tft, tff, ftt, ftf, fft, fff
(b) 1/8   (c) 1/8   (d) 3/8

3-29   (a) rrr, rrr, rrw, rwr, rwr, rww, rwr, rwr, rww, wrr, wrr, wrw, wwr, wwr, www, wwr, wwr, www, brr, brr, brw, bwr, bwr, bww, bwr, bwr, bww
(b) 2/27   (c) 11/27   (d) 5/27   (e) 16/27

3-31   2/7

3-33   1

3-35   Mutually exclusive: a, f, h, i

3-37   4/5

3-39   30/43

3-41   7/12

3-43   2/5

3-45   35/57

3-47   4/5

3-49   63/100

3-51   845/1000

3-53   15/22

3-55   29/35

3-57   61/87

3-59   1

3-61   22/27

3-63   17/60

3-65 $0.4 \leq P(B) \leq 0.8$ and they may or may not be mutually exclusive.

3-67 $P(A \text{ or } B) = P(A) + P(B) - 2P(A \text{ and } B)$

3-69   Independent events: a, c, d

3-71   1/81

3-73   0.0289

3-75   55/96

3-77   0.961

3-79   0.0356

3-81   1/5

3-83   6/57

3-85   (a) 3/10   (b) 9/25

3-87   (a) 9/25   (b) 87/245

3-89   (a) 0.779   (b) 0.777   (c) Sample without replacement

3-91   0.0000225

3-93   0.431

3-95   Answer varies

3-97   3/10, 7/10

3-99   1/2, 1/2

3-101   2/7, 5/7

3-103   3/5, 2/5

3-105   7/8

3-107   0.226

3-109   0.151

3-111   0.401

3-113   0.998

3-115   0.00996

3-117   0.779

3-119   0.0268

3-121   (a) $P(B) = 0$ (b) $P(B) = 1$ (c) $P(B) \le 0.3$

3-123   (a) $P(\overline{A \text{ or } B}) = 1 - P(A) - P(B) + P(A \text{ and } B)$
(b) $P(\overline{A} \text{ or } \overline{B}) = 1 - P(A \text{ and } B)$ (c) Different

3-125   0.970

3-127   720

3-129   40

3-131   20

3-133   120

3-135   1

3-137   1

3-139   6

3-141   5040

3-143   646,646

3-145   362,880

3-147   (a) 5040   (b) 1/5040

3-149   1,000,000,000

3-151   43,680

3-153   (a) 2002   (b) 0.675

3-155   (a) 3,268,760   (b) 1/3,268,760

3-157 6.20×10^{23}

3-159   120

3-161   Calculator: 3.0414093×10^{64}
            Approximation: 3.0363452×10^{64}

3-163   8/15

3-165   2/5

3-167   1/4

3-169   7/15

3-171   1/84

3-173   (a) 9/400   (b) 51/2380

3-175   0.9999999757

3-177   0.999

3-179   1/8

3-181   (a) 0   (b) 1   (c) 0.78   (d) No   (e) Yes

3-183   (a) 40,320   (b) 20,160   (c) 45   (d) 3160

3-185   1/5040

3-187   11,880

3-189   768

3-191   792

3-193   336

CHAPTER 4

4-1   No, sum of $P(x)$ is not 1.

4-3   No, sum of $P(x)$ is not 1.

4-5   Yes

4-7   No, sum of $P(x)$ is not 1.

4-9   Yes.

4-11   Yes.

4-13   (a)   0, 1, 2, 3
     (b)   0.125, 0.375,
          0.375, 0.125
     (c)

x	$P(x)$
0	0.125
1	0.375
2	0.375
3	0.125

     (d)

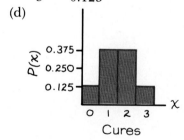

     (e)   Same as part b.

4-15   (a)   0, 1, 2
     (b)   1/3, 1/3, 1/3
     (c)

x	$P(x)$
0	1/3
1	1/3
2	1/3

(continued)

(d)

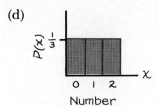

Number

(e) Same as part b.

4-17 (a) 0, 1, 2, 3

(b) 1/8, 3/8,
3/8, 1/8

(c)

x	$P(x)$
0	1/8
1	3/8
2	3/8
3	1/8

(d)

Boys

(e) Same as part b.

4-19 (a) 2, 3, 4, 5, 6,
7, 8, 9, 10,
11, 12

(b) 1/36, 2/36, 3/36,
4/36, 5/36, 6/36,
5/36, 4/36, 3/36,
2/36, 1/36

(c)

x	$P(x)$
2	1/36
3	2/36
4	3/36
5	4/36
6	5/36
7	6/36
8	5/36
9	4/36
10	3/36
11	2/36
12	1/36

(d)

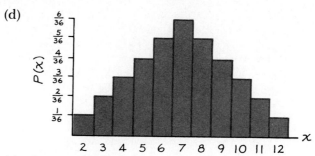

(e)   Same as part b.

4-21   No, sum exceeds 1.

4-23

x	$P(x)$
0	0.410
1	0.410
2	0.154
3	0.026
4	0.002

	Mean	Variance	Standard deviation
4-25	1.1	0.5	0.7
4-27	2.3	0.6	0.8
4-29	21.3	304.7	17.5
4-31	6.0	5.9	2.4
4-33	1.7	1.0	1.0
4-35	3.7	2.2	1.5
4-37	0.4	0.3	0.6
4-39	5.9	1.3	1.2

4-41 $275

4-43   −$70

4-45 $122,500

4-47 $60 per day

4-49   (b)   1/32, 5/32, 10/32, 10/32, 5/32, 1/32

(continued)

(c)

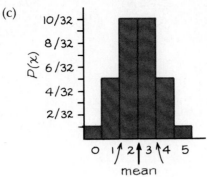

(d) 2.5   (e) 1.25   (f) 1.12

4-51 $\mu = \Sigma x \cdot P(x) = \left(1 \cdot \frac{1}{n}\right) + \left(2 \cdot \frac{1}{n}\right) + \cdots + \left(n \cdot \frac{1}{n}\right)$

$= \frac{1}{n}(1 + 2 + \cdots + n) = \frac{1}{n} \cdot \frac{n(n+1)}{2} = \frac{n+1}{2}$

$\sigma^2 = \Sigma x^2 \cdot P(x) - \mu^2 = \left(1^2 \cdot \frac{1}{n}\right) + \left(2^2 \cdot \frac{1}{n}\right) + \cdots + \left(n^2 \cdot \frac{1}{n}\right) - \left(\frac{n+1}{2}\right)^2$

$= \frac{1}{n}\left(1^2 + 2^2 + \cdots + n^2\right) - \left(\frac{n+1}{2}\right)^2$

$= \frac{1}{n}\frac{n(n+1)(2n+1)}{6} - \frac{n^2+2n+1}{4}$

$= \frac{2n^2+3n+1}{6} - \frac{n^2+2n+1}{4}$

$= \frac{4n^2+6n+2-(3n^2+6n+3)}{12} = \frac{n^2-1}{12}$

4-53   a, b, c, e, f, g, i, j

4-55   (a) 10   (b) 1   (c) 8   (d) 56   (e) 1

4-57 $n = 10$, $x = 4$, $p = 0.5$, $q = 0.5$, $P(4) = 0.205$

4-59 $n = 8$, $x = 3$, $p = 0.6$, $q = 0.4$, $P(3) = 0.124$

4-61 $n = 7$, $x = 4$, $p = 0.7$, $q = 0.3$, $P(4) = 0.227$

4-63 $n = 10$, $x = 7$, $p = 0.6$, $q = 0.4$, $P(7) = 0.215$

4-65 $n = 20$, $x = 1$, $p = 0.02$, $q = 0.98$, $P(1) = 0.272$

4-67   0.093

4-69   0.133

4-71   0.108

4-73   0.000297

4-75

x	$P(x)$
0	0.296
1	0.444
2	0.222
3	0.037

4-77   0.994

4-79   0.885

4-81   0.052

4-83   0.417

	Mean	Variance	Standard deviation
4-85	8.0	4.0	2.0
4-87	3.0	2.1	1.4
4-89	18.0	1.8	1.3
4-91	62.4	25.0	5.0
4-93	6.0	4.0	2.0
4-95	225.0	56.3	7.5
4-97	2.5	1.3	1.1
4-99	6.0		2.0
4-101	0.03		0.17
4-103	4.4		1.7
4-105	338.0		17.1
4-107	88.8		7.5
4-109	6.9		2.4
4-111	127.6		6.8

4-113 $\mu = 4.0$, $\sigma = 1.5$, $\bar{x} = 4.0$, $s = 1.5$

4-115   No

4-117   (a) 0.5   (b) 0.8   (c) 0.6

4-119   (a) 0   (b) 0.4   (c) 0.5   (d) 1509/5000   (e) 999/5000

4-121   (a) 26.5 in.   (b) 23.5 in.   (c) 2/3

4-123   (a) 52.1 cubic cm.   (b) 50.9 cubic cm.   (c) 5/6

4-125 17.2   18.6 19.3 20.1   26.3

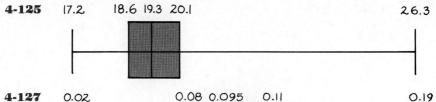

4-127 0.02   0.08 0.095 0.11   0.19

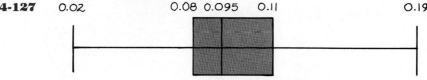

4-129 130   159   171.5   181   198

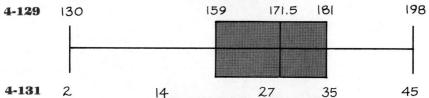

4-131 2   14   27   35   45

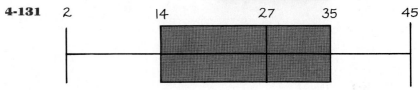

4-133 (a) 9/25   (b) 12/25

4-135 $Q_1 = 42$, $Q_3 = 73$; hinges are also 42 and 73.

4-137 No, the sum
of $P(x)$ is
not 1.

4-139 No, the sum
of $P(x)$ is
not 1.

4-141 (a) 0.3
(b) 8.0
(c) 21.0

4-143 (a)

x	$P(x)$
1	0.2
2	0.2
3	0.2
4	0.2
5	0.2

(b) 3.0   (c) 2.0

4-145 (a) 0.010   (b) 0.230   (c) 2.0   (d) 1.2
(e) 1.1

4-147 (a) 0.196   (b) 8.0   (c) 4.0

4-149   (a) 0.051   (b) 2.30

4-151   (a) 0.682   (b) 1.1   (c) 1.0

4-153   5/8

4-155   31/40

4-157

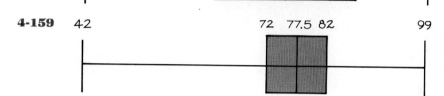

4-159   42   72 77.5 82   99

CHAPTER 5

5-1   0.1915

5-3   0.4192

5-5   0.3413

5-7   0.4793

5-9   0.1587

5-11   0.0143

5-13   0.1587

5-15   0.2546

5-17   0.8413

5-19   0.8790

5-21   0.8185

5-23   0.9104

5-25   0.1359

5-27   0.0233

5-29   0.1464

5-31   0.3128

5-33   0.5

5-35   0.3753

5-37   99.74%

5-39   2.33, −2.33

5-41

x	y
−4	0.0001
−3	0.0046
−2	0.0549
−1	0.2434
0	0.4000
1	0.2434
2	0.0549
3	0.0046
4	0.0001

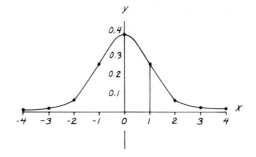

5-43   0.1587

5-45   0.1985

5-47   0.1772

5-49   8.38%

5-51   0.0668

5-53   0.4743

5-55   (a) 77.45%   (b) 30.85%

5-57   24.97%

5-59   22.67%

5-61   0.3202

5-63   0.1743

5-65   7.30%

5-67   0.0734

5-69   (a) 50.5   (b) 2.2   (c) 46.67%   (d) 35.31%

5-71   Uniform: 0.5769; normal: 0.6568.

5-73   1.645

5-75   −1.645

5-77   −1.645,   1.645

5-79   −1.645, 1.28

5-81   0.25

5-83   0.67

5-85   208.87

5-87   273.89

5-89   63.29 in.

5-91   4.35

5-93   75.3, 124.7

5-95   0.154

5-97   24.87 mm, 25.40 mm

5-99   3.470, 5.756

5-101   29.6 min.

5-103   (a) 68.1   (b) 10.0   (c) 82   (d) 84.6

	Is approx. suitable?	μ	σ
5-105	Yes	6.25	2.17
5-107	No	4.32	2.03
5-109	No	2.50	1.57
5-111	No	79.8	2.00

5-113   (a) 0.121   (b) 0.1173

5-115   (a) 0.114   (b) 0.1215

5-117   (a) 0.194   (b) 0.1922

5-119   (a) 0.227   (b) 0.2327

5-121   0.0287

5-123   0.0119

5-125   0.9949

5-127   0.8461

5-129   0.0329

5-131   0.0485; yes

5-133   0.8365

5-135   0.0222

5-137   (a) 0.001   (b) 0.0012   (c) 0.0011

5-139   262

5-141   0.4772

5-143   0.4641

5-145   0.4772

5-147   0.1056

5-149   0.4878

5-151   0.8294

5-153   0.9970

5-155   0.1112

5-157   It is halved.

5-159   Use the correction factor in b, c, d, f, h, i, j.
(b) 0.957   (c) 0.936   (d) 0.791   (f) 0.967
(h) 0.957   (i) 0.894   (j) 0.942

5-161   0.9887

5-163   0.8805

5-165   0.0456

5-167 $\mu = 28.4$; $\sigma = 9.4$;   0.1151.

5-169   (a) 0.4332   (b) 0.5987   (c) 0.3161   (d) 30.35 tons   (e) 0.1469

5-171   (a) 0.0398   (b) 0.6591   (c) 11.22 years   (d) 18.05 years   (e) 0.9920

5-173   0.0146

5-175   0.9656

5-177   (a) 0.2704   (b) 0.0392   (c) 0.8599

CHAPTER 6

6-1   (a) H_0: $\mu \leq 30$
               H_1: $\mu > 30$
       (c) H_0: $\mu \geq 100$
               H_1: $\mu < 100$
       (e) H_0: $\mu = 3271$
               H_1: $\mu \neq 3271$
       (g) H_0: $\mu \leq 26,000$
               H_1: $\mu > 26,000$
       (i) H_0: $\mu \leq 3.2$
               H_1: $\mu > 3.2$

6-2   (a)   Type I error: Reject the claim that the mean age of professors is 30
years or less when their mean age is actually 30 years or less.
Type II error: Fail to reject the claim that the mean age of professors
is 30 years or less when that mean is actually greater than 30 years.

(c)   Type I error: Reject the claim that the mean I.Q. of college students is at least 100 when it really is at least 100.

Type II error: Fail to reject the claim that the mean I.Q. of college students is at least 100 when it is really less than 100.

(e)   Type I error: Reject the claim that the mean monthly cost equals \$3271 when it does equal that amount.

Type II error: Fail to reject the claim that the mean monthly cost equals \$3271 when it does not equal that amount.

(g)   Type I error: Reject the claim that the mean annual salary is less than or equal to \$26,000 when it is less than or equal to \$26,000.

Type II error: Fail to reject the claim that the mean annual salary is less than or equal to \$26,000 when it is actually greater than that amount.

(i)   Type I error: Reject the claim that the mean weight of girls at birth is at most 3.2 kg when it really is at most 3.2 kg.

Type II error: Fail to reject the claim that the mean weight of girls at birth is at most 3.2 kg when it really is greater than 3.2 kg.

6-3   Right-tailed: a, b, g, i, j
Left-tailed: c, d, h
Two-tailed: e, f

6-4   (a)   1.645   (c) -1.96, 1.96   (e) -1.645 (g) -1.645, 1.645   (i) 1.96

6-5   Test statistic: $z = 1.44$. Critical value: $z = 2.33$. Fail to reject H_0: $\mu \leq 100$.

There is not sufficient evidence to warrant rejection of the claim that the mean is less than or equal to 100.

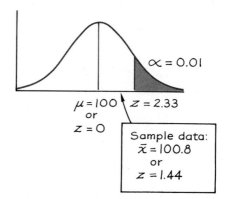

6-7   Test statistic: $z = -4.33$. Critical value: $z = -1.645$. Reject H_0: $\mu \geq 20$. There is sufficient evidence to warrant rejection of the claim that the mean is at least 20.

6-9   Test statistic: $z = 1.73$. Critical values: $z = -1.645$ and $z = 1.645$. Reject H_0: $\mu = 500$.

There is sufficient evidence to warrant rejection of the claim that the mean is equal to 500.

6-11   Test statistic: $z = 1.77$. Critical value: $z = 1.645$. Reject H_0: $\mu \leq 40$. The sample evidence does support the claim that the mean is greater than 40.

6-13   Test statistic: $z = 1.76$. Critical values: $z = 1.96$ and $z = -1.96$. Fail to reject H_0: $\mu = 13.20$.
It appears that the mean for the region is the same as the mean of the population.

6-15   Test statistic: $z = 1.30$. Critical value: $z = 1.645$. Fail to reject H_0: $\mu \leq 500$.
There is not sufficient sample evidence to support the principal's claim that her graduates have better than average scores.

6-17   Test statistic: $z = -4.22$. Critical values: $z = -2.575$ and $z = 2.575$. Reject H_0: $\mu = 600$ mg.
Reject the claim that the mean equals 600 mg. The sample results indicate a mean significantly less than 600 mg.

6-19   Test statistic: $z = 4.74$. Critical value: $z = 2.33$. Reject H_0: $\mu \leq 50$.
It appears that the new model does have an increase in productivity.

6-21   Test statistic: $z = -1.18$. Critical value: $z = -1.28$. Fail to reject H_0: $\mu \geq 0.610$.
There is not sufficient sample evidence to support the instructor's claim of faster reaction times. (Note that "faster" times correspond to lower values.)

6-23   Test statistic: $z = -4.22$. Critical values: $z = 1.96$ and $z = -1.96$. Reject H_0: $\mu = 22.6$.
Reject the claim that the new spark plug does not change fuel consumption. The new spark plug seems to consume more fuel since the mpg rating appears to be significantly lower.

6-25 $\bar{x} = 104.3$ and $s = 13.5$. Test statistic: $z = 1.86$. Critical value: $z = 1.88$. Fail to reject H_0: $\mu \leq 100$.
There is not sufficient sample evidence to warrant support of the claim that the mean is above 100.

6-27 $\bar{x} = 28.4$, $s = 9.5$. Test statistic; $z = -1.57$. Critical values: $z = -2.05$, 2.05. Fail to reject H_0: $\mu = 30.0$.
There is not sufficient sample evidence to warrant rejection of the claim that the mean equals 30.0. The sample mean does not differ significantly from the claimed mean of 30.0.

6-29   Reject the null hypothesis.

6-31   Fail to reject the null hypothesis.

6-33   P-value: 0.0808. Fail to reject the null hypothesis.

6-35   P-value: 0.0262. Reject the null hypothesis.

6-37   P-value: 0.0970. Fail to reject the null hypothesis.

6-39   P-value: 0.0524. Fail to reject the null hypothesis.

6-41   Test statistic: $z = -2.24$. P-value: 0.0125. Reject H_0: $\mu \geq 100$. The mean appears to be significantly less than 100.

6-43   Test statistic: $z = 2.40$. P-value: 0.0164. Fail to reject H_0: $\mu = 75.6$. The sample mean does not differ significantly from the claimed value of 75.6.

6-45   Test statistic: $z = -0.96$. P-value: 0.1685. Fail to reject H_0: $\mu \geq 10.00$. There is not sufficient sample evidence to warrant rejection of the claim that the mean is at least 10.00 minutes.

6-47   Test statistic: $z = 2.59$. P-value: 0.0096. Reject H_0: $\mu = 25$. Reject the claim that the mean equals 25 years. It appears to be greater than 25 years.

6-49   The null hypothesis is the claim that the mean is equal to some value.

6-51   13.52

6-53   (a) $-2.056, 2.056$ (b) 1.337 (c) -3.365 (d) $-4.032, 4.032$
(e) -2.467

6-55   Test statistic: $t = 1.500$. Critical value: $t = 1.860$. Fail to reject H_0: $\mu \leq 10$.
There is not sufficient sample evidence to warrant rejection of the claim that the mean is 10 or less.

6-57   Test statistic: $t = -2.121$. Critical value: $t = -2.110$. Reject H_0: $\mu \geq 98.6$.
There is sufficient sample evidence to warrant rejection of the claim that the mean is 98.6 or more.

6-59   Test statistic: $t = 2.014$. Critical values: $t = 2.145$ and $t = -2.145$. Fail to reject H_0: $\mu = 75$.
There is not sufficient sample evidence to warrant rejection of the claim that the mean equals 75.

6-61   Test statistic: $t = -0.712$. Critical value: $t = -3.250$. Fail to reject H_0: $\mu \geq 3000$.
There is not sufficient sample data to warrant rejection of the claim that the mean is at least 3000 pounds.

6-63   Test statistic: $t = 1.090$. Critical value: $t = 1.734$. Fail to reject H_0: $\mu \leq 77$.
There is insufficient evidence to support the claim that method B is better.

6-65   Test statistic: $t = 2.615$. Critical values: $t = 2.552$ and $t = -2.552$. Reject H_0: $\mu = 1.50$.
There is sufficient evidence to warrant rejection of the claim that the mean voltage level equals 1.50 V.

6-67   Test statistic: $t = -2.549$. Critical values: $t = 2.776$ and $t = -2.776$. Fail to reject H_0: $\mu = 17,850$.
There is not sufficient evidence to warrant rejection of the claim that the mean equals 17,850 pounds.

6-69   Test statistic: $t = -3.162$. Critical value: $t = -1.833$. Reject H_0: $\mu \geq 100$.
The sample data support the claim that the mean level is below 100 decibels.

6-71   Test statistic: $t = 3.969$. Critical values: $t = 2.052$ and $t = -2.052$. Reject H_0: $\mu = 60$.
The mean appears to be significantly different from 60.

6-73   Test statistic: $t = -3.078$. Critical value: $t = -1.328$. Reject H_0: $\mu \geq 30$.
The sample data support the claim that the mean was below 30.

6-75   Test statistic: $t = -8.485$. Critical value: $t = -2.998$. Reject H_0: $\mu \geq 0.88$.
The sample data support the claim that the miss distance is decreased.

6-77   The sample yields a mean of 78.1 and a standard deviation of 5.6. Test statistic: $t = 2.476$. Critical value: $t = 1.729$. Reject H_0: $\mu \leq 75$.
The sample data support the claim that the class is above average.

6-79   (a) The P-value is between 0.025 and 0.05.
(b) The P-value is between 0.01 and 0.02.
(c) The P-value is between 0.10 and 0.25.

6-81   Test statistic: $z = 3.27$. Critical values: $z = -1.96$ and $z = 1.96$. Reject H_0: $p = 0.3$.
There is sufficient sample evidence to warrant rejection of the claim that the proportion of defects is equal to 0.3. It seems to be significantly higher.

6-83   Test statistic: $z = -1.23$. Critical value: $z = -2.33$. Fail to reject H_0: $p \geq 0.7$.
There is not sufficient sample evidence to support the claim that the proportion of voters who favor nuclear disarmament is less than 0.7.

6-85   Test statistic: $z = 1.84$. Critical value: $z = 1.645$. Reject H_0: $p \leq 0.03$.
Reject the manager's claim that production is not out of control; it appears that production is out of control.

6-87   Test statistic: $z = -2.61$. Critical value: $z = -1.645$. Reject H_0: $p \geq 0.12$.
The sample data support the claim that the failure rate has been lowered from the 12% level.

6-89   Test statistic: $z = 2.77$. Critical values: $z = -2.33$ and $z = 2.33$. Reject H_0: $p = 1/8$.
The Mendelian law does not appear to be working since the sample results differ significantly from the $1/8$ proportion that was expected.

6-91   Test statistic: $z = 1.41$. Critical value: $z = 2.33$. Fail to reject H_0: $p \leq 1/3$.
There is not sufficient sample evidence to support the claim that more than $1/3$ of all adults smoke.

6-93   Test statistic: $z = -5.40$. Critical values: $z = -2.575$ and $z = 2.575$. Reject H_0: $p = 0.95$.
There is sufficient sample evidence to warrant rejection of the claim that 95% recognize Columbus. In fact, the rate appears to be lower than 95%.

6-95   Test statistic: $z = -5.13$. Critical value: $z = -1.645$. Reject H_0: $p \geq 0.082$.
Reject the labor leader's claim that the rate was at least 8.2%. The sample data indicate a rate that is significantly less than 8.2%. (The sample rate is 7.57%.)

6-97   Test statistic: $z = 8.26$. Critical values: $z = -1.96$ and $z = 1.96$. Reject H_0: $p = 3/4$.
The sample rate of 91% differs significantly from the claimed rate of $3/4$. Reject the claim that $3/4$ of all such accidents will result in fatalities.

6-99   Test statistic: $z = -4.40$. Critical value: $z = -2.33$. Reject H_0: $p \geq 1/2$.
The sample rate of 39% is significantly less than the claimed rate of $1/2$ or 50%. There is sufficient sample evidence to support the claim that fewer than $1/2$ of San Francisco residential telephones have unlisted numbers.

6-101   Divide numerator and denominator by n. Replace q by $(1 - p)$. Also note that when dividing the square root of npq by n, the n goes under the radical as the square of n.

6-103   1.977%

6-105   (a) 8.907, 32.852   (b) 10.117   (c) 8.643, 42.796   (d) 40.289   (e) 4.075

6-107   Test statistic: $\chi^2 = 50.44$. Critical value: $\chi^2 = 38.885$. Reject H_0: $\sigma^2 \leq 100$.
The sample data support the claim that the variance is greater than 100.

6-109   Test statistic: $\chi^2 = 9.500$. Critical value: $\chi^2 = 7.962$. Fail to reject H_0: $\sigma^2 \geq 416$.
There is not sufficient sample evidence to warrant rejection of the claim that the variance is at least 416.

6-111   Test statistic: $\chi^2 = 23.109$. Critical values: $\chi^2 = 2.603, 26.757$. Fail to reject H_0: $\sigma^2 = 2.38$. (continued)

There is not sufficient sample evidence to warrant rejection of the claim that the variance is equal to 2.38.

6-113   Test statistic: $\chi^2 = 35.743$. Critical value: $\chi^2 = 5.697, 35.718$. Reject H_0: $\sigma = 10.0$.

There is sufficient sample evidence to warrant rejection of the claim that the variance is equal to 10.0.

6-115   Test statistic: $\chi^2 = 108.889$. Critical value: $\chi^2 = 101.879$. Reject H_0: $\sigma^2 \leq 9.00$.

There is sufficient sample evidence to warrant rejection of the claim that the variance is equal to or less than 9.00.

6-117   Test statistic: $\chi^2 = 22.337$. Critical value: $\chi^2 = 21.920$. Reject H_0: $\sigma^2 \leq 2.00$.

The sample data support the claim that the standard deviation is greater than 2.00 hours.

6-119   Test statistic: $\chi^2 = 44.8$. Critical value: $\chi^2 = 51.739$. Reject H_0: $\sigma^2 \geq 0.0225$.

The sample data support the claim that the new machine produces less variance.

6-121   Test statistic: $\chi^2 = 24.000$. Critical values: $\chi^2 = 24.433, 59.342$. Reject H_0: $\sigma^2 = 225$.

There is sufficient sample evidence to warrant rejection of the claim that the variance equals 225.

6-123   Test statistic: $\chi^2 = 10.222$. Critical values: $\chi^2 = 11.689, 38.076$. Reject H_0: $\sigma = 15$.

There is sufficient sample evidence to warrant rejection of the claim that the standard deviation is equal to 15.

6-125   Test statistic: $\chi^2 = 35.490$. Critical value: $\chi^2 = 35.479$. Reject H_0: $\sigma \leq 50$.

The sample data support the claim that the standard deviation of the hardness indices is greater than 50.0.

6-127   (a) Estimated values: 73.772, 129.070.
       Table A-5 values: 74.222, 129.561.
    (b) 116.643, 184.199

6-129 $n = 12$, $\bar{x} = 33.05$, $s = 1.13$. Test statistic: $\chi^2 = 3.511$. Critical value: $\chi^2 = 3.816$. Reject H_0: $\sigma \geq 2.0$.

The sample data support the claim that the standard deviation is less than 2.0 mg.

6-131   (a) $z = -1.645$ (b) $z = 2.33$ (c) $t = -3.106, 3.106$
    (d) $\chi^2 = 10.856$ (e) $\chi^2 = 16.047, 45.722$

6-133   (a) $H_0: \mu \geq 20.0$.
    (b) Left-tailed.

(c) Rejecting the claim that the mean is at least 20.0 minutes when it really is at least 20.0 minutes.

(d) Failing to reject the claim that the mean is at least 20.0 minutes when it is really less than 20.0 minutes.

(e) 0.01

6-135 Test statistic: $z = -3.00$. Critical values: $z = -1.96$ and $z = 1.96$. Reject H_0: $\mu = 10.0$.
There is sufficient sample evidence to warrant rejection of the claim that the mean equals 10.0 seconds.

6-137 Test statistic: $z = -2.73$. Critical value: $z = -2.33$. Reject H_0: $\mu \geq 5.00$.
The sample data support the claim that the mean radiation dosage is below 5.00 milliroentgens.

6-139 Test statistic: $t = 2.432$. Critical value: $t = 2.492$. Fail to reject H_0: $\mu \leq 25.5$.
There is not sufficient sample evidence to warrant rejection of the claim that the mean is less than or equal to 25.5 feet.

6-141 $\bar{x} = 98.0$, $s = 8.4$, $n = 15$.
Test statistic: $t = -0.922$. Critical value: $t = 2.977$ and $t = -2.977$. Fail to reject H_0: $\mu = 100$.
There is not sufficient sample evidence to warrant rejection of the claim that the mean is equal to 100.

6-143 Test statistic: $z = 1.22$. Critical values: $z = -1.96$ and $z = 1.96$. Fail to reject H_0: $p = 0.20$.
There is not sufficient sample evidence to warrant rejection of the claim that the percentage who can't read is equal to 20%.

6-145 Test statistic: $z = 2.47$. Critical value: $z = 1.645$. Reject H_0: $p \leq 1/5$.
The sample data support the claim that more than 1/5 believe that birth control pills prevent venereal disease.

6-147 Test statistic: $\chi^2 = 9.310$. Critical values: $\chi^2 = 8.907, 32.852$. Fail to reject H_0: $\sigma = 4.0$.
There is not sufficient sample evidence to warrant rejection of the claim that the standard deviation is equal to 4.0 grams.

6-149 Test statistic: $\chi^2 = 77.906$. Critical value: $\chi^2 = 74.397$. (Interpolated critical value is 73.274.) Reject H_0: $\sigma^2 \leq 6410$.
The sample data support the counselor's claim that the current group has more varied aptitudes.

CHAPTER 7

7-1 (a) 1.96   (b) 2.33   (c) 2.05
(d) 2.093   (e) 2.977

7-3   2.94

7-5   20.6

7-7   6.192

7-9   3.8625

7-11 $68.77 < \mu < 72.03$

7-13 $98.09 < \mu < 99.12$

7-15 $72.70 < \mu < 78.90$

7-17 $0.593 < \mu < 0.627$

7-19 $297.23 < \mu < 347.57$

7-21 $0.9943 < \mu < 0.9977$

7-23   110

7-25   5968

7-27 $34.81 < \mu < 37.59$

7-29 $12.85 < \mu < 16.55$ (Use $z(\alpha/2)$ since σ is known.)

7-31 $66.23 < \mu < 70.77$

7-33 $13.88 < \mu < 18.52$

7-35   185

7-37   61

7-39   Multiply both sides by the square root of n, then divide both sides by E, then square both sides.

7-41   52

7-43 $\$42,980 < \mu < \$62,000$ (assuming 95% confidence interval)

7-45   3254

7-47

	$\hat{p}$	$\hat{q}$	Point estimate
a.	0.450	0.550	0.450
b.	0.608	0.392	0.608
c.	0.050	0.950	0.050
d.	0.880	0.120	0.880
e.	0.466	0.534	0.466

7-49 $0.208 < p < 0.292$

7-51 $0.553 < p < 0.654$

7-53   423

7-55   9604

7-57 $0.545 < p < 0.595$

7-59 $0.729 < p < 0.858$ (Assuming a 95% confidence int.)

7-61   456

7-63   1522

7-65 $0.828 < p < 0.852$; yes

7-67 $0.647 < p < 0.733$

7-69 $0.177 < p < 0.373$

7-71 $0.121 < p < 0.199$

7-73   0.0307 or 3.07%

7-75 $E = 0.98/\sqrt{n}$

7-77 $p = 0.5$, $q = 0.5$

7-79   (a) 12.5   (b) 234.09   (c) 1.44   (d) 3.61

7-81   9.886, 45.559

7-83   3.247, 20.483

7-85 $58.5 < \sigma^2 < 208.5$

7-87 $13.6 < \sigma^2 < 41.3$

7-89 $406.9 < \sigma < 781.4$

7-91 $43.400 < \sigma^2 < 223.651$

7-93 $5.11 < \sigma^2 < 26.01$

7-95 $6.73 < \sigma < 12.09$

7-97 $0.064 < \sigma < 0.406$ ($s^2 = 0.0136$)

7-99 $31.8 < \sigma < 50.9$

7-101   (a) 1.645   (b) 2.700, 19.023   (c) 2.262   (d) 0.25

7-103 $63.46 < \mu < 78.94$

7-105   (a) 83.2   (b) $82.16 < \mu < 84.24$

7-107   (a) 83.2   (b) $81.01 < \mu < 85.39$

7-109 $15.082 < \mu < 17.718$

7-111   2401

7-113   271

7-115   601

7-117 8687

7-119 $0.033 < p < 0.087$

7-121 $0.437 < p < 0.503$

CHAPTER 8

8-1 Test statistic: $F = 2.000$. Critical value: $F = 4.0260$.
Fail to reject H_0: $\sigma_1^2 = \sigma_2^2$.
There is not sufficient evidence to reject the claim that the variances are equal.

8-3 Test statistic: $F = 7.7931$. The critical value of F is between 8.6565 and 8.5599.
Fail to reject H_0: $\sigma_1^2 = \sigma_2^2$.
There is not sufficient evidence to reject the claim that the variances are equal.

8-5 Test statistic: $F = 2.4091$. Critical value: $F = 2.4523$.
Fail to reject H_0: $\sigma_1^2 = \sigma_2^2$.
There is not sufficient evidence to reject the claim that the variances are equal.

8-7 Test statistic: $F = 169.000$. Critical value: $F = 3.8919$. Reject H_0: $\sigma_1^2 = \sigma_2^2$.
Reject the claim of equal variances.

8-9 Test statistic: $F = 4.0000$. Critical value: $F = 3.1789$. Reject H_0: $\sigma_1^2 \leq \sigma_2^2$.
The variance of population A does appear to exceed that of population B.

8-11 Test statistic: $F = 1.4603$. Critical value: $F = 1.6664$. Fail to reject H_0: $\sigma_1^2 \leq \sigma_2^2$.
There is not sufficient evidence to support the claim that the variance of population A exceeds that of population B.

8-13 Test statistic: $F = 1.7333$. Critical value: $F = 2.5848$. Fail to reject H_0: $\sigma_1^2 = \sigma_2^2$.
There is not sufficient evidence to support the claim of unequal variances.

8-15 Test statistic: $F = 1.8526$. Critical value: $F = 1.8055$. Reject H_0: $\sigma_1^2 \leq \sigma_2^2$.
The second scale does appear to produce greater variance.

8-17 Sample A: $s^2 = 56.9$ and $n = 10$. Sample B: $s^2 = 101.9$ and $n = 14$. Test statistic: $F = 1.7909$. The critical F value is between 3.8682 and 3.7694.
Fail to reject H_0: $\sigma_1^2 = \sigma_2^2$.
There is not sufficient evidence to reject the claim of equal variances.

8-19   First: $s^2 = 81.6$, $n = 61$. Second: $s^2 = 146.1$, $n = 41$. Test statistic: $F = 1.7904$. Critical value: $F = 1.7440$. Reject H_0: $\sigma_1^2 = \sigma_2^2$. There is sufficient evidence to reject the claim of equal standard deviations.

8-21   120

8-23 $\bar{x} = 4.0$ and $s_x = 2.2$. $\bar{y} = 3.6$ and $s_y = 2.5$.
F test results: Test statistic is $F = 1.2913$. Critical F value is between 3.1532 and 3.0527. We conclude that $\sigma_1 = \sigma_2$.
Test of means: Test statistic is $t = 0.449$.
Critical values: $t = -2.056, 2.056$.
Fail to reject: H_0: $\mu_1 = \mu_2$.
There is not sufficient evidence to reject the claim that the samples come from populations with equal means.

8-25   Test statistic: $z = 1.12$. Critical values: $z = -1.96, 1.96$. Fail to reject H_0: $\mu_1 = \mu_2$.
There is not sufficient evidence to reject the claim that the mean volumes put out by both machines are equal.

8-27 F test results: Test statistic is $F = 3.7539$. Critical value of F is between 3.2497 and 3.1532. We conclude that $\sigma_1 \neq \sigma_2$.
Test of means: Test statistic is $t = -1.916$. Critical values: $t = -2.201, 2.201$.
Fail to reject H_0: $\mu_1 = \mu_2$.
There is not sufficient evidence to reject the claim that the mean down times are equal.

8-29 $\bar{d} = 11.9$ and $s_d = 11.2$.
Test statistic: $t = 3.360$. Critical values: $t = -2.262, 2.262$. Reject H_0: $\mu_1 = \mu_2$.
We reject the claim that the pill was ineffective. It appears to have an effect of lowered values.

8-31 $\bar{d} = -1.67$ and $s_d = 3.14$.
Test statistic: $t = -1.837$. Critical values: $t = -2.201, 2.201$. Fail to reject H_0: $\mu_1 - \mu_2$.
There is not sufficient evidence to reject the claim that the diet has no effect. That is, it appears to have no effect.

8-33 F test results: Test statistic is $F = 3.8118$. Critical value: $F = 3.5879$. We conclude that $\sigma_1 \neq \sigma_2$.
Test of means: Test statistic is $t = -3.074$. Critical values: $t = -2.262, 2.262$.
Reject H_0: $\mu_1 = \mu_2$.
The two models appear to have different means.

8-35   Test statistic: $z = 3.10$. Critical values: $z = -1.96, 1.96$. Reject H_0: $\mu_1 = \mu_2$.
We reject the claim that the means for the two locations are equal.

8-37 Test statistic: $z = -13.23$. Critical values: $z = -1.96$, 1.96. Reject H_0: $\mu_1 = \mu_2$.
There is sufficient evidence to reject the claim that the means are equal.

8-39 F test results: Test statistic is $F = 1.6198$. The critical F value is close to 2.0677. We conclude that $\sigma_1 = \sigma_2$.
Test of means: Test statistic is $t = -1.412$. Critical values: $t = -1.960$, 1.960.
Fail to reject H_0: $\mu_1 = \mu_2$.
There is not sufficient evidence to reject the claim that the means are equal. Both brands appear to have the same mean nicotine content.

8-41 A: $n = 9$, $\bar{x} = 26.9$, $s = 4.7$. B: $n = 12$, $\bar{x} = 18.8$, $s = 3.4$.
F test results: Test statistic is $F = 1.9109$. Critical value is $F = 3.6638$. We conclude that $\sigma_1 = \sigma_2$.
Test of means: Test statistic is $t = 4.593$. Critical values: $t = -2.093$, 2.093 (assuming a 0.05 significance level).
Reject H_0: $\mu_1 = \mu_2$.
The two types of string do appear to have different mean breaking points.

8-43 F test results: Test statistic is $F = 1.0331$. Critical value of F is close to 2.0667. We conclude that $\sigma_1 = \sigma_2$.
Test of means: Test statistic is $t = -2.104$. Critical values: $t = -1.960$, 1.960. Reject H_0: $\mu_1 = \mu_2$.
We reject the claim that there is no difference between the two brands. They appear to have different means.

8-45 $\bar{d} = -4.5$ and $s_d = 3.6$.
Test statistic: $t = -3.953$. Critical values: $t = -2.821$, 2.821. Reject H_0: $\mu_1 = \mu_2$.
The two tests appear to have significantly different means.

8-47 Test statistic: $z = -19.19$. Critical value: $z = -1.645$. Reject H_0: $\mu_1 \geq \mu_2$.
The data support the claim that theaters are warmer than stores.

8-49 F test results: Test statistic is $F = 2.6406$. Critical F value is between 2.2989 and 2.3567. We conclude that $\sigma_1 \neq \sigma_2$.
Test of means: Test statistic is $t = 2.327$. Critical value: $t = -1.714$. Fail to reject H_0: $\mu_1 \geq \mu_2$.
The evidence does not support the claim that System 2 has a greater mean than System 1.
(Note: With $\bar{x}_1 > \bar{x}_2$, we can never conclude that $\mu_1 < \mu_2$.)

8-51 $3.50 < (\mu_1 - \mu_2) < 4.30$.

8-53 A: $n = 150$, $\bar{x} = 42.7$, $s = 11.2$. B: $n = 167$, $\bar{x} = 48.4$, $s = 13.7$. Test statistic: $z = -4.07$. Critical values: $z = -1.96$, 1.96. Reject H_0: $\mu_1 = \mu_2$.
Reject the claim that the populations have the same mean age.

8-55 15.11; s_{x+y}^2 is approx. $s_x^2 + s_y^2$.

8-57

	n_1	n_2	x_1	x_2	$\hat{p}$	$\hat{q}$	$\bar{p}$	$\bar{q}$
(a)	200	400	67	148	0.335	0.370	0.358	0.642
(b)	250	300	95	138	0.380	0.460	0.424	0.576
(c)	250	250	50	60	0.200	0.240	0.220	0.780
(d)	300	400	159	212	0.530	0.530	0.530	0.470
(e)	50	100	21	57	0.420	0.570	0.520	0.480

8-59   (a) -2.05 (b) $-1.96, 1.96$
(c) Reject H_0: $p_1 = p_2$.

8-61   Test statistic: $z = 0.71$. Critical values (assuming a 0.05 significance level): $z = -1.96, 1.96$. Fail to reject H_0: $p_1 = p_2$.
There is not a significant difference between the two sample proportions.

8-63   Test statistic: $z = -0.87$. Critical values: $z = -1.96, 1.96$. Fail to reject H_0: $p_1 = p_2$.
There is not sufficient evidence to reject the claim that the two proportions are equal.

8-65   Test statistic: $z = 0.83$. Critical value: $z = 2.33$. Fail to reject H_0: $p_1 \leq p_2$.
There is not sufficient evidence to support the claim that the younger age group has a greater proportion of fatal accidents.

8-67   Test statistic: $z = -1.19$. Critical values: $z = -1.96, 1.96$. Fail to reject H_0: $p_1 = p_2$.
There is not sufficient evidence to reject the claim that the two professors have the same failure rate.

8-69   Test statistic: $z = -12.50$. Critical values: $z = -2.575, 2.575$. Reject H_0: $p_1 = p_2$.
Reject the claim that the proportions of audits from the two groups are equal.

8-71   Test statistic: $z = -6.74$. Critical value: $z = -2.33$. Reject H_0: $p_1 \geq p_2$.
The Salk vaccine appears to be effective.

8-73   Test statistic: $z = 0.92$. Critical value: $z = 1.645$. Fail to reject H_0: $p_1 \leq p_2$.
There is not sufficient evidence to reject the claim that the percentage of defective units is not higher than the corresponding percentage for models made by competitors.

8-75   Test statistic: $z = 2.03$. Critical values: $z = -1.96, 1.96$. Reject H_0: $p_1 = p_2$.
Reject the claim that the proportions are equal.

8-77   Test statistic: $z = -9.13$. Critical values: $z = -2.575, 2.575$. Reject H_0: $p_1 = p_2$.
There is sufficient evidence to support the claim that there is a difference between the proportion of males who are Democrats and the proportion of females who are Democrats.

8-79   Test statistic: $z = -2.40$. Critical values: $z = -1.96, 1.96$. Reject H_0: $p_1 - p_2 = 0.25$.
Reject the claim that the California percentage exceeds the New York percentage by an amount equal to 25%.

8-81   Test statistic: $z = -1.23$. Critical values: $z = -2.33, 2.33$. Fail to reject H_0: $p_1 = p_2$.
There is not sufficient evidence to reject the claim that both plants have the same rate of defects.

8-83   Test statistic: $F = 1.8225$. Critical F value is between 3.9639 and 3.8682. Fail to reject H_0: $\sigma_1 = \sigma_2$.
There is not sufficient evidence to reject the claim that the standard deviations are equal.

8-85   Test statistic: $z = 10.24$. Critical values: $z = -1.96, 1.96$. Reject H_0: $\mu_1 = \mu_2$.
Reject the claim that there is no difference between the mean scores from both states. The means appear to be different.

8-87 $\bar{d} = -1.21$ and $s_d = 1.23$.
Test statistic: $t = -2.95$. Critical values: $t = -2.306, 2.306$. Reject H_0: $\mu_1 = \mu_2$.
Reject the claim that both means are equal.

8-89 F test results: see 8-88. We conclude that $\sigma_1 \neq \sigma_2$.
Test of means: Test statistic is $t = -4.843$. Critical values: $t = -2.093$, 2.093.
Reject H_0: $\mu_1 = \mu_2$.
Reject the claim that the means are equal.

8-91   Test statistic: $z = -2.45$. Critical values: $z = -1.645, 1.645$. Reject H_0: $p_1 = p_2$.
The sample proportions appear to be different.

8-93   Test statistic: $F = 3.4490$. The critical F value is between 2.9222 and 2.8249. Reject H_0: $\sigma_1 = \sigma_2$.
Reject the claim that the two production methods yield batteries whose lives have equal standard deviations.

8-95 F test results: Test statistic is $F = 1.0374$. The critical F value is close to 2.0677. We conclude that $\sigma_1 = \sigma_2$.
Test of means: Test statistic is $t = -1.008$. Critical values: $t = -1.96$, 1.96.
Fail to reject H_0: $\mu_1 = \mu_2$.
There is not sufficient evidence to reject the claim that both means are the same.

8-97 $\bar{d} = -234.0$ and $s_d = 199.5$.
Test statistic: $t = -3.709$. Critical value: $t = -1.833$. Reject H_0: $\mu_1 = \mu_2$.
The program appears to be effective.

8-99 F test results: see 8-98. Conclude that $\sigma_1 \neq \sigma_2$.
Test of means: Test statistic is $t = 36.567$. Critical values: $t = -2.064$, 2.064.
Reject H_0: $\mu_1 = \mu_2$.
There is sufficient evidence to reject the claim that the means are equal.

CHAPTER 9

9-1   (a) Positive correlation   (b) Positive correlation
     (c) Positive correlation   (d) Negative correlation
     (e) No correlation

9-3   (a) Significant positive linear correlation.
     (b) Significant negative linear correlation.
     (c) No significant linear correlation.
     (d) No significant linear correlation.
     (e) No significant linear correlation.

9-5   (b) 4   (c) 7   (d) 15   (e) 49   (f) 20   (g) -0.191

9-7   (b) 5   (c) 9   (d) 31   (e) 81   (f) 47   (g) 0.917

9-9   (b) 0.997   (c) 0.878
     (d) Significant positive linear correlation.

9-11   (b) 0.946   (c) 0.878
     (d) Significant positive linear correlation.

9-13   (b) 0.983   (c) 0.878
     (d) Significant positive linear correlation.

9-15   (b) 0.994   (c) 0.754
     (d) Significant positive linear correlation.

9-17   (b) 0.891   (c) 0.632
     (d) Significant positive linear correlation.

9-19   (b) 0.999   (c) 0.632
     (d) Significant positive linear correlation.

9-21   (b) 0.310   (c) 0.514
     (d) No significant linear correlation.

9-23   (b) 0.600   (c) 0.707
     (d) No significant linear correlation.

9-25   In attempting to calculate r we get a denominator of zero, so a real value of r does not exist. However, it should be obvious that the value of x is not at all related to the value of y.

9-27 r decreases from 0.997 to 0.912. The effect of an extreme value can be minimal or severe, depending on the other data.

9-29   Same results as in 9-6.

9-31   10

9-35 $y = -0.36x + 3.64$

9-37 $y = 0.62x + 3.08$

9-39 $y = 0.027x - 0.00048$

9-41 $y = 2.84x - 7.65$

9-43 $y = 1.08x + 21.27$

9-45 $y = 0.21x - 1.37$

9-47 $y = 0.51x + 39.66$

9-49 $y = 0.26x + 37.66$

9-51 $y = 0.034x + 3.880$

9-53 $y = 0.000124x - 0.210$

9-55   (a) 6   (b) 17   (c) −13   (d) 6   (e) 6

9-57 $y = -55.0x + 2306.6$; the effect can be substantial.

9-59   They are equal.

9-61   Note that s_x and s_y are never negative.

9-63   The sum of the squares using the regression line ($y = -2.00x + 7.25$) is 0.75.
The sum of the squares using the line $y = -x + 6$ is 5.0.

9-65   0.111; 11.1%

9-67   0.640; 64.0%

9-69   Zero. There is no error between the sample points and the regression line since all sample points lie on the regression line.

9-71   0.48

9-73   (a) 14   (b) Since the standard error of estimate is zero, $E = 0$ and there is no "interval" estimate.

9-75   (a) 18.94   (b) $17.12 < y < 20.76$

9-77   (a) 36.8   (b) 34.3   (c) 0.932   (d) −0.949

9-79 $m = 0$

9-81   (a) Correlation   (b) Regression   (c) Regression
(d) Correlation   (e) Correlation

9-83   (a) 0.919
(b) 0.878

(c) Significant positive linear correlation.
(d) $y = 0.70x + 32.33$

9-85   (a) 0.687
(b) 0.811
(c) No significant linear correlation.
(d) $y = 0.22x + 5.41$

9-87   (a) 0.988
(b) 0.707
(c) Significant positive linear correlation.
(d) $y = 1.80x + 1.16$

9-89   (a) 0.887
(b) 0.707
(c) Significant positive linear correlation.
(d) $y = 0.95x + 7.01$

9-91   (a) 0.506
(b) $7.86 < y < 11.61$

CHAPTER 10

10-1   The expected frequencies are 20, 20, 20, 20, 20.
(a) 5.900   (b) 13.277
(c) Fail to reject the claim that absences occur on the five days with equal frequency.

10-3   Test statistic: $\chi^2 = 4.800$. Critical value: $\chi^2 = 14.067$. Fail to reject the claim that the aspirins are equally effective.

10-5   Test statistic: $\chi^2 = 9.474$. The critical value depends on the significance level. With a 0.05 level of significance, $\chi^2 = 16.919$ and we fail to reject the claim that students enroll in the various sections with equal frequencies.

10-7   Test statistic: $\chi^2 = 14.470$. Critical value: $\chi^2 = 11.071$. Reject the manufacturer's claim that the outcomes occur with the stated percentages.

10-9   Test statistic: $\chi^2 = 5.957$. Critical value: $\chi^2 = 9.488$. Fail to reject the manager's claim that the stated percentages are correct.

10-11   Test statistic: $\chi^2 = 4.556$. Critical value: $\chi^2 = 5.991$. Fail to reject the claim that teachers have the same accident rate.

10-13   Test statistic: $\chi^2 = 2.911$. Critical value: $\chi^2 = 7.815$. Fail to reject the claim that the actual number of classes agreed with the teacher's expectation of equal amounts of time.

10-15   180 rolls; reject the claim of equal frequencies.

10-17   Combining A with B, E with F, G with H, and I with J, we get a test statistic of $\chi^2 = 4.118$ and a critical value of $\chi^2 = 11.071$ with a significance level of 0.05. Fail to reject the claim that the observed and expected frequencies are compatible.

10-19   (a) 0.296, 0.444, 0.222, 0.037
   (b) 88.8, 133.2, 66.6, 11.1
   (c) Test statistic: $\chi^2 = 23.202$. Critical value: $\chi^2 = 7.815$. Reject the claim that the observed frequencies fit the given binomial distribution.

10-21   (a) 27.778, 22.222, 22.222, 17.778
   (b) 11.025   (c)   3.841
   (d) Reject the claim that voting is independent of party.

10-23   Test statistic: $\chi^2 = 18.726$. Critical value: $\chi^2 = 6.635$. Reject the claim that adjustment is independent of sex.

10-25   Test statistic: $\chi^2 = 4.776$. Critical value: $\chi^2 = 9.210$. Fail to reject the claim that grade category is independent of sex.

10-27   Test statistic: $\chi^2 = 11.825$. The critical value depends on the significance level. With a 0.05 level of significance, the critical value is $\chi^2 = 9.488$ and we reject the claim that day of the week is independent of the number of defects.

10-29   Test statistic: $\chi^2 = 2.723$. Critical value: $\chi^2 = 9.488$. Fail to reject the claim that grade distribution is independent of the subject.

10-31   Test statistic: $\chi^2 = 17.344$. Critical value: $\chi^2 = 13.362$. Reject the claim that the subject and grade are independent.

10-33   (Use categories of East, Central, and West.) Test statistic: $\chi^2 = 10.362$. Critical value: $\chi^2 = 9.488$. Reject the claim that region and opinion are independent.

10-35   It is multiplied by the same constant.

10-37   Test statistic: $F = 1.6373$. Critical value: $F = 3.2317$. Fail to reject the claim of equal means.

10-39   Test statistic: $F = 1.5430$. Critical value: $F = 2.6060$. Fail to reject the claim of equal means.

10-41   Test statistic: $F = 1.6500$. Critical value: $F = 3.0556$. Fail to reject the claim of equal means.

10-43   Test statistic: $F = 8.2086$. Critical value: $F = 3.8853$. Reject the claim of equal means.

10-45   Test statistic: $F = 15.8140$. Critical value: $F = 3.6823$. Reject the claim of equal means.

10-47   Test statistic: $F = 9.8683$. The critical value depends on the significance level. With a 0.05 level of significance, the critical value is $F = 3.6823$ and we fail to reject the claim of equal means.

10-49   Test statistic: $F = 3.7239$. Critical value: $F = 6.2262$. Fail to reject the claim of equal means.

10-51   Test statistic: $F = 11.6744$. Critical value: $F = 4.6755$. Reject the claim of equal means.

10-53   (a)   With test statistic $z = -8.41$ and critical values $z = -1.96, 1.96$ we reject the claim that $\mu_1 = \mu_2$.
            (b)   With test statistic $z = 8.58$ and critical values $z = -1.96, 1.96$ we reject the claim that $\mu_2 = \mu_3$.
            (c)   With test statistic $z = 1.21$ and critical values $z = -1.96, 1.96$ we fail to reject the claim that $\mu_1 = \mu_3$.
            (d)   With test statistic $F = 48.2442$ and critical value $F = 3.0000$ (approx.) we reject the claim of equal means.

10-55   (a)   The test statistic does not change.
            (b)   The test statistic is multiplied by the square of the constant.

10-57   Test statistic: $\chi^2 = 27.325$. Critical value: $\chi^2 = 12.592$. Reject the claim that customers arrive on the different days with equal frequencies.

10-59   Test statistic: $\chi^2 = 10.000$. Critical value: $\chi^2 = 11.345$. Fail to reject the claim that the characteristics are equally likely.

10-61   Test statistic: $\chi^2 = 3.620$. Critical value: $\chi^2 = 13.277$. Fail to reject the claim that the responses are independent of the student group.

10-63   Test statistic: $\chi^2 = 25.008$. Critical value: $\chi^2 = 13.277$. Reject the claim that opinion is independent of region.

10-65   Test statistic: $F = 10.8266$. Critical value: $F = 3.0718$. Reject the claim of equal means.

10-67   Test statistic: $F = 4.3124$. Critical value: $F = 4.8740$. Fail to reject the claim of equal means.

CHAPTER 11

11-1 $x = 4$ (the smaller of 4 and 8) and the critical value is 2. Since $x = 4$ is not less than or equal to the critical value, fail to reject the null hypothesis that the x and y samples come from the same population. They appear to come from the same population.

11-3 $x = 1$ (the smaller of 1 and 8) and the critical value is 1. Since $x = 1$ is less than or equal to the critical value, reject the null hypothesis of no effect. The pill appears to lower blood pressure.

11-5 $x = 2$ (the smaller of 2 and 8), $n = 10$, and the critical value is 1. Since $x = 2$ is not less than or equal to the critical value, fail to reject the null hypothesis that the diet has no effect. The diet appears to be ineffective.

11-7   The test statistic $x = 4$ is not less than or equal to the critical value of 0. Fail to reject the null hypothesis of no difference. There appears to be no difference between the ages of husbands and wives.

11-9   The test statistic $x = 5$ is less than or equal to the critical value of 5. Reject the null hypothesis that half (or fewer) of the students feel that they understand the purpose of their student senate. The data support the claim that most understand.

11-11   The test statistic $x = 7$ is not less than or equal to the critical value of 5. Fail to reject the null hypothesis that both parties are preferred equally. There is not a significant difference.

11-13   The test statistic $x = 32$ is converted to $z = -1.15$. The critical value is $z = -1.645$. Since the test statistic is not less than or equal to the critical value, fail to reject the null hypothesis that the median life is at least 40 hours.

11-15   The statistic $x = 18$ is converted to the test statistic $z = -1.84$. The critical value is $z = -1.645$. Since the test statistic is less than or equal to the critical value, reject the null hypothesis that production was unchanged or lowered by the new fertilizer. It appears that production was increased. '

11-17   The statistic $x = 6$ is converted to the test statistic $z = -2.83$. The critical values are $z = -2.575$ and $z = 2.575$. Since the test statistic is less than or equal to the critical values, reject the null hypothesis that the drink had no effect.

11-19   The statistic $x = 22$ is converted to the test statistic $z = -0.71$. The critical value is $z = -1.28$. Since the test statistic is not less than or equal to the critical value, fail to reject the null hypothesis that at most half of the voters favor the bill. There is not sufficient evidence to support the claim that the majority favor the bill.

11-21   With $k = 3$ we get $6 < M < 17$.

11-23   78

11-25   The given entries 5, 8, 10, 12, 15 correspond to ranks 1, 2, 3, 4, 5.

11-27   The given entries 50, 100, 150, 200, 400, 600 correspond to ranks 1, 2, 3, 4, 5, 6.

11-29   The given entries 6, 8, 8, 9, 12, 20 correspond to ranks 1, 2.5, 2.5, 4. 5, 6.

11-31   The given entries 12, 13, 13, 13, 14, 15, 16, 16, 18 correspond to ranks 1, 3, 3, 3, 5, 6, 7.5, 7.5, 9.

11-33   (a)   3, -7, -2, -4, -1, 0 (discard), -10, -15
      (b)   3, 5, 2, 4, 1, 6, 7
      (c) $+3$, -5, -2, -4, -1, -6, -7
      (d) $T = 3$ (the smaller of 3 and 25)

11-35   (a)   1, -1, 2, -3, -6, 5, -10, -20, 0 (discard)
      (b)   1.5, 1.5, 3, 4, 6, 5, 7, 8
      (c) $+1.5$, -1.5, $+3$, -4, -6, $+5$, -7, -8
      (d) $T = 9.5$ (the smaller of 9.5 and 26.5)

11-37   (a)   25; reject H_0
      (b)   25; reject H_0
      (c)   25; fail to reject H_0
      (d)   81; reject H_0
      (e)   35; fail to reject H_0

11-39 $T = 6.5, n = 9$. The test statistic $T = 6.5$ is greater than the critical value of 6, so fail to reject the null hypothesis of equal results.

11-41 $T = 4.5, n = 8$. The test statistic of $T = 4.5$ is greater than the critical value of 4, so fail to reject the null hypothesis of equal perceptions of depth.

11-43 $T = 11.5, n = 14$. The test statistic of $T = 11.5$ is less than or equal to the critical value of 21, so reject the null hypothesis of equal anxiety levels.

11-45 $T = 3, n = 10$. The test statistic of $T = 3$ is less than or equal to the critical value of 8, so reject the null hypothesis of equal test results.

11-47 $T = 31, n = 12$. The test statistic $T = 31$ is greater than the critical value of 14, so fail to reject the null hypothesis of no difference in attitudes towards the two groups.

11-49 $T = 51.5, n = 36$. Test statistic $z = -4.42$ is less than or equal to the critical value of $z = -2.575$, so reject the null hypothesis that both systems require the same times. It appears that the scanner system is faster.

11-51   (a)   0, 18   (b)   0, 27.5   (c)   0, 637.5

11-53 $\mu_R = 150, \sigma_R = 17.32, R = 166, z = 0.92$. The test statistic $z = 0.92$ is not in the critical region bounded by $z = -1.96, 1.96$, so fail to reject the null hypothesis of equal performances.

11-55 $\mu_R = 144, \sigma_R = 16.25, R = 163.5, z = 1.20$. The test statistic of $z = 1.20$ is not in the critical region bounded by the critical values $z = -1.96$ and $z = 1.96$, so fail to reject the null hypothesis of equal ratings.

11-57 $\mu_R = 203$, $\sigma_R = 21.76$, $R = 184$, $z = -0.87$. The test statistic $z = -0.87$ is not in the critical region bounded by the critical values of $z = -1.96$ and $z = 1.96$, so fail to reject the null hypothesis of equal times.

11-59 $\mu_R = 201.5$, $\sigma_R = 23.89$, $R = 163$, $z = -1.61$. The   test   statistic $z = -1.61$ is not in the critical region bounded by the critical values of $z = -2.575$ and $z = 2.575$, so fail to reject the null hypothesis of equality.

11-61 $\mu_R = 224$, $\sigma_R = 20.26$, $R = 270.5$, $z = 2.29$. The test statistic $z = 2.29$ is in the critical region bounded by the critical values $z = -1.96$, 1.96, so reject the null hypothesis of equal attitudes.

11-63 $\mu_R = 162$, $\sigma_R = 19.44$, $R = 86$, $z = -3.91$. The test statistic $z = -3.91$ is in the critical region bounded by the critical values $z = -1.96$, 1.96, so reject the null hypothesis of no difference between the two treatments.

11-65   (a) $\mu_R = 232.5$, $\sigma_R = 24.11$, $R = 120$, $z = -4.67$. The   test   statistic $z = -4.67$ is in the critical region bounded by the critical values $z = -1.96$, 1.96, so reject the null hypothesis that both groups come from the same population.

         (b) $\mu_R = 232.5$, $\sigma_R = 24.11$, $R = 225$, $z = -0.31$. The   test   statistic $z = -0.31$ is not in the critical region bounded by the critical values of $z = -1.96$, 1.96, so fail to reject the null hypothesis that both groups come from the same population.

11-67   (a)

ABAB	4
ABBA	5
BBAA	7
BAAB	5
BABA	6

         (b)

R	p
3	1/6
4	1/6
5	2/6
6	1/6
7	1/6

         (c)   No, the most extreme rank distribution has a probability of at least 1/6 and we can never get into a critical region with a probability of 0.10 or less.

11-69   0

11-71   The test statistic $H = 2.435$ is less than the critical value of 5.991, so fail to reject the null hypothesis that the samples come from identical populations.

11-73   The test statistic of $H = 12.167$ exceeds the critical value of 5.991, so reject the null hypothesis that the samples come from identical populations.

11-75   The test statistic of $H = 0.775$ is less than the critical value of 9.210, so fail to reject the null hypothesis of equally effective methods.

11-77   The test statistic is $H = 8.756$. The critical value depends on the significance level chosen. With a 0.05 level of significance, the critical value is 9.488 and we fail to reject the null hypothesis that the methods are equally effective.

11-79   The test statistic of $H = 31.086$ exceeds the critical value of 9.488, so reject the null hypothesis that the samples come from identical populations.

11-81 $H = \dfrac{1}{1176}(R_1^2 + R_2^2 + \cdots + R_8^2) - 147$

11-83   12.204

11-85   (a) ± 0.450 (b) ± 0.280 (c) ± 0.373
       (d) ± 0.475 (e) ± 0.342

11-87

x	1	2	4	3	5
y	5	4	3	2	1

11-89 $d = 3, 0, 0, 3; \ \Sigma d^2 = 18; \ n = 4; \ r_s = -0.8$

11-91 $d = 3, 3, 1, 2, 2, 5; \ \Sigma d^2 = 52; \ n = 6; \ r_s = -0.486$

11-93 $d = 2, 1, 0, 1, 2, 0, 0, 1, 3, 2; \ \Sigma d^2 = 24; \ n = 10; \ r_s = 0.855$; critical values: $r_s = \pm 0.648$.
       There is a positive correlation.

11-95 $r_s = 0.818$; critical values: $r_s = \pm 0.525$.
       There is a positive correlation.

11-97 $r_s = -0.077$; critical values: $r_s = \pm 0.591$.
       There is no correlation.

11-99 $r_s = 0.748$; critical values: $r_s = \pm 0.648$.
       There is a positive correlation.

11-101 $r_s = 0.339$; critical values: $r_s = \pm 0.525$.
       There is no correlation.

11-103 $r_s = -0.821$; critical values: $r_s = \pm 0.683$.
       There is a negative correlation.

11-105   (a) 123, 132, 213, 231, 312, 321
       (b) 1, 0.5, 0.5, -0.5, -0.5, -1
       (c) 1/6 or 0.167

11-107   The rank correlation coefficient, because the trend is not linear.

11-109 $n_1 = 5, \ n_2 = 3, \ G = 3$, critical values: 1, 8.

11-111 $n_1 = 4$, $n_2 = 6$, $G = 2$, critical values: 2, 9.

11-113 $n_1 = 7$, $n_2 = 2$, $G = 2$, critical values: 1, 6.

11-115 $n_1 = 10$, $n_2 = 10$, $G = 9$, critical values: 6, 16

11-117 $\bar{x} = 11.0$; BBBBBABAAA; $n_1 = 6$, $n_2 = 4$, $G = 4$.
The critical values are 2 and 9. Fail to reject randomness.

11-119 $\bar{x} = 9.2$, $n_1 = 6$, $n_2 = 8$, $G = 2$.
Critical values are 3, 12. Reject randomness.

11-121   Median is 9.5, $n_1 = 5$, $n_2 = 5$, $G = 2$.
Critical values are 2, 10. Reject randomness.

11-123   Median is 9 so delete the three 9s to get BBBBBBBAAAAAAA. $n_1 = 7$,
$n_2 = 7$, $G = 2$.
Critical values are 3, 13. Reject randomness.

11-125 $n_1 = 9$, $n_2 = 18$, $G = 4$. Critical values: 8, 18.
Reject randomness.

11-127 $n_1 = 30$, $n_2 = 15$, $G = 14$, $\mu_G = 21$, $\sigma_G = 2.94$.
Test statistic is $z = -2.38$. Critical values are $z = -1.96$, 1.96. Reject
randomness.

11-129   Median is 834. $n_1 = 8$, $n_2 = 8$, $G = 2$.
Critical values are 4, 14. Reject randomness. There appears to be a
downward trend.

11-131 $n_1 = 17$, $n_2 = 28$, $G = 12$. $\mu_G = 22.16$, $\sigma_G = 3.11$, and the test statistic is
$z = -3.27$. Critical values are $z = -1.96$, 1.96. Reject randomness.

11-133 $n_1 = 30$, $n_2 = 30$, $G = 8$. $\mu_G = 31$, $\sigma_G = 3.84$.
The test statistic is $z = -5.99$. Critical values are $z = -1.96$, 1.96. Reject
randomness.

11-135 $n_1 = n_2 = 40$ since the median is used. Also, $G = 6$, $\mu_G = 41$, $\sigma_G = 4.44$.
The test statistic is $z = -7.88$ and the critical values are $z = -1.96$, 1.96.
Reject randomness.

11-137   Minimum is 2, maximum is 4. Critical values of 1 and 6 can never be
realized so that the null hypothesis of randomness can never be rejected.

11-139   The eighty-four sequences yield two runs of 2, seven runs of 3, twenty
runs of 4, twenty-five runs of 5, twenty runs of 6, and ten runs of 7 so
that $P(2 \text{ runs}) = 2/84$, $P(3 \text{ runs}) = 7/84$, $P(4 \text{ runs}) = 20/84$, $P(5 \text{ runs}) = 25/84$, $P(6 \text{ runs}) = 20/84$, $P(7 \text{ runs}) = 10/84$. From this we
conclude that the G values of 3, 4, 5, 6, 7 can easily occur by chance while
$G = 2$ is unlikely since $P(2 \text{ runs})$ is less than 0.025. The lower critical G
value is therefore 2 and this agrees with Table A-11. The table lists 8 as
the upper critical value, but it is impossible to get 8 runs using the given
elements.

11-141 $\bar{x} = 72.5$, $n_1 = 10$, $n_2 = 12$, $G = 6$.
Critical values: 7, 17. Reject randomness.

11-143 $r_s = -0.105$. Assuming a 0.05 level of significance, the critical values are $r_s = \pm 0.591$.
Fail to reject the null hypothesis of no correlation. There does not appear to be a correlation between age and amount billed.

11-145 $\mu_R = 162$, $\sigma_R = 19.44$, $R = 154$, $z = -0.41$.
The test statistic $z = -0.41$ is not in the critical region bounded by $z = -1.96$, 1.96, so fail to reject the null hypothesis of no difference. There appears to be no difference between the speeds in the two states.

11-147 $T = 4$, $n = 12$. The test statistic $T = 4$ is less than the critical value of 14, so reject the null hypothesis that the tune-up has no effect.

11-149   (Use the Kruskal-Wallis test.) $N = 31$, $n_1 = 11$, $R_1 = 220.5$, $n_2 = 13$, $R_2 = 136.5$, $n_3 = 7$, $R_3 = 139$. The test statistic is $H = 8.195$ and the critical value is 5.991. Reject the null hypothesis of identical populations.

11-151 $\mu_R = 137.50$, $\sigma_R = 17.26$, $R = 177.5$, $z = 2.32$.
The test statistic $z = 2.32$ is in the critical region bounded by $z = -1.96$, 1.96. Reject the null hypothesis of no difference.

11-153   (Use the sign test.) Discard the zero to get 10 negative signs and 9 positive signs so that $x = 9$ which is not less than or equal to the critical value of 4 found from Table A-8. Fail to reject the null hypothesis that the median is 38.

11-155 $\mu_R = 126.5$, $\sigma_R = 15.23$, $R = 155.5$, $z = 1.90$.
The test statistic $z = 1.90$ is not in the critical region bounded by $z = -1.96$, 1.96. Fail to reject the null hypothesis that both brands are the same.

11-157 $T = 12$, $n = 7$. The test statistic of $T = 12$ is greater than the critical value of 2, so fail to reject the null hypothesis of no difference. There does not appear to be a difference.

Index

abuse of statistics, 7
accuracy, 11, 28
addition rule, 108–115
airlines, 5
algebra, 5
alternative hypothesis, 273
American Automobile Association, 8
analysis of variance, 483–491
Apgar rating scale, 563
area, 9
arithmetic mean, 51
average, 8, 50–51
axiom, 5

bar graph, 8, 9
baseball, 484
Bennett, William, 113
Berman, Ronald, 190
Better Business Bureau, 302
bibliography, A30, A31
Bills of Mortality, 4
bimodal distribution, 197
binomial approximated by normal, 239–247
binomial distribution, 173–184
binomial distribution mean, 189
binomial distribution standard deviation, 189
binomial distribution variance, 189
binomial experiment, 173
binomial probability formula, 176
binomial probability table, A1–A9
bivariate data, 418
box-and-whisker, 199
boxplot, 199
Brooks, Juanita, 122
Bureau of Census, 18, 25
Bureau of Labor Statistics, 520

Campbell, John, 270, 291
cancer, 425, 548
Cardan, Jerome, 101
census, 25
Census Bureau, 18, 25
central limit theorem, 249–258
central tendency, measures of, 51
centroid, 426
Chebyshev's theorem, 78
chi-square distribution, 316–317, 464–465, 472–473
chi-square table, A12–A13
cigarettes, 548
class, 25
class boundaries, 24
classical approach to probability, 97
class marks, 24, 25

class size paradox, 56
class width, 24, 25
clusters, 202
cluster sampling, 576
Coca-Cola, 528
coefficient of determination, 450
coefficient of variation, 77
Collins case, 122
Columbus, 190
combinations, 140
complementary events, 130–133
completely randomized design, 491
compound event, 109
confidence interval, 334
confidence interval for mean, 334
confidence interval for proportion, 347
confidence interval for standard deviation, 358
confidence interval for variance, 357
confidence interval limits, 335
confidential survey, 576
consistency, 355
consumer price index, 80, 561
contingency table, 472–478
continuity correction, 242, 247
continuous random variable, 158
correlation, 418–428
correlation coefficient tables, A20, A23
countable, 157
counting, 136–143
counting formula, 175
crime, 386
critical region, 273
critical value, 273
cumulative frequency polygon, 38
cumulative frequency table, 29
cycles, 561

Darwin, Charles, 433
deciles, 80
decisions, 332
decision theory, 167
deductive, 5
degree of confidence, 334
degrees of freedom, 298, 317, 372, 380, 465, 475, 485, 533
dependent, 122
dependent samples, 378
descriptive statistics, 22
discrete random variable, 157
disease, 202
disobedience, 14
dispersion, 64
Disraeli, Benjamin, 7

distribution, 22, 43
distribution-free tests, 504
distribution shapes, 195
draft, 557
drug, 273

economic indicators, 436
economics, 561
efficiency, 505, 525
empirical approximation of probability, 97
estimates, 332
event, 96
exclusive or, 109, 118
exit poll, 372
expectation, 168
expected frequency, 463–464, 473
expected value, 167–168
experiment, 96
explained deviation, 449
explained variation, 450
exploratory data analysis, 40

factorial, 138
factorial rule, 138
failure, 174
farm, 70
F distribution, 371, 484
F distribution tables, A14–A19
Federal Aviation Administration, 155
Federalist papers, 28
Federal Trade Commission, 302
finite population correction factor, 255–256
Firestone, 302
Fisher, R. A., 464
frequency, 23
frequency polygon, 38, 39
frequency table, 23
Freud, 190
fundamental counting rule, 137

Gallup, George, 491, 577
Galton, Sir Francis, 422, 442
gender gap, 378
General Motors, 302
genius, 235
geometric distribution, 187
geometric mean, 63–64
geometry, 6
glossary, A25–A29
goodness-of-fit, 467
Gosset, William, 297
graph, 8, 9
Graunt, John, 4, 16

Hamilton, Alexander, 28
Hamlet, 113
harmonic mean, 63
Harris, Lou, 491
heights, 442, 476
hemline index, 436
Hertz, 8
hinge, 199
histogram, 37–38, 156, 196
H test, 533
Huff, Darrell, 11
hypergeometric distribution, 188
hypothesis, 268
hypothesis test, 268–321

imputation, 25
inclusive or, 109, 118
independent, 122
index numbers, 80
inductive, 5
inference, 22
inferential statistics, 22
infinity, 157
inflation, 561
interquartile range, 82
interval, 14
interval estimate, 334
interval estimate of mean, 334
interval estimate of predicted value, 453

Jay, John, 28

Keyes, Ralph, 476
Kruskal-Wallis test, 533–536

Landon, Alfred, 3
least squares property, 442
left-tailed test, 278–279, 291
leukemia, 202
levels of measurement, 12
linear correlation coefficient, 420
linear correlation coefficient table, A20
Listerine, 302
Literary Digest, 3
lottery, 159, 168, 557
lower class limits, 24

Madison, James, 28
Mann-Whitney U test, 525
maximum error of estimate of mean, 334, 339
maximum error of estimate of proportion, 347
McClellan, David, 450
mean, 51, 165
mean deviation, 67, 71
mean of discrete random variable, 165
measures of central tendency, 51

measures of dispersion, 64–83
measures of position, 79–84
median, 8, 53, 506
Mendel, Gregor, 464
midquartile, 82
midrange, 55
Milgram, Stanley, 14
mode, 8, 54
Mt. St. Helens, 10
multimodal distribution, 54, 197
multinomial distribution, 188
multinomial experiments, 462–468
multiplication rule, 119–126
mutually exclusive, 112

Napoleon, 190
negative correlation, 421, 424, 544, 545
negotiations, 8
Nielsen television ratings, 335
nominal, 12
nonparametric, 504–561
nonstandard normal distributions, 225–229
normal approximation to binomial, 239–247
normal distribution, 216–258
nuclear power, 120, 310
null hypothesis, 272

observed frequency, 464, 473
odds, 132
ogive, 38–39
one-way analysis of variance, 491
ordinal, 13

parachuting, 174
parameter, 5
Pearson, Karl, 426
Pearson's product moment, 426
Pepsi, 528
percentiles, 80
permutations, 139
phone surveys, 351, 575
pie chart, 35
plane geometry, 5
point estimate, 332, 346, 355
point estimate of mean, 332
point estimate of proportion, 346
point estimate of variance, 355–356
Poisson distribution, 187
police, 386
polio, 404
poll, 10
pollution, 9
pooled estimate of proportion, 400
population, 5
population size, 255
position, measures of, 79–84
positive correlation, 421, 424, 544, 545

precision, 11
predicted value, 448
probability:
    addition rule, 108–115
    complementary events, 130–133
    defined, 97
    distributions, 155–207
    multiplication rule, 119–126
    nature of, 95
    review, 147–149
    value, 289
product testing, 279
proportions, tests of, 307–312
P-value, 289–293, 303, 311

quadratic mean, 64
quartiles, 79, 199

random, 553, 574
random sample, 574
random variable, 156
range, 65
rank correlation, 540–549
rank correlation coefficient table, A23
ranked data, 506
ratio, 15
redundancy, 109
regression, 418, 433–443
regression analysis, 433
regression line, 433
Reichard, Robert, 11
relative frequency, 97
reliability, 252
report, 577
right-tailed test, 278–279, 291
rigorously controlled design, 491
robust, 355
Roosevelt, Franklin D., 3
root mean square, 64
rounding, 55, 103
run, 554
runs test, 553–562
runs test table, A24

Salk vaccine, 404
sample, 5
sample proportion, 346
sample size, 337, 338, 349, 350, 358, 359
sample size for estimating mean, 337–338
sample size for estimating proportion, 348–352
sample size for estimating standard deviation, 358–359
sample size for estimating variance, 358, 359
sample space, 96
S.A.T. scores, 98, 450
scatter diagram, 418, 419–421

Schwartz, Noel, 279
seat belts, 536
semi-interquartile range, 82
Shakespeare, 113, 190
sigma, 52
significance level, 273
sign test, 506–513
sign test table, A21
simple event, 96
simulation, 153
single factor analysis of variance,
    491
skewed distribution, 197
slope, 434
smoking, 548
Spearman, Charles, 542
standard deviation:
    calculation of, 70–73
    definition, 70
    of discrete random variable, 166
    units, 71
standard error of estimate, 452
standard error of the mean, 252
standard normal distribution,
    217–223
standard normal distribution table,
    A10
standard score, 79, 225
STATDISK, 82, 84, 182, 198, 243,
    289, 292, 303, 340, 381,
    388–389, 402, 443, 477, 478,
    490
statistic, 6
statistics, 4
stem-and-leaf plot, 40–43
stratified sampling, 575
student *t* distribution, 295–297
student *t* distribution table, A11
success, 174

summarizing data, 23–30
Super-Bowl omen, 436
survey medium, 575
syphilis, 183
systematic sampling, 575

tables, A1–A24
tally marks, 28–29
taxes, 9–10
t distribution, 297–299
t distribution table, A11
teacher ratings, 422
telephone starts, 575
telephone surveys, 351, 575
television, 42, 333, 541
testing hypotheses, 267–321
tests of proportions, 307–312
tests of significance, 272
tests of variances, 315–321
test statistic, 273
total deviation, 448
total variation, 450
tree diagram, 125–126, 136
trends, 561
triangle, 5
t test, 295–303
Tukey, John, 40
two-tailed test, 278–279, 291
two-way tables, 472
type I error, 273
type II error, 273

unbiased estimator, 332, 346, 356
unemployment, 520
unexplained deviation, 449
unexplained variation, 450
uniform distribution, 201–202
upper class limits, 24

validity, 252
variability, 65
variable, 52
variance:
    calculation of, 66
    calculator, 69–70
    definition, 66
    of discrete random variable, 166
    for frequency table, 69
    notation, 66
    units, 71
variance between samples, 485
variances, tests of, 315–321
variance within samples, 485
variation, 448–453
Vega, 302
volume, 10
voting power, 319

weather, 94, 227
weighted mean, 55
Wickramashinghe, N. C., 130
Wilcoxon rank-sum test, 525–529
Wilcoxon signed-ranks test, 517–521
Wilcoxon signed-ranks test table,
    517–521

Yankelovich, Daniel, 339
y-intercept, 434

ZIP codes, 486
z score, 79, 225

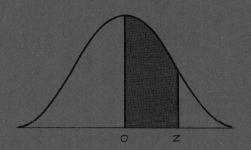

Table A-3 The Standard Normal (z) Distribution

z	.00	.01	.02	.03	.04	.05	.06	.07	.08	.09
0.0	.0000	.0040	.0080	.0120	.0160	.0199	.0239	.0279	.0319	.0359
0.1	.0398	.0438	.0478	.0517	.0557	.0596	.0636	.0675	.0714	.0753
0.2	.0793	.0832	.0871	.0910	.0948	.0987	.1026	.1064	.1103	.1141
0.3	.1179	.1217	.1255	.1293	.1331	.1368	.1406	.1443	.1480	.1517
0.4	.1554	.1591	.1628	.1664	.1700	.1736	.1772	.1808	.1844	.1879
0.5	.1915	.1950	.1985	.2019	.2054	.2088	.2123	.2157	.2190	.2224
0.6	.2257	.2291	.2324	.2357	.2389	.2422	.2454	.2486	.2517	.2549
0.7	.2580	.2611	.2642	.2673	.2704	.2734	.2764	.2794	.2823	.2852
0.8	.2881	.2910	.2939	.2967	.2995	.3023	.3051	.3078	.3106	.3133
0.9	.3159	.3186	.3212	.3238	.3264	.3289	.3315	.3340	.3365	.3389
1.0	.3413	.3438	.3461	.3485	.3508	.3531	.3554	.3577	.3599	.3621
1.1	.3643	.3665	.3686	.3708	.3729	.3749	.3770	.3790	.3810	.3830
1.2	.3849	.3869	.3888	.3907	.3925	.3944	.3962	.3980	.3997	.4015
1.3	.4032	.4049	.4066	.4082	.4099	.4115	.4131	.4147	.4162	.4177
1.4	.4192	.4207	.4222	.4236	.4251	.4265	.4279	.4292	.4306	.4319
1.5	.4332	.4345	.4357	.4370	.4382	.4394	.4406	.4418	.4429	.4441
1.6	.4452	.4463	.4474	.4484	.4495	.4505	.4515	.4525	.4535	.4545
1.7	.4554	.4564	.4573	.4582	.4591	.4599	.4608	.4616	.4625	.4633
1.8	.4641	.4649	.4656	.4664	.4671	.4678	.4686	.4693	.4699	.4706
1.9	.4713	.4719	.4726	.4732	.4738	.4744	.4750	.4756	.4761	.4767
2.0	.4772	.4778	.4783	.4788	.4793	.4798	.4803	.4808	.4812	.4817
2.1	.4821	.4826	.4830	.4834	.4838	.4842	.4846	.4850	.4854	.4857
2.2	.4861	.4864	.4868	.4871	.4875	.4878	.4881	.4884	.4887	.4890
2.3	.4893	.4896	.4898	.4901	.4904	.4906	.4909	.4911	.4913	.4916
2.4	.4918	.4920	.4922	.4925	.4927	.4929	.4931	.4932	.4934	.4936
2.5	.4938	.4940	.4941	.4943	.4945	.4946	.4948	.4949	.4951	.4952
2.6	.4953	.4955	.4956	.4957	.4959	.4960	.4961	.4962	.4963	.4964
2.7	.4965	.4966	.4967	.4968	.4969	.4970	.4971	.4972	.4973	.4974
2.8	.4974	.4975	.4976	.4977	.4977	.4978	.4979	.4979	.4980	.4981
2.9	.4981	.4982	.4982	.4983	.4984	.4984	.4985	.4985	.4986	.4986
3.0	.4987	.4987	.4987	.4988	.4988	.4989	.4989	.4989	.4990	.4990

Frederick Mosteller and Robert E. K. Rourke, *Sturdy Statistics* Table A-1. (Reading, Mass.: Addison-Wesley, 1973.) Reprinted with permission.

NOTE: For values of z above 3.09, use 0.4999.